W9-CBC-094

TCM MANUAL - $115.00

To Order, Contact: Timberweld Mfg.
P.O. Box 21,000
Billings, MT 59104
(406) 652-3600

TIMBER CONSTRUCTION MANUAL

TIMBER CONSTRUCTION MANUAL

Fifth Edition

AMERICAN INSTITUTE OF TIMBER CONSTRUCTION

JOHN WILEY & SONS, INC.

This book is printed on acid-free paper. ♾

Copyright © 2005 by American Institute of Timber Construction. All rights reserved

Published by John Wiley & Sons, Inc., Hoboken, New Jersey
Published simultaneously in Canada

No part of this publication may be reproduced, stored in a retrieval system, or transmitted in any form or by any means, electronic, mechanical, photocopying, recording, scanning, or otherwise, except as permitted under Section 107 or 108 of the 1976 United States Copyright Act, without either the prior written permission of the Publisher, or authorization through payment of the appropriate per-copy fee to the Copyright Clearance Center, Inc., 222 Rosewood Drive, Danvers, MA 01923, (978) 750-8400, fax (978) 750-4470, or on the web at www.copyright.com. Requests to the Publisher for permission should be addressed to the Permissions Department, John Wiley & Sons, Inc., 111 River Street, Hoboken, NJ 07030, (201) 748-6011, fax (201) 748-6008, e-mail: permcoordinator@wiley.com.

Limit of Liability/Disclaimer of Warranty: While the publisher and author have used their best efforts in preparing this book, they make no representations or warranties with respect to the accuracy or completeness of the contents of this book and specifically disclaim any implied warranties of merchantability or fitness for a particular purpose. No warranty may be created or extended by sales representatives or written sales materials. The advice and strategies contained herein may not be suitable for your situation. You should consult with a professional where appropriate. Neither the publisher nor author shall be liable for any loss of profit or any other commercial damages, including but not limited to special, incidental, consequential, or other damages.

For general information on our other products and services or for technical support, please contact our Customer Care Department within the United States at (800) 762-2974, outside the United States at (317) 572-3993 or fax (317) 572-4002.

Wiley also publishes its books in a variety of electronic formats. Some content that appears in print may not be available in electronic books. For more information about Wiley products, visit our web site at www.wiley.com.

Library of Congress Cataloging-in-Publication Data:

Timber construction manual/American Institute of Timber Construction—5th ed.
p. cm.
Includes bibliographical references and index.
ISBN 0-471-23687-X (cloth)
1. Building, Wooden—Handbooks, manuals, etc. I. American Institute of Timber Construction.
TA666.T47 2004
694—dc22 2004000225

Printed in the United States of America

10 9 8 7 6 5 4 3 2 1

CONTENTS

PREFACE

The American Institute of Timber Construction (AITC) has developed this *Timber Construction Manual* for convenient reference by architects, engineers, contractors, teachers, the laminating and fabricating industry, and all others having a need for reliable up-to-date technical data and recommendations on engineered timber construction. The information and recommendations herein are based on the most reliable technical data available and reflect the commercial practices found to be most practical. Their application results in structurally sound construction.

The American Institute of Timber Construction, established in 1952, is a nonprofit industry association for the structural glued laminated timber industry. Its members design, plant-laminate, fabricate, assemble, and erect structural timber systems utilizing both sawn and structural glued laminated timber components. These systems are used in homes, schools, churches, commercial and industrial buildings, and for other structures, such as bridges, towers, and marine installations. Institute membership also includes engineers, architects, building officials, and associates from other industries related to timber construction.

The first edition of the *Timber Construction Manual* was published in 1966. Changes in the wood products industry, technological advances, and improvements in the structural timber fabricating industry necessitated revisions of the manual. New lumber sizes and revisions in grading requirements for lumber and glued laminated timber were reflected in the second edition published in 1974. The third edition was published in 1985 to reflect new information on timber design methods. The fourth edition of the manual was published in 1994 and contained updated design procedures used for timber construction. This fifth edition contains updated information for timber design, including a new section on timber rivet fasteners and a chapter on load and resistance factor design. The manual has also been restructured. Design stresses, which were included in earlier editions, have been removed from the main text and are available separately as supplemental materials. In this way the design procedures will remain relevant while design stresses are revised periodically.

Preparation of the *Timber Construction Manual* was guided by the AITC Technical Advisory Committee and was carried out by AITC staff, the engineers and technical representatives of AITC member firms, and private consultants. Suggestions for its improvement will be welcomed and will receive consideration in the preparation of future editions.

The *Timber Construction Manual* has been adopted by the American Institute of Timber Construction as its official recommendation in glued laminated timber design.

While the information herein has been prepared in accordance with recognized engineering principles and is based on the most accurate and reliable technical data available, it should not be used or relied upon for any general or specific application without competent professional examination and verification of its accuracy, suitability, and applicability by a licensed professional engineer, designer, or architect. By the publication of this manual, AITC intends no representation or warranty, expressed or implied, that the information contained herein is suitable for any general or specific use or is free from infringement of any patent or copyright. Any user of this information assumes all risk and liability arising from such use.

GENERAL NOMENCLATURE

The following abbreviations and symbols are in general use throughout this book. Other abbreviations and symbols and deviations from the following are noted as they occur.

ABBREVIATIONS

AASHTO	American Association of State Highway and Transportation Officials
AITC	American Institute of Timber Construction
ANSI	American National Standards Institute
AREMA	American Railway Engineering and Maintenance-of-Way Association
ASCE	American Society of Civil Engineers
ASD	Allowable Stress Design
ASME	American Society of Mechanical Engineers
ASTM	American Society for Testing and Materials
AWPA	American Wood-Preservers' Association
BF	Board foot
BFM	Board footage, board-foot measure
Btu	British thermal unit
DL	Dead load
EMC	Equilibrium moisture content
°F	Degree(s) Fahrenheit
ft	Foot

GLB	Glulam beam
Glulam	Structural glued laminated timber
hr	Hour
in.	Inch
k	Kip (thousand pounds)
lb	Pound (weight or force)
LL	Live load
LRFD	Load and resistance factor design
MBF	Thousand board feet
MC	Moisture content
MEL	Machine-evaluated lumber
MMBF	Million board feet
MSR	Machine stress rated
MRI	Mean recurrence interval
NDS®	*National Design Specification® for Wood Construction*
o.c.	On center (spacing)
pcf	Pound(s) per cubic foot
plf	Pound(s) per lineal foot
psf	Pound(s) per square foot
psi	Pound(s) per square inch
PTC	Pitched and tapered curved beam
SL	Snow load
SP	Southern pine
s.w.	Self-weight
TL	Total load
WL	Wind load

SYMBOLS

A	Area of cross section
A	Coefficient or factor
A_m	Gross cross-sectional area of main member in connection
A_n	Net area considering section loss due to holes and/or notches
A_s	Gross cross-sectional area of side member or sum of gross cross-sectional areas of side members in connection
a	Dimension of member
a	Distance to load for column bracket
B	Coefficient or factor
b	Width (breadth) of rectangular member
b	Width of member at specific horizontal shear plane
b	Width of column flange
C	Coefficient or factor
C_b	Bearing area factor

C_c	Curvature factor
C_D	Load duration factor
C_d	Penetration depth factor
C_{di}	Diaphragm factor
C_{dt}	Constant for tapered beam deflection
C_{eg}	End grain factor
C_F	Size factor
C_{fu}	Flat use factor
C_g	Group action factor
C_I	Stress interaction factor
C_L	Beam stability factor
C_M	Wet service factor
C_P	Column stability factor
C_r	Repetitive member factor
C_r	Reduction factor for pitched and tapered curved beams
C_{st}	Metal side plate factor
C_t	Temperature factor
C_{tn}	Toenail factor
C_V	Volume factor
C_x	Spaced column fixity factor
C_y	Factor for tapered beam deflection
C_Δ	Geometry factor
$C_{\Delta e}$	Edge distance (geometry) factor
$C_{\Delta n}$	End distance (geometry) factor
$C_{\Delta s}$	Spacing (geometry) factor
c	Distance from neutral axis to extreme fiber (outer surface)
c	Camber
D	Dead load
D	Diameter
D_r	Root diameter (of fastener)
d	Depth of member
d	Pennyweight of nail or spike
d_c	Depth of member at midspan
d_e	Depth of member at end
d_e	Effective depth at connection
d_t'	Depth of curved beam at tangent point measured perpendicular to bottom face
d_x'	Depth of beam at distance x measured perpendicular to sloped bottom face tangent point
d_1, d_2	Compression member cross-section dimensions
E	Earthquake load
E, E'	Design value and adjusted modulus of elasticity
E_m	Modulus of elasticity of main member in connection
E_s	Modulus of elasticity of side members in connection

E_x, E'_x Design value and adjusted modulus of elasticity for glued laminated timber loaded perpendicular to wide faces of laminations
E_y, E'_y Design value and adjusted modulus of elasticity for glued laminated timber loaded parallel to wide faces of laminations
E_{05} Lower fifth percentile modulus of elasticity
e Eccentricity of load
e_1, e_2 Eccentricity of column load
F Fluid pressure load
$F_b, F'_b, F_{bx}, F'_{bx}, F_{by}, F'_{by}, F_{b1}, F'_{b1}, F_{b2}, F'_{b2}$ Design values and adjusted (allowable) stress values for bending
F_{bE} Critical buckling design value for bending member
F_c, F'_c Design value and adjusted (allowable) compressive stress parallel to the grain
F_{cE} Critical buckling design value for compression member
$F_{c\perp}, F'_{c\perp}$ Design value and adjusted (allowable) compressive stress perpendicular to the grain
F_e Dowel bearing strength
F_{em} Dowel bearing strength of main member
F_{es} Dowel bearing strength of side member(s)
$F_{e\|}$ Dowel bearing strength parallel to the grain
$F_{e\perp}$ Dowel bearing strength perpendicular to the grain
$F_{e\theta}$ Dowel bearing strength at angle θ with respect to the grain
F_{rt}, F'_{rt} Design value and adjusted (allowable) radial tension stress
F_t, F'_t Design value and adjusted (allowable) tension stress
F_{vt}, F'_{vt} Design value and adjusted (allowable) torsional stress
f_b Bending stress (extreme fiber)
f_c Compressive stress parallel to the grain
$f_{c\perp}$ Compressive stress perpendicular to the grain
f_o Reference stress for pitched and tapered curved beams
f_r Radial stress
f_{rc} Radial compressive stress
f_{rt} Radial tension stress
f_{vt} Maximum torsional stress of rectangular member
G Specific gravity
G Shear modulus (modulus of rigidity)
G_m Specific gravity at specified moisture content
G_0 Specific gravity based on oven-dry weight and oven-dry volume
g Gauge of screw
H Lateral earth pressure load
h Rise in middepth from end to apex
h_a Height of apex
h_s Height of soffit at midspan

I	Initial moisture content
I	Moment of inertia
K	Factor applied to dead load for deflection criteria
K_{bE}	Euler buckling coefficient for bending
K_{cE}	Euler buckling coefficient for columns
K_D	Diameter coefficient for dowel fasteners
K_e	Equivalent length coefficient for column buckling
K_r	Radial stress factor for pitched and tapered curved beams
K_θ	Bending stress coefficient for pitched and tapered curved beams
k	Thermal conductivity
L	Live load
L	Span or overall length
L_r	Roof live load
ℓ	Span or length
ℓ_b	Bearing length
ℓ_c	Length of curved portion of beam
ℓ_e	Effective length of compression or bending member
ℓ_t	Length of tapered end portion of pitched and tapered curved beam
ℓ_1,ℓ_2	Distance between points of lateral support in planes 1 and 2 of compression member
ℓ_3	Distance from center of spacer block to centroid of group of connectors in spaced column
M	Bending moment
M'	Adjusted moment resistance (LRFD)
M_u	Factored moment (LRFD)
M_x,M_y	Bending moments with respect to x and y directions (typically, load perpendicular to wide faces of laminations and load parallel to wide faces of laminations, respectively)
m	Moisture content
m_f,m_i	Final and initial moisture content values
MF	Magnification factor for ponding
N,N'	Nominal and allowable lateral fastener design value at an angle to the grain
n	Number of fasteners in a row
n_C	Number of rivets per row parallel to direction of load
n_R	Number of rows of rivets
P	Axial load or total concentrated load
P,P'	Design or nominal and adjusted or allowable fastener load parallel to the grain
P,P'	Nominal and allowable timber rivets joint capacity parallel to the grain
P_r,P_w	Nominal timber rivets joint capacity parallel to the grain based on rivets and wood, respectively

P_s	Assumed side load used for columns with side bracket loads
p	Penetration (effective) of fastener
Q	Moment of area about neutral axis
Q,Q'	Design or nominal and adjusted or allowable fastener load perpendicular to the grain
Q,Q'	Nominal and allowable timber rivets joint capacity perpendicular to the grain
Q_r,Q_w	Nominal timber rivets joint capacity perpendicular to the grain based on rivets and wood, respectively
R	Thermal resistance
R,R_V	Vertical reaction
R	Radius of curvature of inside face of lamination
R	Soffit radius of curved section of pitched and tapered curved beam
R	Required spacing along connector axis
R	Rain load
R_B	Slenderness ratio for bending member
R_d	Reduction term in fastener design
R_H	Horizontal reaction
R_m	Radius of curvature (at middepth) of curved member
r	Radius of gyration
S	Snow load
S,S_x,S_y	Section modulus
S	Tributary width for bending member (ponding analysis)
S	Stringer spacing
S_C	Curved length (at middepth) of pitched and tapered curved beam
S_m	Percent shrinkage from initial to final moisture content
S_0	Total percent shrinkage from green to oven-dry conditions
s	Spacing of fasteners
s	Effective bridge deck span
T	Temperature
T	Tensile force
T	Torque
T	Self-straining load
t	Thickness (lamination, column flange, bridge deck)
t_m	thickness of main member in connection
t_s	Thickness of side member(s) in connection
V	Shear force
V'	Adjusted shear resistance (LRFD)
V_u	Factored shear (LRFD)
W	Wind load
x	Horizontal distance or location
y	Vertical distance or location or distance from neutral axis
Z,Z'	Design value and adjusted (allowable) load for single fastener

Z'_{GT}	Allowable load for group of fasteners based on group tear-out
Z'_{RT}	Allowable load for row of fasteners based on row tear-out
$Z_{\parallel}, Z'_{\parallel}$	Design value and adjusted (allowable) load for single fastener parallel to the grain
$Z_{\perp}, Z'_{\perp}$	Design value and adjusted (allowable) load for single fastener perpendicular to the grain in both members
$Z_{m\perp}, Z'_{m\perp}$	Design value and adjusted (allowable) load for single fastener perpendicular to the grain in main member and parallel to the grain in side member(s)
$Z_{s\perp}, Z'_{s\perp}$	Design value and adjusted (allowable) load for single fastener perpendicular to the grain in side member(s) and parallel to the grain in main member
Z'_{α}	Allowable load on fastener with load at angle α with wood surface

Greek Symbols

α	Angle; angle of sloped end with respect to grain
α	Coefficient or factor
β	Coefficient or factor
γ	Specific weight
γ_m	Specific weight of wood at specified moisture content
γ_w	Specific weight of water
θ	Angle of load to the grain
θ	Angle of tapered cut on compression face to end of beam
λ	Ratio of long-term deflection to immediate deflection of sustained loads
λ	Time effect factor (LRFD)
π	Pi (3.1416)
σ_b, σ_Δ	Allowable area loads based on bending and deflection
ϕ	Angle measure
ϕ	Resistance factor (LRFD)
ϕ_B, ϕ_T	Bottom (soffit) and top (roof) slopes of pitched and tapered beam
ω	Distributed (line) load

CHAPTER 1

TIMBER CONSTRUCTION

1.1 INTRODUCTION

The American Institute of Timber Construction (AITC) has developed this *Timber Construction Manual* to provide state-of-the-art technical information and recommendations on engineered timber construction. The first chapters include basic information related to the characteristics and use of wood in modern construction. Topics of the first chapter include economy, permanence, seasoning, handling, storage, erection, and fire safety. With an understanding of these topics, the designer can more effectively utilize the advantages of wood construction. General and specific design information and recommendations for timber construction are covered in subsequent chapters, with actual design examples. Additional design information is provided in the Appendix. References to supplemental information available from AITC are also cited in the text.

This book applies to two types of engineered timber construction: sawn lumber and structural glued laminated timber (*glulam*). Sawn lumber is the product of lumber mills and is produced from many species. Glued laminated timber members are produced in laminating plants by gluing together dry lumber, normally of 1 or 2 in. nominal thickness, under controlled conditions of temperature and pressure. Members with a wide variety of sizes, profiles, and lengths can be produced that have superior characteristics of strength, serviceability, and appearance. Glued laminated timber members are manufactured from a number of species and grades of lumber, resulting in optimum use of the timber resource. The book does not cover the use of proprietary products such as wood I-joists, laminated veneer lumber, and other structural composite lumber.

1.2 ECONOMY

The economic success of the construction of a project may be influenced greatly by design. Important elements of design include, but are not limited to, the layout of the framing, proper selection of materials, proper design of all components, ease of construction, serviceability for the intended use, and durability. The best economy in timber construction is generally realized when standard-size members can be utilized in a repetitious arrangement. However, timber framing, especially glued laminated timber, can be custom fabricated to provide a nearly infinite variety of unique but cost-effective architectural forms and arrangements.

1.2.1 Standard Sizes

The selection of standard sizes and grades in timber construction will result in maximum economy. Standard sizes of glued laminated timber, sawn lumber, and heavy timber are given in Tables 1.1 to 1.3. The member length of glued laminated timber is limited only by transportation and handling restrictions. Standard lengths for sawn lumber are generally available in even 2-ft increments, with the maximum length practically limited by what is available from lumber suppliers.

TABLE 1.1 Standard Sizes for Glued Laminated Timber

	Nominal Width (in.)							
	3[a]	4	5	6	10	12	14	16
Western species	$2\frac{1}{8}$ or $2\frac{1}{2}$	$3\frac{1}{8}$	$5\frac{1}{8}$	$6\frac{3}{4}$	$8\frac{3}{4}$	$10\frac{3}{4}$	$12\frac{1}{4}$	$14\frac{1}{4}$
Southern pine	$2\frac{1}{8}$ or $2\frac{1}{2}$	3 or $3\frac{1}{8}$	5 or $5\frac{1}{8}$	$6\frac{3}{4}$	$8\frac{1}{2}$	$10\frac{1}{2}$	12	14

	Net Depth of Member (in.)		
		Nominal 2-in. Laminations[b]	
Number of Laminations	Nominal 1-in. Laminations	$1\frac{1}{2}$ in.	$1\frac{3}{8}$ in.
4	3	6	$5\frac{1}{2}$
5	$3\frac{1}{4}$	$7\frac{1}{2}$	$6\frac{7}{8}$
6	$4\frac{1}{2}$	9	$8\frac{1}{4}$
7	$5\frac{1}{4}$	$10\frac{1}{2}$	$9\frac{5}{8}$
8	6	12	11
etc.	etc.	etc.	etc.

[a] $2\frac{1}{2}$-in. width may not be available in architectural or premium appearance grades. Consult the manufacturer for availability.

[b] $1\frac{1}{2}$-in.-thick laminations are normal for western softwoods; $1\frac{3}{8}$-in.-thick laminations are normal for southern pine.

1.2.2 Board-Foot Measure

The volume of structural timbers is typically measured in terms of board feet (BF). Large volumes of wood are typically measured in thousands of board feet (MBF) and millions of board feet (MMBF). Reducing the board-foot measure of timbers in a structure generally results in improved economy; therefore, it is important to understand how board footage is measured and to be able to convert into other measures of volume. Board-foot measure is calculated based on nominal dimensions; therefore, board-foot measure must be converted to actual volume for calculations other than for pricing. For example, weights of quantities of wood must be calculated based on actual volume.

One board foot of sawn lumber or timber is equal to 144 in^3 based on nominal thickness, nominal width, and actual length. For example, a 1-ft length of nominal 2 in. × 6 in. or 1 in. × 12 in. lumber measures 1 board foot. Similarly, a 4-in. length of nominal 6 in. × 6 in. timber measures 1 board foot. Structural glued laminated timber is measured based on the nominal dimensions of the input lumber. For example, a $5\frac{1}{8}$ in. × 12 in. Douglas fir glulam beam is manufactured from eight laminations of nominal 2 in. × 6 in. lumber. Each 2 in. × 6 in. lamination measures 1 board foot per lineal foot, resulting in a total board foot measure of 8 board feet per lineal foot

TABLE 1.2 Standard Sizes for Sawn Lumber

	Thickness (in.)			Face Widths (in.)		
		Minimum Dressed			Minimum Dressed	
Item	Nominal	Dry	Green	Nominal	Dry	Green
Boards	$\frac{3}{4}$	$\frac{5}{8}$	$\frac{11}{16}$	2	$1\frac{1}{2}$	$1\frac{9}{16}$
	1	$\frac{3}{4}$	$\frac{25}{32}$	3	$2\frac{1}{2}$	$2\frac{9}{16}$
	$1\frac{1}{4}$	1	$1\frac{1}{32}$	4	$3\frac{1}{2}$	$3\frac{9}{16}$
	$1\frac{1}{2}$	$1\frac{1}{4}$	$1\frac{9}{32}$	5	$4\frac{1}{2}$	$4\frac{5}{8}$
				6	$5\frac{1}{2}$	$5\frac{5}{8}$
				7	$6\frac{1}{2}$	$6\frac{5}{8}$
				8	$7\frac{1}{4}$	$7\frac{1}{2}$
				9	$8\frac{1}{4}$	$8\frac{1}{2}$
				10	$9\frac{1}{4}$	$9\frac{1}{2}$
				12	$11\frac{1}{4}$	$11\frac{1}{2}$
				14	$13\frac{1}{4}$	$13\frac{1}{2}$
				16	$15\frac{1}{4}$	$15\frac{1}{2}$
Dimension lumber	2	$1\frac{1}{2}$	$1\frac{9}{16}$	2	$1\frac{1}{2}$	$1\frac{9}{16}$
	$2\frac{1}{2}$	2	$2\frac{1}{16}$	3	$2\frac{1}{2}$	$2\frac{9}{16}$
	3	$2\frac{1}{2}$	$2\frac{9}{16}$	4	$3\frac{1}{2}$	$3\frac{9}{16}$
	$3\frac{1}{2}$	3	$3\frac{1}{16}$	5	$4\frac{1}{2}$	$4\frac{5}{8}$
	4	$3\frac{1}{2}$	$3\frac{9}{16}$	6	$5\frac{1}{2}$	$5\frac{5}{8}$
	$4\frac{1}{2}$	4	$4\frac{1}{16}$	8	$7\frac{1}{4}$	$7\frac{1}{2}$
				10	$9\frac{1}{4}$	$9\frac{1}{2}$
				12	$11\frac{1}{4}$	$11\frac{1}{2}$
				14	$13\frac{1}{4}$	$13\frac{1}{2}$
				16	$15\frac{1}{4}$	$15\frac{1}{2}$
Timbers	5 and thicker	—	$\frac{1}{2}$ off	5 and wider	—	$\frac{1}{2}$ off

TABLE 1.3 Heavy Timber Sizes and Minimum Equivalent Glued Laminated Timber for Heavy Timber Construction (Width × Depth)

Minimum Nominal Size (in.)	Minimum Glued Laminated Timber Net Size (in.)	
	$1\frac{1}{2}$-in.-Thick Laminations	$1\frac{3}{8}$-in.-Thick Laminations
8×8	$6\frac{3}{4} \times 9$	$6\frac{3}{4} \times 8\frac{1}{4}$
6×10	$5\frac{1}{8} \times 10\frac{1}{2}$	5 or $5\frac{1}{8} \times 11$
6×8	$5\frac{1}{8} \times 9$	5 or $5\frac{1}{8} \times 8\frac{1}{4}$
6×6	$5\frac{1}{8} \times 6$	5 or $5\frac{1}{8} \times 6\frac{7}{8}$
4×6	$3\frac{1}{8} \times 7\frac{1}{2}$	3 or $3\frac{1}{8} \times 6\frac{7}{8}$

for the beam. Similarly, a $5\frac{1}{8}$ in. × $12\frac{3}{8}$ in. southern pine beam, manufactured from nine laminations of nominal 2 in. × 6 in. lumber, measures 9 board feet per lineal foot.

1.2.3 Standard Connection Details

A great variety of fasteners and connection hardware is readily available for wood construction. Where possible, code-approved fasteners and hardware should be used and will generally result in the best economy. In applications requiring custom design, typical construction details are provided in AITC 104 [1] and Chapter 8 of this book to assist in the design of safe and permanent connections.

1.2.4 Framing Systems

There is a great variety of structural timber framing systems. The relative economy of any one system over another will depend on the particular requirements of a specific job. Consideration of the overall structure, intended use, geographical location, required configuration, and other factors play an important part in determining the framing system to be used on a job. Table 1.4 may be used for preliminary design purposes to determine the economical span ranges for various timber framing systems. It must be emphasized that the table is to be used for preliminary planning purposes only. Any particular project will require a more extensive analysis for selection of the final framing system.

Additional considerations, when applied to timber framing system design, tend to reduce costs. Connections (joints) should be simple and as few as practically possible. Splices can be used to minimize design, fabrication, and erection problems. Unnecessary variations in members should be avoided; that is, identical members should be used repetitively where practical and the number of variations kept to a minimum. Certain roof profiles will affect the amount and type of load on a structure and may therefore affect economy. Continuous spans and cantilever systems may be used to balance positive and negative moments and reduce deflections, and may therefore result in lower costs.

1.2.5 Structural Grades

For projects using sawn lumber and/or stock glued laminated members, suppliers should be consulted for the relative availability and economy of various grades and species. For large jobs utilizing numerous glued laminated timbers, and for custom members, better economy may be obtained by specifying the required design stresses for the members, allowing the manufacturer flexibility

TABLE 1.4 Economic Span Ranges for Various Timber Framing Systems

Type of System	Economical Span Range (ft)	Considerations
A. Primary Framing Systems		
	Roof Framing Systems	
Beams		Beam systems are frequently used where a low-pitched roof shape is desired
Simple spans		
Straight beams		
Glued laminated	10–100	
Sawn	6–32	
I Joists	12–40	
Tapered beams	25–100	
Double tapered-pitched beams	25–100	
Curved beams	25–100	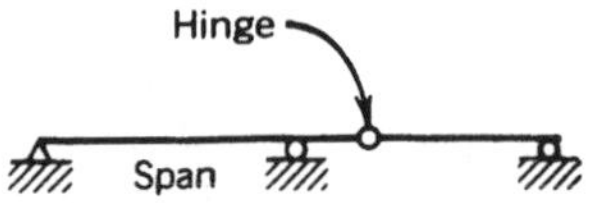 Hinge
Cantilevered systems		Span
Glued laminated	Up to 90	
Sawn	Up to 24	Usually more economical than simple spans when span is over 40 ft
Continuous spans		 Span
Glued laminated	10–32	
Sawn	Up to 16	
Girders	40–80	
Arches		
Three-hinged arches		For relatively high rise applications
Gothic	40–90	
Tudor	20–120	Provides required vertical wall frame
A-Frame	20–100	
Three centered	40–250	
Parabolic	40–250	
Radial	40–250	
Two-hinged arches		For relatively low rise applications
Radial	50–200	
Parabolic	50–200	

TABLE 1.4 (*Continued*)

Type of System	Economical Span Range (ft)	Considerations
Trusses (Heavy)		Provide openings for passage of wiring, piping, etc.
Flat or parallel chord	50–150	Low roof profile
Triangular or pitched	50–90	For pitched roofs requiring flat surfaces
Bowstring (continuous chord)	50–200	Provide greatest clearance with least wall height
Carrying	40–60	
Trusses (light)		Most light trusses commonly used within these ranges are based on proprietary connections and fabrication methods
Flat or parallel chord	20–50	
Triangular or pitched	20–75	
Tied arches	50–200	Good where no ceiling is wanted; give clear open appearance for low-rise curve; normally more expensive than bowstring; buttress not required
Dome structures	50–500+	
	Floor Framing Systems	
Beams		
Simple span		
Glued laminated	6–40	
Sawn	6–20	
I Joists	12–30	
Continuous	25–40	
B. Secondary Framing Systems		
	Roof Framing Systems	
Sheathing and decking		
1-in. sheathing applied directly to primary system	1–4	Check deflection on spans greater than 32 in.
2-in. roof deck applied directly to primary system	6–10	Check deflection on spans greater than 8 ft
3-in. roof deck applied directly to primary system	8–15	2-, 3-, and 4-in. decking provide good insulation, fire resistance, appearance; easy to erect

TABLE 1.4 (*Continued*)

Type of System	Economical Span Range (ft)	Considerations
4-in. roof deck applied directly to primary system	12–20	
Plywood or structural panel sheathing applied directly to primary system	1–4	
Stressed skin panels	8–40	
Joists with sheathing	16–24	
Purlins with sheathing	16–36	
Beams	20–40	
	Floor Framing Systems	
Plank decking		Floor and ceiling in one
Edge to edge	4–16	
Wide face to wide face	4–16	
Joists with sheathing	10–24	

in use of laminate materials. A guide for specifying glued laminated timber is provided in Chapter 8.

1.2.6 Appearance Grades for Glued Laminated Timber

An additional consideration related to economy is appearance. AITC has developed specifications for four standard appearance grades of glued laminated members (framing, industrial, architectural, and premium). Details of these grades are given in *Standard Appearance Grades for Structural Glued Laminated Timber,* AITC 110 [2]. Appearance grades are not related to strength. It is generally more economical to specify the appearance grade best suited for each job, or perhaps different members in a particular job, than to require the best appearance grade for all jobs or members.

1.3 PERMANENCE

Timber structures may be built not only to be structurally adequate but also to be practically permanent, with a minimum of maintenance. With proper design details, construction procedures, and use, wood is a permanent construction material. Certain conditions affect permanence and required maintenance. If proper consideration is given to these conditions in the design phases of a project, there will be greater assurance that the structure will be permanent and that maintenance will be minimal.

1.3.1 Wood Decay, Mold, and Stain

Decay of wood is caused by fungi that grow from microscopic spores. These spores are present wherever wood is used. The fungi use wood substance as their source of nutrition. If deprived of any one of the four essentials of life (nutrients, air, moisture, and favorable temperature), decay growth is prevented or stopped and the wood remains sound, retaining its existing strength with no further deterioration. Wood will not be attacked by fungi if it is submerged in water (thereby excluding air), kept continuously below 20% moisture content (excluding sufficient free moisture), or maintained at temperatures below freezing or much above 100°F. Growth can begin or resume whenever conditions are favorable. In the design of wood structures, fungal decay is typically prevented by assuring a low moisture content or by making the wood toxic to the fungi by pressure preservative treatment (where low moisture content cannot be assured). The *Standard for Preservative Treatment of Structural Glued Laminated Timber,* AITC 109 [3], provides information about preservative treatment of glued laminated timber. The American Wood-Preservers' Association [4] provides information on the treatment of sawn lumber.

Wood has a proven performance of indefinitely long service without special treatments if it is kept below 20% moisture content. When wood is exposed to the weather and not properly protected by a roof, eave overhang, or similar covering, or is subjected to other conditions of free water or high relative humidity, either preservative treatment is required or wood that is naturally decay resistant must be used. Woods naturally resistant to decay include the heartwood of Alaska cedar and redwood. (*Wood Handbook* [5] lists additional domestic woods that are naturally resistant to decay.) The need for preservative treatment (or decay resistance) is a design consideration based on the conditions intended for the wood in service. AITC Technical Note 12 [6] provides additional information related to permanence.

Wood may also experience mold and stain. Molds and stains are confined largely to sapwood and are of various colors. Molds generally do not stain the wood but produce surface blemishes varying from white or light colors to black that can often be brushed off. Fungal stains may penetrate the wood and normally cannot be removed by scraping or sanding. The presence of molds and stains are not necessarily signs of decay, as stain-producing fungi do not attack the wood substance appreciably. For most uses in which appearance is not a factor, stains alone are not necessarily unacceptable, as wood strength is practically unaffected. Ordinarily, the only effects of stains and mold are confined to those properties that determine shock resistance or toughness. The only danger is that the early stages of decay may also be hidden in the discolored areas of molds or stain.

Under conditions that favor their growth, decay-producing fungi, may attack both heartwood and sapwood. Heartwood is more resistant than sapwood to attack. Fresh surface growths are usually fluffy or cottony, although seldom powdery like the surface growths of molds. The early stages of decay are often accompanied by discoloration of the wood, which is more evident on freshly exposed surfaces of unseasoned wood than on dry wood. However, many fungi produce early stages of decay that are similar in color to that of normal wood or give the wood a water-soaked appearance. Later stages of decay produce observable changes in color and texture and wood volume. Decayed wood has reduced strength and fire resistance. In the extreme, the wood appears rotten and crumbly and reduces the member section. *Dry rot* is wood that has decayed in the presence of moisture and has subsequently become dry.

Prevention of decay in wood structures is best accomplished by keeping wood members at low moisture content. Proper construction and connection details are important in decay prevention. Storage, handling, and erection considerations related to decay prevention are also important and are covered in the following section. Rainwater and melted snow must be directed away from wood members by adequately sloped framing and appropriate flashing. The building envelope should include an appropriate moisture barrier. The American Society of Heating, Refrigeration, and Air-Conditioning Engineers' *Fundamentals Handbook* [7] may be consulted, as well as the local building authority, for guidance in providing an effective moisture barrier. Roof overhangs with gutters and downspouts are advised. Finish grade around a structure should be sloped to direct surface and subsurface runoff away from the structure. Untreated wood must be placed above finish grade (typically, a minimum of 6 in.). Wood in contact with soil or concrete must be pressure treated or be wood recognized for its resistance to decay in moist conditions and should be separated from the concrete or soil by a sealing membrane. Girder and joist openings in masonry and concrete walls should be large enough to assure that there will be an airspace around the sides and ends of these wood members. If the members are below the outside soil level, moisture proofing of the outer face of the walls is essential. Enclosed spaces such as attics and crawl spaces must be adequately ventilated. Moisture from soil floors in crawl spaces can be inhibited from moving into the crawl space by covering the soil with a vapor barrier. Special consideration should be taken for structures containing significant sources of moisture, such as pools. Special consideration should also be taken for parts of the structure that are outside the building envelope, such as decks. Prescriptive requirements for wood construction in model building codes generally take into consideration decay potential and generally reflect many of the safeguards noted above. Periodic inspection of wood structures is recommended to identify signs of excess moisture, decay, or damage to the wood–moisture separation.

1.3.2 Insects

In terms of economic loss, the most destructive insect to attack wood buildings is the subterranean termite. In certain localities, aboveground termites are also very destructive. Other insects attack timber buildings, but ordinarily, these occurrences are rather rare and their damage is slight. In many cases, these insects can be controlled by the methods used for termites.

1.3.2.1 Subterranean Termites The extent of the occurrence of damage from subterranean and nonsubterranean termites is shown in Figure 1.1. Damage occurrence in general is much greater in southern states, where temperature conditions are more favorable. However, damage to individual buildings may be just as great in northern states. Subterranean termites develop and maintain colonies in the ground from which they build tunnels through the earth and around obstructions to get at the wood they need for food. The worker members of the colony cause the destruction of wood. At certain seasons of the year, male and female winged forms swarm from the colony, fly a short time, lose their wings, mate, and if they succeed in locating suitable places, start new colonies. The occurrence of flying termites (similar in appearance to flying ants) or their shed wings may be an indication of a nearby colony.

Figure 1.1 Termite damage in the continental United States. A, northern limit of recorded damage done by subterranean termites; B, northern limit of damage done by dry-wood or nonsubterranean termites. (From *Wood Handbook* [5].)

Subterranean termites do not establish themselves in buildings by being carried there in lumber, but by entering from ground nests after the building has been constructed. Termites must have continual access to moisture, such as from the soil. Signs of the presence of termites are the earthen tubes, or runways, built by these insects from the ground over the surfaces of foundation walls to reach the wood above. In wood, the termites make galleries that follow the grain, often concealed by a shell of sound wood. Because the galleries seldom show on the wood surface, probing may be necessary to identify termite infestation and damage.

Where subterranean termites are prevalent, the best protection is to build so as to prevent their gaining access to the building from the ground. Construction details prescribed by model building codes separating wood from soil and moisture to prevent decay are also useful for preventing termite infestation. In general, foundations must be of preservative-treated wood, concrete, or other material through which the termites cannot penetrate. Cement mortar should be used with masonry foundations, as termites can work through some other types of mortar. Wood that is not impregnated with an effective preservative must be kept away from the ground. Basement floors should preferably be concrete slab on grade. In general, wood floor framing must be treated if within 18 in. of the ground below the floor. Untreated posts must stand off at least 1 in. above concrete slabs unless the slab has been protected from moisture and pest infestation. Moisture condensation on the floor joists and subfloor, which may cause conditions favorable to decay and thus make the wood more attractive to termites, can be avoided by covering the soil with a waterproof membrane. Expansion joint material for slabs must be pest-resistant or treated to resist termites.

All concrete forms, stakes, stumps, waste wood, and other potential sources of infestation must be removed from a building site at the time of construction. Where protection is needed in addition to that obtained by physical methods, the soil adjacent to foundation walls and piers beneath a building may be thoroughly treated with a recognized insecticide.

1.3.2.2 Aboveground Termites Nonsubterranean or dry-wood termites have been found only in a narrow strip of territory extending from central California around the southern edge of the continental United States to Virginia and also in the West Indies and Hawaii (Figure 1.1). Their damage is confined to an area in southern California, to parts of southern Florida, notably Key West, and to Hawaii. Nonsubterranean termites are fewer in number, and their depredations are not rapid, but if they are allowed to work unchecked for a few years, they can occasionally ruin timbers with their tunneling. In the principal damage areas, careful examination of wood is needed to avoid the occurrence of infestations during construction of a building. All exterior wood can be protected by placing fine-mesh screen over all holes in the walls or roof of the building. If a building is found to be infested by dry-wood termites, badly damaged wood must be replaced. Further termite activity can be arrested by approved chemical treatments applied under proper supervision, which will provide for safety of people, domestic animals, and wildlife. Where practical, fumigation is another method of destroying insects.

1.3.2.3 Other Insects Large wood-boring beetles and wood wasps may infect green wood and complete their development in seasoned wood. The borers can be killed by heating the wood to a center temperature of 130°F for 1 hour or by fumigation. Once seasoned wood has been cleared of borers, it will not be reinfested. If infested wood is used in a building, emerging adults may bore $\frac{1}{8}$- to $\frac{1}{2}$-in. holes to the surface, penetrating insulation, vapor barriers, siding, or interior surface materials. Infested members may be damaged and require structural or cosmetic repair. If the remaining structure has been built with seasoned wood without initial infestation, infestation of the other members will generally not take place.

Powder-post beetles can infest and reinfest dry wood, reducing it to floury sawdust. *Lyctus* powder-post beetles, which are encountered most frequently, attack large-pored hardwoods. Their attacks may be recognized by tunnels packed with floury sawdust and numerous emergence holes $\frac{1}{32}$ to $\frac{1}{8}$ in. in diameter. Heat or fumigation treatments will kill the beetles but will not prevent reinfestation. Infestation can be prevented by a surface application of an approved insecticide in a light-oil solution. Any finishing material that plugs the surface holes of wood will also protect the wood from *Lyctus* attack. Usually, infestations in buildings result from the use of infested wood, and insecticidal treatment or fumigation may be needed to eliminate them.

Carpenter ants chew nesting galleries in wood. The principal species are large dark-colored ants, and individuals in the colony may be $\frac{1}{2}$ in. long. They exist throughout the United States. Because the ants require a nearly saturated atmosphere in their nest, an ant infestation may indicate a moisture problem in the wood that could also result in decay damage. Ant infestations can be controlled by insecticides or by keeping the wood dry.

1.3.3 Marine Borers

Fixed or floating wood structures in salt or brackish water are subject to attack by marine borers. Marine borers include shipworms such as *Teredo* and *Bankia,* the pholads *Martesia* and *Zylophaga,* and *Limnoria* and *Sphaeroma*. Almost all attack wood as free-swimming organisms in the early part of their lives. Shipworms and pholads bore an entrance hole generally at the waterline, attach themselves, and grow in size as they bore tunnels into the wood. *Limnoria* and *Sphaeroma* generally burrow just below the surface of the water.

1.3.3.1 Protection and Control For areas where shipworm and pholad attack are known or expected and where *Limnoria* attack is not expected, the wood should be pressure treated with a creosote and/or creosote–coal tar solution. For areas where *Limnoria* and pholad attack are known or expected, dual treatment of waterborne salts and creosote is recommended. Where *Limnoria* attack is known or expected and where pholads are absent, either a dual treatment or waterborne salt preservatives may be used.

1.3.4 Checking

Wood naturally expands and shrinks with increases and decreases in moisture content. Unless wood is by some means sealed from the atmosphere or other sources of contact with moisture, the wood will expand and shrink as it gains or loses water. In the case of rapid drying, wood at the surface tends to shrink faster than the inner wood, and the surface wood fibers may separate, causing *drying checks*. Drying checks are common and may be expected in sawn lumber. Drying checks may also be expected in glued laminated members, although to a lesser extent, due to the manufacturing process of the members. Glued laminated members are manufactured using wood with controlled moisture content (lamination to lamination) and at moisture contents generally close to the expected end-use conditions of the members. Proper storage, handling, and final construction details of wood members will minimize checking of structural significance. In cases where checking is considered severe, AITC Technical Notes 11 [8] and 18 [9] may be used to evaluate the significance of checking in glued laminated members and ASTM D245 [10] for sawn members.

1.3.5 Temperature

Wood may be exposed temporarily to temperatures up to 150°F without loss of strength. Where exposed to sustained temperatures in excess of 100°F, wood suffers loss in strength and stiffness as the wood substance is degraded. Strength loss is greater in members subject simultaneously to high moisture content. Design adjustment factors for wood construction take into consideration strength loss at sustained high temperature as well as high temperature and moisture content. Wood at low temperatures tends to have greater strength than at normal temperatures.

1.3.6 Chemical Environments

Wood is superior to many other common construction materials in its resistance to chemical attack. For this reason, wood is used in storage buildings and for containers for many chemicals and in processing plants in which structural members are subjected to spillage, leakage, or condensation of chemicals. Wooden tanks are commonly employed for the storage of water or chemicals that deteriorate other materials and have the unique feature of being self-sealing due to the expansion of wood where exposed to moisture. Experience has shown that the heartwood of cypress, Douglas Fir–Larch,

Southern Pine, and California redwood is most suitable for water tanks and that the heartwood of the first three of these species is most suitable for tanks when resistance to chemicals in appreciable concentrations is an important factor. More information on the use of wood in chemical environments is provided in Chapter 2.

1.4 SEASONING, HANDLING, STORAGE, AND ASSEMBLY

Seasoning, in general, is the drying of wood from its wet or green condition when first cut to the end condition in which it is in equilibrium with its surroundings. Since wood shrinks as it dries (as discussed previously and in more detail in Chapter 2), it is preferable that wood used during construction be predried to a low moisture content, because the equilibrium moisture content for most end uses in buildings is generally low. Excessive shrinkage of green wood after installation may cause structural and serviceability problems. In its end use, wood used in construction will experience changes in moisture content as surrounding temperature and humidity conditions change. In structures with a good building envelope, these changes tend to be slow, and the dimensional changes associated with them are generally accommodated by the wood, fastenings, and properly detailed connections themselves.

If green wood is used in construction (although not generally recommended), or wood with a significantly higher moisture content than end use, physical accommodations may need to be made for the dimensional changes associated with shrinkage to equilibrium. This is particularly true if the wood members are large and attached to relatively rigid, structural or nonstructural, elements. Equations for estimating wood shrinkage amounts are provided in Chapter 2. Where temperature and moisture conditions during construction differ significantly from the building end use (e.g., construction during cold seasons), and where large timber members are exposed to view, it is recommended that a building be brought to end-use conditions gradually, to reduce the occurrence of checking.

Wood is not a particularly hard material, so care must be taken during handling not to damage the wood surface or members themselves. Forklifts or cranes should be used to lift wood from delivery vehicles to a construction site. Upon delivery, wood members should be inspected for damage and tally. At the job site, larger members should be handled with fabric slings, as chains and cables tend to mark the wood surfaces. Wood may also be marked or scuffed easily at the site if not protected from soil and traffic.

Dimension lumber delivered to a job site should be kept off the ground and protected from sunlight and moisture. Wrapped dimension lumber and glued laminated members (generally wrapped) should also be stored off the ground and the wrapping maintained until used, except that the underside should be punctured to allow drainage of excess moisture or condensation. The wrapping of glued laminated members should not be removed until the building has been enclosed. Wood tends to creep under load, so it should not be stored haphazardly or in ways that will introduce undesirable twists or bends.

Timber trusses are usually shipped partially or completely disassembled and are assembled on the ground at the site before erection. Arches, which are generally shipped in halves, may be assembled on the ground, or connections may be made after the half arches are in position. When trusses and arches are assembled on the ground at the site, they should be assembled on level blocking to permit connections to be fitted properly and tightened securely without damage. The end compression joints should be brought into full bearing and compression plates installed where specified. Before erection, the assembly should be checked for prescribed overall dimensions, prescribed camber, and accuracy of anchorage connections. Erection should be planned and executed in such a way that the close fit and neat appearance of joints and the structure as a whole will not be impaired.

Anchor bolts should be checked prior to start of erection. Before erection begins, all supports and anchors should be complete, accessible, and free of obstructions. The weights and balance points of the structural timber framing should be determined before lifting begins so that proper equipment and lifting methods may be employed. When long members or timber trusses of long span are raised from a flat to a vertical position preparatory to lifting, stresses entirely different from the normal design stress may be introduced. The magnitude and distribution of these stresses will vary depending on such factors as the weight, dimensions, and type of member. A competent rigger should consider these factors in determining how much suspension and stiffening, if any, is required, and where it should be located.

All framing must be true and plumb. Temporary erection bracing must be provided to hold the framing in a safe position until sufficient permanent bracing is in place. Proper and adequate erection bracing must accommodate all loads to which the structure may be subjected during erection, including loads from equipment and its operation. Final tightening of alignment bolts should not be completed until the structure has been properly aligned. Temporary erection bracing should be removed only after diaphragms and permanent bracing are installed, the structure has been aligned properly, and connections and fastenings have been finally tightened. Retightening of connections prior to final completion or closing in of inaccessible connections is recommended. Field welding of structural connections must be performed in accordance with accepted standards for steel construction and welding [11,12].

All field cuts in timbers should be coated with an approved moisture seal if the member was initially coated, unless otherwise specified. If timber framing has been pressure treated, field cutting after treatment must be avoided or at least, insofar as possible, held to a minimum. When field cuts in pressure-treated material are unavoidable, additional treatment should be provided in accordance with AWPA Standard M4 [13].

During erection operations, all timber framing that requires moisture content control, whether sawn or glued laminated timbers, should be protected against moisture pickup. Any fabricated structural materials to be stored for an extended period of time before erection should, insofar as is practicable, be assembled into subassemblies for storage purposes. Additional information on handling, storage, and erection of glued laminated timbers is provided in AITC 111 [14].

1.5 FIRE SAFETY

Fire safety of structures is a function of building materials, fire detection, exiting conditions, fire suppression, and the use of the structure, as well as proximity of other structures or sources of fire. Although wood itself is a combustible material, it can be used in construction in ways that are acceptably safe. The fire and life safety provisions of model building codes, adopted and made law by local building authorities, place restrictions on all construction, established by history and research, to provide reasonable safety to building occupants. These provisions generally do not produce fire-proof structures, but instead, structures that endure the occurrence of fires in ways that allow the occupants to exit safely.

Depending on the use and occupancy of a structure, construction may be required to have a certain fire rating. Wood stud walls, for example, covered (protected) by fire-rated gypsum, may have a fire rating of, for example, 1 hour. A great number of wall and floor assemblies have been fire tested and have accepted fire ratings. Many of these assemblies are already recognized in the model building codes. Certain fire ratings for structural wood members may be achieved by protecting the members with fire-rated gypsum, or the natural fire-enduring characteristics of the wood may be used alone. When exposed to heat and/or flame, wood forms a self-insulating surface layer of char. Although the surface chars, the undamaged inner wood below the char retains its strength and may support loads equivalent to the capacity of the uncharred section. Very often, heavy timber members will retain their structural integrity through long periods of fire exposure and remain serviceable after the surface has been cleaned and refinished. The fire endurance and excellent performance of heavy timber is attributable to the size of the wood members and to the slow rate at which the charring penetrates. AITC Technical Note 7 [15] and AF&PA Technical Report 10 [16] may be used to calculate the fire resistance of glued laminated members. In cases where calculated fire resistance is inadequate, members may be oversized to provide the necessary resistance.

Unlike most other construction materials, timber is not distorted appreciably by high temperatures; therefore, it is not likely that walls will be displaced or pushed over by expanding members, as often happens with some other materials. For fire-resistance classification purposes, buildings of wood construction are generally classified as heavy timber construction, ordinary construction, or wood frame construction. In heavy timber construction, fire resistance is attained by placing limitations on the minimum size, thickness, or composition of all load-carrying wood members; by avoiding concealed spaces under floors or roofs; by providing the required degree of fire resistance in the exterior and interior walls; and by other construction details. Ordinary construction has exterior masonry walls and wood framing members of sizes smaller than heavy timber sizes. Wood frame construction has wood-framed walls and structural framing of sizes smaller than heavy timber sizes. Depending on the occupancy of a building or the hazard of the operations within it, a building of wood frame or ordinary construction may have its members covered with fire-resistive coverings.

All the nationally recognized building codes and most other codes recognize heavy timber construction by allowing larger building areas and uses for which ordinary construction and wood frame construction are not permitted. The requirements for heavy timber construction are contained in *Standard for Heavy Timber Construction,* AITC 108 [17].

CHAPTER 2

PHYSICAL PROPERTIES OF WOOD

2.1 INTRODUCTION

Wood is a cellular organic material made up primarily of cellulose, which comprises the structural units (*cells*), and *lignin,* which bonds the structural units together. Wood cells are hollow and vary from about 0.04 to 0.33 in. in length and from 0.0004 to 0.0033 in. in diameter. Most cells are elongated and are oriented vertically in the growing tree, but some, called *rays,* are oriented horizontally and extend from the bark toward the center or pith of the tree.

2.1.1 Hardwoods and Softwoods

Trees are divided into two broad classes: *hardwoods,* which have broad leaves, and *softwoods* or *conifers,* which have needlelike or scalelike leaves. Most hardwoods shed their leaves annually at the end of each growing season, and most softwoods shed only damaged or unused needles and are thus termed *evergreens.* The notable conifer exceptions are the tamarack and larch, whose needles turn yellow in the fall and are lost, leaving the tree bare of needles through the winter and early spring. The terms *hardwood* and *softwood* are often misleading, because they do not directly indicate the hardness or softness of wood. Some hardwoods are softer than certain softwoods, and some softwoods are harder than some hardwoods.

2.1.2 Heartwood and Sapwood

The cross section of a tree generally shows several distinct zones: the *bark,* which protects the tree and typically has a rough surface; a light-colored zone called *sapwood;* and an inner zone, generally of darker color, called *heartwood.* Trees grow by adding new layers of cells to the outside of the sapwood. The sapwood functions to conduct sap and store nutrients, as well as to support the tree. As the tree continues to grow, the inner layers of the sapwood stop conducting sap and storing nutrients, becoming heartwood, which acts only to support the tree. In some species, heartwood is significantly more resistant to decay than is sapwood. There is no consistent difference between the weight and strength properties of heartwood and sapwood.

2.1.3 Earlywood and Latewood

In climates where temperature limits the growing season of a tree, each annual increment of growth usually is readily distinguishable. Such an increment is known as an *annual growth ring* or *annual ring.* In many wood species, large, thin-walled cells are formed in the spring when growth is fastest, whereas smaller, thicker-walled cells are formed later in the year. The areas of fast growth, called *earlywood* or *springwood,* form the lighter band of the annual ring. The areas of slower growth, called *latewood* or *summerwood,* form the darker band in each ring. Because latewood typically contains more solid wood substance than does earlywood, it is stronger and more dense than earlywood. The proportion of the width of the latewood to the width of the annual ring and the number of rings per inch are used as measures of the quality and strength of wood for some species.

2.1.4 Grain and Texture

The terms *grain* and *texture* are used in many ways to describe the characteristics of wood and, in fact, do not have a definite meaning. Terms such as *close-grained* and *coarse-grained* refer to the width of the annual rings, while *straight-grained* and *cross-grained* indicate whether the fibers are parallel or at an angle to the sides of a particular piece. The terms *parallel to grain* and *perpendicular to grain* are generally structural terms that indicate the direction of load or stress in relation to the longitudinal axis of the wood cells (or fibers). *Texture* generally refers to the fineness of wood structure rather than to the annual rings. When these terms are used in connection with wood, the intended meanings should be clearly defined.

2.1.5 Moisture Content

Water is found in wood as free water in the cell cavities, water absorbed into the cell walls, and bound water. *Moisture content* (MC) is generally defined as the weight of water in a piece of wood expressed as a percentage of the oven-dry weight of the same piece. The moisture content of wood is important in both the design and construction of timber structures. Wood freshly sawn from living trees, termed *green*, may have a moisture content greater than 100%, as green wood in some species contains more water than wood substance itself. The *Wood Handbook* [1] contains average values for the moisture content of heartwood and sapwood for various species. Once cut, wood generally loses moisture, first from the cell cavities (*lumens*), and then from the cell walls. The condition in which the free water has left the cell cavities, but in which the cell walls are still saturated, is termed the *fiber saturation point*. Below the fiber saturation point, wood shrinks and swells with changes in moisture content. Although the moisture content at fiber saturation varies from species to species and from piece to piece within a species, a value of 30% moisture content is generally associated with the fiber saturation point.

Wood generally continues to lose moisture, or *season,* until an *equilibrium moisture content* (EMC) is reached between the wood moisture content and the surrounding air relative humidity and temperature. Equilibrium moisture content values for wood in the temperature and humidity ranges of practical structural importance are shown in Table 2.1. The EMC values shown are essentially independent of species. As the surrounding temperature or humidity conditions change, wood takes on or loses moisture to achieve the corresponding new EMC.

When control of the moisture content of the wood is to be achieved or the seasoning process accelerated, which is often the case with structural wood,

TABLE 2.1 Moisture Content of Wood (%) in Equilibrium with Stated Dry-Bulb Temperature and Relative Humidity[a]

Temperature		Relative Humidity (%)																		
°C	°F	5	10	15	20	25	30	35	40	45	50	55	60	65	70	75	80	85	90	95
−1	30	1.4	2.6	3.7	4.6	5.5	6.3	7.1	7.9	8.7	9.5	10.4	11.3	12.4	13.6	14.9	16.5	18.5	21.0	24.3
4	40	1.4	2.6	3.7	4.6	5.5	6.3	7.1	7.9	8.7	9.5	10.4	11.3	12.4	13.5	14.9	16.5	18.5	21.0	24.4
10	50	1.4	2.6	3.6	4.6	5.5	6.3	7.1	7.9	8.7	9.5	10.3	11.2	12.3	13.4	14.8	16.4	18.4	20.9	24.3
16	60	1.3	2.5	3.6	4.6	5.4	6.3	7.0	7.8	8.6	9.4	10.2	11.1	12.1	13.3	14.6	16.2	18.2	20.7	24.1
21	70	1.3	2.5	3.5	4.5	5.4	6.2	6.9	7.7	8.5	9.2	10.1	11.0	12.0	13.1	14.4	16.0	18.0	20.5	23.9
27	80	1.3	2.4	3.5	4.4	5.3	6.1	6.8	7.6	8.3	9.1	9.9	10.8	11.8	12.9	14.2	15.7	17.7	20.2	23.6
32	90	1.2	2.4	3.4	4.3	5.1	5.9	6.7	7.4	8.1	8.9	9.7	10.6	11.5	12.6	13.9	15.4	17.4	19.9	23.3
38	100	1.2	2.3	3.3	4.2	5.0	5.8	6.5	7.2	7.9	8.7	9.5	10.3	11.2	12.3	13.6	15.1	17.0	19.5	22.9
43	110	1.1	2.2	3.2	4.0	4.9	5.6	6.3	7.0	7.7	8.5	9.2	10.0	11.0	12.0	13.2	14.7	16.6	19.1	22.5
49	120	1.1	2.1	3.0	3.9	4.7	5.4	6.1	6.8	7.5	8.2	8.9	9.8	10.7	11.7	12.9	14.4	16.2	18.6	22.0
54	130	1.0	2.0	2.9	3.7	4.5	5.2	5.9	6.6	7.3	7.9	8.7	9.5	10.3	11.3	12.5	14.0	15.8	18.2	21.5
60	140	0.9	1.9	2.8	3.6	4.3	5.0	5.7	6.3	7.0	7.7	8.4	9.1	10.0	11.0	12.2	13.6	15.4	17.7	21.0
66	150	0.9	1.8	2.6	3.4	4.1	4.8	5.5	6.1	6.7	7.4	8.1	8.8	9.7	10.6	11.8	13.2	14.9	17.2	20.5

[a] Values were calculated using Equation (3.3), *Wood Handbook* [1].

high-temperature kilns are often used to dry the wood to a specified moisture content. Framing lumber is typically *kiln dried* to a moisture content of 19% or less. Lumber used for laminating is typically kiln dried to 16% or less moisture content, with an average moisture content of about 12%. Wood pieces (particularly larger pieces) are sometimes *surface dried,* in which case allowance is made for additional drying of the interior of the pieces. *Oven-dry* refers to the condition in which wood is dried in a laboratory oven until no further moisture can be driven from the wood.

The rate at which wood reaches equilibrium moisture content varies depending on the degree to which the wood has been enclosed or sealed, environmental conditions, and the wood itself. Rapid change in moisture content may cause checking in the wood, as discussed in Chapter 1. Once in use in a structure, the rate of moisture content change is comparable to seasonal or monthly changes in environmental conditions, and as such, monthly or seasonal values of EMC are generally considered. For structures in many locations, the EMC is significantly lower than the fiber saturation point. The *Wood Handbook* contains 30-year average monthly EMC values for approximately 50 U.S. cities.

2.1.6 Growth Characteristics

Wood contains certain natural growth characteristics such as knots, slope of grain, compression wood, and shakes, which may, depending on their size, number, and location in a structural member, affect the strength properties of that member adversely. These characteristics are discussed in detail in the *Wood Handbook* [1]. Structural grading rules take into account the effects of

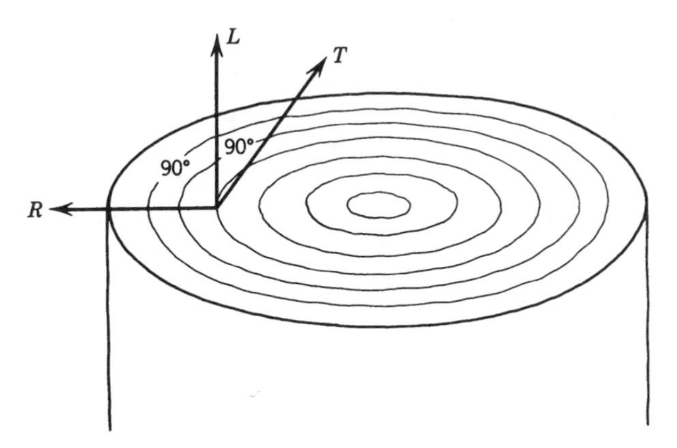

Figure 2.1 Three principal axes of wood: *L*, longitudinal (parallel to the grain); *R*, radial (perpendicular to the grain, radial to annual rings); *T*, tangential (perpendicular to the grain, tangential to annual rings).

these growth characteristics on the strength of wood in establishing design values for lumber and glued laminated timber.

2.1.7 Directional Properties

Wood is anisotropic because of its cellular structure. The structure of wood in any particular log is generally considered to have three axes of symmetry: longitudinal, radial, and tangential (Figure 2.1). Wood pieces sawn from logs are typically oriented with their long axis (or faces) approximately parallel to the longitudinal axis of the log, but with the other faces indiscriminate with respect to the radial and tangential directions. As such, for practical purposes, directional properties of wood are generally distinguished between parallel to grain (longitudinal) and perpendicular to grain. Values perpendicular to grain accommodate both radial and tangential properties.

2.2 SPECIFIC GRAVITY AND SPECIFIC WEIGHT OF COMMERCIAL LUMBER SPECIES

Specific gravity and specific weight values for various lumber species or species combinations are shown in Table 2.2. Detailed descriptions of each species and species combinations can be found in the *Wood Handbook* [1] and the *National Design Specification® for Wood Construction* [2]. Since the weight of a particular piece of lumber may vary with moisture content, specific gravity value is used in timber construction, which is the oven-dry weight of a piece divided by the volume at a specified moisture content, normalized by the specific weight of water. Typical moisture content values for determining specific gravity are green, 12% MC, and oven dry. Table 2.2 shows specific gravity values based on oven-dry weight and 12% MC and oven-dry volumes. The values based on oven-dry volume are used in the *National Design Specification®* for fastener design.

The specific weight values in Table 2.2 include the weight of the water in the wood and are for a moisture content of 12%. Specific weight values at 12% MC are customarily used in design for computing self-weight and dead loads. Specific weight values at other moisture contents up to 30% may be determined as follows:

$$G_m = \frac{G_0}{1 + 0.265\ G_0\ (m/30\%)} \tag{2.1}$$

and

TABLE 2.2 Specific Gravity and Weight of Commercial Lumber Species

Species or Species Combination	Specific Gravity		Specific Weight, 12% MC (pcf)[c]
	Oven Dry Volume[a]	MC 12% Volume[b]	
Aspen	0.39	0.37	26
Balsam Fir	0.36	0.35	24
Beech–Birch–Hickory	0.71	0.66	46
Coast Sitka Spruce	0.39	0.37	26
Cottonwood	0.41	0.39	27
Douglas Fir–Larch	0.50	0.47	33
Douglas Fir–Larch (North)	0.49	0.47	33
Douglas Fir (South)	0.46	0.44	31
Eastern Hemlock	0.41	0.39	27
Eastern Hemlock–Balsam Fir	0.36	0.35	24
Eastern Hemlock–Tamarack	0.41	0.39	27
Eastern Hemlock–Tamarack (North)	0.47	0.45	31
Eastern Softwoods	0.36	0.35	24
Eastern Spruce	0.41	0.39	27
Eastern White Pine	0.36	0.35	24
Engelmann Spruce–Lodgepole Pine	0.38	0.37	26
Engelmann Spruce–Lodgepole Pine (higher grades)	0.46	0.44	31
Hem–Fir	0.43	0.41	29
Hem–Fir (North)	0.46	0.44	31
Mixed Maple	0.55	0.52	36
Mixed Oak	0.68	0.63	44
Mixed Southern Pine	0.51	0.48	34
Mountain Hemlock	0.47	0.45	31
Northern Pine	0.42	0.40	28
Northern Red Oak	0.68	0.63	44
Northern Species	0.35	0.34	24
Northern White Cedar	0.31	0.30	21
Ponderosa Pine	0.43	0.41	29
Red Maple	0.58	0.55	38
Red Oak	0.67	0.63	44
Red Pine	0.44	0.42	29
Redwood, close grain	0.44	0.42	29
Redwood, open grain	0.37	0.36	25
Sitka Spruce	0.43	0.41	29
Southern Pine	0.55	0.52	36
Spruce–Pine–Fir	0.42	0.40	28
Spruce–Pine–Fir (higher grades)	0.50	0.47	33
Spruce–Pine–Fir (South)	0.36	0.35	24
Western Cedars	0.36	0.35	24
Western Cedars (North)	0.35	0.34	24
Western Hemlock	0.47	0.45	31

(*continues*)

TABLE 2.2 (*Continued*)

Species or Species Combination	Specific Gravity: Oven Dry Volume[a]	Specific Gravity: MC 12% Volume[b]	Specific Weight, 12% MC (pcf)[c]
Western Hemlock (North)	0.46	0.44	31
Western White Pine	0.40	0.38	27
Western Woods	0.36	0.35	24
White Oak	0.73	0.68	47
Yellow Poplar	0.43	0.41	29

[a] Specific gravity values from the *National Design Specification® for Wood Construction* [2] based on oven-dry weight and oven-dry volume.
[b] Specific gravity based on volume at 12% moisture content and oven-dry weight, calculated using Equation (3-5) of the *Wood Handbook* [1] and specific gravity values from the *NDS®*.
[c] Specific weight at 12% moisture content, including weight of water.

$$\gamma_m = G_m \left(1 + \frac{m}{100\%}\right)\gamma_w \tag{2.2}$$

where

G_m = the specific gravity at a specified moisture content, m,
G_0 = the specific gravity based on oven-dry weight and oven-dry volume,
m = the moisture content (between 0 and 30%),
γ_m = the specific weight of wood at the specified moisture content m,
γ_w = the specific weight of water.

Occasionally, the weight of wood may need to be calculated at moisture contents above the fiber saturation point. Because the volume and oven-dry weight do not change above the fiber saturation point (30% moisture content), the specific gravity calculated will be equal to that calculated for 30% moisture content. However, the specific weight includes the weight of the water in the wood and will change with moisture content above 30%. The following equation may be used to calculate specific weight for pieces above 30% moisture content:

$$\gamma_m = \gamma_{30} \left(1 + \frac{m - 30\%}{30\%}\right) \tag{2.3}$$

where γ_{30} = the specific weight at 30% moisture content.

Example 2.1 Estimating Shipping Weight of Lumber

Given: 10,000 board feet (10 MBF) of Douglas Fir–Larch (DF-L) 2 × 6 dimension lumber.

Determine: The shipping weight at 5%, 19%, and 30% MC.

Approach: The specific gravity of DF-L at oven-dry weight and oven-dry volume will be obtained from Table 2.2. The specific gravity values at 5%, 19%, and 30% will be obtained using equation (2.1). Equation (2.2) and will then be used to calculate specific weights at the three MC levels. Using the definition of board feet from Chapter 1 and the specific weight of water, the total weights of 10 MBF will be calculated at the MC levels stated.

Solution: First, considering 5% MC, from Table 2.2, the value of G_0 for DF-L is 0.50. From equation (2.1),

$$G_5 = \frac{0.50}{1 + (0.265)(0.50)(5\%/30\%)} = 0.49$$

From equation (2.2),

$$\gamma_5 = 0.49\left(1 + \frac{5\%}{100\%}\right) 62.4 \text{ pcf} = 32 \text{ pcf}$$

From Chapter 1, 1 board foot of 2 × 6 is equal to 1 lineal foot of 2 × 6, or 1.5 in. × 5.5 in. × 12 in. = 99 in^3 of wood; therefore, 10,000 BF of 2 × 6 has a volume of 990,000 in^3 or 573 ft^3. The weight of 10 MBF at 5% MC is then 573 ft^3 × 32 pcf = 18,300 lb.

Similarly, for 19% MC, from equation (2.1), $G_{19} = 0.46$; from equation (2.2), $\gamma_{19} = 34$ pcf. The weight of 10 MBF at 19% MC is then 573 ft^3 × 34 pcf = 19,500 lb.

And for 30% MC, from equation (2.1), $G_{30} = 0.44$; from equation (2.2), $\gamma_{19} = 36$ pcf. The weight of 10 MBF at 30% MC is then 573 ft^3 × 36 pcf = 20,600 lb.

Answer: The weight of 10 MBF of DF-L at 5%, 19%, and 30% MC is 18,300, 19,500, and 20,600 lb, respectively.

Discussion: As shown above, the weight of typical structural wood decreases by about 5% from fiber saturation to 19% MC, and another 5% from 19% MC to 5% MC.

2.2.1 Effect of Adhesives and Preservative Treatments

Adhesives used in laminating do not have an appreciable effect on specific weights. Some preservative treatments—in particular, oil-borne treatments at high retention levels—may change the specific weight significantly. The American Wood-Preservers' Association [3] should be consulted for more detailed information regarding preservative treatment and specific weight changes.

2.3 DIMENSIONAL CHANGES DUE TO MOISTURE AND TEMPERATURE

Dimensional changes in wood may come about by temperature and moisture content changes in the wood. Wood containing moisture responds to varying temperature differently than other typical building materials. As wood is heated, it tends to expand due to temperature but also shrinks due to an accompanying loss of moisture. Unless the wood is very dry initially (3 or 4% MC or less), shrinkage due to moisture loss on heating will exceed thermal expansion, resulting in a negative net dimensional change due to heating. In timber design, the dimensional changes due to temperature are considered small in comparison with dimensional changes due to changing moisture content. As such, proper detailing of connections to accommodate wood shrinkage is also generally considered adequate to accommodate dimensional changes due to temperature.

2.3.1 Moisture Content and Shrinkage

Between the fiber saturation point and zero moisture content, wood shrinks as it loses moisture and swells as it absorbs moisture. Above the fiber saturation point, there is no dimensional change with variation in moisture content. The amount of shrinkage and swelling differs in the tangential, radial, and longitudinal directions of a piece of wood or lumber. Good engineering design considers shrinkage and swelling in the detailing and use of wood members. Shrinking and swelling are expressed as percentages based on the green dimensions of the wood. Wood shrinks most in a direction tangent to the annual growth rings and somewhat less in the radial direction or across these rings. In general, shrinkage amounts are greater with greater wood density.

Table 2.3 gives the average tangential, radial, and volumetric shrinkage values for various species drying from the green condition to 0% MC and ranges of values for species groups. Because the faces of pieces of lumber are seldom oriented so that the annual growth rings are exactly tangent and radial to the faces of the piece, it is customary in estimating cross-sectional dimensional changes to use an intermediate or average value between the

TABLE 2.3 Average Shrinkage Values for Wood Based on Green[a] Dimensions

	Percent Dimension Change (Shrinkage) from Green to Oven-Dry Moisture Content		
Species or Species Combination	Radial	Transverse	Volumetric
Aspen	3–4	7–8	11–12
Balsam Fir	3	7	11
Beech–Birch–Hickory	5–8	9–13	14–19
Coast Sitka Spruce	4	8	11–12
Coast Species	4–5	7–9	11–14
Cottonwood	3–4	7–9	10–14
Douglas Fir–Larch	4–5	7–9	11–14
Eastern Hemlock	3	7	10
Eastern Hemlock–Balsam Fir	3–4	7	10–14
Eastern Hemlock–Tamarack	3–4	7	10–14
Eastern Hemlock–Tamarack (North)	3–4	7	10–14
Eastern Softwoods	2–4	6–8	8–14
Eastern Spruce	4	7–8	11–12
Eastern White Pine	2	6	8
Hem–Fir	3–5	7–9	10–13
Mixed Maple	3–5	7–10	12–15
Mixed Oak	4–7	8–13	13–19
Mixed Southern Pine	4–5	7–8	11–12
Mountain Hemlock	4	7	11
Northern Pine	4	7	10–11
Northern Red Oak	4	9–11	14–15
Northern White Cedar	2	5	7
Ponderosa Pine	4	6	10
Red Maple	4	8	13
Red Oak	4–5	9–11	14–19
Red Pine	4	7	11
Redwood	2–3	4–5	7
Sitka Spruce	4	8	12
Southern Pine	5	7–8	12
Spruce–Pine–Fir	3–4	7–8	10–12
Western Cedars	2–5	5–7	7–10
Western Cedars (North)	2–3	5–6	7–9
Western Hemlock	4	8	12
Western Hemlock (North)	4	8	12
Western White Pine	4	7	12
Western Woods	—	—	—
White Oak	4–7	9–13	13–16
Yellow Poplar	5	8	13

[a] Values presented are from the *Wood Handbook* [1] and represent average values for a single species or a range of average values for a species group. Additional information may be found in the *Wood Handbook*.

tangential and radial values. The values in Table 2.3 can be used to estimate shrinkage using

$$S_m = S_0 \frac{m_i - m_f}{30\%} \tag{2.4}$$

where S_m = shrinkage from initial moisture condition to final moisture content m (%),
S_0 = total shrinkage from Table 2.3 (%),
m_f = final moisture content (at or below 30%), and
m_i = initial moisture content (at or below 30%).

Values for longitudinal shrinkage with change in moisture content are not tabulated in Table 2.3 because they are ordinarily negligible. The total longitudinal shrinkage of commonly used species from fiber saturation to oven-dry condition usually ranges from 0.1 to 0.2% of the green dimension. Abnormal longitudinal shrinkage may occur in compression wood, wood with steep slope of grain, and exceptionally lightweight wood of any species. Because there is considerable variation in shrinkage for individual pieces within any species, it is difficult to predict the exact shrinkage of an individual piece of wood. If the species of wood is known, the values given in Table 2.3 or the values in the *Wood Handbook* [1] may be used to estimate the average shrinkage of a quantity of pieces, such as a stack of lumber, or the accumulative shrinkage of a the wood in a wood structure.

A rule of thumb commonly used to estimate shrinkage perpendicular to the grain in glued laminated timber is to assume that 1% shrinkage occurs for every 5% change in moisture content. Lumber used in laminating tends to be *flat grained* (wide faces more closely parallel with the growth rings); therefore, shrinkage perpendicular to the glue lines is more comparable to radial shrinkage than tangential shrinkage. As such, the shrinkage of a typical glued laminated beam across the glue lines (decrease in beam depth) tends to be less than the values predicted using the rule of thumb.

Example 2.2 Glulam Beam Shrinkage

Given: Douglas fir $5\frac{1}{8}$ in. ×24 in. glued laminated beam.

Determine: The decrease in depth (shrinkage) for a change in moisture content from 12% to 5%.

Approach: The decrease in the depth dimension will be perpendicular to the wide faces or across the glue lines, which from the discussion above, will be assumed to be somewhere between radial and tangential shrinkage, but closer to the radial value. It will also be assumed that the dimension at 12% MC is a true 24 in.

Solution: From Table 2.3, a value of 6% is chosen for S_0. From equation (2.4),

$$S_5 = 6.0\% \left(\frac{12\% - 5\%}{30\%} \right) = 1.4\%$$

The shrinkage is therefore, 1.4% of 24 in., or 0.34 in.

Answer: The expected change in depth of a 24-in. DF glulam beam from 12% MC to 5% MC is 0.34 in. The dimension change of 1.4% agrees with the rule-of-thumb amount of approximately 1% change in dimension per 5% change in MC.

Accommodation of shrinkage and swelling is very important in the proper design, detailing, and construction of wood structures and wood structural elements, particularly where constraint introduces stresses perpendicular to the grain. The effects of dimensional changes due to a change in moisture content have been considered in the development of the details shown in *Typical Construction Details,* AITC 104 [4], included in Chapter 8.

2.3.2 Dimensional Changes due to Temperature

The coefficient of linear thermal expansion differs in wood's three structural directions. In the longitudinal direction (parallel to the grain), the coefficient appears to be independent of specific gravity and species and ranges from about 1.7×10^{-6} to 2.5×10^{-6} in./in. per °F for oven-dry wood for both hardwoods and softwoods. This value is about one-tenth to one-third of those for other common structural materials. Coefficients of linear thermal expansion across the grain (radial and tangential) are proportional to and range from about 5 to more than 10 times greater than coefficients parallel to the grain. Equations relating the coefficient of thermal expansion for oven-dry wood in the across-grain (radial and tangential) directions as functions of specific gravity are published in the *Wood Handbook* [1].

2.4 THERMAL INSULATING PROPERTIES

Due to its physical structure, wood is considered to be a very good insulator, because the cells contain voids that retard heat transfer. The thermal conductivity of wood varies with (1) the direction of grain, (2) specific gravity, (3) moisture content, (4) extractives present in the wood, and (5) growth characteristics such as knots, slope of grain, seasoning checks, and growth rings.

Thermal conductivity is approximately the same in the radial and tangential directions but is about 2.5 times greater along the grain. Cross-grain thermal conductivity values for various wood species at 12% MC may be calculated using [1]

$$k = 1.676G + 0.129 \tag{2.5}$$

where k = thermal conductivity (Btu-in./hr-ft^2-°F), and
G = specific gravity at 12% MC (Table 2.2).

Thermal conductivity of individual wood pieces may vary considerably with calculated values. Thermal conductivity also increases with moisture content. Thermal conductivity values at moisture content levels different from 12% may be calculated with equations in the *Wood Handbook*.

The thermal resistance, R, for a piece of material with one-dimensional heat flow may be calculated as the reciprocal of the thermal conductivity, k, multiplied by the piece thickness, t, or $R = (1/k)t$. Using equation (2.5), the thermal conductivity of Douglas Fir–Larch, at 12% MC, is $k = (1.676)(0.47) + 0.129 = 0.92$ Btu-in./hr-ft^2-°F, where $G = 0.47$ from Table 2.2. The thermal resistance of a 2-in. (nominal) plank, therefore, is $R = (1/0.92)(1.5) = 1.6$ hr-ft^2-°F/Btu-in. Since wood is commonly used in combination with other building materials for insulating assemblies, consideration must be made for heat flow through the assembly as a whole. The *Fundamentals Handbook* of the American Society of Heating, Refrigerating, and Air-Conditioning Engineers [5] provides information and example calculations of resistance values for various wood assemblies. Heat capacity and thermal diffusivity values for various wood species at various moisture content levels may be found in the *Wood Handbook* [1].

2.5 WOOD IN CHEMICAL ENVIRONMENTS

As discussed in Chapter 1, wood is naturally resistant to chemical action, making it a favorable building material in chemically adverse environments. Chemical actions of three general types may affect the strength of wood. The first causes swelling and a resulting weakening of the wood. Liquids such as water, alcohols, and some other organic liquids cause wood to swell. This action is almost completely reversible; hence, if the swelling liquid or solution is removed by evaporation or by extraction followed by evaporation of the solvent, the original dimensions and strength are practically restored. Liquids such as petroleum oils and creosote do not cause wood to swell. The second type of action brings about permanent changes in the wood, such as hydrolysis of the cellulose by acids or acid salts. The third type of action, which is also permanent, involves delignification of the wood and dissolving of hemicelluloses by alkalis.

Experience and available data indicate species and conditions where wood is equal or superior to other materials in resisting the degradative action of chemicals. In general, heartwood of such species as cypress, Douglas fir, larch, southern pine, California redwood, maple, and white oaks are quite resistant to attack by dilute mineral and organic acids. Oxidizing acids, such as nitric acid, have a greater degradative action than that of nonoxidizing acids. Alkaline solutions are more destructive than acidic solutions, and hardwoods are more susceptible than softwoods to attack by both acids and alkalis.

Highly acidic salts tend to hydrolyze wood when present in high concentrations. Even relatively low concentrations of such salts have shown signs that the salt may migrate to the surface of railroad ties, which are occasionally wet and dried in a hot, arid region. This migration, combined with the high concentrations of salt relative to the small amount of water present, causes an acidic condition sufficient to make wood brittle.

Iron salts, which develop at points of contact with plates, bolts, and other fasteners, have a degradative action on wood, especially in the presence of moisture. In addition, iron salts may precipitate toxic extractives and thus lower the natural decay resistance of wood. The softening and discoloration of wood around corroded iron fastenings is a commonly observed phenomenon; it is especially pronounced in acidic woods, such as oak, and in woods such as California redwood, which contain considerable tannin and related compounds. The oxide layer formed on iron is transformed through reaction with wood acids into soluble iron salts, which not only degrade the surrounding wood but may catalyze the further corrosion of the metal. The action is accelerated by moisture; oxygen may also play an important role in the process. This effect is not encountered with well-dried wood used in dry locations. Under damp-use conditions, it can be avoided or minimized by using corrosion-resistant fastenings.

Many substances have been employed as impregnants to enhance the natural resistance of wood to chemical degradation. One of the more economical treatments involves pressure impregnation with a viscous coke-oven coal tar to retard liquid penetration. Acid resistance of wood is increased by impregnation with phenolic resin solutions followed by appropriate drying and curing. Treatment with furfuryl alcohol has been used to increase resistance to alkaline solutions. Another procedure involves massive impregnation with a monomeric resin, such as methyl methacrylate, followed by polymerization.

2.6 ACOUSTICAL PROPERTIES

The acoustical properties of a material or composite construction are determined by its sound insulation and sound absorption abilities. Sound insulation abilities are measured in terms of the reduction in intensity of sound when it passes through a barrier. Sound absorption refers to the amount of incident sound on a surface that is not reflected by the surface. Sound insulating values for materials of construction are related to the sound transmission loss for the construction measured in decibels at various frequencies. Like most common construction materials, wood alone does not provide good sound insulation, but when properly combined with other materials in typical constructions, will provide a structural unit of satisfactory sound-insulating ability. Sound-absorption values for wood vary with moisture content, direction of the grain, and density.

2.7 ELECTRICAL PROPERTIES

The most important electrical properties of wood are its resistance to the passage of an electric current and its dielectric properties. The electrical resistance of wood is utilized in electric moisture meters used to determine moisture content. The dielectric properties of wood are utilized in the high-frequency curing of adhesives in glued laminated members and in the seasoning of wood.

The electrical resistance of wood varies with moisture content, density, direction of travel of the current with respect to the direction of the grain, and temperature. It varies greatly with moisture content, especially below the fiber saturation point, decreasing with an increase in moisture content. At low moisture content, wood is considered an insulator. Electrical resistance varies inversely with the density of wood, although this effect is slight compared to the variance due to moisture content and is greater across the grain than along it. The electrical resistance of wood approximately doubles for each drop in temperature of 22.5°F, but this relationship varies considerably with the level of moisture content. There is also a variation in electrical resistance between species, which is possibly caused by minerals or electrolytes in the wood itself or dissolved in the water present in the wood. Metallic salts such as those used in preservative and fire-retardant treatments, may lower the electrical resistance of wood considerably. Use of wood containing such salts should be avoided for applications where electrical resistance is critical and in processes involving dielectric heating. Electric moisture meters may give erroneous readings for such wood.

2.8 COEFFICIENT OF FRICTION

Coefficients of static friction for wood in contact with wood and for wood in contact with other materials depend on the moisture content of the wood and surface roughness. The coefficients vary little with species except for those species that contain abundant oily or waxy extractives. Coefficients of static friction for wood have been reported to be approximately 0.7 for dry wood on unpolished steel and 0.4 for green wood on steel. Coefficients of static friction for smooth wood on smooth wood are reported to be 0.6 for dry wood and 0.8 for green wood. It is not usual practice to use friction to resist forces in the design of timber connections.

CHAPTER 3

TIMBER DESIGN

3.1 INTRODUCTION

The design of timber structures requires all components of the structure to be of sufficient size and strength to withstand the forces to which the structure might reasonably be subjected during its useful life. Proper design also requires sufficient stiffness in members and structural systems to prevent deformations associated with reasonable loads from impairing the serviceability of nonstructural systems, or causing the structure to feel or appear unsafe. The selection of the timber elements, connections, and other structural systems to ensure the strength and serviceability described above serves as the context of design for this book.

Timber design includes proper evaluation of timber or wood elements, such as individual planks, joists, beams, posts, or columns. Design also includes the connections of elements to other elements and the foundation, and the performance of the elements and connections making up structural systems. Design in general is covered in this chapter; subsequent chapters are devoted specifically to elements, connections, and systems.

The design approach used in this manual is *allowable stress design* (ASD), also known as *working stress design*. The allowable stress design approach requires the designer to specify structural members (size, species, and grade) and fasteners so that the material stresses and fastener loads at design loading conditions do not exceed allowable values for the structural materials. In ASD, factors of safety accounting for uncertainties in loads, materials, construction, and failure mechanisms are incorporated into allowable stress and load values.

The loads and stresses used in allowable stress design are covered in the remainder of this chapter. Successive chapters cover the use of ASD in the design of individual timber members, connections, and structural systems. An alternative design methodology, *load and resistance factor design* (LRFD), is presented in Chapter 7.

3.2 LOADS AND FORCES

The loads or forces that act upon structures may be divided into two broad categories: applied loads and dead loads. *Dead loads* are generally the weight forces associated with a structure itself, and include the weight of structural and nonstructural materials permanently attached to the structure. *Applied loads* generally include other forces that might reasonably be experienced by a structure during its design life. Applied loads include live loads, such as the weights of building occupants and the forces generated by their activities, and include the weights of materials stored in or on the structure. Applied loads also include loads or forces brought on by the environment, such as wind, earthquakes, snow, floods, and earth pressures. The design of any structure must include the deliberate determination of all of the appropriate loads and forces.

Minimum loads and forces for design are generally prescribed by building codes. Local building codes are generally derived from model building codes adopted and amended by local governing bodies. Building codes also stipulate how the loads and forces are to be resisted, as well as stress or strength values that may be assigned to building materials. Prior to the design of any structure, the applicable building code requirements should be obtained from the governing building authority. In the absence of a governing building authority, model building codes may be used, such as the *BOCA National Building Code* [1], the *International Building Code* [2], the *Standard Building Code* [3], and the *Uniform Building Code* [4]. The design of any structure, regardless of governing codes, must be done in accordance with accepted engineering practice. Historically, the methods presented in this book have been considered accepted engineering practice.

The loads and forces acting on timber structures are described in greater detail in the following sections of this chapter. The typical stresses resulting from design loading conditions and corresponding allowable stresses are also described in this chapter. Design methods for individual timber members, connections and fasteners, and building systems used in timber design are covered in successive chapters. Wood's strength depends on the duration of time in which a particular load is applied. Hence, for timber design, the duration of an applied load may be as important as the magnitude of applied load. Building codes may therefore stipulate load duration values in addition to load magnitudes for the design of timber structures. Adjustments for load duration are discussed in greater detail in Section 3.4.

3.2.1 Dead Loads

Dead loads may be defined as the weights of all permanent structural and nonstructural components of a building or structure, such as walls, floors, roofs, partitions, stairways, and fixed service equipment. These loads include the weights of materials as a part of the original construction and should include materials that are expected to be added later. For example, if additional roofing materials are expected to be added later (re-roofing), the original design of structural framing for the roof should take into consideration the eventual total roof material weights (dead loads), so that the framing system need not be retrofitted or reinforced when the additional load is applied.

The actual weights of various materials and systems or assemblies should be used in design if this information is available. Minimum dead load values for common construction materials and assemblies as recommended by the American Society of Civil Engineers [5] are shown in Tables 3.1 and 3.2. The weight of any structural member under consideration, often referred to as the *self-weight,* is also a dead load. Care should be taken to ensure that self-weights of members are considered properly. Care must also be taken to ensure that material weights on sloped or curved surfaces are considered properly, in that the plan area weights are greater than the surface area weights.

3.2.2 Live Loads

For the purposes of this book, *live loads* are considered to be loads arising from occupancy or use of a structure, and include the weight of occupants, vehicles, and materials placed in or on the structure during normal use, including stored materials and furniture. Other loads or forces, such as those arising from wind, rain, snow, and earthquakes, will be considered separately. Minimum live loads recommended by the American Society of Civil Engineers [5], listed in Table 3.3, may be used as a guide in the absence of local code requirements. Conditions of any particular project may require the use of greater loads than those in the tables or provided by local codes. The loads used in design should be clearly shown on design and construction documents.

3.2.2.1 Floor Live Loads Loads arising from the activities of the occupants on floors of buildings are of two types: distributed loads and concentrated loads. Minimum loads of both types are listed in Table 3.3. Floor systems must be designed to carry both types of loads, although not necessarily simultaneously. Where floor support members have multiple spans, the conditions of all-span, alternate-span, and adjacent-span loading must be con-

TABLE 3.1 Minimum Design Dead Loads (psf)

Component	Load (psf)
CEILINGS	
Acoustical Fiber Board	1
Gypsum board (per mm thickness)	0.55
Mechanical duct allowance	4
Plaster on tile or concrete	5
Plaster on wood lath	8
Suspended steel channel system	2
Suspended metal lath and cement plaster	15
Suspended metal lath and gypsum plaster	10
Wood furring suspension system	2.5
COVERINGS, ROOF, AND WALL	
Asbestos-cement shingles	4
Asphalt shingles	2
Cement tile	16
Clay tile (for mortar add 10 psf)	
Book tile, 2-in.	12
Book tile, 3-in.	20
Ludowici	10
Roman	12
Spanish	19
Composition:	
Three-ply ready roofing	1
Four-ply felt and gravel	5.5
Five-ply felt and gravel	6
Copper or tin	1
Corrugated asbestos-cement roofing	4
Deck, metal, 20 gage	2.5
Deck, metal, 18 gage	3
Decking, 2-in. wood (Douglas fir)	5
Decking, 3-in. wood (Douglas fir)	8
Fiberboard, 1/2-in.	0.75
Gypsum sheathing, 1/2-in.	2
Insulation, roof boards (per inch thickness)	
Cellular glass	0.7
Fibrous glass	1.1
Fiberboard	1.5
Perlite	0.8
Polystyrene foam	0.2
Urethane foam with skin	0.5
Plywood (per 1/8-in. thickness)	0.4
Rigid insulation, 1/2-in.	0.75
Skylight, metal frame, 3/8-in. wire glass	8
Slate, 3/16-in.	7
Slate, 1/4-in.	10
Waterproofing membranes:	
Bituminous, gravel-covered	5.5
Bituminous, smooth surface	1.5
Liquid applied	1
Single-ply, sheet	0.7
Wood sheathing (per inch thickness)	3
Wood shingles	3
FLOOR FILL	
Cinder concrete, per inch	9
Lightweight concrete, per inch	8
Sand, per inch	8
Stone concrete, per inch	12
FLOORS AND FLOOR FINISHES	
Asphalt block (2-in.), 1/2-in. mortar	30
Cement finish (1-in.) on stone-concrete fill	32
Ceramic or quarry tile (3/4-in.) on 1/2-in. mortar bed	16
Ceramic or quarry tile (3/4-in.) on 1-in. mortar bed	23
Concrete fill finish (per inch thickness)	12
Hardwood flooring, 7/7-in.	4
Linoleum or asphalt tile, 1/4-in.	1
Marble and mortar on stone-concrete fill	33
Slate (per mm thickness)	15
Solid flat tile on 1-in. mortar base	23
Subflooring, 3/4-in.	3
Terrazzo (1-1/2-in.) directly on slab	19
Terrazzo (1-in.) on stone-concrete fill	32
Terrazzo (1-in.), 2-in. stone concrete	32
Wood block (3-in.) on mastic, no fill	10
Wood block (3-in.) on 1/2-in. mortar base	16

TABLE 3.1 ***(Continued)***

Component	12-in. spacing (lb/ft^2)	16-in. spacing (lb/ft^2)	24-in. spacing (lb/ft^2)	Load (psf)
FLOORS, WOOD-JOIST (NO PLASTER) DOUBLE WOOD FLOOR				
Joist sizes (inches):				
2 × 6	6	5	5	
2 × 8	6	6	5	
2 × 10	7	6	6	
2 × 12	8	7	6	
FRAME PARTITIONS				
Movable steel partitions				4
Wood or steel studs, 1/2-in. gypsum board each side				8
Wood studs, 2 × 4, unplastered				4
Wood studs, 2 × 4, plastered one side				12
Wood studs, 2 × 4, plastered two sides				20
FRAME WALLS				
Exterior stud walls:				
2 × 4 @ 16-in., 5/8-in. gypsum, insulated, 3/8-in. siding				11
2 × 6 @ 16-in., 5/8-in. gypsum, insulated, 3/8-in. siding				12
Exterior stud walls with brick veneer				48
Windows, glass, frame and sash				8
Clay brick wythes:				
4 in.				39
8 in.				79
12 in.				115
16 in.				155

Component			Load (psf)		
Hollow concrete masonry unit wythes:					
Wythe thickness (in inches)	4	6	8	10	12
Density of unit (16.49 kN/m^3)					
No grout	22	24	31	37	43
48″ o.c.		29	38	47	55
40″ o.c. grout		30	40	49	57
32″ o.c. spacing		32	42	52	61
24″ o.c.		34	46	57	67
16″ o.c.		40	53	66	79
Full Grout		55	75	95	115
Density of unit (125 pcf):					
No grout	26	28	36	44	50
48″ o.c.		33	44	54	62
40″ o.c. grout		34	45	56	65
32″ o.c. spacing		36	47	58	68
24″ o.c.		39	51	63	75
16″ o.c.		44	59	73	87
Full Grout		59	81	102	123
Density of unit (21.21 kN/m^3)					
No grout	29	30	39	47	54
48″ o.c.		36	47	57	66
40″ o.c. grout		37	48	59	69
32″ o.c. spacing		38	50	62	72
24″ o.c.		41	54	67	78
16″ o.c.		46	61	76	90
Full Grout		62	83	105	127
Solid concrete masonry unit wythes (incl.					
Wythe thickness (in mm)	4	6	8	10	12
Density of unit (105 pcf):	32	51	69	87	105
Density of unit (125 pcf):	38	60	81	102	124
Density of unit (135 pcf):	41	64	87	110	133

Source: Reprinted with Permission from ASCE 7-02 *Minimum Design Loads for Buildings and Other Structures,* Copyright © 2003, American Society of Civil Engineers. This information is extracted from ASCE 7-02; for further information, the complete text of the manual should be referenced (http://www.pubs.asce.org/ASCE7.html?9991330).

sidered. Floor live loads are generally considered to have *normal* load duration of 10 years. The live loads listed are treated as static loads but are assumed to accommodate dynamic (impact) effects of normal building use and occupancy. Crane loads and cyclic loads associated with machinery are considered separately.

TABLE 3.2 Minimum Design Dead Loads (pcf)

Material	Load (lb/ft^3)	Material	Load (lb/ft^3)
Aluminum	170	Earth (submerged)	
Bituminous products		Clay	80
Asphaltum	81	Soil	70
Graphite	135	River mud	90
Paraffin	56	Sand or gravel	60
Petroleum, crude	55	Sand or gravel and clay	65
Petroleum, refined	50	Glass	160
Petroleum, benzine	46	Gravel, dry	104
Petroleum, gasoline	42	Gypsum, loose	70
Pitch	69	Gypsum, wallboard	50
Tar	75	Ice	57
Brass	526	Iron	
Bronze	552	Cast	450
Cast-stone masonry (cement, stone, sand)	144	Wrought	48
Cement, portland, loose	90	Lead	710
Ceramic tile	150	Lime	
Charcoal	12	Hydrated, loose	32
Cinder fill	57	Hydrated, compacted	45
Cinders, dry, in bulk	45	Masonry, Ashlar stone	
Coal		Granite	165
Anthracite, piled	52	Limestone, crystalline	165
Bituminous, piled	47	Limestone, oolitic	135
Lignite, piled	47	Marble	173
Peat, dry, piled	23	Sandstone	144
Concrete, plain		Masonry, brick	
Cinder	108	Hard (low absorbtion)	130
Expanded-slag aggregate	100	Medium (medium absorbtion)	115
Haydite (burned-clay aggregate)	90	Soft (high absorbtion)	100
Slag	132	Masonry, concrete*	
Stone (including gravel)	144	Lightweight units	105
Vermiculite and perlite aggregate, non–load-bearing	25–50	Medium weight units	125
Other light aggregate, load-bearing	70–105	Normal weight units	135
Concrete, reinforced		Masonry grout	140
Cinder	111	Masonry, rubble stone	
Slag	138	Granite	153
Stone (including gravel)	150	Limestone, crystalline	147
Copper	556	Limestone, oolitic	138
Cork, compressed	14	Marble	156
Earth (not submerged)		Sandstone	137
Clay, dry	63	Mortar, cement or lime	130
Clay, damp	110	Particleboard	45
Clay and gravel, dry	100	Plywood	36
Silt, moist, loose	78	Riprap (Not submerged)	
Silt, moist, packed	96	Limestone	83
Silt, flowing	108	Sandstone	90
Sand and gravel, dry, loose	100	Sand	
Sand and gravel, dry, packed	110	Clean and dry	90
Sand and gravel, wet	120	River, dry	106

TABLE 3.2 (*Continued*)

Material	Load (lb/ft³)	Material	Load (lb/ft³)
Slag		Tin	459
Bank	70	Water	
Bank screenings	108	Fresh	62
Machine	96	Sea	64
Sand	52	Wood, seasoned	
Slate	172	Ash, commercial white	41
Steel, cold-drawn	492	Cypress, southern	34
Stone, quarried, piled		Fir, Douglas, coast region	34
Basalt, granite, gneiss	96	Hem fir	28
Limestone, marble, quartz	95	Oak, commercial reds and whites	47
Sandstone	82	Pine, southern yellow	37
Shale	92	Redwood	28
Greenstone, hornblende	107	Spruce, red, white, and Stika	29
Terra cotta, architectural		Western hemlock	32
Voids filled	120	Zinc, rolled sheet	449
Voids unfilled	72		

Source: Reprinted with Permission from ASCE 7-02 *Minimum Design Loads for Buildings and Other Structures,* Copyright © 2003, American Society of Civil Engineers. This information is extracted from ASCE 7-02; for further information, the complete text of the manual should be referenced (http://www.pubs.asce.org/ASCE7.html?9991330).

3.2.2.2 Roof Live Loads Roof live loads generally arise from construction and re-roofing activities and various materials or equipment placed on or supported by the roof. Where there is potential for significant snow accumulations on roofs, snow loads may govern roof design. In addition, flat and relatively flat roofs must also be designed for the potential of rain or melted snow to pond on the roof. Roof loads (live or snow) are determined by the building official. Consideration of ponding must be made by the designer. Minimum roof live loads from the *International Building Code* [2] are calculated using the following equation:

$$L_r = 20R_1R_2 \qquad \text{psf} \tag{3.1}$$

where $12 \leq L_r \leq 20$ psf and where

$$R_1 = \begin{cases} 1 & \text{for } A_f \leq 200 \text{ ft}^2 \\ 1.2 - 0.001A_f & \text{for } 200 \leq A_f \leq 600 \text{ ft}^2 \\ 0.6 & \text{for } A_f \geq 600 \text{ ft}^2 \end{cases}$$

and where

A_f = Tributary area (span length multiplied by effective width) supported by any structural member (ft²) and

$$R_1 = \begin{cases} 1 & \text{for } F \leq 4 \\ 1.2 - 0.05F & \text{for } 4 \leq F \leq 12 \\ 0.6 & \text{for } F \geq 12 \end{cases}$$

where F for a sloped roof is the inches of rise per foot, and for an arch or dome is the rise/span ratio multiplied by 32. These loads are assumed to arise from construction and repair activities and are generally assumed to have a maximum duration of seven days. Snow loads and ponding are considered in later sections. Roofs to be used for special purposes must be designed for appropriate anticipated loads.

TABLE 3.3 Minimum Design Live Loads

Occupancy or Use	Uniform psf (kN/m^2)	Conc. lbs (kN)
Apartments (see residential)		
Access floor systems		
Office use	50 (2.4)	2000 (8.9)
Computer use	100 (4.79)	2000 (8.9)
Armories and drill rooms	150 (7.18)	
Assembly areas and theaters		
Fixed seats (fastened to floor)	60 (2.87)	
Lobbies	100 (4.79)	
Movable seats	100 (4.79)	
Platforms (assembly)	100 (4.79)	
Stage floors	150 (7.18)	
Balconies (exterior)	100 (4.79)	
On one- and two-family residences only, and not exceeding 100 ft.^{2} (9.3 m^2)	60 (2.87)	
Bowling alleys, poolrooms, and similar recreational areas	75 (3.59)	
Catwalks for maintenance access	40 (1.92)	300 (1.33)
Corridors		
First floor	100 (4.79)	
Other floors, same as occupancy served except as indicated		
Dance halls and ballrooms	100 (4.79)	
Decks (patio and roof)		
Same as area served, or for the type of occupancy accommodated		
Dining rooms and restaurants	100 (4.79)	
Dwellings (see residential)		
Elevator machine room grating (on area of 4 in.^{2} (2580 mm^2))		300 (1.33)
Finish light floor plate construction (on area of 1 in.^{2} (645 mm^2))		200 (0.89)
Fire escapes	100 (4.79)	
On single-family dwellings only	40 (1.92)	
Fixed ladders		See Section 4.4
Garages (passenger vehicles only)	40 (1.92)	Note (1)
Trucks and buses	Note (2)	

TABLE 3.3 ***(Continued)***

Occupancy or Use	Uniform psf (kN/m²)	Conc. lbs (kN)
Grandstands (see stadium and arena bleachers)		
Gymnasiums, main floors, and balconies	100 (4.79) Note (4)	
Handrails, guardrails, and grab bars	See Section 4.4	
Hospitals		
Operating rooms, laboratories	60 (2.87)	1000 (4.45)
Private rooms	40 (1.92)	1000 (4.45)
Wards	40 (1.92)	1000 (4.45)
Corridors above first floor	80 (3.83)	1000 (4.45)
Hotels (see residential)		
Libraries		
Reading rooms	60 (2.87)	1000 (4.45)
Stack rooms	150 (7.18) Note (3)	1000 (4.45)
Corridors above first floor	80 (3.83)	1000 (4.45)
Manufacturing		
Light	125 (6.00)	2000 (8.90)
Heavy	250 (11.97)	3000 (13.40)
Marquees and canopies	75 (3.59)	
Office buildings		
File and computer rooms shall be designed for heavier loads based on anticipated occupancy		
Lobbies and first floor corridors	100 (4.79)	2000 (8.90)
Offices	50 (2.40)	2000 (8.90)
Corridors above first floor	80 (3.83)	2000 (8.90)
Penal institutions		
Cell blocks	40 (1.92)	
Corridors	100 (4.79)	
Residential		
Dwellings (one- and two-family)		
Uninhabitable attics without storage	10 (0.48)	
Uninhabitable attics with storage	20 (0.96)	
Habitable attics and sleeping areas	30 (1.44)	
All other areas except stairs and balconies	40 (1.92)	
Hotels and multifamily houses		
Private rooms and corridors serving them	40 (1.92)	
Public rooms and corridors serving them	100 (4.79)	
Reviewing stands, grandstands, and bleachers	100 (4.79) Note (4)	
Roofs	See Sections 4.3 and 4.9	

TABLE 3.3 ***(Continued)***

Occupancy or Use	Uniform psf (kN/m^2)	Conc. lbs (kN)
Schools		
Classrooms	40 (1.92)	1000 (4.45)
Corridors above first floor	80 (3.83)	1000 (4.45)
First floor corridors	100 (4.79)	1000 (4.45)
Scuttles, skylight ribs, and accessible ceilings		200 (9.58)
Sidewalks, vehicular driveways, and yards subject to trucking	250 (11.97) Note (5)	8000 (35.60) Note (6)
Stadiums and arenas		
Bleachers	100 (4.79) Note (4)	
Fixed Seats (fastened to floor)	60 (2.87) Note (4)	
Stairs and exit-ways	100 (4.79)	Note (7)
One- and two-family residences only	40 (1.92)	
Storage areas above ceilings	20 (0.96)	
Storage warehouses (shall be designed for heavier loads if required for anticipated storage)		
Light	125 (6.00)	
Heavy	250 (11.97)	
Stores		
Retail		
First floor	100 (4.79)	1000 (4.45)
Upper floors	75 (3.59)	1000 (4.45)
Wholesale, all floors	125 (6.00)	1000 (4.45)
Vehicle barriers	See Section 4.4	
Walkways and elevated platforms (other than exit-ways)	60 (2.87)	
Yards and terraces, pedestrians	100 (4.79)	

Notes

(1) Floors in garages or portions of building used for the storage of motor vehicles shall be designed for the uniformly distributed live loads of Table 4-1 or the following concentrated load: (1) for garages restricted to passenger vehicles accommodating not more than nine passengers, 3000 lb (13.35 kN) acting on an area of 4.5 in. by 4.5 in. (114 mm by 114 mm, footprint of a jack); (2) for mechanical parking structures without slab or deck which are used for storing passenger car only, 2250 lb (10 kN) per wheel.

(2) Garages accommodating trucks and buses shall be designed in accordance with an approved method, which contains provisions for truck and bus loadings.

(3) The loading applies to stack room floors that support nonmobile, double-faced library bookstacks subject to the following limitations:

a. The nominal bookstack unit height shall not exceed 90 in. (2290 mm);

b. The nominal shelf depth shall not exceed 12 in. (305 mm) for each face; and

c. Parallel rows of double-faced bookstacks shall be separated by aisles not less than 36 in. (914 mm) wide.

(4) In addition to the vertical live loads, the design shall include horizontal swaying forces applied to each row of the seats as follows: 24 lbs/ linear ft of seat applied in a direction parallel to each row of seats and 10 lbs/ linear ft of seat applied in a direction perpendicular to each row of seats. The parallel and perpendicular horizontal swaying forces need not be applied simultaneously.

(5) Other uniform loads in accordance with an approved method, which contains provisions for truck loadings, shall also be considered where appropriate.

(6) The concentrated wheel load shall be applied on an area of 4.5 in. by 4.5 in. (114 mm by 114 mm, footprint of a jack).

(7) Minimum concentrated load on stair treads (on area of 4 in.^{2} (2580 mm^2)) is 300 lbs (1.33 kN).

Source: Reprinted with Permission from ASCE 7-02 *Minimum Design Loads for Buildings and Other Structures,* Copyright © 2003, American Society of Civil Engineers. This information is extracted from ASCE 7-02; for further information, the complete text of the manual should be referenced (http://www.pubs.asce.org/ASCE7.html?9991330).

3.2.3 Snow Loads

Snow loads vary widely from region to region. In mountainous areas, snow loads may vary widely in short distances. Factors affecting snow load accumulation on structures include climatic variables, geographic location, roof exposure, roof slope, roof thermal condition, snow drifting, and sliding snow. Snow loads are generally provided by the building official. In mountainous regions, snow loads may be site-specific and may be determined as a service by the building official, or it may be required that the snow load be determined by the designer. Structural engineers' associations of a number of states publish information helpful for calculating site-specific snow loads. Such information generally consists of ground snow load maps and methods to determine corresponding roof snow loads.

In the absence of specific snow load information, the amount of snow on the ground at a site at an accepted mean recurrence interval (MRI) must be determined by accepted hydrological methods. The mean recurrence interval typically used is 50 years. Structures containing essential facilities (emergency-related facilities) and those containing particularly hazardous materials may be designed to withstand snow (and other) loads associated with greater recurrence intervals. Snow loads provided by the building official or determined from snow load maps are generally for shallow sloping roofs. Adjustments may be used (if approved by the building official) for reduced snow on steep roofs and on roofs with unobstructed slippery surfaces.

Consideration must also be made for snow accumulations. Examples to be considered include the following.

1. Snow drifting and accumulating in valleys between parallel and nonparallel ridges
2. Snow drifting behind parapets or other roof projections
3. Snow accumulating and ice damming on eaves
4. Snow drifting onto lower roofs (or balconies or decks)
5. Snow drifting and sliding onto lower roofs (or balconies or decks
6. Snow drifting from one side of a predominant ridge and accumulating on the other side

The American Society of Civil Engineers [5] and various structural engineers' associations provide guidance for the adjustments above. It may be beneficial for a professional charged with designing for snow loads to visit areas with structures loaded with snow to better understand snow accumulations. Snow loads for mountainous areas, especially in the western states, may be enormous. Mt. Baker, in western Washington, recently received over 90 ft of snow in a single season. Actual snow packs of over 700 psf have been recorded in isolated regions, and some inhabited regions have snow loads ranging up to 300 psf. Roof snow loads are generally prescribed assuming heated structures. Unheated structures, structures with particularly effective insulation, or structures used for cold storage should be given special consideration and the snow loads increased accordingly.

For snow loads, a duration of two months is customarily used. Although snow may remain on a structure for periods exceeding two months, the amounts of snow for longer periods of time are generally less than the design-load. In cases where design- or near-design-level amounts of snow may be expected for longer periods of time, the use of a longer load duration is recommended.

3.2.4 Wind Loads

Wind forces on structures are generally calculated from surface pressures. Surface pressures may be positive or negative. Negative (suction) pressures on roofs may produce significant uplift forces, in some cases equal to or greater than gravity loads. Both positive and negative pressures load structural members and act to overturn or slide a structure. Pressures used for wind design are generally derived from basic or design wind speeds for a given geographical area. Adjustments are made for height above ground, the conditions of the surrounding terrain (*exposure*), and for the part of the structure under consideration. Local pressures for individual framing members or pieces of sheathing are higher than the pressures assumed over larger surfaces of the structure. Similarly, surfaces near discontinuities (roof eaves, rakes and ridges, and wall corners) experience higher pressures than surfaces distant from the discontinuities. Importance factors in wind design are associated with the mean recurrence interval of the design wind speeds. An importance factor of 1.00 generally corresponds to an MRI of 50 years; higher importance factors correspond to greater mean recurrence intervals.

The design wind speed and exposure condition for a structure should be determined by the building official and indicated on the design and construction documents. Wind pressures and resulting forces on the structure should be calculated according to the locally adopted code. In the absence of a locally adopted code, procedures in a model building code, or *Minimum Design Loads for Buildings and Other Structures,* SEI/ASCE-7 [5], should be used. Special consideration should be made for increased wind speeds in valleys, gorges, and mountainous areas. In areas of extremely high wind (e.g., hurricane-prone regions), consideration must also be made for windborne debris.

3.2.5 Earthquake Loads

The ground movements from earthquakes produce significant forces on structures. The *seismic zone or seismic design category* for a proposed structure should be obtained from the building official as well as from the governing building code stipulating seismic design procedures and requirements. In areas of high seismic risk, site location with respect to known faults is important. In some cases, a site-specific study of seismic ground motions may be necessary. *Minimum Design Loads for Buildings and Other Structures,* SEI/ASCE-7 [5], may be used in the absence of local code requirements, with careful attention to site-specific hazards.

With regard to earthquakes, building codes have been developed with the interest of public safety and not for the prevention of all damage to structures. As a minimum, buildings are designed *not to collapse* during the design earthquake event. The degree of damage deemed acceptable during the design earthquake is reflected in building codes, with the importance category or seismic design category of the structure as determined by the building official. The expected performance of a structure during a design earthquake should be agreed upon clearly between designer and owner. The design of a structure to sustain less damage than that allowed by the minimum requirements of the code will generally increase the cost of the structure as well as design costs.

3.2.6 Highway Loads

Timber bridges can and have been designed to satisfy the requirements of modern vehicle transportation. Load and other design requirements for highway loads should be obtained from the *Standard Specifications for Highway Bridges,* adopted by the American Association of State Highway and Transportation Officials (AASHTO) [6]. Due to the increased strength of wood to withstand impact loads, wheel loads used in the design of timber bridges are generally treated as live loads with a two-month load duration without special consideration of dynamic effects. Dynamic effects must be considered, however, in metal-to-metal connections and in railing design. *Glued Laminated Timber Bridge Systems* [7], available from AITC, provides assistance and examples for timber bridge design. Timber bridges have historically performed very well, are aesthetically pleasing, and are relatively environmentally friendly.

3.2.7 Railway Loads

Timber railway bridges and trestles have been used extensively, and with the advent of pressure preservative treatments, a service life of 50 years is commonplace. For the design of timber bridges and trestles, the recommendations in the *Manual for Railway Engineering* of the American Railway Engineering and Maintenance-of-Way Association (AREMA) [8] are ordinarily used. As with highway bridge design, rail loads are treated as live loads with normal load duration without additional consideration for impact effects for wood members. Dynamic effects must be taken into consideration for metal-to-metal connections.

3.2.8 Crane Loads

Timber is often used for crane beams and girders because of its ability to absorb impact forces. In designing crane beams, wheel loads recommended by crane manufacturers should be used. At a minimum, loads should be taken to be at least equal to the rated capacity of the crane plus the weight of the crane and trolley. In general, crane loads are considered live loads with normal load duration but are increased a certain percentage to account for impact effects. SEI/ASCE-7 [5] prescribes the minimum impact, lateral, and longitudinal force effects that must be considered in design. Beams or girders supporting cranes should also be designed to prevent undue vertical and lateral deflection in accordance with crane manufacturers' recommendations.

3.2.9 Dynamic Loads

Wood by nature performs well under dynamic loads, as discussed earlier with regard to bridge and highway loads. As discussed in more detail in Section 3.4.1, certain design stresses may be increased significantly for short-duration loads. This is particularly true of stresses induced parallel to the grain, such as in bending. Loading or service conditions that produce tension perpendicular to the grain (notches, or fasteners located close to and loaded toward member edges) should be avoided, particularly in repeated load applications.

3.2.9.1 Cyclic Loads Research and experience indicate that wood performs well in situations that may cause fatigue failures in other structural materials. This is true particularly with stresses parallel to the grain, such as in bending. In applications where shear stresses are relatively high and more than 1 million cycles are expected, the design shear stresses should be adjusted as recommended in ASCE Paper 2470, *Design Considerations for Fatigue in Timber Structures* [9] and/or USDA Forest Products Laboratory Report 2236, *Fatigue Resistance of Quarter-Scale Bridge Stringers in Flexure and Shear* [10]. Whereas special consideration for timber in cyclic loading conditions may not be necessary, accompanying mechanical fasteners and connections may require special consideration.

3.2.9.2 Vibration Loads The effects of vibration may generally be neglected in timber structures, except for cases in which the vibration forces are critical or in which vibrations may be objectionable to human occupancy. Vibrating equipment should be installed with isolation devices and manufacturer recommendations should be followed. In the absence of detailed analysis or information from the manufacturer, it is recommended that dead load of the equipment itself be doubled for design of the supporting members. Excess vibration, although not necessarily of structural concern with timber members, may tend to loosen threaded fasteners.

3.2.9.3 *Impact Loads* Design stresses may generally be increased 100% for impact loads, as shown in greater detail in Section 3.4.1. Loads for mechanical fasteners and connections, however, are generally increased by a lesser amount, as indicated in the *National Design Specification® for Wood Construction* [11] and/or local building codes.

3.2.9.4 *Blast Loads* Data on the design of structures resistant to blast loading such as may be caused by nuclear weapon explosions are beyond the scope of this book. Such data may be found in the American Society of Civil Engineers Manual 42, *Design of Structures to Resist Nuclear Weapons Effects* [12].

3.2.10 Load Combinations

Design must take into consideration the most severe realistic distribution, concentration, and combination of loads and forces. The following example load combinations for allowable stress design are from the *International Building Code* [2]:

$$D$$

$$D + L$$

$$D + L + (L_r \text{ or } S \text{ or } R)$$

$$D + (W \text{ or } 0.7E) + L + (L_r \text{ or } S \text{ or } R)$$

$$0.6D + W$$

$$0.6D + 0.7E$$

where D = dead load,
L = live load,
L_r = roof live load,
S = snow load,
R = rain load,
W = wind load, and
E = earthquake load.

Load combinations generally take into account that some design loads may occur simultaneously—for example, floor live load and roof snow load—or that a particular design load may occur simultaneous with some fraction of another load—for example, wind and snow. The combinations also recognize that some loads will in all likelihood not occur simultaneously: for example, wind and earthquake.

3.2.11 Design Stresses

Structural members must be selected such that critical stresses arising from all appropriate load combinations are less than or equal to the corresponding allowable stress values for the load combination and service conditions in question. Typical stresses analyzed in timber stresses are as follows:

f_b = Extreme fiber bending stress,
f_v = Shear stress parallel to grain,
f_c = Compression parallel to grain,
$f_{c\perp}$ = Compression perpendicular to grain,
f_t = Tension parallel to grain,
f_{rt} = Radial tension stress in curved members.

For any particular structural member, the location of one critical stress may be different from another critical stress, and the two may be examined independently. For example, the critical shear stress in a bending member is at the neutral axis near a support, whereas the location of critical bending stress is at the extreme fiber (top or bottom of the member) near or at the center of the span for a simply supported member. Other critical stresses may occur at the same location, and their combined stress or stress interaction must be considered: for example, bending and axial compression or tension.

Where curved members are subject to bending, radial stresses are developed. Bending loads that tend to straighten a curved member produce radial stresses, f_{rt}, that act in tension across the grain and are of particular concern. Curved members must be sized so that such radial stresses are less than allowable radial tension stresses. Chapter 4 provides design examples, including all of the stresses listed above and appropriate combined stresses.

3.3 DESIGN VALUES

In ASD, structural member sizes and grades are selected such that the stresses due to design loading conditions are not greater than the allowable stresses for the same loading and service conditions:

$$f_{()} \leq F'_{()} \tag{3.2}$$

where $f_{()}$ is the applied or design stress (Section 3.2.11),

() denotes the type of stress (shear, flexure, compression, etc.), and
$F'_{()}$ is the allowable stress determined from the published design value, $F_{()}$, modified by appropriate adjustment factors.

Allowable stresses are obtained from design values and adjustment factors published by grading agencies for sawn lumber, the American Institute of Timber Construction for glued laminated timber, and are summarized in the *National Design Specification®* [11]. Design values are generally published by species or species group and grade (or species and combination or stress class for glulam) for all of the stresses listed in Section 3.2.11. Adjustment factors take into consideration the end use of the structural member (wet use versus dry, load duration, etc.). Design values are generally denoted by F; and the adjusted or allowable stress value, by F'. Design values are discussed in greater detail in this section.

For glued laminated timber, structural properties vary across the beam section, so additional distinction in design values is made for tension and compression zones as well as orientation of the laminations with respect to load. Design values represent the ultimate strength of the wood material for the type of stress specified divided by an appropriate factor of safety, except compression perpendicular to the grain, which is based on an accepted deformation limit. Design values for modulus of elasticity are also published. Design values are not published for tension perpendicular to the grain in timber design, because this condition is to be avoided.

3.3.1 Grading

Wood as it is sawn from a log is quite variable in its mechanical properties. Individual pieces may differ in strength by as much as several hundred percent. For simplicity and economy in use, pieces of lumber of similar mechanical properties are placed in classes known as *grades*. The structural properties of a particular grade depend on the sorting criteria used, species or species group, and other factors. Rules for determining lumber grades are written by rules-writing agencies authorized by the American Lumber Standards Committee [13]. Four such agencies are the Southern Pine Inspection Bureau [14], the West Coast Lumber Inspection Bureau [15], the Western Wood Products Association [16], and the National Lumber Grades Authority [17]. Lumber grading is also certified by agencies authorized by the American Lumber Standards Committee. Generally, the designer of timber structures is not charged with grading but, instead, with selecting commercially available grades that meet necessary structural requirements.

Lumber grading agencies also establish design values and adjustment factors for each grade. Design values provided by the grading agencies are published in the *National Design Specification® (NDS®)* [11]. These values and factors are generally accepted by model and/or local building codes but are sometimes adopted with amendments particular to the jurisdiction. The *NDS®* is a recommended supplement to this book. Grading is accomplished by sorting according to visually observable characteristics (visual grading) and/or by measurable mechanical and visual properties (mechanical grading). Both grading methods relate visually observable characteristics or mechanically observed properties to expected strength.

3.3.1.1 Visual Grading In visual grading, the wood pieces are sorted into grades according to species or species group, size and presence of knots, slope of grain, growth rate, and other characteristics. Design values are assigned to each grade by one of two procedures. In the first procedure, strength properties are assigned to clear pieces of wood in each species or species group, and then the strength values are reduced for the presence of strength-reducing characteristics such as knots and slope of grain. The strength characteristics of the clear wood are established by *Standard Test Methods for Establishing Clear Wood Strength Values,* ASTM D2555 [18]. The reduced strength values within each grade are determined according to *Standard Practice for Establishing Structural Grades and Related Allowable Properties for Visually Graded Lumber,* ASTM D245 [19]. This method is used for dimensional lumber of thickness up to 4 in. and various depths as well as for sawn timbers (5 in. × 5 in. and larger). In the second procedure, full-size pieces within each grade are assigned design values by *Standard Practice for Establishing Allowable Properties for Visually-Graded Dimension Lumber from In-Grade Tests of Full-Size Specimens,* ASTM D1990 [20]. This procedure is generally applied only to pieces of dimension lumber.

3.3.1.2 Mechanical Grading In addition to visual grading, structural lumber may also be graded mechanically. Two common mechanical grading classifications are machine-stress-rated (MSR) lumber and machine evaluated lumber (MEL). Mechanical grading is based on the relationship between the modulus of elasticity or specific gravity of a piece of wood and other structural characteristics. Pieces are graded or sorted according to mechanically measured modulus of elasticity or specific gravity, and then are checked further to meet visual requirements. Randomly selected pieces are also subjected regularly to destructive tests to ensure adequate performance. The use of both mechanical and visual sorting criteria tends to produce more consistent strength predictions than either visual or mechanical evaluation alone. Procedures for mechanical grading are given in *Standard Practice for Assigning Allowable Properties for Mechanically-Graded Lumber,* ASTM D6570 [21]. Mechanical grading is generally applied to pieces of 2 in. nominal thickness or less.

3.3.1.3 Glued Laminated Timber The term *structural glued laminated timber* refers to an engineered, stress-rated product of a timber laminating plant comprising assemblies of suitably selected and prepared wood laminations bonded together with adhesives. The grain of all laminations is approximately parallel longitudinally. Individual laminations do not exceed 2 in. net thickness. Individual lamination pieces may be joined end to end to produce laminated timbers much longer than the laminating stock itself. Pieces may also be placed or glued edge to edge to make timbers wider than the input stock. As such, glued laminated timbers (glulam) may be made to almost any size; however, shipping considerations generally limit the size of glulam normally produced. Standard dimensions for glulam are discussed in Chapter 1. Glued laminated timbers may also be manufactured into curved shapes, adding to their appeal for use in institutional and commercial structures.

In addition to versatile member size and length and shape, glued laminated timbers are an efficient utilization of wood resource. Higher-grade laminations are placed in the parts of the beam expected to experience higher stresses under load; lower-grade laminations are placed in the parts of the beam subjected to lesser stress. Pieces used as lamination stock in glulam may be graded either visually or mechanically; however, *lamination grades* are typically used instead of *structural joists and planks* grades.

The arrangement or layup of the pieces making up various glulam combinations is described in AITC 117, *Standard Specifications for Structural Glued Laminated Timber of Softwood Species* [22]. Quality control and other manufacturing requirements are governed by ANSI/AITC A190.1 [23]. Design values for the various layups or combinations are determined by *Standard Practice for Establishing Allowable Properties for Structural Glued Laminated Timber (Glulam)*, ASTM D3737 [24]. In the past, glued laminated timbers have been specified by species and combination symbol. A stress classification system for glulam has also been developed that provides more flexibility for the manufacturer as well as simpler glulam specification for the designer.

The laminations in the outer 5% of the beam depth that are located in the tension zone of glued laminated timbers are required to satisfy stringent requirements. The special tension laminations are located in the bottom 5% of beam depth in unbalanced beams (beams manufactured to be used as simply supported members) and in the bottom and top 5% of beam depth for balanced beams (beams intended for use as cantilevered or continuous members). Beams manufactured without special tension laminations are assigned reduced strength values.

Joining laminations end-to-end is achieved using either scarf or finger joints, although finger joints are the most common. End joints in glulam manufactured in accordance with ANSI/AITC A190.1 [23] are subject to strict quality control, which ensures that full design values for the member may be used without adjustment for the joints.

Glued laminated timbers are generally manufactured with the intent that the loads be applied perpendicular to the wide faces of the member. Since the members are used primarily for resisting gravity loads, such members are referred to as *horizontally laminated*. Glued laminated timbers may also be used to resist axial loads (tension and compression) and combined loads (axial and bending). Laminated timbers may also be placed so that the bending loads are applied parallel to the wide faces of the laminations (vertically laminated members). Design values for the various loading conditions are determined by the procedures in ASTM D3737 [24] and are published in AITC 117 [22] and the *National Design Specification*® [11].

3.4 ADJUSTMENT FACTORS

Design values are published for standard conditions and adjusted for end use and loading conditions by the designer. Adjustment factors may also be applied for member size and orientation. Many of the adjustment factors are common to both sawn lumber and glued laminated timber. Others are for use with only one of the types of wood products. Adjustment factors are published in this book and in the *National Design Specification*® [11]. Some adjustment factors may be prescribed or amended by local building codes; so the local building authority should be contacted and all applicable requirements obtained as part of the design process. Most adjustment factors are cumulative. In general, the allowable stress is determined by the design values multiplied by all appropriate adjustment factors.

3.4.1 Load Duration Factor, C_D

Design values for timber construction are published for a *normal* duration of load, assumed to be 10 years. For loads taken to have normal load duration, no adjustment of the design value is necessary for load duration. For loads of shorter duration, the design values for bending, shear parallel to the grain, compression parallel to the grain, and tension parallel to the grain are adjusted in accordance with Table 3.4. The values in the table reflect the increased strength of wood for a short duration of load. Load duration factors for durations not shown in the table may be obtained from Figure 3.1. Although the load duration factor for dead loads is less than unity (thus decreasing the

TABLE 3.4 Frequently Used Load Duration Factors

Load Duration[a]	C_D	Typical Design Loads
Permanent	0.9	Dead loads
10 years	1.0	Occupancy live loads
2 months	1.15	Snow loads
7 days	1.25	Construction loads
10 minutes	1.6	Wind and earthquake loads
Impact[b]	2.0	Impact loads

Source: National Design Specification® [11].

[a]Load duration factors do not apply to modulus of elasticity, E, nor to design values for compression perpendicular to the grain, $F_{c\perp}$, based on a deformation limit.

[b]The impact load duration factor does not apply to structural members pressure-treated with waterborne preservatives, nor to structural members pressure-treated with fire retardant chemicals. The impact load duration factor does not apply to connections.

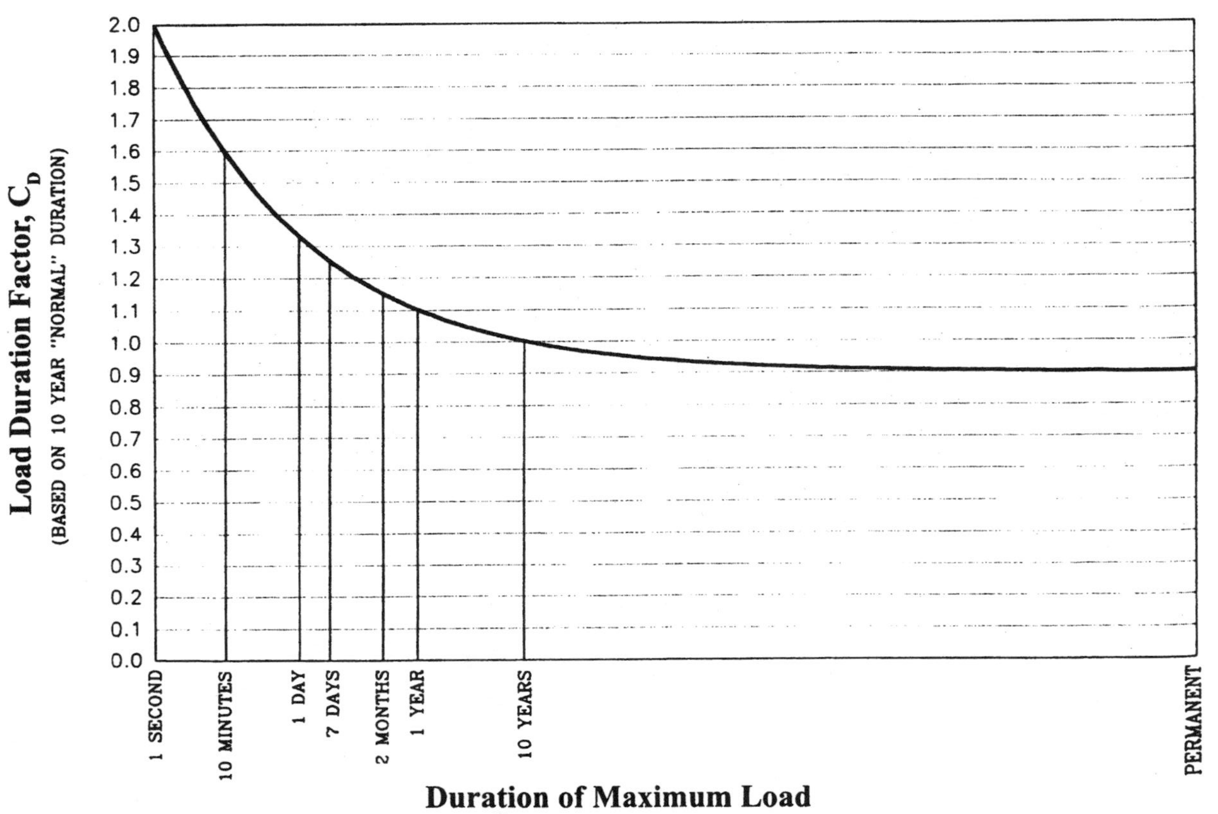

Figure 3.1 Load duration factors for various load durations. (From *National Design Specification*® [11].

design value), it seldom governs design except in cases of heavy permanent loads (such as storage uses) or where the dead loads are in excess of 90% of the total load. Compression perpendicular to the grain is not adjusted for load duration, nor is the modulus of elasticity. Some building codes require use of 1.33 for wind and earthquake loads. Load duration factors for connections differ from those listed and are discussed in greater detail in Chapter 5.

3.4.2 Wet Service Factor, C_M

Design values assume an in-service moisture content of 19% or less for dimension lumber and 16% or less for glued laminated timber. If wood members are subject to conditions that may cause sustained moisture content values higher than those listed, the wet service factor is to be used. Wet service factors account for reduced strength of wood at higher moisture content. Wet service factors are summarized in Table 3.5. Southern pine design values for decking and sawn timbers 5 in. × 5 in. and larger require no adjustment for wet service.

3.4.3 Factors Related to Size, Shape, and Length

Research has indicated that design values are affected by the size and shape of a member as reflected by its dimensions: thickness, width, and length for

TABLE 3.5 Wet Service Factor, C_M

	Strength Property						
	F_b	F_t	F_v	$F_{c\perp}$	F_c	E	F_{rt}
	Glued Laminated Timber						
All species, MC >16%	0.80	0.80	0.875	0.53	0.73	0.833	0.875
	Sawn Lumber						
Visually graded dimension lumber, MC > 19%	0.85[a]	1.00	0.97	0.67	0.80[b]	0.90	
Visually graded timbers (5 in. × 5 in. and larger) MC > 19%, except southern pine and mixed southern pine[c]	1.00	1.00	1.00	0.67	0.91	1.00	
Visually graded decking MC > 19%, except southern pine[c]	0.85[b]			0.67		0.90	

[a] For visually graded dimension lumber of all species with $F_bC_F \leq 1150$ psi, $C_M = 1.0$.
[b] Where F_bC_F of all visually graded species except southern pine ≤ 750 psi, $C_M = 1.0$; when F_c for visually graded southern pine ≤ 750 psi, $C_M = 1.0$.
[c] Design values for southern pine and mixed southern pine are used without further adjustment.

sawn lumber; width, depth, and length for glued laminated timber. Adjustments for the effects of size and shape are described in greater detail below.

3.4.3.1 *Size Factor, C_F (Sawn Lumber)* Visually graded dimension lumber except for southern pine is graded such that the size factor, C_F, is to be applied to the design values for bending, tension parallel to the grain, and compression parallel to the grain for various sizes within each grade. Size factors for use with dimension lumber are published with the design values in the *NDS*® [11]. The size factor is not applied to visually graded timbers (5 in. × 5 in. and larger) except for timbers with depths exceeding 12 in. or where subject to loads applied to the wide face. The size factor for timbers of depth greater than 12 in. is determined by

$$C_F = \left(\frac{12\text{ in.}}{d}\right)^{1/9} \tag{3.3}$$

where d = member depth (in.). Size factors for 5 in. × 5 in. and larger timbers loaded perpendicular to the wide faces are tabulated with the design values in the *NDS*® [11].

Design values for visually graded southern pine are published for different size ranges within each range, and thus no size factor adjustment is necessary except for particular grades with widths greater than 12 in. No size factor adjustment is necessary for mechanically graded lumber (MSR and MEL). Examples of the use of size factors are included in design examples in the following chapters. Sawn lumber tends to be limited in length, so design values are established such that no length adjustment is necessary.

3.4.3.2 *Volume Factor, C_V (Structural Glued Laminated Timber)* Design values for bending, F_b, in horizontally laminated glued laminated timbers are based on standard beam dimensions of $5\frac{1}{8}$ in. width × 12 in. depth and 21 ft length. Adjustment for other sizes is made using the volume factor, C_V:

$$C_V = \left(\frac{5.125\text{ in.}}{b}\right)^{1/x}\left(\frac{12\text{ in.}}{d}\right)^{1/x}\left(\frac{21\text{ ft}}{L}\right)^{1/x} \leq 1.0 \tag{3.4}$$

where

b = member width (in.),
d = member depth (in.),
L = length of member between points of zero moment (ft), and
x = 20 for southern pine and 10 for other species.

The volume factor should in no case be greater than unity.

Example 3.1 Volume Factor

Given: $6\frac{3}{4}$ in. × $25\frac{1}{2}$ in. × 30 ft western species glulam beam.

Determine: The volume factor.

Approach: Equation (3.4) will be used with $x = 10$.

Solution:

$$C_V = \left(\frac{5.125 \text{ in.}}{6.75 \text{ in.}}\right)^{1/10}\left(\frac{12 \text{ in.}}{25.5 \text{ in.}}\right)^{1/10}\left(\frac{21 \text{ ft}}{30 \text{ ft}}\right)^{1/10} = 0.87$$

Answer: The volume factor for the beam is 0.87.

3.4.3.3 Flat Use Factor, C_{fu} (Dimension Lumber) Design values for bending, f_b, for dimension lumber are based on pieces loaded on edge (loaded parallel to the wide face). Where these members are placed flat and loaded perpendicular to the wide face, the design values for bending may be increased using the flat use factor, C_{fu}. Flat use factors are published with the design values [11]. Design values for visually graded decking [11] generally take into consideration flat use with no further adjustment to be taken. The flat use factor is not applied to visually graded timber (5 in. × 5 in. and larger).

3.4.3.4 Flat Use Factor, C_{fu} (Glued Laminated Timber) The flat use factor should be applied to the design value for bending for glued laminated timber loaded parallel to the wide faces of the laminations (F_{by}) for wide-face lamination widths other than 12 in. The flat use factor for glulam is calculated by

$$C_{fu} = \left(\frac{12 \text{ in.}}{d}\right)^{1/9} \tag{3.5}$$

Note that the flat use factor produces an increase in bending stress for depths (widths of the wide faces of the laminations) less than 12 in., and a reduced bending stress for lamination widths greater than 12 in.

3.4.3.5 Curvature Factor, C_c (Glued Laminated Timber) The curvature factor is applied to the design value in bending, F_b, for curved glued laminated members only. It takes into account the prestress induced in laminations bent into curved shapes during the manufacturing process, thereby lowering the allowable bending stress for the design loading. The curvature factor is calculated using equation (3.6). The curvature factor is significant for glulam timbers manufactured to relatively short curvature radii. For large-radius glued laminated timbers, such as cambered beams, the adjustment is not significant and is typically not considered. The reduction of bending strength for a glulam with $1\frac{1}{2}$-in. laminations manufactured to a curvature of radius of 56 ft is only 1%.

$$C_c = 1 - 2000\left(\frac{t}{R}\right)^2 \tag{3.6}$$

where t = thickness of laminations and R = radius of curvature of the inside face of lamination. The ratio t/R may not exceed $\frac{1}{100}$ for hardwoods and southern pine or $\frac{1}{125}$ for softwoods other than southern pine. The curvature factor is not applied to design values in the straight portion of a member regardless of curvature in other portions. Curved members also develop radial stresses, which must be considered separately, as shown in Chapter 4.

3.4.3.6 Other Factors Related to Size, Shape, and Length: C_L, C_I, and C_P Member stability in bending and compression is related to size, shape, and length as well as to distribution of load, member stiffness, and lateral support. The bending design value F_b for sawn lumber and glued laminated timber should be adjusted for stability by multiplying by the beam stability factor, C_L. The beam stability factor is taken to be unity if the beam depth is less than or equal to the beam width, or if the compression side of the beam is laterally supported along its entire length. In all other cases, the beam stability factor should be calculated as shown in Chapter 4. The beam stability factor is not cumulative with the volume factor in glued laminated timbers but is cumulative with the size factor for sawn lumber.

The stress interaction factor, C_I, should be applied to the bending design values for glued laminated timbers with taper cuts. Examples of the use and calculation of C_I are included in Chapter 4. The stress interaction factor is not cumulative with the volume factor, C_V, when the taper is on the compression side of the beam. Taper cuts on the tension side of a beam are not recommended. The stress interaction factor is not applied to arches or to curved portions of pitched and tapered curved beams.

For members subject to axial compression, the design value for compression parallel to the grain should be multiplied by the column stability factor, C_P, which is related to size, shape, length, load, stiffness, and lateral support conditions. Examples including the calculation of C_P are included in Chapter 4.

3.4.4 Repetitive Member Factor, C_r

Where three or more parallel members of dimension lumber are spaced 24 in. on center (o.c.) or less, the repetitive member factor, C_r, of 1.15 may be applied to the design value for bending. The members must be joined to one another mechanically (e.g., built-up beams) or must be joined by floor, roof, or other load-distributing elements capable of supporting the design load. The repetitive member factor takes into consideration the variability of lumber and that stiffer members also tend to be stronger. Where several such members are configured in parallel (as described above), the stiffer member(s) take a greater portion of the applied load. The repetitive member factor is not used with decking because the factor has already been included in establishing the design values for decking. Also, it does not apply to glued laminated timber.

3.4.5 Temperature Factor, C_t

Most wood is used under conditions where temperature effects are not significant. However, as discussed in Chapter 1, when an elevated temperature of a member is anticipated for an extended period of time or when elevated temperatures are expected to occur simultaneously with maximum design loads, design values should be adjusted by the temperature factors of Table 3.6.

3.4.6 Bearing Area Factor, C_b

For any bearing lengths of less than 6 in. and not less than 3 in. at the end of a member, the design values for compression perpendicular to the grain, $F_{c\perp}$, may be multiplied by the bearing area factor, C_b, which may be obtained from

$$C_b = \frac{\ell_b + 0.375 \text{ in.}}{\ell_b} \tag{3.7}$$

where ℓ_b = the length in bearing parallel to the grain of the wood (in.).

The bearing area factor effectively increases the compression perpendicular to grain value for bearing conditions away from the ends of a member such

TABLE 3.6 Temperature Factor, C_t

Design Values	In-Service Moisture Conditions[a]	C_t		
		$T \leq 100°F$	$100°F < T \leq 125°F$	$125°F < T \leq 150°F$
F_t, E	Wet or dry	1.0	0.90	0.90
F_b, F_v, F_c, and $F_{c\perp}$	Dry	1.0	0.80	0.70
	Wet	1.0	0.70	0.50

[a] Wet-service condition for sawn lumber is defined as an in-service moisture content greater than 19%. Wet-service condition for glued laminated timber is defined as an in-service moisture content of 16% or greater.

as may arise from bearing plates and washers. For round bearing areas, the bearing length is taken to be the diameter of the bearing area.

Example 3.2 Bearing Area Factor

Given: 3 in. × 3 in. timber washer bears against wood perpendicular to the grain in a connection.

Determine: The appropriate bearing area factor. Assume that the washer is not near the end of the member.

Approach: Use equation (3.7).

Solution: Since the bearing condition is away from the end of the member, equation (3.7) may be used:

$$C_b = \frac{3\text{ in.} + 0.375\text{ in.}}{3\text{ in.}} = 1.13$$

Answer: The design value for compression perpendicular to the grain may be increased by 13% along with all other appropriate adjustments. The area to be used to determine the bearing stress should be the actual bearing area (the area of the washer).

3.4.7 Adjustments for Treatments

The design values published for sawn dimension lumber and glued laminated timber may be applied to pressure-preservatively treated members when treated by an approved process as recommended by the American Wood-Preservers' Association (AWPA) [25]. For dimension lumber incised to facilitate preservative treatment, the incising factor, C_i, of Table 3.7 should be used. For timber treated with waterborne preservative salts or fire retardant chemicals, the duration-of-load factor for impact does not apply. It is recommended that the effects of fire retardant chemical treatments on strength be considered and that information on these effects be obtained from the company providing the treating and redrying service. Research indicates that fire retardant treatments may reduce bending strength by 10–25%. Because

TABLE 3.7 Incising Factor, C_i, for Sawn Dimension Lumber

Design Value	C_i
E	0.95
F_b, F_t, F_c, and F_v	0.80
$F_{c\perp}$	1.00

Source: National Design Specification® [11].

of the excellent fire performance characteristics of structural glued laminated timber, fire retardant treatments are not typically applied.

3.4.8 Other Adjustment Factors

The preceding discussion covers adjustment factors usually applicable to timber design. For special cases, such as poles and piles, other adjustment factors may be necessary.

3.4.9 Summary and Application of Adjustment Factors

Adjustment factors for timber design and their application are summarized in Table 3.8. Where adjustment factors are not applied cumulatively (e.g., C_L and C_V for glulam), the smaller factor applies.

3.5 SERVICEABILITY AND DEFLECTION

Serviceability concerns with timber design generally involve deflections of the framing members. Deflection considerations are also important for the prevention of ponding as discussed in Section 3.7. Deflections should be

TABLE 3.8 Adjustment Factor Application Summary

Loading Condition and Member	Design Value and Adjustment Factors
Bending	
Visually graded dimension lumber	$F'_b = F_b C_D C_M C_F C_{fu} C_L C_r C_t C_i$
Mechanically graded dimension lumber	$F'_b = F_b C_D C_M C_L C_r C_t C_i$
Sawn timbers 5 in. × 5 in. and larger	$F'_b = F_b C_D C_M C_F C_L C_t$
Glulam (load perpendicular to wide faces of lams)	$F'_b = F_b C_D C_M (C_V \text{ or } C_L) C_C C_I C_t$
Glulam (load parallel to wide faces of lams)	$F'_b = F_b C_D C_M C_L C_{fu} C_C C_I C_t$
Decking	$F'_b = F_b C_D C_M C_t$
Tension parallel to the grain	
Dimension lumber	$F'_t = F_t C_D C_M C_F C_t$
Glulam	$F'_t = F_t C_D C_M C_t$
Compression parallel to the grain	
Dimension lumber	$F'_c = F_c C_D C_M C_F C_P$
Sawn timbers 5 in. × 5 in. and larger	$F'_c = F_c C_D C_M C_P C_t$
Glulam	$F'_c = F_c C_D C_M C_P$
Compression perpendicular to the grain	$F'_{c\perp} = F_{c\perp} C_M C_t$
Shear parallel to the grain	$F'_v = F_v C_D C_M C_t$
Modulus of elasticity	$E' = E C_M C_t$
Radial tension	$F'_{rt} = F_{rt} C_D C_M C_t$

limited to prevent a deflected part from interfering with another part of a structure, to reduce vibration, and to prevent the appearance or feel of an unsafe or sagging structure. Both vertical deflection and horizontal deflection may need to be considered by the designer. Deflection limitations must be in compliance with the governing building code. The deflection limitations listed in the tables in this section are recommendations of the American Institute of Timber Construction to provide greater serviceability and may be more restrictive than required by codes.

Deflections are generally calculated for spanning members such as joists, beams, and trusses and may also be calculated for columns, frames, shear walls, and diaphragms. Beam deflections are discussed in this section, with examples given in Chapter 4. Deflections of trusses and frames are covered in Chapter 6, except for deflections of metal plate connection trusses, which are typically proprietary and provided as a service by the truss manufacturer. Deflections of shear walls and diaphragms are covered in the *Plywood Design Specification* [26].

Wood members experience immediate elastic deformation under load, as well as long-term plastic deformation (or creep) for loads that are sustained. The ratio of the long-term deflection to the initial deflection of a sustained load is approximated as 1.5 for seasoned wood and 2.0 for green wood members. Where actual deflections are calculated, such as for camber and ponding, both immediate- and long-term effects must be considered. Deflection limitation criteria, however, as shown in the following tables and as stipulated in typical building codes, may not require consideration of plastic deflection.

Deflection limitations are generally specified as some fraction of the member span. Two calculation checks are generally performed for beams and other spanning members. The first check is the deflection due to applied load (generally, live or snow load) only; the second check is for deflection due to applied load plus all or some fraction of dead load. Deflection criteria are generally more restrictive for the applied load only, although either case may govern, depending on the relative magnitudes of dead and applied loads on the member. Table 3.9 provides recommended deflection limitations for applied and total loads for roof and floor beams and highway and railway stringers. Table 3.10 provides deflection limitations for floor beams where additional stiffness is desired. For special uses, such as beams supporting vibrating machinery or carrying moving loads, more severe limitations may be required.

Example 3.3 Deflection Criteria Check for a Floor Beam

Given: the deflections of a $5\frac{1}{8}$ in. × 12 in. × 20 ft floor beam are calculated to be 0.62 in. under live load and 0.93 in. under live load plus dead load.

Determine: Whether or not the deflection criteria of Table 3.9 are met for ordinary use.

TABLE 3.9 Recommended Deflection Criteria

Use Classification	Applied Load Only	Applied Load + Dead Load
Roof beams		
Industrial	ℓ/180	ℓ/120
Commercial and institutional		
Without plaster ceiling	ℓ/240	ℓ/180
With plaster ceiling	ℓ/360	ℓ/240
Floor beams, ordinary usage[a]		
Highway bridge stringers	ℓ/300	
Railway bridge stringers	ℓ/300 to ℓ/400	

[a]Ordinary use classification for floors is intended for construction in which walking comfort and minimized plaster cracking are the main considerations. These recommended deflection limits may not eliminate all objectionable vibrations, such as in long spans approaching the maximum limits, nor may the limitations be suitable for some office and institutional applications where increased floor stiffness is desired. The deflection limitations in Table 3.10 may be used where additional stiffness or reduction of vibration is desired.

TABLE 3.10 Deflection Criteria for Increased Floor Stiffness

Use Classification	Live Load Only	Live Load + K(Dead Load)[a]
Floor beams		
Commercial, office, and institutional		
Floor joists, spans to 26 ft[b]		
LL[c] $\leq$ 60 psf	ℓ/480	ℓ/360
60 psf $<$ LL $\leq$ 80 psf	ℓ/480	ℓ/360
LL $\geq$ 80 psf	ℓ/420	ℓ/360
Girders, spans to 36 ft[b]		
LL $\leq$ 60 psf	ℓ/480[d]	ℓ/360
60 psf $<$ LL $\leq$ 80 psf	ℓ/420[d]	ℓ/300
LL $\geq$ 80 psf	ℓ/360[d]	ℓ/240

[a]$K = 1$ except for seasoned members, where $K = 0.5$. Seasoned members for this use are defined as having a moisture content of less than 16% at the time of installation.

[b]For girder spans greater than 36 ft and joist spans greater than 26 ft, special design considerations may be required, such as more restrictive deflection limits and vibration considerations that include the total mass of the floor.

[c]LL, live load.

[d]Based on reduction of live load where permitted by the code.

Approach: The appropriate ratios from Table 3.9 will be used with the total beam length to calculate allowable deflection for each case. The given deflections will then be compared with the allowable deflections to evaluate suitability with regard to the recommended criteria.

Solution: Applied load only: The recommended limit is $\ell/360$ or (20×12) in./360 = 0.67 in.; 0.62 in. $\leq$ 0.67 in.; good. Applied load plus dead load: The recommended limit is $\ell/240 = (20 \times 12)$ in./240 = 1.0 in.; 0.93 in. $\leq$ 1.00 in.; good.

Answer: The stated deflections satisfy the ordinary floor beam use recommendations of Table 3.9.

3.6 CAMBER

Camber is built into structural glued laminated timber members by introducing a curvature, either circular or parabolic, opposite to the anticipated long-term deflection. Camber recommendations are provided in Table 3.11. Recommendations for camber depend on whether the member is of simple, continuous, or cantilever span. The intent of the camber is typically to provide a straight or nearly straight structural member under the long-term effects of

TABLE 3.11 Recommended Minimum Camber for Glued Laminated Timber Beams

Roof beams[a]	$1\frac{1}{2}$ times the immediate dead load deflection
Floor beams[b]	$1\frac{1}{2}$ times the immediate dead load deflection
Bridge beams[c]	
Long span	2 times the immediate dead load deflection
Short span	2 times the immediate dead load deflection + $\frac{1}{2}$ of applied (live) load deflection

[a]The minimum camber of $1\frac{1}{2}$ times the immediate dead load deflection will produce a nearly level member under dead load alone after plastic deformation has occurred. Additional camber is usually provided to improve appearance and/or provide necessary roof drainage. Roof beams should have a positive slope or camber equivalent to $\frac{1}{4}$ in. per foot of horizontal distance between the level of the drain and the high point of the roof, in addition to the minimum camber to avoid the ponding of water. In addition, on long spans, level roof beams may not be desirable because of the optical illusion that the ceiling sags. This condition may also apply to floor beams in multistory buildings.

[b]The minimum camber of $1\frac{1}{2}$ times the immediate dead load deflection will produce a nearly level member under dead load alone after plastic deformation has occurred. For warehouse or similar floors where live load may remain for long periods, additional camber should be provided to give a level floor under the permanently applied load.

[c]Bridge members are normally cambered for dead load only on multiple spans to obtain acceptable riding qualities.

dead load, including plastic deformation or creep. Camber may also be specified to ensure adequate roof drainage.

For simple beams, a single radius member is typically selected or specified. For continuous members, either a straight member or custom camber is used. Reverse cambers may be necessary for cantilever spans and continuous spans under some loading conditions but should be specified cautiously, as they may also create drainage or ponding problems.

Camber is usually designated in custom members by specifying the amount of camber required in inches and the location along the member. For a simple-span beam, the camber is usually specified for the midpoint. The location and camber amount for cantilevered or continuous beams depends on the shape of the deflected structure.

Camber for stock glulam beams is usually designated by the manufacturer in terms of radius of curvature. Typically, long billets are made with constant radius and cut and sold in shorter pieces. Radius of curvature values of 1600 and 2000 ft have commonly been used, with 3500 ft becoming more accepted to prevent excessive camber interfering with other framing components during construction.

Constant-radius shapes have been found to provide suitable camber for large-radius (shallow) calculated deflections. The following equation may be used to calculate the camber associated with a constant radius of curvature and beam length:

$$c = R - \tfrac{1}{2}\sqrt{4R^2 - L^2} \tag{3.8}$$

where

c = camber at midspan assuming a circular shape,
R = radius of curvature,
L = span.

Alternatively, equation (3.8) may be cast to calculate the required radius of curvature for a given camber:

$$R = \frac{L^2}{8c} - \frac{c}{2} \tag{3.8a}$$

Example 3.4 Glulam Beam Camber

Given: The dead load deflection for a 32-ft, simple span, glued laminated timber is calculated to be 0.49 in., without consideration for creep.

Determine: The minimum recommended camber and associated radius of curvature for the beam.

Approach: Table 3.11 will be used to determine the minimum camber based on deflection and use (floor beam); equation (3.8) will be used to calculate the associated radius of curvature.

Solution: From Table 3.11, camber, c = 1.5 in. × 0.49 in. = 0.735 in. (or $\frac{3}{4}$ in.).

To be dimensionally consistent with the other terms in equation (3.8a), the camber is converted to feet:

$$\frac{0.735 \text{ in.}}{12 \text{ in./ft}} = 0.0613 \text{ ft (minimum recommended camber)}$$

From equation (3.8a),

$$R = \frac{(32 \text{ ft})^2}{(8)(0.0613 \text{ ft})} - \frac{0.0613 \text{ ft}}{2} = 2090 \text{ ft}$$

A smaller radius of curvature will produce more camber or a slight crown after the long-term effect of dead load; use R = 2000 ft.

Answer: The minimum recommended camber is $\frac{3}{4}$ in. and the associated radius of curvature is 2000 ft.

Example 3.5 Evaluation of Given Camber

Given: A glued laminated timber with camber radius 3500 ft has been recommended for a roof beam with simple span of 24 ft . The deflection from the dead load, including creep, is calculated to be 0.21 in. using commercially available design software.

Determine: Whether or not the 3500-ft radius is suitable for the simple roof beam application.

Approach: Since the calculated deflection already includes the effect of creep, the dead load deflection will not be modified further by the factors of Table 3.11, and the value 0.21 in. will be used directly with equation (3.8) to determine the radius of curvature. The radius of curvature calculated will then be compared with the stated curvature.

Solution: Converting 0.21 in. into feet gives (0.21 in.)/(12 in./ft) = 0.0175 ft. From equation (3.8),

$$R = \frac{(24 \text{ ft})^2}{(8)(0.0175 \text{ ft})} - \frac{0.0175 \text{ ft}}{2} = 4114 \text{ ft}$$

The stated curvature is 3500 ft, which will produce more camber or a slight crown under long-term load. The slight crown should also satisfy the appearance considerations stated in footnote [a] of Table 3.11.

Answer: The recommended radius of curvature of 3500 ft should be suitable for the stated application.

3.7 PONDING

Ponding of water on roofs or other flat elevated surfaces can cause catastrophic structural collapse if not properly prevented in design. Collapses from ponding are generally the result of insufficient roof stiffness and/or inadequate roof drainage, where the deflection due to water accumulation creates a progressively deeper pond or pool until the supporting structural elements fail. Ponding problems have been observed to be more prevalent on roofs with relatively light design loads, as these systems tend to be more flexible, and the weight of ponded water is in higher relative proportion to the design loads.

The best design against ponding is to provide adequate roof slope and drainage. A minimum roof surface slope of $\frac{1}{4}$ in. per foot is recommended. The minimum slope should remain after long-term deflection (creep) has occurred, thus requiring camber of roof members, framed roof slopes greater than $\frac{1}{4}$ in. per foot, or other accommodations. Drains should be selected or designed properly to discharge the design rainfall without creating undue hydraulic head on the roof surface. Consideration must also be made for gravel stops, scuppers, and other appurtenances that may cause water to pond on the roof surface. Consideration should also be made for uneven roof surfacing materials. Drainage systems should be selected to be as free as possible from debris or ice accumulation and should be maintained regularly.

For flat roofs or roofs with less than the recommended minimum slope, the supporting members must be sufficiently stiff and strong to prevent progressive deflection once water begins accumulating. Design for sufficient strength and stiffness is discussed below and included in Chapter 4. Where roof features may dam water, the weight of the dammed water must be included with the other design loads for the roof system. Careful consideration should be made for potential combined effects of ponding due to rain or snowmelt in combination with some or all of other design loads, such as rain and snow.

Roof systems in which the secondary framing members, sheathing or decking, and other framing are relatively stiff compared to the primary framing members are regarded as one-way systems for the purposes of investigating ponding. Design procedures for one-way systems are discussed below. For systems where the secondary and other members may themselves experience significant deflection or ponding, the combined effects of the deflections of primary and the other members must be investigated. Such systems may be regarded as two-way systems and may be investigated methodically, taking into account the accumulated effect of ponding for all members, or may be investigated using the provisions in AF&PA/ASCE 16-95 [27].

3.7.1 Simplified Approach

For flat roofs or roofs with insufficient slope for drainage, and where water is not dammed or caused to pond by others means, the members are generally considered stiff enough to resist ponding if the deflection from a 5-psf load on the member will cause not more than $\frac{1}{2}$ in. of deflection (5 psf corresponds to approximately 1 in. of standing water).

Example 3.6 Ponding (Simplified Approach)

Given: A $5\frac{1}{8}$ in. × 12 in. glulam beam has been selected to span 21 ft on a nearly flat roof. The beam supports a total load of 40 psf (25 psf of live load and 15 psf of dead load) with a tributary width of 8 ft. The self-weight of the beam is found to be 15 plf. The modulus of elasticity of the beam is 1.8 million psi, and the total load deflection, including self-weight but without consideration for creep, is estimated to be 1.10 in. (0.66 in. due to live load plus 0.44 in. due to dead load and self-weight).

Determine: The suitability of the beam stiffness with regard to ponding.

Approach: Assuming that the secondary and other framing members are relatively stiff (one-way action), the deflection for 5 psf will be calculated and compared to the simplified criteria of $\frac{1}{2}$ in. or less deflection for 5 psf of applied load.

Solution: The equivalent total area load for the beam, including self-weight, is 40 psf + 15 plf/8 ft = 42 psf. The deflection of the beam under 5 psf load may be calculated from the linear elastic stiffness of the beam determined by the total elastic deflection and total load by

$$\Delta_{5\ \text{psf}} = \Delta_{42\ \text{psf}} \left(\frac{5\ \text{psf}}{42\ \text{psf}}\right) = 1.10\ \text{in.} \left(\frac{5\ \text{psf}}{42\ \text{psf}}\right) = 0.13\ \text{in.}$$

$$0.13\ \text{in.} \leq 0.5\ \text{in.} \qquad \text{good}$$

Answer: Since the deflection of the beam for 5 psf is not greater than $\frac{1}{2}$ in., the beam is assumed to be stiff enough to resist ponding problems by the simplified approach.

It should be noted that the simplified approach does not consider additional stresses produced by ponding and thus assumes that the design stresses for the member are not critical or near critical in the design. Evaluation of the effect of ponding on member stresses may be performed using a magnification factor, as discussed in the next section.

3.7.2 Magnification Factor

The effect of ponding on member stresses and deflections may be determined using the magnification factor, MF. The magnification factor for a single span, simply supported beam is applied to the design stresses and deflections simultaneous with the ponding event:

$$\mathrm{MF} = \frac{1}{1 - \lambda\gamma SL^4/\pi^4 EI} \tag{3.9}$$

where

MF = the magnification factor to account for ponding,
λ = the ratio of long-term dead load deflection to immediate dead load deflection (typically, 1.5 for glulam and seasoned lumber and 2.0 for unseasoned lumber),
γ = the specific weight of the ponding liquid (typically, water),
S = the tributary width for the member under consideration,
L = the span of the member under consideration,
E = the modulus of elasticity of the member.

The modulus of elasticity, E, discussed thus far in this book has been assumed to be the average modulus of elasticity for the particular species and grade under consideration. For ponding considerations it is customary to use the lower fifth percentile modulus of elasticity, E_{05}. Values of E_{05} may be calculated from published design values for E and coefficient of variation (COV_E) in the *National Design Specification*® [11]. Using a normal distribution of E values within a particular species and grade, E_{05} may be computed using

$$E_{05} = E - 1.645E(\mathrm{COV}_E) \tag{3.10}$$

where

E_{05} = the lower fifth percentile modulus of elasticity,
E = the published (average) modulus of elasticity design value,
COV_E = the coefficient of variation for E published with the design values.

Coefficient of variation values for E for various types of members are shown in Table 3.12.

TABLE 3.12 Coefficient of Variation for Modulus of Elasticity for Various Wood Products

Wood Product	COV_E
Visually graded sawn lumber	0.25
Machine-evaluated lumber (MEL)	0.15
Machine-stress-rated lumber (MSR)	0.11
Glued laminated timber	
Six or more laminations	0.10
Fewer than six laminations	0.15

Example 3.7 Modulus of Elasticity

Given: Visually graded sawn lumber beam with a modulus of elasticity design value of 1,800,000 psi.

Determine: The fifth percentile modulus of elasticity.

Approach: The published design value is the mean for the grade and species; Equation (3.10) with the appropriate coefficient of variation will be used to compute the lower fifth percentile E.

Solution: From Table 3.12, $COV_E = 0.25$.

$$E_{05} = 1{,}800{,}000 \text{ psi} - (1.645)(1{,}800{,}000 \text{ psi})(0.25) = 1{,}060{,}000 \text{ psi}$$

Answer: The appropriate modulus of elasticity for ponding calculations for this grade is 1,060,000 psi.

A conservative approach to the use of the magnification factor of equation (3.9) is to apply it to all loads acting on the member. Strictly speaking, it is applied to the deflections and stresses associated with loads causing a curved shape that allows water or some other liquid to pool, resulting in an additional load of pooled liquid, additional deflection, additional pooling, and so on. If, for example, a member is cambered to offset all dead load deflection, the magnification factor applies only to the additional loads overcoming the camber and producing a shape capable of pooling water or other ponding liquid. In such cases the magnification factor is applied only to the additional deflections and stresses. The magnified values must then be added to the unmagnified deflections and stresses to compute the total deflection and stress conditions.

The magnification factor of equation (3.9) also assumes that all load and deflection effects are long term by inclusion of the long-term deflection factor λ. The short-term effects of ponding coincident with short-term loads would be expected to be less. However, while the event producing the ponding liquid may be of short duration (e.g., a rainstorm), the resulting pond may be required to endure for a significantly longer period of time. As such, it is recommended that all effects be considered long term and that λ be used with all magnified loads. The magnification factor of equation (3.9) applies to simply supported beams. The deflections and magnification factors for continuously framed beams tend to be less than those for simple members.

Example 3.8 Magnification Factor

Given: The beam considered in Example 3.6 supports a portion of roof where water may potentially pond to a depth of 2 in., due to the arrangement of the roof drainage system. The beam is to be cambered to offset 1.5 times the dead load (plus self-weight) deflection.

Determine: The magnified load condition for the beam, taking the 2-in. surcharge and ponding effects into consideration.

Approach: The beam is cambered to offset the long-term dead load deflection; thus, the live load and water surcharge loads only will be magnified. The fifth percentile modulus of elasticity will be calculated using equation (3.10) with the coefficient of variation for E from Table 3.12. The duration of the surcharge and live loads are not stated; thus, the magnification factor will include the ratio $\lambda = 1.5$ for glulam.

Solution: From Example 3.6, the modulus of elasticity is 1,800,000 psi. From Table 3.12, COV_E is 0.10.

$$E_{05} = 1{,}800{,}000 \text{ psi} - (1.645)(1{,}800{,}000 \text{ psi})(0.10) = 1{,}500{,}000 \text{ psi}$$

$$\lambda = 1.5 \text{ (glulam)}$$

For dimensional consistency, the specific weight of water converted into lb/in^3 is (62.4 pcf)/(12 in./ft)^3 = 0.0361 lb/in^3.

$$S = (8 \text{ ft})(12 \text{ in./ft}) = 96 \text{ in.}$$

$$L = (21 \text{ ft})(12 \text{ in./ft}) = 252 \text{ in.}$$

$$I = \tfrac{1}{12}(5.125 \text{ in.})(12 \text{ in.})^3 = 738 \text{ in}^4$$

$$\text{MF} = \frac{1}{1 - (1.5)(0.0361 \text{ lb/in}^3)(96 \text{ in.})(252 \text{ in.})^4/\pi^4(1{,}500{,}000 \text{ psi})(738 \text{ in}^4)} = 1.24$$

Two inches of water pooled on the roof will add (2 in.)/(12 in./ft)(62.4 pcf) = 10.4 psf of water surcharge. The total loading condition on the beam considering the 2-in. surcharge plus the magnification factor is thus in terms of area load:

$$15 \text{ psf (dead load)} + 2 \text{ psf (self-weight equivalent)} + 1.24(25 \text{ psf} + 10.4 \text{ psf}) = 61 \text{ psf}$$

Answer: The total area load on the beam considering self-weight, surcharge, and ponding magnification is 61 psf.

Discussion: This example illustrates that where conditions create a physical pond above the structural member, the surcharge load due to the pond must also be calculated. The magnification factor itself [equation (3.9)] relates only to the potential of progressive deflection (thus, the term *ponding instability*).

The suitability of the beam with regard to bending and shear stresses under the magnified loads must be investigated using the methods of Chapter 4. The deflection due to the magnified loads should be evaluated by criteria acceptable to the owner and approved by the building official. The magnification factor presented herein is based on a derivation by Kuenzi and Bohannan [28].

CHAPTER 4

STRUCTURAL MEMBERS

4.1 INTRODUCTION

This chapter covers the design of timber beams; columns; tension and compression members in timber trusses; curved, tapered, and pitched beams; and timber arches. Emphasis is placed on the use of structural glued laminated timber members, although solid sawn and/or engineered lumber products are used in some cases. Design of connecting or supporting hardware and fasteners is covered in Chapter 5. Design of structural systems that may be comprised of members from this chapter is covered in Chapter 6.

The design methodology used in this chapter is allowable stress design (ASD), where critical stresses in the members are calculated and compared to allowable stresses. The allowable stresses shown in examples are based on the design values published in AITC 117 [1] and the *National Design Specification® for Wood Construction* [2] adjusted according to the factors of Chapter 3 and in some cases adjusted further by the methods described in this chapter. In all examples, complete hand calculations are shown. Many of the calculations shown may be carried out using commercially available design software or may be solved by programming the equations into spreadsheet software.

4.2 BENDING MEMBERS

Bending members are structural components loaded perpendicular to their long axes, such as beams, joists, and planks (or decking). Beams and joists carry loads more efficiently when the direction of the load is parallel to their greater cross-section dimension (Figure 4.1). Beams with narrow cross sections (Figure 4.1 $d > b$) tend to be the most efficient; however, they are also less stable and must either be adequately braced or their bending strength reduced so that the member may also resist buckling. Planks or decking pieces are generally oriented flat, which is less efficient for resisting bending loads; however, unlike joists and beams, they serve the additional function of providing a floor or roof surface.

The determination of a suitable bending member size is most commonly governed by bending stress or deflection limitations based on given load, span, and bracing conditions. The design of bending members with relatively short

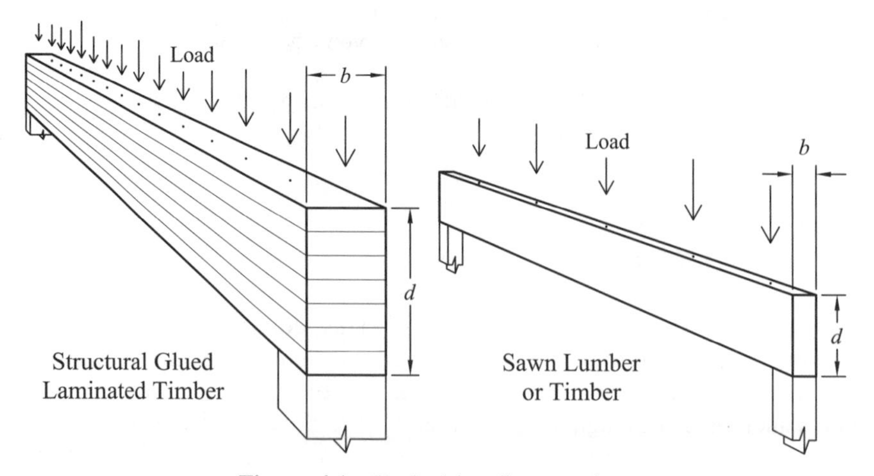

Figure 4.1 Typical bending member.

spans may, however, be governed by shear stress parallel to the grain. Examples of the foregoing cases are included in this chapter. Bearing stresses at supports or connections seldom govern the selection of bending member size and are covered in Chapter 5. The examples in this chapter illustrate the design process for individual structural members and move generally from relatively simple to more complex.

4.2.1 Bending (Flexure)

Loads acting perpendicular to the axis of spanning members produce flexural tension on one face of the member and flexural compression on the other. Wood products in general tend to better resist flexural compression than flexural tension; hence flexural tension is often the only bending stress checked (particularly for simple beams). Bending stresses are checked for the extreme fiber at the section with greatest bending moment. The applied bending stress is given by

$$f_b = \frac{M}{S} \tag{4.1}$$

where

f_b = extreme fiber bending stress,
M = applied bending moment,
S = section modulus, I/c,
I = area moment of inertia with respect to the neutral axis, and
c = distance from the neutral axis to the extreme fiber.

The design check (as with other stresses) is given by

$$f_b \leq F'_b \tag{3.2}$$

where

F'_b = the allowable extreme fiber bending stress obtained by multiplying the design value in bending by all applicable adjustment factors.

The compression zone of a member subject to bending may become unstable and buckle laterally if not braced. The design value in bending is adjusted by the beam stability factor, C_L, to account for potential lateral buckling. Bending members are considered to be adequately braced when the requirements of Table 4.1 are satisfied, and the beam stability factor may be taken to equal 1.0.

TABLE 4.1 Bracing Rules for Lateral Stability of Bending Members

Member	Rule
	Sawn Lumber
Depth/Breadth (d/b) (nominal dimensions)	
$d/b \leq 2$	No lateral bracing is required.
$2 < d/b \leq 4$	The ends are to be held in position, as by full-depth solid blocking, bridging, hangers, nailing, or bolting to other framing members, or other acceptable means.
$4 < d/b \leq 5$	The compression edge is to be held in line for its entire length to prevent lateral displacement, as by adequate sheathing or subflooring, and ends at points of bearing are to be held in position to prevent rotation and/or lateral displacement.
$5 < d/b \leq 6$	Bridging, full-depth sold blocking, or diagonal cross bracing is to be installed at intervals not exceeding 8 ft, the compression edge of the member is to be held in line for its entire length to prevent lateral displacement, as by adequate sheathing or subflooring, and ends at points of bearing are to be held in position to prevent rotation and/or lateral displacement.
$6 < d/b \leq 7$	Both edges of the member are to be held in line for their entire length, and ends at points of bearing are to be held in position to prevent rotation and/or lateral displacement.
Combined bending and axial compression	The depth/breadth ratio may be as much as 5 if one edge is held firmly in line. If under all conditions of load the unbraced edge is in tension, the depth/breadth ratio may be as much as 6.
	Glued Laminated Timber
Depth/Breadth (d/b)	
$d/b \leq 1$	No lateral support is required.
$d/b > 1$	The compression edge is supported throughout its length to prevent lateral displacement, and the ends at points of bearing have lateral support to prevent rotation.

Some of the means of preventing lateral rotation of a beam at its end and/or bearing points are as follows:

1. Attachment of the bottom of the beam to a wall, column, or pilaster that prevents movement of the bottom of the beam in a direction perpendicular to the beam axis, and attachment of the top of the beam to the wall or parapet, or to the roof diaphragm, provided that the roof is adequately attached to the wall
2. Full-depth blocking between the ends and bearing points of parallel beams
3. Cross bracing between parallel beams at ends and bearing points
4. Attachment of top and bottom of the ends of beams to girts or rim members
5. Suitable and approved connection hardware

Means of providing support against lateral movement of the beam at intermediate points or continuously along the beam include:

1. Attachment of structural roof or floor sheathing or deck material directly to the beam
2. Suitable attachment of roof and floor framing to the beam (girders, joists, trusses)
3. Intermediate full-depth blocking
4. Bridging or bracing at intermediate points
5. Proper attachment of suitable ceiling materials

Deck planks alone may not be sufficient to prevent lateral movement unless individual courses are nailed to one another or are made into a diaphragm with the additional application of structural sheathing.

Where the conditions of Table 4.1 are not satisfied, the beam stability factor must be calculated by

$$C_L = \frac{1 + F_{bE}/F_b^*}{1.9} - \sqrt{\left(\frac{1 + F_{bE}/F_b^*}{1.9}\right)^2 - \frac{F_{bE}/F_b^*}{0.95}} \tag{4.2}$$

where

F_{bE} = critical bucking design value for bending members, and
F_b^* = bending design value multiplied by all applicable adjustment factors except C_{fu}, C_L, and C_V.

The critical buckling design value for bending members, F_{bE}, is calculated using

$$F_{bE} = \frac{K_{bE}E'}{R_B^2} \tag{4.3}$$

where

K_{bE} = Euler buckling coefficient for bending members,
E' = allowable modulus of elasticity, and
R_B = slenderness ratio of the bending member.

Since lateral buckling of the beam is being considered, the modulus of elasticity, E, for lateral bending should be used. Design values of E for lumber are generally the same for both directions of bending but generally vary for glued laminated timber.

The Euler buckling coefficient for bending members is a function of the member modulus of elasticity coefficient of variation and is determined by

$$K_{bE} = 0.745 - 1.225(\text{COV}_E) \tag{4.4}$$

Using the coefficients of variation from Table 3.12, the values for K_{bE} are shown in Table 4.2.

The slenderness ratio of the bending member is determined by

$$R_B = \sqrt{\frac{\ell_e d}{b^2}} \tag{4.5}$$

where

ℓ_e = the effective length in bending,
d = the member depth, and
b = the width (breadth). The bending slenderness ratio is not to exceed 50 for wood members.

TABLE 4.2 Euler Buckling Coefficients for Bending Members

Product	K_{bE}
Visually graded lumber	0.439
Products with $0.11 < \text{COV}_E \leq 0.15$ [machine-evaluated lumber (MEL) and glued laminated timber with less than six laminations]	0.561
Products with $\text{COV}_E \leq 0.11$ (MSR lumber and glued laminated timber with six or more laminations)	0.610

The effective length in bending, ℓ_e, is given in Table 4.3 and is a function of the loading conditions, support conditions, member cross section, and unsupported length, ℓ_u.

4.2.2 Shear Parallel to the Grain

In wood members subject to bending, shear stresses develop both parallel and perpendicular to the grain. Due to its cellular structure, wood is much stronger in resisting shear perpendicular to the grain; therefore, shear stresses parallel to the grain (sometimes referred to as *horizontal shear stresses*) always govern shear design. For this reason, shear design values are published only for shear

TABLE 4.3 Effective Length, ℓ_e, for Bending Members

Cantilever[1]	when $\ell_u/d < 7$	when $\ell_u/d \geq 7$
Uniformly distributed load	$\ell_e = 1.33\ \ell_u$	$\ell_e = 0.90\ \ell_u + 3d$
Concentrated load at unsupported end	$\ell_e = 1.87\ \ell_u$	$\ell_e = 1.44\ \ell_u + 3d$
Single Span Beam[1,2]	when $\ell_u/d < 7$	when $\ell_u/d \geq 7$
Uniformly distributed load	$\ell_e = 2.06\ \ell_u$	$\ell_e = 1.63\ \ell_u + 3d$
Concentrated load at center with no intermediate lateral support	$\ell_e = 1.80\ \ell_u$	$\ell_e = 1.37\ \ell_u + 3d$
Concentrated load at center with lateral support at center	$\ell_e = 1.11\ \ell_u$	
Two equal concentrated loads at 1/3 points with lateral support at 1/3 points	$\ell_e = 1.68\ \ell_u$	
Three equal concentrated loads at 1/4 points with lateral support at 1/4 points	$\ell_e = 1.54\ \ell_u$	
Four equal concentrated loads at 1/5 points with lateral support at 1/5 points	$\ell_e = 1.68\ \ell_u$	
Five equal concentrated loads at 1/6 points with lateral support at 1/6 points	$\ell_e = 1.73\ \ell_u$	
Six equal concentrated loads at 1/7 points with lateral support at 1/7 points	$\ell_e = 1.78\ \ell_u$	
Seven or more equal concentrated loads, evenly spaced, with lateral support at points of load application	$\ell_e = 1.84\ \ell_u$	
Equal end moments	$\ell_e = 1.84\ \ell_u$	

1. For single span or cantilever bending members with loading conditions not specified in Table 3.3.3:
   $\ell_e = 2.06\ \ell_u$ when $\ell_u/d < 7$
   $\ell_e = 1.63\ \ell_u + 3d$ when $7 \leq \ell_u/d \leq 14.3$
   $\ell_e = 1.84\ \ell_u$ when $\ell_u/d > 14.3$
2. Multiple span applications shall be based on table values or engineering analysis.

Reprinted with permission from National Design Specification® for Wood Construction. Copyright © 2001. American Forest & Paper Association, Inc.

parallel to the grain. Shear stress parallel to the grain must be checked for all bending members. The horizontal shear stress acting on any plane in a wood bending member may be determined by

$$f_v = \frac{VQ}{Ib} \tag{4.6a}$$

where

f_v = the horizontal shear stress or shear parallel to the grain,
V = the shear force,
Q = the first moment about the neutral axis of the area of the section between the plane of interest and outside edge of the cross section on the same side of the neutral axis,
I = moment of inertia, as defined previously, and
b = the width of the plane across which the shear stress is calculated.

The shear stress at a section is maximum at the neutral axis. Many structural wood members are rectangular in cross section. For a rectangular section, the maximum shear stress (at the neutral axis or middepth), from equation (4.6a), becomes

$$f_v = \frac{3V}{2A} = \frac{3V}{2bd} \tag{4.6b}$$

where

A = the cross-sectional area.

Shear forces are generally greatest near beam supports. In the computation of shear stress, it is permissible to neglect distributed loads applied within a distance d from the edge of the support in cases where the loads are applied on one face and support is provided by bearing on the opposite face (Figure 4.2). In practice, instead of omitting such loads, for example, in shear and bending moment analysis, the critical shear force is simply taken at a distance d from the support. Concentrated loads within a distance d from the support are permitted to be multiplied by x/d, where x is the distance from the beam support face to the load.

In applications including moving loads, the moving load should be placed at a distance d from the support for shear stress analysis. Moving loads must be considered in combination with the other applicable nonmoving loads. In cases where the magnitudes of concentrated loads are known but a point or points of application are not necessarily specified (e.g., wall framing on headers), loads placed a distance d from the supports generally produce the

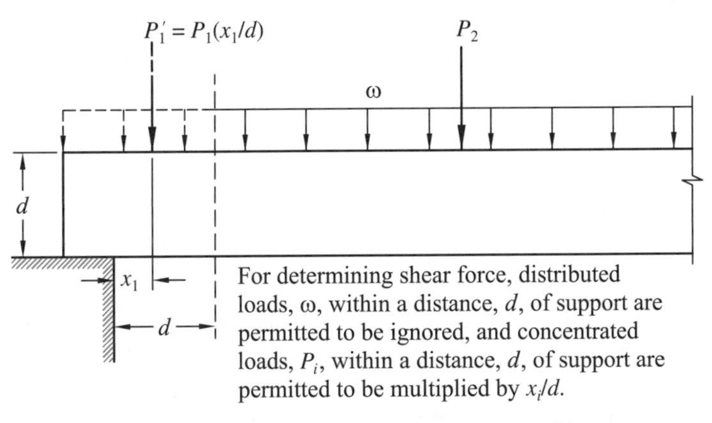

Figure 4.2 Loads and shear force at ends of beam.

greatest shear values. In both cases, placement of loads at or near midspan (away from the supports) will produce the critical bending moments and deflections.

4.2.3 Deflection

Deflections may be determined using principles of mechanics, energy-based methods, or finite element computer analysis. The Appendix contains mechanics-based formulas for shear, bending moment, and deflections for commonly encountered support and loading conditions for beams. Deflection limitation criteria are given in Chapter 3. In many applications, both the immediate (instantaneous) deflection due to live and other applied loads and the total deflection (including creep) due to all loads are considered.

4.2.4 Summary of Design Checks

The basic design checks covered in Sections 4.2.1 to 4.2.3 must be performed for all bending members, and any one of the checks may govern selection of the member. As a rule of thumb, shear parallel to the grain tends to govern the selection of short members, bending stress tends to govern the selection of medium-length members, and deflection considerations govern the selection of long members. Additional design checks are required for curved, tapered, arched, and other members. The additional design checks are discussed in later sections of the chapter.

4.2.5 Simple Beams

One of the most common members in wood framing is the *simple beam,* a member that is supported only at its ends and resists loads across a single span. The beam ends are idealized as being allowed to rotate freely, referring to simple or *pinned* end conditions. For simple beams, the following three design checks are made:

1. Bending stress (flexural tension)
2. Shear stress parallel to the grain (horizontal shear)
3. Deflection

Any one of the design checks may govern selection of the appropriate member. Bearing stress and other connection conditions must also be checked, although they seldom govern selection of the member. Bearing stress and other connection conditions are discussed in Chapter 5.

Example 4.1 Simple Beam with Continuous Lateral Support

Given: A structural glued laminated timber is to be used as a beam to support a residential floor load. The basic tributary width for the beam is 16 ft and the beam will span 12 ft from the centerline of one support to the centerline of the other. Floor joists at 16 in. o.c. will span over and bear on top of the beam. Douglas fir 24F-1.8E is found to be a readily available species and grade beam member for the project. A floor dead load of 15 psf is assumed, not including the weight of the beam.

Determine: An appropriate structural glued laminated timber section.

Approach: Unless the plans specify otherwise, it will be assumed that the floor joists span continuously over the beam. In the condition of two equal spans, the joists will deliver 25% more load over the center support than would be calculated using straight tributary areas alone. The joists will be assumed to provide continuous support of the compression side (top) of the beam. Design values and deflection criteria will be obtained from AITC 117[1] and Chapter 3, respectively.

Solution: The floor joists will be assumed to deliver the floor live and dead loads uniformly along the beam. The floor live load, from Table 3.1, is 40 psf. Thus, the loads are calculated as follows.

$$\text{Floor live load:} \quad 16 \text{ ft} \times 1.25 \times 40 \text{ psf} = 800 \text{ plf}$$
$$\text{Floor dead load:} \quad 16 \text{ ft} \times 1.25 \times 15 \text{ psf} = 300 \text{ plf}$$

The beam self-weight is calculated based on an assumed size of $5\frac{1}{8}$ in. $\times$ 12 in. An assumed moisture content of 12% will be used; a specific weight of 33 lb/ft^3 is obtained from Table 2.2:

$$\text{Beam self-weight:} \quad \left(\frac{5.125 \text{ in.}}{12 \text{ in./ft}}\right)\left(\frac{12 \text{ in.}}{12 \text{ in./ft}}\right)\left(\frac{33 \text{ lb}}{\text{ft}^3}\right) = 14 \text{ lb/ft}$$

The total load on the beam is 800 lb/ft + 300 lb/ft + 14 lb/ft = 1114 lb/ft.

The maximum bending moment for a simple beam under uniform load is given in Section A.2.1, condition 1, as $M = \omega\ell^2/8$, where ω is the uniform load and ℓ is the span. Therefore,

$$M = \left(1114 \frac{\text{lb}}{\text{ft}}\right)\left[\frac{(12 \text{ ft})^2}{8}\right] = 20{,}050 \text{ lb-ft}$$

$$= (20{,}050 \text{ lb-ft})\left(\frac{12 \text{ in.}}{\text{ft}}\right)$$

$$= 240{,}600 \text{ lb-in.}$$

The bending stress is $f_b = M/S$, where $S = I/c$.

The moment of inertia, I, may be obtained from Section A.1.2 or the *NDS®* [2], or calculated as follows:

$$I = \frac{bd^3}{12} = \frac{(5.125 \text{ in.})(12 \text{ in.})^3}{12} = 738 \text{ in}^4$$

$$S = \frac{I}{c} = \frac{I}{d/2} = \frac{2I}{d} = \frac{(2)(738 \text{ in}^4)}{12 \text{ in.}} = 123 \text{ in}^3$$

Therefore,

$$f_b = \frac{240{,}600 \text{ lb-in.}}{123 \text{ in}^3} = 1956 \text{ psi}$$

The bending stress of 1956 psi must be checked against the allowable bending stress, which is the bending stress design value multiplied by all applicable adjustment factors, $f_b \le F_b'$. From AITC 117 [1] the following design values are obtained for the 24F-1.8E stress class: $F_b = 2400$ psi, $F_v = 265$ psi, and $E_x = 1{,}800{,}000$ psi. The applicable adjustment factors are obtained from Table 3.8:

$$F_b' = F_b C_D C_M (C_V \text{ or } C_L) C_c C_I C_t$$

The load duration factor for occupancy live load is 1.0, from Table 3.4. The beam will be assumed to be in dry conditions of use with normal temperatures and will be straight or slightly cambered (not curved), and without tapered cuts; therefore, all the adjustment factors will be 1.00 except for C_V and C_L. Factors C_V and C_L will both be calculated and the smaller value used. (The factors C_V and C_L do not act accumulatively.) From equation (3.3),

$$C_V = \left(\frac{5.125 \text{ in.}}{5.125 \text{ in.}}\right)^{\frac{1}{10}} \left(\frac{12 \text{ in.}}{12 \text{ in.}}\right)^{\frac{1}{10}} \left(\frac{21 \text{ ft}}{12 \text{ ft}}\right)^{\frac{1}{10}} = 1.06 \leq 1.00$$

where the exponent of $\frac{1}{10}$ is used for western species. The volume factor cannot be taken greater than unity; hence, C_V is taken to be 1.00. Since the beam is assumed to be supported continuously along the compression edge, $C_L = 1.00$. In this example, therefore, all the adjustment factors are unity; hence, $F'_b = (2400 \text{ psi})\ (1.0)(1.0) \cdots = 2400$ psi. It is customary in timber design *not* to show the adjustment factors that are unity. Therefore, the design check for bending becomes

$$f_b = 1956 \text{ psi} \leq F'_b = 2400 \text{ psi} \qquad \text{good}$$

The section selected is adequate with regard to bending.

The reactions for the beam are each $R = \omega\ell/2 = (1114 \text{ lb/ft})\ (12 \text{ ft})/2 = 6684$ lb. Since the loads are applied to the top of the beam and the beam supports will be assumed to provide bearing to the bottom, the shear force to be checked may exclude the distributed loads applied within d of the end, or

$$V = R - \omega d = 6684 \text{ lb} - (1114 \text{ lb/ft})\ (1 \text{ ft}) = 5570 \text{ lb}$$

The shear stress to be checked is therefore

$$f_v = \frac{3V}{2A} = \frac{(3)(5570 \text{ lb})}{(2)(5.125 \text{ in.} \times 12 \text{ in.})} = 136 \text{ psi}$$

The applicable adjustment factors for shear are obtained from Table 3.8 as C_D, C_M, and C_t, which in this example are all 1.00. Therefore, the design check for shear is

$$f_v = 135 \text{ psi} \leq F'_v = 265 \text{ psi} \qquad \text{good}$$

The deflection under live load, using the simple beam formula of condition 1, Section A.2.1, is

$$\Delta_{\text{LL}} = \frac{5}{384}\frac{\omega\ell^4}{EI} = \frac{(5)(800 \text{ lb/ft})(1 \text{ ft}/12 \text{ in.})[12 \text{ ft}(12 \text{ in./ft})]^4}{(384)(1{,}800{,}000 \text{ lb/in}^2)(738 \text{ in}^4)} = 0.28 \text{ in.}$$

From Table 3.9, the live load deflection should be limited to $\frac{1}{360}$ of the span, in this case 144 in./360 = 0.40 in. Since 0.28 in. is less than 0.40 in., the selected beam is satisfactory with regard to live load deflection. Table 3.9 also states that the deflection due to applied load plus dead load should be limited to $\frac{1}{240}$ of the span. The dead load deflection in this case will be

$$0.28 \text{ in.} \left(\frac{300 \text{ plf} + 14 \text{ plf}}{800 \text{ plf}}\right) = 0.11 \text{ in.}$$

Thus the total deflection is 0.28 in. + 0.11 in. = 0.39 in. The limiting deflection is 144 in./240 = 0.60 in. Since 0.39 in. is less than 0.60 in., the selected beam is satisfactory with regard to both deflection criteria.

Answer: For the stated conditions, a $5\frac{1}{8}$ in. × 12 in. Douglas fir (DF) 24F-1.8E beam is satisfactory with regard to bending, shear, and deflection. Connection and bearing requirements are examined in Chapter 5.

Discussion: It is common to express the results of the design checks for bending, shear, and deflection as ratios of actual to allowable values, which in this case are as follows:

$$\text{Shear:} \quad \frac{f_v}{F'_v} = \frac{136 \text{ psi}}{265 \text{ psi}} = 0.51 \leq 1.00 \qquad \text{good}$$

$$\text{Bending:} \quad \frac{f_b}{F'_b} = \frac{1956 \text{ psi}}{2400 \text{ psi}} = 0.82 \leq 1.00 \qquad \text{good}$$

$$\text{Deflection (LL):} \quad \frac{\Delta_{\text{LL}}}{\ell/360} = \frac{0.28 \text{ in.}}{0.40 \text{ in.}} = 0.70 \leq 1.00 \qquad \text{good}$$

$$\text{Deflection (LL + DL):} \quad \frac{\Delta_{\text{LL+DL}}}{\ell/240} = \frac{0.39 \text{ in.}}{0.60 \text{ in.}} = 0.65 \leq 1.00 \qquad \text{good}$$

From the foregoing ratios, it is seen that bending controls (giving the ratio closest to but not exceeding unity). Since the joists frame over the beam, the beam selected should be checked with regard to clearance. It can be shown that $3\frac{1}{8}$ in. × 15 in., or $6\frac{3}{4}$ in. by $10\frac{1}{2}$ in., 24F-1.8E members would also be satisfactory.

Glued laminated timber is generally priced by cost per board feet. Nominal dimensions of the input lumber are used to calculate the board footage (BFM) of the glulam beams. The $3\frac{1}{8}$ in. × 15 in. beam is made from ten 2 × 4's, the $5\frac{1}{8}$ in. × 12 in. beam is made from eight 2 × 6's, and the $6\frac{3}{4}$ in. × $10\frac{1}{2}$ in. beam is made from seven 2 × 8's. The board footage calculations (per foot of beam) follow.

$$\text{For the } 3\tfrac{1}{8}\text{-in. beam:} \quad \frac{\text{BFM}}{\text{ft}} = \frac{(10)(2 \text{ in.})(4 \text{ in.})}{144 \text{ in}^3/\text{BF}} \left(\frac{12 \text{ in.}}{\text{ft}}\right) = 6.67 \frac{\text{BF}}{\text{ft}}$$

$$\text{For the } 5\tfrac{1}{8}\text{-in. beam:} \quad \frac{\text{BFM}}{\text{ft}} = \frac{(8)(2 \text{ in.})(6 \text{ in.})}{144 \text{ in}^3/\text{BF}} \left(\frac{12 \text{ in.}}{\text{ft}}\right) = 8 \frac{\text{BF}}{\text{ft}}$$

$$\text{For the } 6\tfrac{3}{4}\text{-in. beam:} \quad \frac{\text{BFM}}{\text{ft}} = \frac{(7)(2 \text{ in.})(8 \text{ in.})}{144 \text{ in}^3/\text{BF}} \left(\frac{12 \text{ in.}}{\text{ft}}\right) = 9.33 \frac{\text{BF}}{\text{ft}}$$

The $3\frac{1}{8}$ in. × 15 in. section would probably be more economical but may also have clearance or headroom problems.

In Example 4.1, the loads from the joists were assumed to be distributed uniformly over the beam span. It can be shown that using concentrated loads from each joist (if the exact joist locations are known) produces shear, bending, and deflection values that vary at most by a few percent from the values obtained from assuming uniformly distributed loads in this example. It was also assumed that the joists provided continuous support of the top (compression) edge of the beam. The following example demonstrates that for practical purposes, typical joist (or truss) framing can be assumed to provide continuous lateral support. The relative values in Example 4.1 also illustrate that exact knowledge of the moisture content of the member is not critical with regard to self-weight (14 lb/ft compared to the total load of 1114 lb/ft).

Example 4.2 Simple Beam with Support Provided by Framing

Given: The $5\frac{1}{8}$ in. × 12 in. 24F-1.8E DF glued laminated timber beam of Example 4.1. The modulus of elasticity for lateral buckling is E_y = 1,600,000 psi.

Determine: The beam stability factor, C_L, assuming that the joists provide lateral support every 16 in.

Approach: Use equation (4.2).

Solution: The unsupported length, ℓ_u, is 16 in. or 1.33 ft. Assuming that the joists frame over the supports and at increments of 16 in., there are (12 ft) (12 in./ft) × 1 joist/16 in. − 1 = 8 equal magnitude and equally spaced concentrated loads.

From Table 4.3, the effective length ℓ_e is $1.84 \times \ell_u$, or 29.4 in.

Using equation (4.5), $R_B = \sqrt{(29.4 \text{ in.})(12 \text{ in.})/(5.125 \text{ in.})^2} = 3.67$, which is less than 50; good.

From Table 3.8, $E' = EC_M C_t$, = (1,600,000 psi)(1.0)(1.0) = 1,600,000 psi.

From Table 4.2, $K_{bE} = 0.610$.

From equation (4.3), $F_{bE} = (0.610)(1.6)(10^6 \text{ psi})/(3.67)^2 = 72{,}600$ psi.

F_b^* is F_b multiplied by all applicable adjustment factors except C_{fu}, C_V, and C_L; in this case, $F_b^* = (2400 \text{ psi})C_D C_c C_M C_I C_t = 2400$ psi, as all the factors are unity. Therefore,

$$C_L = \frac{1 + 72{,}600 \text{ psi}/2400 \text{ psi}}{1.9} - \sqrt{\left(\frac{1 + 72{,}600 \text{ psi}/2400 \text{ psi}}{1.9}\right)^2 - \frac{72{,}600 \text{ psi}/2400 \text{ psi}}{0.95}} = 0.998$$

Answer: The beam stability factor, C_L, assuming that the joists provide lateral support every 16 in., is 0.998, nearly unity.

Discussion: This example shows that where joists or other members provide lateral support at relatively short intervals, the beam may be assumed to be fully supported laterally. In this example it can be shown that the reduction of the allowable bending stress using the beam stability factor is less than the reduction of applied bending stress that would have been obtained by considering concentrated loads at 16-in. intervals instead of the uniformly distributed load.

Example 4.3 Simple Beam with Unsupported Length

Given: A 60-ft western species 24F-1.8E beam is loaded with varying loads along its length that develop a moment of 1,500,000 lb-in. The beam is braced only at a point 25 ft from one end. Assume snow load duration and dry and normal temperature conditions.

Determine: An appropriate section (based on bending).

Approach: A trial beam section will be selected, volume and beam stability factors will be calculated, and the size will be adjusted as necessary such that the bending stress is not greater than the design value properly adjusted for volume or stability effects. The modulus of elasticity for bending in the y–y direction will be used for the beam stability calculations.

Solution: A trial beam size may be obtained from equation (4.1) and the design value for bending adjusted by all factors except C_V or C_L. From equation (4.1), let $f_b = F'_b$ to give $S_{\text{required}} = M/F'_b$, where F'_b will be estimated by F_bC_D = (2400 psi) (1.15) = 2760 psi. Thus,

$$S_{\text{required}} = 1{,}500{,}000 \text{ lb-in.}/2760 \text{ psi} = 543 \text{ in}^3$$

From the Appendix (or *NDS®*), some 24F-1.8E western species beam sections that have the required section modulus are $5\frac{1}{8}$ in. × 25.5 in. (S = 555 in^3), $6\frac{3}{4}$ in. × 22.5 in. (S = 570 in^3), and $8\frac{3}{4}$ in. × 19.5 in. (S = 555 in^3). In anticipation that the volume or beam stability factors will cause some reduction in the bending design value, a slightly larger section will be selected, for example, $6\frac{3}{4}$ in. × 24 in. (S = 648 in^3). The design stress is, therefore,

$$f_b = \frac{M}{S} = \frac{1{,}500{,}000 \text{ lb-in.}}{648 \text{ in}^3} = 2315 \text{ psi}$$

The volume factor is, from equation (3.4),

$$C_V = \left(\frac{5.125 \text{ in.}}{6.75 \text{ in.}}\right)^{\frac{1}{10}} \left(\frac{12 \text{ in.}}{24 \text{ in.}}\right)^{\frac{1}{10}} \left(\frac{21 \text{ ft}}{60 \text{ ft}}\right)^{\frac{1}{10}} = 0.817$$

The beam stability factor, from equation (4.2), is

$$C_L = \frac{1 + F_{bE}/F_b^*}{1.9} - \sqrt{\left(\frac{1 + F_{bE}/F_b^*}{1.9}\right)^2 - \frac{F_{bE}/F_b^*}{0.95}}$$

where, for F_{bE},

$$R_B = \sqrt{\frac{\ell_e d}{b^2}}.$$

For Table 4.3, $\ell_u = 60 \text{ ft} - 25 \text{ ft} = 35 \text{ ft}$; $d = 24 \text{ in.} = 2 \text{ ft}$ and $\ell_u/d = 35 \text{ ft}/2 \text{ ft} = 17.5$. From Table 4.3, assuming that the loads are distributed more or less uniformly,

$$\ell_e = 1.63\ell_u + 3d = (1.63)(35 \text{ ft})(12 \text{ in./ft}) + (3)(24 \text{ in.}) = 757 \text{ in.}$$

$$R_B = \sqrt{\frac{(757 \text{ in.})(24 \text{ in.})}{6.75 \text{ in}^2}} = 19.96 \leq 50 \qquad \text{good}$$

$$E' = E'_y = (1{,}600{,}000 \text{ psi})C_M C_t = 1{,}600{,}000 \text{ psi}$$

$$F_{bE} = \frac{K_{bE}E'}{R_B^2} = \frac{(0.610)[(1.6)(10^6) \text{ psi}]}{(19.96)^2} = 2449 \text{ psi}$$

So

$$C_L = \frac{1 + 2449 \text{ psi}/2760 \text{ psi}}{1.9} - \sqrt{\left(\frac{1 + 2449 \text{ psi}/2760 \text{ psi}}{1.9}\right)^2 - \frac{2449 \text{ psi}/2760 \text{ psi}}{0.95}} = 0.764.$$

The smaller of C_L and C_V is chosen ($C_L = 0.764$), because they are not taken cumulatively. The adjusted bending design value becomes

$$F'_b = (2400 \text{ psi})(1.15)(0.764) = 2109 \text{ psi}$$

Since $f_b = 2315 \text{ psi} > F'_b = 2109 \text{ psi}$, the selected section is not suitable; a larger section is required. Try $6\frac{3}{4}$ in. × $25\frac{1}{2}$ in., $S = 732 \text{ in}^3$ and

$$\frac{\ell_u}{d} = \frac{(35 \text{ ft})(12 \text{ in./ft})}{25.5 \text{ in.}} = 16.5$$

The volume factor is

$$C_V = \left(\frac{5.125 \text{ in.}}{6.75 \text{ in.}}\right)^{\frac{1}{10}} \left(\frac{12 \text{ in.}}{25.5 \text{ in.}}\right)^{\frac{1}{10}} \left(\frac{21 \text{ ft}}{60 \text{ ft}}\right)^{\frac{1}{10}} = 0.812$$

From Table 4.3,

$$\ell_e = (1.63)(35 \text{ ft})(12 \text{ in./ft}) + (3)(25.5 \text{ in.}) = 761 \text{ in.}$$

$$R_B = \sqrt{\frac{(761 \text{ in.})(25.5 \text{ in.})}{(6.75 \text{ in.})^2}} = 20.6 \leq 50 \qquad \text{good}$$

$$E' = (1{,}600{,}000 \text{ psi})C_M C_t = 1{,}600{,}000 \text{ psi}$$

$$F_{bE} = \frac{K_{bE}E'}{R_B^2} = \frac{(0.610)(1{,}600{,}000 \text{ psi})}{(20.6)^2} = 2290 \text{ psi}$$

So

$$C_L = \frac{1 + 2290 \text{ psi}/2760 \text{ psi}}{1.9} - \sqrt{\left(\frac{1 + (2290 \text{ psi}/2760 \text{ psi})}{1.9}\right)^2 - \frac{2290 \text{ psi}/2760 \text{ psi}}{0.95}} = 0.731$$

The beam stability factor controls (0.731 is less than 0.812); therefore,

$$F_b' = (2400 \text{ psi})(1.15)(0.731) = 2015 \text{ psi}$$

The bending stress is 1,500,000 lb-in./732 in^3 = 2049 psi. Since f_b = 2049 psi > F_b' = 2015 psi, the $6\frac{3}{4}$ in. × $25\frac{1}{2}$ in. section is not adequate in bending.

A larger section will be considered. Try $6\frac{3}{4}$ in. × 27 in. with S = 820 in^3:

$$C_V = \left(\frac{5.125 \text{ in.}}{6.75 \text{ in.}}\right)^{\frac{1}{10}} \left(\frac{12 \text{ in.}}{27 \text{ in.}}\right)^{\frac{1}{10}} \left(\frac{21 \text{ ft}}{60 \text{ ft}}\right)^{\frac{1}{10}} = 0.808$$

From Table 4.3, for ℓ_u/d = (35 ft)(12 in./ft)/27 in. = 15.56,

$$\ell_e = (1.63)(35 \text{ ft})(12 \text{ in./ft}) + (3)(27 \text{ in.}) = 766 \text{ in.}$$

$$R_B = \sqrt{\frac{(766 \text{ in.})(25.5 \text{ in.})}{(6.75 \text{ in.})^2}} = 20.7 \leq 50 \qquad \text{good}$$

$$E' = (1{,}600{,}000 \text{ psi})C_M C_t = 1{,}600{,}000 \text{ psi}$$

$$F_{bE} = \frac{K_{bE}E'}{R_B^2} = \frac{0.610(1{,}600{,}000 \text{ psi})}{(20.7)^2} = 2277 \text{ psi}$$

So

$$C_L = \frac{1 + 2277\ \text{psi}/2760\ \text{psi}}{1.9} - \sqrt{\left(\frac{1 + 2277\ \text{psi}/2760\ \text{psi}}{1.9}\right)^2 - \frac{2277\ \text{psi}/2760\ \text{psi}}{0.95}} = 0.728$$

The beam stability factor controls (0.728); therefore,

$$F_b' = (2400\ \text{psi})(1.15)(0.728) = 2009\ \text{psi}$$

The bending stress is f_b = 1,500,000 lb-in./820 in^3 = 1829 psi. Since f_b = 1829 psi $\leq F_b'$ = 2009 psi, the $6\frac{3}{4}$ in. × 27 in. section is adequate in bending.

Answer: A Western Species $6\frac{3}{4}$ in. × 27 in. 24F-1.8E glued laminated timber beam is adequate in bending for the stated conditions.

Discussion: The beam stability factor controlled over the volume effect in this example. Although deeper beams have greater section modulus, they are also less stable. A wider and shallower section could be investigated. Complete design of the beam would require consideration of shear and deflection as well.

4.2.6 Biaxial Bending (Bending about Both Axes)

Bending members are commonly intended to be loaded primarily in their strong direction (loaded and supported on edge with the load direction parallel to the wide face of the member). Many cases exist, however, where bending members are loaded in both strong and weak directions or are loaded obliquely. A common example of an obliquely loaded member is a purlin on a sloped roof where the top of the member is framed flush with or parallel to the roof surface (Figure 4.3). The components of the loads parallel to the wide and narrow faces must be found; these loads are used to compute flexural

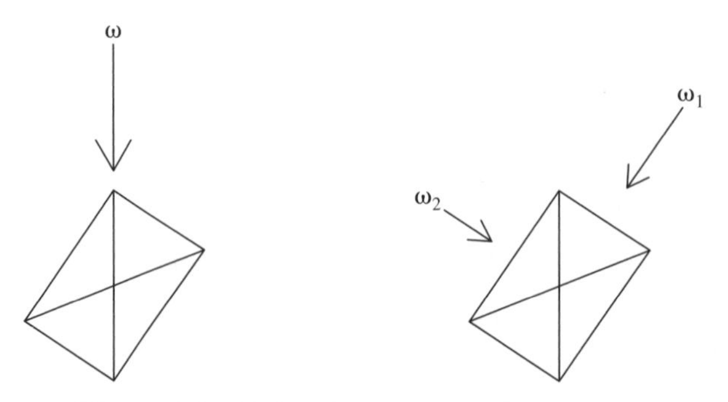

Figure 4.3 Obliquely loaded bending member.

stresses with respect to loading in these directions; and the combined effects are evaluated as follows.

The stress, strength, stiffness, and other properties or conditions of a bending member loaded on edge (in the member's strong orientation) may be denoted by the subscript 1. The conditions or properties associated with the member being loaded flatwise (weak orientation) are denoted by the subscript 2. Design properties for glued laminated timber are published with respect to x and y orientation, where x refers to the load direction perpendicular to the wide faces *of the laminations,* and y indicates the load direction parallel to the wide faces of the laminations. Thus, for horizontally laminated members with member depth greater than member width, the 1 and x orientations are coincident; similarly for the 2 and y orientations.

When members are subject to bending in both directions, the following equation must be satisfied:

$$\frac{f_{b1}}{F'_{b1}} + \frac{f_{b2}}{F'_{b2}[1 - (f_{b1}/F_{bE})^2]} \leq 1.00 \qquad (4.7)$$

where

f_{b1} = actual bending stress about the 1 or strong axis,
f_{b2} = actual bending stress about the 2 or weak axis,
F'_{b1} = bending design value for the strong axis multiplied by all applicable adjustment factors,
F'_{b2} = bending design value for the weak axis multiplied by all applicable adjustment factors, and
F_{bE} = $K_{bE}E'/R_{bE}^2$ [equation (4.3)].

Example 4.4 Biaxial Bending

Given: A 24F-1.8E glued laminated, Southern Pine beam 30 ft long is to be used to support a uniform dead load of 400 plf and a vertical roof load of 800 plf. In addition, the beam is subjected to a horizontal wind load P of 4000 lb located at the midpoint of the member which can act in either direction (Figure 4.4). The member is to be used under normal temperature and dry service conditions and is not laterally braced between the end supports. The vertical roof load is assumed to have construction load duration.

Determine: An acceptable beam section assuming 24F-1.8E southern pine.

Approach: The stated loads will be examined according to the load combinations from the *International Building Code* [3] listed in Section 3.2.10. This code requires that the member be adequate to resist the dead load plus either transient load as well as the dead load plus 75% of the transient loads acting simultaneously. The load duration factors to be used are obtained from Table

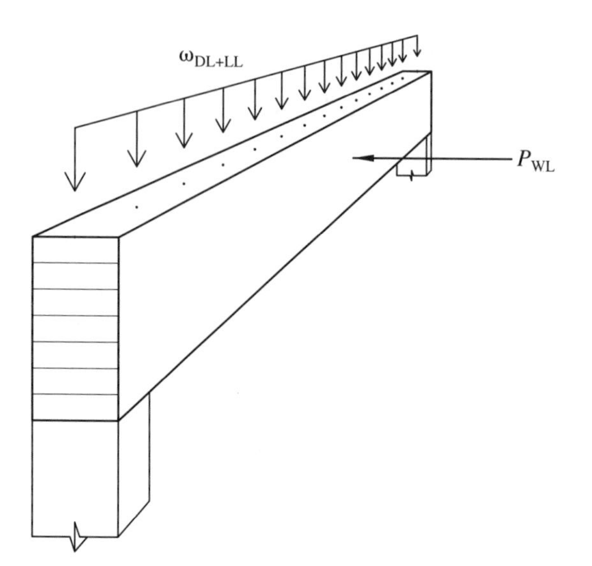

Figure 4.4 Biaxially loaded bending member—Example 4.4.

3.4 as 1.25 for roof load (construction) and 1.60 for wind. A trial beam size will be investigated and the size increased or decreased as appropriate.

Solution: A trial size will be determined from bending due to the gravity loads (dead and roof/construction). The bending moment about the x–x axis is

$$M_{\text{DL}} = \frac{\omega \ell^2}{8} = \frac{(400 \text{ plf})(30 \text{ ft})^2 \ (12 \text{ in./ft})}{8} = 540{,}000 \text{ lb-in.}$$

$$M_{\text{LL}} = \left(\frac{800 \text{ plf}}{400 \text{ plf}}\right)(540{,}000 \text{ lb-in.}) = 1{,}080{,}000 \text{ lb-in.}$$

$$M_x = M_{\text{DL}} + M_{\text{LL}} = 1{,}620{,}000 \text{ lb-in.}$$

Unless a very narrow profile is used, it will be assumed that the size will be determined by the deadload plus the full roof load; therefore, the load duration factor of 1.25 will be used for the trial size. A volume factor of 0.85 will be assumed. From the *NDS*®, $F_{bx} = 2400$ psi, $E_x = 1{,}800{,}000$ psi, $F_{by} = 1600$ psi, and $E_y = 1{,}600{,}000$ psi. Letting $f_b = F'_b$, where $F'_b = F_b C_D C_V = (2400 \text{ psi})\ (1.25)(0.85) = 2550$ psi, gives

$$S_{x(\text{required})} = \frac{M}{F'_{bx}} = \frac{1{,}620{,}000 \text{ lb-in.}}{2550 \text{ psi}} = 635 \text{ in}^3$$

For the moment about the y–y axis, the full wind load will be considered:

$$M_{\mathrm{WL}} = \frac{PL}{4} = \frac{(4000 \text{ lb})(30 \text{ ft})(12 \text{ in./ft})}{4} = 360{,}000 \text{ lb-in.}$$

It will be assumed that the section will be deeper than it is wide; therefore, the beam stability factor will be unity. Assuming a width of 8.5 in., a flat use factor of $C_{fu} = (12 \text{ in.}/8.5 \text{ in.})^{1/9} = 1.04$ is calculated from equation (3.5). The section modulus required for horizontal bending is therefore

$$S_{y(\text{required})} = \frac{M}{F'_{by}} = \frac{M}{F_{by}C_D C_{fu}} = \frac{360{,}000 \text{ lb-in.}}{(1600 \text{ psi})(1.6)(1.04)} = 135 \text{ in}^3$$

The beam will have to resist the dead load plus 75% of the roof and wind loads acting simultaneously. Since the vertical load is largely roof load, a trial section will be selected that has approximately 175% of the section modulus values calculated above:

$$S_x = (1.75)(635 \text{ in}^3) = 1111 \text{ in}^3 \quad \text{and} \quad S_y = (1.75)(135 \text{ in}^3) = 237 \text{ in}^3$$

From Section A.1.2, the following sections have equal or greater section modulus values:

$6\frac{3}{4}$ in. × $31\frac{5}{8}$ in.: $S_x = 1125 \text{ in}^3$, $S_y = 240 \text{ in}^3$

$8\frac{1}{2}$ in. × $28\frac{7}{8}$ in.: $S_x = 1181 \text{ in}^3$, $S_y = 348 \text{ in}^3$

$8\frac{1}{2}$ in. × $30\frac{1}{4}$ in.: $S_x = 1296 \text{ in}^3$, $S_y = 364 \text{ in}^3$

$10\frac{1}{2}$ in. × $26\frac{1}{8}$ in.: $S_x = 1194 \text{ in}^3$, $S_y = 480 \text{ in}^3$

The 6-$\frac{3}{4}$ in.-wide section appears excessively narrow, and the $10\frac{1}{2}$-in.-wide section appears excessively wide. Try the $8\frac{1}{2}$ in. × $28\frac{7}{8}$ in. section with $S_x = 1181 \text{ in}^3$, $I_x = 17{,}050 \text{ in}^4$, $S_y = 348 \text{ in}^3$, and $I_y = 1478 \text{ in}^4$,

$$\begin{aligned} f_{bx} &= \frac{M_{\mathrm{DL}}}{S_x} + 0.75\frac{M_{\mathrm{RL}}}{S_x} = \frac{540{,}000 \text{ lb-in.}}{1181 \text{ in}^3} + 0.75\left(\frac{1{,}080{,}000 \text{ lb-in.}}{1181 \text{ in}^3}\right) \\ &= 457 \text{ psi} + 686 \text{ psi} \\ &= 1143 \text{ psi} \end{aligned}$$

$$f_{by} = 0.75\frac{M_{\mathrm{WL}}}{S_y} = 0.75\left(\frac{360{,}000 \text{ lb-in.}}{348 \text{ in}^3}\right) = 776 \text{ psi}$$

$$F_{bx} = F_{bx}C_D C_V \quad \text{or} \quad F_{bx}C_D C_L$$

$$C_D = 1.60$$

$$C_V = \left(\frac{5.125\text{ in.}}{b}\right)^{1/x}\left(\frac{12\text{ in.}}{d}\right)^{1/x}\left(\frac{21\text{ ft}}{L}\right)^{1/x} \quad \text{where } x = 20 \text{ for southern pine}$$

$$= \left(\frac{5.125\text{ in.}}{8.5\text{ in.}}\right)^{1/20}\left(\frac{12\text{ in.}}{28.875\text{ in.}}\right)^{1/20}\left(\frac{21\text{ ft}}{30\text{ ft}}\right)^{1/20} = 0.917$$

For beam stability, the beam is supported at the ends to prevent rotation;

$$\ell_u/d = (30\text{ ft})(12\text{ in./ft})/28.875\text{ in.} = 12.5\text{; thus from Table 4.3,}$$

$$\ell_e = (1.63)(30\text{ ft})(12\text{ in./ft}) + (3)(28.875\text{ in.}) = 673\text{ in.} = 56.1\text{ ft}$$

$$R_B = \sqrt{\frac{\ell_e d}{b^2}} = \sqrt{\frac{(673\text{ in.})(28.875\text{ in.})}{(8.5\text{ in.})^2}} = 16.4 \le 50 \qquad \text{good}$$

$$K_{bE} = 0.610$$

$$E' = E'_y = 1{,}600{,}000\text{ psi}$$

$$F_{bE} = \frac{K_{bE}E'}{R_B^2} = \frac{(0.610)(1{,}600{,}000\text{ psi})}{16.4^2} = 3626\text{ psi}$$

$$F^*_{bx} = (2400\text{ psi})(1.6) = 3840\text{ psi}$$

$$C_L = \frac{1 + F_{bE}/F^*_{bx}}{1.9} - \sqrt{\left(\frac{1 + F_{bE}/F^*_{bx}}{1.9}\right)^2 - \frac{F_{bE}/F^*_{bx}}{0.95}}$$

$$= \frac{1 + 3626\text{ psi}/3840\text{ psi}}{1.9}$$

$$- \sqrt{\left(\frac{1 + 3626\text{ psi}/3840\text{ psi}}{1.9}\right)^2 - \frac{3626\text{ psi}/3840\text{ psi}}{0.95}} = 0.793$$

The beam stability factor controls (0.793 versus 0.917); therefore,

$$F'_{bx} = (2400\text{ psi})(1.60)(0.793) = 3045\text{ psi}$$

In considering loading about the y–y axis, $b_y = 28.875$ in. and $d_y = 8.5$ in.; therefore, $b/d = 28.875$ in./8.5 in. $= 3.4 > 1.0$; therefore, $C_L = 1.0$. C_V is not applicable to members loaded parallel to the wide faces of the laminations, but C_{fu} applies. Therefore, $F'_{by} = (1600\text{ psi})(1.60)(1.04) = 2662$ psi.

Using equation (4.7),

$$\frac{f_{b1}}{F'_{b1}} + \frac{f_{b2}}{F'_{b2}[1 - (f_{b1}/F_{bE})^2]} \leq 1.00$$

$$\frac{1143 \text{ psi}}{3045 \text{ psi}} + \frac{776 \text{ psi}}{(2662 \text{ psi})[1 - (1143 \text{ psi}/3626 \text{ psi})^2]} = 0.70 \leq 1.00 \qquad \text{good}$$

The $8\frac{1}{2}$ in. $\times$ $28\frac{7}{8}$ in. beam is adequate.

Investigate the use of an $8\frac{1}{2}$ in. $\times$ $26\frac{1}{8}$ in. section.

$$S_x = 967 \text{ in}^3$$

$$S_y = 315 \text{ in}^3$$

$$f_{bx} = \frac{M_{\text{DL}}}{S_x} + 0.75\,\frac{M_{\text{RL}}}{S_x} = \frac{540{,}000 \text{ lb-in.}}{967 \text{ in}^3} + 0.75\left(\frac{1{,}080{,}000 \text{ lb-in.}}{967 \text{ in}^3}\right)$$

$$= 558 \text{ psi} + 838 \text{ psi}$$

$$= 1396 \text{ psi}$$

$$f_{by} = 0.75\,\frac{M_{\text{WL}}}{S_y} = 0.75\left(\frac{360{,}000}{315}\right) = 857 \text{ psi}$$

$$F'_{bx} = F_{bx}C_DC_V \quad \text{or} \quad F_{bx}C_DC_L$$

$$C_D = 1.60$$

$$C_V = \left(\frac{5.125 \text{ in.}}{b}\right)^{1/x}\left(\frac{12 \text{ in.}}{d}\right)^{1/x}\left(\frac{21 \text{ ft}}{L}\right)^{1/x} \quad \text{where } x = 20 \text{ for southern pine}$$

$$= \left(\frac{5.125 \text{ in.}}{8.5 \text{ in.}}\right)^{1/20}\left(\frac{12 \text{ in.}}{26.125 \text{ in.}}\right)^{1/20}\left(\frac{21 \text{ ft}}{30 \text{ ft}}\right)^{1/20} = 0.921$$

From Table 4.3,

$$\ell_e = (1.63)(30 \text{ ft})(12\text{in./ft}) + (3)(26.125 \text{ in.}) = 665 \text{ in. or } 55.4 \text{ ft}$$

$$R_B = \sqrt{\frac{\ell_e d}{b^2}} = \sqrt{\frac{(665 \text{ in.})(26.125 \text{ in.})}{(8.5 \text{ in.})^2}} = 15.5 \leq 50 \qquad \text{good}$$

$$K_{bE} = 0.610$$

$$E' = E'_y = 1{,}600{,}000 \text{ psi}$$

$$F_{bE} = \frac{K_{bE}E'}{R_B^2} = \frac{(0.610)(1{,}600{,}000 \text{ psi})}{15.5^2} = 4060 \text{ psi}$$

$$F^*_{bx} = 2400 \text{ psi } (1.60) = 3840 \text{ psi}$$

$$C_L = \frac{1 + F_{bE}/F_{bx}^*}{1.9} - \sqrt{\left(\frac{1 + F_{bE}/F_{bx}^*}{1.9}\right)^2 - \frac{F_{bE}/F_{bx}^*}{0.95}}$$

$$= \frac{1 + 4060\ \text{psi}/3840\ \text{psi}}{1.9}$$

$$- \sqrt{\left(\frac{1 + 4060\ \text{psi}/3840\ \text{psi}}{1.9}\right)^2 - \frac{4060\ \text{psi}/3840\ \text{psi}}{0.95}} = 0.839$$

The beam stability factor controls (0.839 versus 0.921); therefore,

$$F'_{bx} = (2400\ \text{psi})(1.60)(0.839) = 3222\ \text{psi}$$

For the y–y axis, $b/d > 1.0$; therefore, $C_L = 1.0$; C_V does not apply, and C_{fu} is again 1.04. Therefore,

$$F'_{by} = (1600\ \text{psi})(1.60)(1.04) = 2662\ \text{psi}$$

From equation (4.7),

$$\frac{f_{b1}}{F'_{b1}} + \frac{f_{b2}}{F'_{b2}[1 - (f_{b1}/F_{bE})^2]} \leq 1.00$$

$$\frac{1396\ \text{psi}}{3222\ \text{psi}} + \frac{857\ \text{psi}}{(2662\ \text{psi})\ [1 - (1396\ \text{psi}/4057\ \text{psi})^2]} = 0.80 \leq 1.00 \qquad \text{good}$$

The $8\frac{1}{2}$ in. × $26\frac{1}{8}$ in. SP 24F-1.8E beam is also adequate.

Answer: The $8\frac{1}{2}$ in. × $26\frac{1}{8}$ in. SP 24F-1.8E is adequate to resist the stated loads.

Discussion: It is possible that even smaller sections will be adequate. Smaller sections may be investigated for the combined transient load case, as above. The sections will also be adequate for the other cases of dead load plus either transient load separately as long as the original minimum section modulus values ($S_x = 635\ \text{in}^3$ and $S_y = 135\ \text{in}^3$) are still satisfied.

4.2.7 Continuous Members

Bending members that are framed continuously across intermediate supports tend to be stiffer, and in many cases may resist greater loads, than simply supported members with the same spans. Considerations for continuous members include the following.

1. Glued laminated timbers should be specified in a balanced grade or layup that utilizes higher-grade tension laminations on both top and

bottom, because both positive and negative moments are generated in continuous framing.

2. The length L used for the volume factor calculation is the distance between points of zero moment.
3. Since the bottom of the member will be in compression in some places, continuous attachment of the top of the member to roof or floor framing, sheathing, or decking will not necessarily prevent buckling of the lower portion of the beam. As such, additional lateral support must be provided or an appropriate beam stability factor calculated.
4. *Unbalanced* or *simple* (beams with tension laminations on the bottom only) may be used if their reduced ability to resist flexural tension on the upper part of the beam is properly taken into consideration.
5. Beams with *stock* camber may no longer be appropriate except for very short spans; straight or custom camber beams should be specified.

Example 4.5 Continuous Beam

Given: A glued laminated timber beam is to span continuously over two 10-ft openings while supporting 300 lb/ft of dead load and 900 lb/ft of snow load.

Determine: Based on bending, the appropriate 24F-1.8E DF sections in both unbalanced and balanced layup.

Approach: It will be assumed that full lateral support of both the top and bottom of the beam may be specified; therefore, C_L will be 1.00 for both positive and negative bending. The volume factor will also be assumed to be near unity, as the loads to be resisted are relatively light. For the unbalanced beam, the design values for the 24F-1.8E DF sections are $F_{bx}^{+} = 2400$ psi, $F_{bx}^{-} = 1450$ psi, and $E_x = 1.8 \times 10^6$ psi. For the balanced beam, $F_{bx}^{+} = F_{bx}^{-} = 2400$ psi and $E_x = 1.8 \times 10^6$ psi. Maximum positive and negative moment information will be obtained from the load and deflection tables in Section A.2.1.

Solution: The maximum positive bending moment is

$$M^{+} = \frac{9\omega\ell^2}{128} = \frac{(9)(300 + 900)\text{lb/ft}(10\text{ ft})^2}{128} = 8438\text{ lb-ft} = 101{,}250\text{ lb-in.}$$

The maximum negative moment is

$$M^{-} = \frac{\omega\ell^2}{8} = -\frac{(300 + 900)\text{lb/ft}(10\text{ ft})^2}{8} = 15{,}000\text{ lb-ft} = 180{,}000\text{ lb-in}$$

Unbalanced Beam Let $f_b = F_b' = M/S$ to determine an appropriate section modulus, assuming tension on the top of the beam controls.

$$S_{\text{required}} = \frac{M}{F'_b} \quad \text{where}$$

$$F'_b = F^{-}_{bx}C_D(C_V \text{ or } C_L) = (1450 \text{ psi})(1.15)(1.00)$$

$$= 1667.5 \text{ psi}$$

$$S_{\text{required}} = \frac{180{,}000 \text{ lb-in.}}{1667.5 \text{ psi}} = 1.08 \text{ in}^3$$

From Section A.2.1, a $3\frac{1}{8}$ in. × 15 in. ($S_x = 117$ in^3) or a $5\frac{1}{8}$ in. × 12 in. ($S_x = 123$ in^3) section appears satisfactory.

Check the $3\frac{1}{8} \times 15$:

$$f^{-}_b = \frac{M^{-}}{S} = \frac{180{,}000 \text{ lb-in.}}{117 \text{ in}^3} = 1538 \text{ psi}$$

$$F^{-\prime}_b = F^{-}_b C_D(C_C \text{ or } C_L)$$

The points of zero moments are at the ends and near the center support. From Section A.2.1, condition 29a, the points of zero moment can be located as follows.

$$M = 0 = R_1 x - \frac{\omega x^2}{2} \qquad \text{where } R_1 = \tfrac{3}{8}\,\omega\ell:$$

$$\tfrac{3}{8}\,\omega\ell x = \frac{\omega\, x^2}{2}$$

$$x = \tfrac{3}{4}\,\ell \quad \text{or} \quad \tfrac{1}{4}\,\ell \text{ from the support}$$

For the negative moment, therefore, the L for the volume effect is $\frac{1}{4} + \frac{1}{4} = \frac{1}{2}$ of 10 ft or 5 ft. The volume factor becomes

$$C_V = \left(\frac{5.125 \text{ in.}}{3.125 \text{ in.}}\right)^{1/10} \left(\frac{12 \text{ in.}}{15 \text{ in.}}\right)^{1/10} \left(\frac{21 \text{ ft}}{10 \text{ ft}}\right)^{1/10} = 1.11$$

but not greater than 1.00, so $C_V = 1.00$. Thus,

$$F^{-\prime}_b = (1450 \text{ psi})\ (1.15) = 1667.5 \text{ psi}$$

Therefore,

$$f^{-}_b = 1538 \text{ psi} \leq F^{-\prime}_b = 1667.5 \text{ psi} \qquad \text{good}$$

Check shear. V at d away from the support is

$$V = \frac{5\omega\ell}{8} - \omega d$$

$$= \frac{(5)(1200\ \text{lb/ft})\ (10\ \text{ft})}{8}$$

$$- (1200\ \text{lb/ft}\ (15\ \text{in.})\left(\frac{1\ \text{ft}}{12\ \text{in.}}\right) = 6000\ \text{lb}$$

$$f_v = \frac{3V}{2bd} = \frac{(3)(6000\ \text{lb})}{(2)(3.125\ \text{in.})(15\ \text{in.})} = 192\ \text{psi}$$

$$F_v = 265\ \text{psi (AITC 117)}$$

$$F'_v = F_v C_D = (265\ \text{psi})(1.15) = 305\ \text{psi}$$

Since $f_v = 192$ psi $\leq F'_b = 305$ psi good

Balanced Beam The negative moment at the support still controls, as the absolute moment magnitude is greatest at that point. $S_{\text{required}} = M/F'_b$; C_V and C_L will be assumed to be unity.

$$F'_b = F^-_{bx} C_D\ (C_V \text{ or } C_L) = (2400\ \text{psi})(1.15)(1.00) = 2760\ \text{psi}$$

$$S_{\text{required}} = \frac{180{,}000\ \text{lb-in.}}{2760\ \text{psi}} = 65\ \text{in}^3$$

From the Appendix, a $3\frac{1}{8}$ in. $\times$ 12 in. section (S_x = 75 in^3) appears satisfactory.

Check the $3\frac{1}{8}$ in. $\times$ 12 in. section:

$$f_b = \frac{180{,}000\ \text{lb-in.}}{75\ \text{in}^3} = 2400\ \text{psi}$$

$$F'_b = F^-_{bx} C_D (C_V \text{ or } C_L)$$

The volume factor becomes

$$C_V = \left(\frac{5.125\ \text{in.}}{3.125\ \text{in.}}\right)^{1/10} \left(\frac{12\ \text{in.}}{12\ \text{in.}}\right)^{1/10} \left(\frac{21\ \text{ft}}{10\ \text{ft}}\right)^{1/10} = 1.13,$$

but not greater than 1.00, so $C_V = 1.00$. Thus, $F'_b = F^-_{bx} C_D (C_V \text{ or } C_L) = (2400\ \text{psi})(1.15)(1.00) = 2760$ psi, unchanged. Therefore,

$$f^-_b = 2400\ \text{psi} \leq F^-_{bx} = 2760\ \text{psi} \qquad \text{good}$$

Check shear. V at d away from the support is 6000 lb, unchanged. The shear stress is

$$f_v = \frac{3V}{2bd} = \frac{(3)(6000 \text{ lb})}{(2)(3.125 \text{ in.})(12 \text{ in.})} = 240 \text{ psi}$$

$$F_v = 265 \text{ psi} \quad \text{(AITC 117)}$$

$$F'_v = F_v C_D = (265 \text{ psi})(1.15) = 305 \text{ psi}$$

Since $f_v = 240 \text{ psi} \leq F'_b = 305 \text{ psi}$ good

Answer: Suitable size and grade for both unbalanced and balanced layup for the stated conditions are:

Unbalanced: $3\frac{1}{8}$ in. × 15 in., 24F-1.8E DF

Balanced: $3\frac{1}{8}$ in. × 12 in., 24F-1.8E DF

The top and bottom of the beam should both be braced to prevent lateral movement for the entire length.

Discussion: A complete analysis would also include deflection checks and camber specifications. In some cases it is necessary to check both positive and negative bending stresses for continuous beams. It would not be appropriate to frame an unbalanced member upside down even though the stronger side of the beam would in this case coincide with the greatest moment.

4.2.8 End-Notched Members

Normally, beams should not be notched or tapered on the tension side. If it becomes necessary to notch a bending member at its end at a support on the tension side, the shear parallel to grain f_v should be computed by the equation

$$f_v = \frac{3R_v}{2bd_e}\left(\frac{d}{d_e}\right)^2 \tag{4.8}$$

where

R_v = vertical reaction (lb),
b = width of the beam (in.),
d = depth of the beam (in.),
d_e = depth of the beam minus the depth of the notch (in.).

The notching of a bending member on the tension side results in a decrease in capacity caused by stress concentrations around the notch as well as a reduction in the area resisting the shear forces. The notch induces tension perpendicular to grain stresses which interact with the shear parallel to the grain, creating a tendency to split. For these reasons, the notching of large glued laminated timbers is not recommended, and it should be limited in smaller wood members. The equation given above is an empirical equation developed for the condition of a square-cornered end notch, and the depth of the notch should be limited to 10% of the beam depth or 3 in., whichever is less. The stress values calculated by equation (4.8) must be compared to the allowable shear stress for nonprismatic members.

The designer should also consider reducing the stress concentration that occurs when a member is notched by using a gradual tapered notch configuration in lieu of a square-cornered notch (Figure 4.5). The designer may also need to consider the use of reinforcement such as full-threaded lag screws to resist the tendency to split at the notch. Notching on the tension side of beams away from the ends is not permitted.

When a beam is notched or beveled on its upper (compression) side at the ends, a less severe condition from the standpoint of stress concentrations is realized. If such a notch is square cornered, as illustrated in Figure 4.6, the shear should be checked by the equation

$$f_v = \frac{3R_v}{2b\{d - [(d - d_e)/d_e]e\}} \tag{4.9}$$

where e is the distance the notch extends inside the inner edge of the support (Figure 4.6). The shear stresses calculated using equation (4.9) must be checked against the allowable shear stress for nonprismatic members.

If e exceeds d_e, equation (4.9) is not used; instead, the shear strength is computed by using only the depth of beam below the notch, d_e. In no case should a notch on the upper side of a beam exceed 40% of the total depth of a beam. If the end of a beam is beveled (as shown by the dashed line in Figure 4.6), d_e is measured from the inner edge of the support to the bevel.

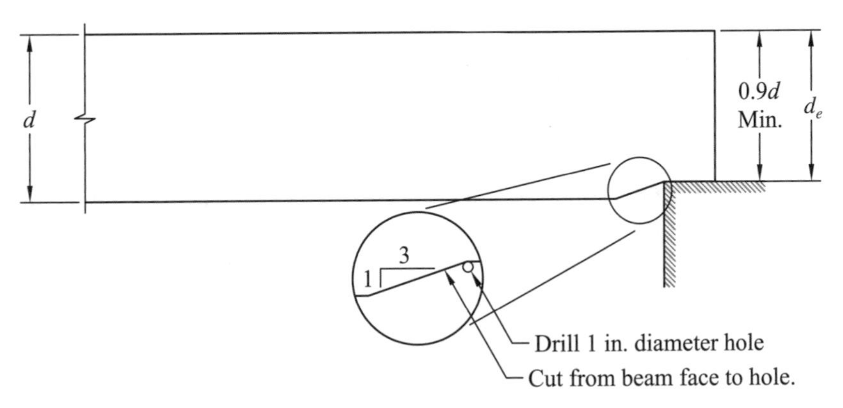

Figure 4.5 End-notched beam.

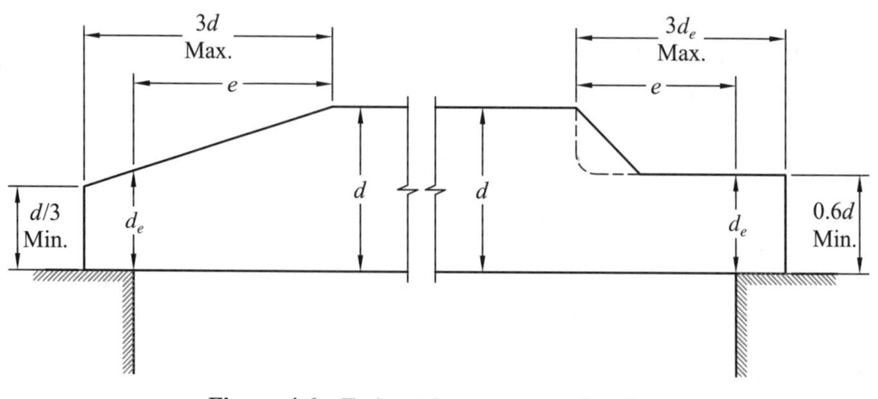

Figure 4.6 End notch on compression face.

The same equation and considerations apply as for beams notched on the compression side.

Generally, continuous span or cantilever span beams should not be notched in the top of the member over the interior support(s) or on the bottom side away from the supports in areas of positive moments. Additional information on notches may be obtained in AITC Technical Note 19 [4]. Under no circumstances should a member be notched on both top and bottom at a support or between supports.

Example 4.6 End-Notched Member

Given: The $3\frac{1}{8}$ in. × 12 in. beam of Example 4.5.

Determine: The suitability of a 1-in. notch on the tension face at one end of the beam.

Approach: Use equation (4.8) to check stress and check the depth against overall size limitations. For the analysis of shear stresses at the notch, the design value for nonprismatic members will be used (0.72 times the tabular value for prismatic members per AITC 117).

Solution: The depth of the notch must not exceed $\frac{1}{10}$ the depth of the member, or 3 in. Since $d/10 = 1.2$ in., the depth of the notch of 1 in. $\leq$ 1.2 in. $\leq$ 3 in.; okay. From Section A2.1, $R_v = \frac{3}{8}\,\omega\ell = (\frac{3}{8})(1200\text{ lb/ft})(10\text{ ft}) = 4500$ lb. Using equation (4.8) yields

$$f_v = \frac{3R_v}{2bd_e}\left(\frac{d}{d_e}\right)^2$$

$$f_v = \frac{(3)(4500)\ \text{lb}}{(2)(3.125\ \text{in.})(11\ \text{in.})}\left(\frac{12\ \text{in.}}{11\ \text{in.}}\right)^2 = 234\ \text{psi}$$

$$F_v = (265\ \text{psi})\ (0.72) = 190\ \text{psi}$$

$$F'_v = (190\ \text{psi})\ (1.15) = 219\ \text{psi}$$

Since $f_v = 234\ \text{psi} > 219\ \text{psi}$, the notch is not permissible.

Answer: The 1-in. notch in the $3\frac{1}{8}$ in. × 12 in. beam produces a stress condition that is not permissible.

4.2.9 Beams with Tapered End Cuts

Taper cuts on the top at the end of beams are sometimes used to improve drainage, to provide extra head for downspouts and scuppers, to facilitate discharge of water, and to reduce the height of a wall (Figure 4.7). The lengths of taper cuts vary depending on span length, roof-framing systems, or other requirements. Sloping cuts from 4 to 8 ft long are commonly used. The depth of the end cut usually depends on the drainage requirements. A sloping cut is commonly made through the compression zone laminations, exposing the lower-strength core laminations and resulting in a member with a lower design value in bending than listed for the stress class or combination in the *NDS®* [2] or AITC 117 [1]. The design values of F_{bx}, E_x, and $F_{c\perp}$ for taper cuts up to two-thirds the depth of the member are provided in Table 4.4. The variations in design values for combinations with the same design value in bending are caused by the use of various grades or species in the core of the combination. The value of F_v should be taken as the nonprismatic shear value. The values of E_x and $F_{c\perp}$ are usually less because of the removal of some of the higher-grade lumber in the compression zone.

Beams with taper cuts confined to the ends of the member are generally selected based on prismatic (uncut) members, and then the effects of the taper at the end are considered. The design considerations at the end are listed below. The shear stress near the end of the beam must be checked with regard

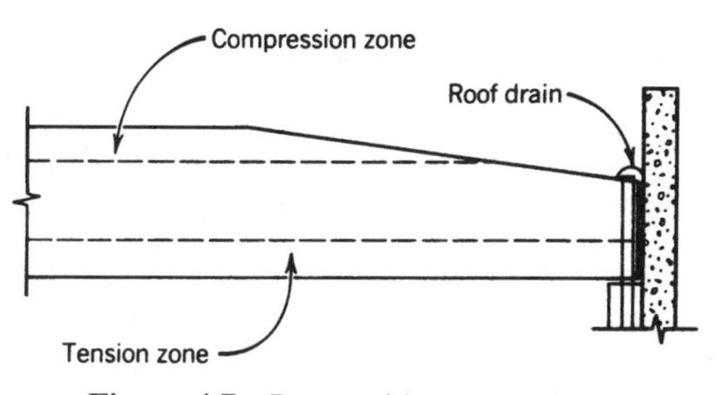

Figure 4.7 Beam with taper end cut.

TABLE 4.4 Allowable Properties and Moduli of Elasticity for Glued Laminated Timber with Tapered Cuts on Compression Face

Stress Class	$F_{bx}{}^{a}$ (psi)	$F_{c\perp x}{}^{b}$ (psi)	$E_x{}^{a}$ (10^6 psi)	$F_{vx}{}^{c}$ (psi)
16F-1.3E	1050	315	1.2	140
20F-1.5E	1250	375	1.4	150
24F-1.7E	1250	375	1.4	150
24F-1.8E	2000	560	1.7	190
26F-1.9E	2000	560	1.7	190
28F-2.1E	2400	650	1.9	215
30F-2.1E	2400	650	1.9	215

[a] Value is applicable to members that have up to two-thirds the depth on the compression side removed by taper cutting. Value is for dry conditions of use.
[b] Design value in compression perpendicular to the grain for the core laminations of the combination.
[c] Design value in horizontal shear for nonprismatic members.

to Section 4.2.7 for notches (and taper cuts) in the compression zone at the end of the beam. The bending stress near the end of the beam must be less than the modified design value of Table 4.4 adjusted by all applicable factors, including the stress interaction factor, C_I, calculated using

$$C_I = \left[\frac{1}{1 + (F_b \tan \theta / F_v)^2 + (F_b \tan^2\theta / F_{c\perp})^2}\right]^{1/2} \tag{4.10}$$

where θ is the angle of the taper and F_b, F_v, and $F_{c\perp}$ are the design values in bending, shear, and compression perpendicular to the grain. Beams manufactured with tapers and beams with taper cuts over significant portions of the span are covered in Section 4.4.

Example 4.7 Taper Cut on Compression Face at End of Beam

Given: A simply supported southern pine beam spans 40 ft carrying 240 lb/ft of roof load (construction load duration) and 300 lb/ft of dead load. The beam is supported at 10-ft intervals along its length to resist buckling. Framing considerations dictate a beam depth of 25 in. overall with a taper cut $7\frac{1}{2}$ in. $\times$ 6 ft as shown in Figure 4.8. The roof is not attached to nonstructural materials subject to damage due to deflection. Since the roof is sloped for drainage, ponding does not need to be investigated.

Determine: A suitable SP 24F-1.7E section, taking into consideration the taper cuts at the ends. Also specify the appropriate camber for the beam.

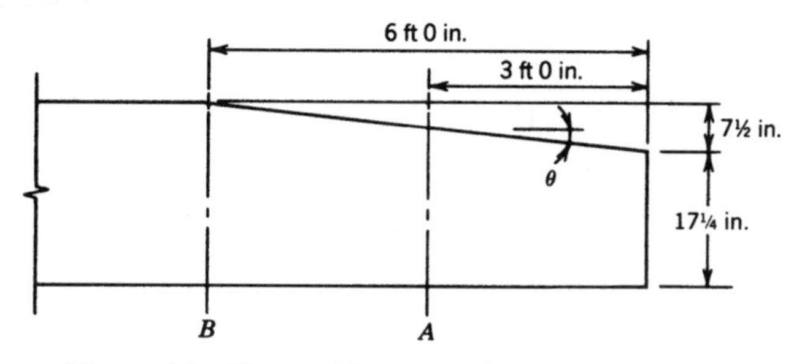

Figure 4.8 Beam with taper end cut—Example 4.7.

Approach: The beam section will be determined based on bending stress and the untapered section. The deflection will be checked. The tapered end will be checked. Once an appropriate section has been verified, the appropriate camber will be calculated and specified. Since the beam is tapered, the shear design value for nonprismatic members will be used.

Solution: Assuming a C_V or C_L value of 0.85, the required section modulus for the beam is $S_{\text{required}} = M/F_b'$:

$$M = \frac{\omega\ell}{8} = \frac{(300 \text{ lb/ft} + 240 \text{ lb/ft})(40 \text{ ft})^2}{8}$$

$$= 108{,}000 \text{ lb-ft} = 1{,}296{,}000 \text{ lb-in.}$$

$$F_b' = (2400 \text{ psi})\,(1.25)(0.85) = 2550 \text{ psi}$$

$$S_{\text{required}} = \frac{1{,}296{,}000 \text{ lb-in.}}{2550 \text{ psi}} = 508 \text{ in}^3$$

From the Appendix, a $5\frac{1}{8}$ in. × $24\frac{3}{4}$ in. section has the following properties: $S_x = 523 \text{ in}^3$, $I_x = 6475 \text{ in}^4$, $A = 126.8 \text{ in}^2$, and

$$C_V = \left(\frac{5.125 \text{ in.}}{5.125 \text{ in.}}\right)^{1/20}\left(\frac{12 \text{ in.}}{24.75 \text{ in.}}\right)^{1/20}\left(\frac{21 \text{ ft}}{40 \text{ ft}}\right)^{1/20} = 0.934$$

For C_L, $\ell_u = 10$ ft, and

$$\frac{\ell_u}{d} = \frac{(10 \text{ ft})\,(12 \text{ in./ft})}{24.75 \text{ in.}} = 4.8 < 7$$

so, from Table 4.3,

$$\ell_e = (2.06)\ell_u = (2.06)(10 \text{ ft}) = 20.6 \text{ ft}$$

$$R_B = \sqrt{\frac{\ell_e d}{b^2}} = \sqrt{\frac{(20.6)\ (12 \text{ in.})\ (24.75 \text{ in.})}{(5.125 \text{ in.})^2}} = 15.26$$

$$E' = E_y \text{ for lateral stability} = 1{,}300{,}000 \text{ psi (AITC 117)}$$

$$F_{bE} = \frac{K_{bE}E'}{R_B^2} = \frac{(0.610)\ (1.3)(10^6) \text{ psi}}{(15.26)^2} = 3404 \text{ psi}$$

$$F_b^* = (2400 \text{ psi})\ (1.25) = 3000 \text{ psi}$$

$$C_L = \frac{1 + F_{bB}/F_b^*}{1.9} - \sqrt{\left(\frac{1 + F_{bE}/F_b^*}{1.9}\right)^2 - \frac{F_{bE}/F_b^*}{0.95}}$$

$$= \frac{1 + 3404 \text{ psi}/3000 \text{ psi}}{1.9}$$

$$- \sqrt{\left(\frac{1 + (3404 \text{ psi}/3000 \text{ psi})}{1.9}\right)^2 - \frac{3404 \text{ psi}/3000 \text{ psi}}{0.95}} = 0.863$$

C_L controls over C_V, and since neither is less than the 0.85 assumed, the section is adequate with regard to bending.

Deflection of the beam is assumed to be little affected by notches near the beam ends; therefore,

$$\Delta_{\text{LL}} = \frac{5}{384}\frac{\omega\ell^4}{EI} = \left(\frac{5}{384}\right)\frac{(240/12 \text{ lb/in.})[(40 \text{ ft})(12 \text{ in./ft})]^4}{(1{,}700{,}000 \text{ psi})(6475 \text{ in}^4)} = 1.26 \text{ in.}$$

From Table 3.9, the live load deflection should be limited to

$$\frac{\ell}{240} = \frac{(40 \text{ ft})(12 \text{ in./ft})}{240} = 2 \text{ in.} \quad \text{(no plaster ceiling)} \qquad \text{good}$$

The total deflection is

$$\Delta_{\text{TL}} = \Delta_{\text{LL}} \frac{240 \text{ plf} + 300 \text{ plf}}{240 \text{ plf}} = 1.26 \text{ in.} \left(\frac{540 \text{ plf}}{240 \text{ plf}}\right) = 2.83 \text{ in.}$$

which should be limited to

$$\frac{\ell}{180} = \frac{(40 \text{ ft})(12 \text{ in./ft})}{180} = 2.67 \text{ in.}$$

The total load deflection is slightly larger than the recommended limit of $\ell/180$. Since the selected section matched the desired framing height, either a wider section should be selected or a member with a higher modulus of elasticity. Considering the 24F-1.8E grade, the deflection under total load becomes

$$\Delta_{\text{TL}} = (2.83 \text{ in.})(1.7/1.8) = 2.66 \text{ in.}$$

which satisfies the recommended deflection limitation. The higher-grade beam also has greater stiffness to resist lateral buckling ($C_L = 0.914$) and will also experience less deflection under live load.

The effects of the taper itself may now be investigated.

1. *Check shear.* From Section 4.2.7, since the distance from the edge of the support to the end of the taper, e, is greater than d_e, the shear stress is computed from equation (4.6b) using the depth d_e for the member depth.

$$f_v = \frac{3V}{2bd_e}$$

where $d_e = 24.75 \text{ in.} - 7.5 \text{ in.} = 17.25 \text{ in.}$ and

$$V = \frac{\omega\ell}{2} = \frac{(540 \text{ lb/ft})(40 \text{ ft})}{2} = 10{,}800 \text{ lb}$$

The shear stress is, therefore,

$$f_v = \frac{3}{2}\left[\frac{10{,}800 \text{ lb}}{2\,(5.125 \text{ in.})(17.25 \text{ in.})}\right] = 183 \text{ psi}$$

From AITC 117,

$$F'_v = 0.72\, F_v C_D = [(0.72)(300 \text{ psi})](1.25) = (215 \text{ psi})(1.25) = 270 \text{ psi}$$

Since 183 psi $\leq$ 270 psi, the member is satisfactory with regard to shear at the tapered end.

2. *Check the bending interaction.* First the angle of the taper, θ, is calculated:

$$\theta = \tan^{-1}\left[\left(\frac{7.5 \text{ in.}}{6 \text{ ft}}\right)\left(\frac{1 \text{ ft}}{12 \text{ in.}}\right)\right] = 5.95°$$

The interaction factor is computed by equation (4.10),

$$C_I = \left[\frac{1}{1 + (F_b \tan \theta / F_v)^2 + (F_b \tan^2\theta / F_{c\perp})^2} \right]^{1/2}$$

where F_b and $F_{c\perp}$ from Table 4.4 are 2000 psi and 560 psi, respectively. F_v = (300 psi) (0.72) = 215 psi from AITC 117, and

$$C_I = \left\{ \frac{1}{1 + [(2000 \text{ psi}) \tan(5.95°)/215 \text{ psi}]^2 + [(2000 \text{ psi}) \tan^2(5.95°)/560 \text{ psi}]^2} \right\}^{1/2}$$

$$= 0.718$$

Considering section A,

$$M_A = \frac{\omega x}{2}(\ell - x) = \frac{(540 \text{ lb/ft})(3 \text{ ft})}{2}(40 \text{ ft} - 3 \text{ ft})$$

$$= 29{,}970 \text{ lb-ft} = 359{,}600 \text{ lb-in.}$$

The depth at A = d_A = 24.75 in. − (7.5 in./2) = 21.0 in.

$$S_A = \frac{bd_A^2}{6} = \frac{(5 \text{ in.})(21.0 \text{ in.})^2}{6} = 367.5 \text{ in}^3$$

$$f_{bA} = \frac{M_A}{S_A} = \frac{359{,}600 \text{ lb-in.}}{367.5 \text{ in}^3} = 979 \text{ psi}$$

The volume factor C_V for section A–A is calculated on the basis of the depth d_A at A–A, the width of the beam, and the entire length of the beam.

$$C_V = \left(\frac{5.125 \text{ in.}}{b}\right)^{1/20} \left(\frac{12 \text{ in.}}{d}\right)^{1/20} \left(\frac{21 \text{ ft}}{L}\right)^{1/20}$$

$$= \left(\frac{5.125 \text{ in.}}{5.125 \text{ in.}}\right)^{1/20} \left(\frac{12 \text{ in.}}{21.0 \text{ in.}}\right)^{1/20} \left(\frac{21 \text{ ft}}{40 \text{ ft}}\right)^{1/20} = 0.942$$

Since 0.942 > 0.718, C_I controls (C_I and C_V are not cumulative):

$$F'_{bx} = F^*_{bx} C_I = (2000 \text{ psi})(1.25)(0.718) = 1794 \text{ psi}$$

Since f_b = 979 psi ≤ F'_b = 1794 psi, the bending stress interaction is okay at A.

Considering section B,

$$M_B = \frac{\omega x}{2}(\ell - x) = \frac{(540 \text{ lb/ft})(6 \text{ ft})}{2}(40 \text{ ft} - 6 \text{ ft}) = 660{,}960 \text{ lb-ft}$$

$$= 359{,}600 \text{ lb-in.}$$

$$S_B = 511 \text{ in}^3 \quad \text{(full section)}$$

$$f_{bB} = \frac{M_B}{S_B} = \frac{660{,}960 \text{ lb-in.}}{511 \text{ in}^3} = 1295 \text{ psi}$$

$$C_V \text{ (for section B)} = \left(\frac{5.125 \text{ in.}}{5.125 \text{ in.}}\right)^{1/20}\left(\frac{12 \text{ in.}}{24.75 \text{ in.}}\right)^{1/20}\left(\frac{21 \text{ ft}}{40 \text{ ft}}\right)^{1/20} = 0.934$$

Since $0.718 < 0.934$, C_I controls. Again,

$$F'_{bx} = F^*_{bx} C_I = (2000 \text{ psi})(1.25)(0.718) = 1794 \text{ psi}$$

Since $f_b = 1295 \text{ psi} \leq F'_b = 1794$ psi, the bending stress interaction is okay at B.

3. *Check bearing*. The bearing condition must be checked using the procedures of Chapter 5.

Now that an acceptable section and grade have been determined, the camber will be calculated. From Table 3.11, the camber should be $1\frac{1}{2}$ times the dead load deflection. From the deflection calculations,

$$\Delta_{DL} = \Delta_{LL} \frac{300 \text{ plf}}{240 \text{ plf}} = 1.26 \text{ in.}\left(\frac{300 \text{ plf}}{240 \text{ plf}}\right) = 1.58 \text{ in.}$$

1.5×1.58 in. = 2.36 in.; specify $2\frac{3}{8}$-in. of camber.

Answer: A $5\frac{1}{8}$ in. $\times$ $24\frac{3}{4}$ in. SP 24F-1.8E beam is suitable for the stated conditions, including considerations for the taper cut at the end.

4.3 TORSION

Torsional stresses are sometimes encountered in timber members. Examples of loading conditions in which torsion may be significant are beams or other members with eccentric transverse loads, bridge stringers resisting guardrail loads, and utility towers. The maximum torsion stress for rectangular members occurs in the middle of the wide face and is computed using [5]

$$f_{vt} = \frac{T(3a + 1.8b)}{a^2b^2} \tag{4.11}$$

where

f_{vt} = the maximum torsion stress under design service load,
T = the applied internal torque under design service load,
a = the dimension of the wide face of the member, and
b = the narrow face dimension.

For glued laminated timber, the recommended allowable torsional stress is given by

$$F'_{vt} = F_{vt}C_DC_MC_t$$

where [6]

$$F_{vt} = \tfrac{2}{3}\,F_{vx} \tag{4.12a}$$

where the value for F_{vx} is for the value for nonprismatic members. For sawn lumber the recommended allowable torsional stress is given by

$$F'_{vt} = F_{vt}C_DC_MC_tC_i$$

where [6]

$$F_{vt} = \tfrac{2}{3}\,F_v \tag{4.12b}$$

Connections to members subject to torsion must be designed to adequately resist the torsional reactions. Consideration should also be made for the effect of shrinkage at connections and providing necessary rigidity. Where torsion stresses may be excessive, design modifications such as additional bracing should be considered to reduce or eliminate torsional effects.

Example 4.8 Timber Member Subject to Torsional Loading

Given: The southern pine glued laminated beam in Figure 4.9 is subject to the load P = 2100 lb from a side bracket producing bending about the strong axis and the torsion of the beam about its longitudinal axis. The bracket is located 3.5 ft from one end of the 12-ft beam and is such that the load P is applied 3 in. from the wide face of the beam. The size shown has been found satisfactory with regard to strong axis bending without consideration of the torsional effect of the applied load P. Normal load duration is to be used.

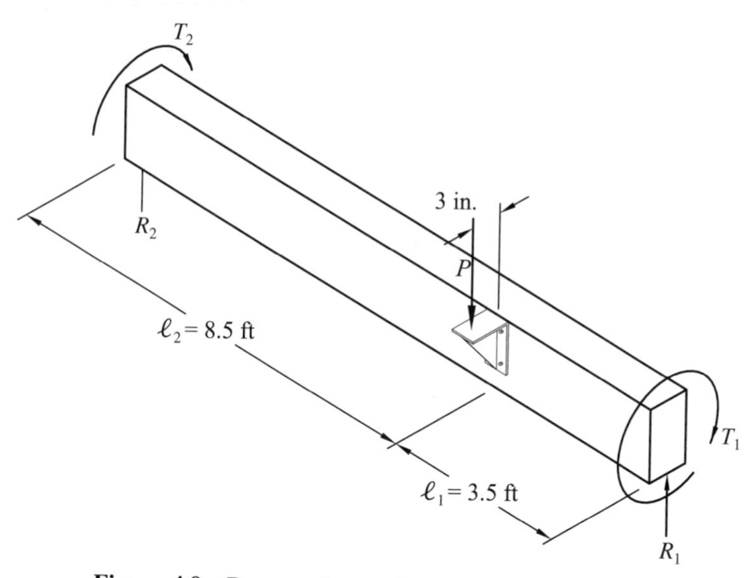

Figure 4.9 Beam under torsional load—Example 4.8.

Temperature conditions are to be considered normal and the beam will be dry in service.

Determine: The acceptability of a $5\frac{1}{8}$ in. × $9\frac{5}{8}$ in. 24F-V4 southern pine glued laminated timber. Recommend an acceptable size if the given size is not suitable.

Approach: The trial size will be investigated with regard to torsion about the longitudinal axis with the torsional stress calculated using equation (4.11). The beam is assumed to be braced adequately at both ends to resist torsion and thus develops torsional reactions at both ends.

Solution: The total torsional load applied to the beam is equal to eP, where e is the amount of eccentricity of the transverse load P. In this case,

$$eP = \left(\frac{5.125 \text{ in.}}{2} + 3 \text{ in.}\right)(2100 \text{ lb}) = (5.563 \text{ in.})\ (2100 \text{ lb}) = 11{,}680 \text{ lb-in.}$$

For a prismatic member supported at its ends and loaded in between, the internal torsional load is greater for the shorter (near) end and is calculated to be

$$T = \frac{eP}{1 + \ell_1/\ell_2} = \frac{11{,}680 \text{ lb-in.}}{1 + 3.5 \text{ ft}/8.5 \text{ ft}} = 8270 \text{ lb-in.}$$

where ℓ_1 is the distance from the load location to the near end and ℓ_2 is the distance from the load location to the far end.

The corresponding maximum torsional stress is

$$f_{vt} = \frac{T(3a + 1.8b)}{a^2b^2} = \frac{(8270 \text{ lb-in.})[(3)(9.625 \text{ in.}) + (1.8)(5.125 \text{ in.})]}{(9.625 \text{ in.})^2 (5.125 \text{ in.})^2}$$

$$= 129 \text{ psi}$$

From AITC 117 [1], the design value for F_{vx} for nonprismatic members for the SP 24F-V4 grade is 150 psi; thus,

$$F_{vt} = \tfrac{2}{3} F_{vx} = \tfrac{2}{3} (150 \text{ psi}) = 100 \text{ psi}$$

The allowable torsional stress is therefore

$$F'_{vt} = F_{vt}C_DC_MC_t = (100 \text{ psi})(1.0)(1.0)(1.0) = 100 \text{ psi}$$

Since $f_{vt} = 129$ psi $> F'_{vt} = 100$ psi, the section is not acceptable with regard to torsional loading.

Considering a $6\frac{3}{4}$ in. $\times$ 11 in. member in the same grade with load P applied 3 in. from the wide face, the eccentricity becomes (6.75 in.)/2 + 3 in. = 6.375 in. The internal torque becomes

$$T = 8270 \text{ lb-in.} \left(\frac{6.375 \text{ in.}}{5.563 \text{ in.}}\right) = 9480 \text{ lb-in.}$$

The maximum torsional stress becomes

$$f_{vt} = \frac{T(3a + 1.8b)}{a^2b^2} = \frac{(9480 \text{ lb-in.})[(3)(11 \text{ in.}) + (1.8)(6.75 \text{ in.})]}{(11 \text{ in.})^2 (6.75 \text{ in.})^2} = 78 \text{ psi}$$

Since $f_{vt} = 78$ psi $\leq F'_{vt} = 100$ psi, the larger section is acceptable with regard to torsional loading.

Answer: The $5\frac{1}{8}$ in. $\times$ $9\frac{5}{8}$ in. 24F-V4 southern pine member shown is not sufficient when torsional effects of the given loading condition are considered. A $6\frac{3}{4}$ in. $\times$ 11 in. member in the same grade is found to be adequate.

4.4 AXIALLY LOADED MEMBERS

Axially loaded members in timber construction are found in trusses, are used as struts and ties, and are commonly used as columns. Many axially loaded members are loaded axially only, such as simple columns and struts, whereas others are loaded in combination with flexural forces. Axial members may be loaded eccentrically, causing flexural (bending) action. Columns and other members subject to concentric axial loads only are the subject of this section.

Members subject to axial tension only are covered, by example, in the following chapter, as they are typically governed by the net section at connections. Members subject to combined or eccentric loads are covered in Section 4.5.

4.4.1 Simple Columns

The term *column* is generally applied to all compression members, including truss members, posts, or other structural components stressed in compression. Columns are divided into three general types, consisting of simple columns, spaced columns, and built-up columns. Simple wood columns consist of a single piece of sawn lumber, post, timber, pole, or glued laminated timber. Spaced columns consist of two or more individual members of sawn lumber or glued laminated timbers with their longitudinal axes parallel, separated at their ends and intermediate points by blocking, and joined at the ends by connectors capable of developing the required shear resistance. Built-up columns consist of two or more pieces of lumber placed side by side in direct contact and joined to one another with mechanical fasteners.

4.4.2 Effective Column Length

The unbraced length of a column or compression member is the distance between two points along its length, between which the member is not prevented from buckling. For a laterally unsupported simple column with assumed pinned ends, the effective length ℓ_e is equal to the total length of the column. Columns with different degrees of fixity at the ends have effective lengths given in Table 4.5. As illustrated in Figure 4.10, the unbraced length may vary with loading direction and dimensions of the compression member. The effective buckling length factors K_e are shown as theoretical K_e values and the recommended values of K_e to use for design. The recommended values of K_e for design take into account the lack of perfect fixity in use. Where compression members depend on the rigidity of other in-plane members entering the joint to provide fixity and the combined stiffness of these members is relatively small, K_e can exceed the values shown in Table 4.5. The effective length is determined by multiplying K_e by the length of the member between points of lateral support. Columns with intermediate bracing points may have different theoretical K_e factors along their lengths. In practice, many columns are supported along their length in one direction (e.g., in the plane of a wall adjoining the column on both sides) and unsupported along their entire length in the other direction (out of the wall plane).

4.4.3 Column Design

Timber members used as columns are oriented with their long axes parallel to the load, producing compression stresses parallel to the grain. These members are designed so that the compression stresses under design load, f_c =

TABLE 4.5 Effective Column Length Factors

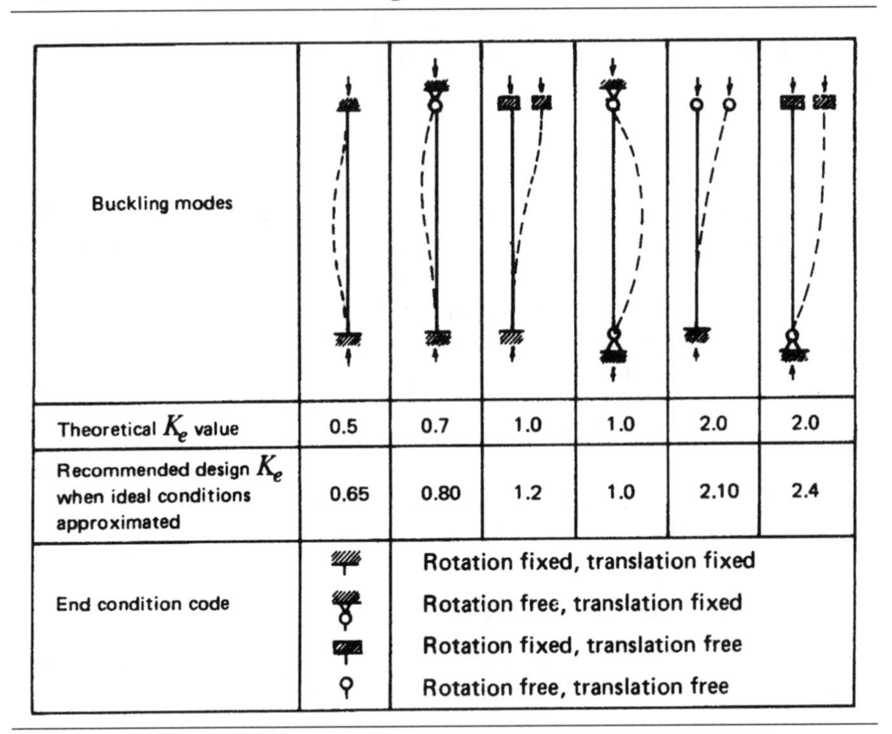

Buckling modes	(a)	(b)	(c)	(d)	(e)	(f)
Theoretical K_e value	0.5	0.7	1.0	1.0	2.0	2.0
Recommended design K_e when ideal conditions approximated	0.65	0.80	1.2	1.0	2.10	2.4

End condition code	
	Rotation fixed, translation fixed
	Rotation free, translation fixed
	Rotation fixed, translation free
	Rotation free, translation free

Source: National Design Specification® [2].

P/A, do not exceed the published design values modified by all applicable adjustment factors. The column stability factor, C_p, adjusts the design value in compression parallel to the grain to account for potential buckling. Where a column is prevented in buckling in all directions, the column stability factor is 1.00. Rectangular section columns typically have a strong direction and weak direction with regard to buckling. In cases where bracing is provided in only one direction, good column design orients the column so that buckling in the otherwise weak direction is totally prevented, resulting in a column stability factor based on the strong direction only. In cases where buckling in both directions is not prevented, the weak direction governs unless the effective lengths differ substantially.

The column stability factor, C_P, is determined as follows:

$$C_P = \frac{1 + F_{cE}/F_c^*}{2c} - \sqrt{\left(\frac{1 + F_{cE}/F_c^*}{2c}\right)^2 - \frac{F_{cE}/F_c^*}{c}} \tag{4.13}$$

where F_c^* = the compression design value multiplied by all applicable adjustment factors except C_P, and

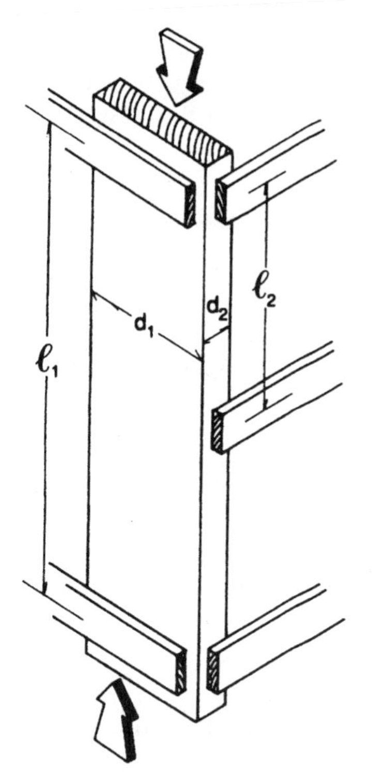

Figure 4.10 Simple column lengths and widths. (From National Design Specification® [2].)

$$F_{cE} = \frac{K_{cE}E'}{(\ell_e/d)^2} \tag{4.14}$$

K_{cE} = 0.510 − 0.839(COV_E)
= 0.3 for visually graded lumber
= 0.384 for products with $0.11 < COV_E \leq 0.15$ such as machine-evaluated lumber (MEL)
= 0.418 for products with $COV_E \leq 0.11$ (glulam and MSR)
c = 0.8 for sawn lumber
= 0.85 for round timber piles
= 0.90 for glued laminated timber
$\ell_e = K_e\ell$,
K_e = effective length coefficient (Table 4.5),
ℓ_e/d = the larger of the ratios ℓ_{e1}/d_1 or ℓ_{e2}/d_2 (Figure 4.10), and
E' = E for the direction in which buckling may take place multiplied by all applicable adjustment factors.

The slenderness ratio ℓ_e/d must in no case be greater than 50.

Glued laminated timbers manufactured in beam or column layups may be used as columns or axial members. Axial loading properties are published for each. Glued laminated beam stock layups with camber, however, should *not* be used, as the camber induces eccentricity with the axial load.

The following examples illustrate the design of sawn lumber and glued laminated timber columns.

Example 4.9 Simple Sawn Column

Given: Solid sawn southern pine, No. 1 dense, 6 × 6 column, dry conditions of use.

Determine: The concentric axial load capacity for 5-, 10-, 15-, and 20-ft lengths assuming pinned end conditions, no intermediate supports resisting buckling, and snow load duration.

Approach: The capacity of the column will be based on the compression design value parallel to the grain published for SP No. 1 dense adjusted by the column stability factor for each of the lengths and by setting the adjusted design value equal to the compression stress under load.

Solution:

1. *Obtain the design values for SP No. 1 dense from the NDS®* [2] (visually graded timbers, 5 × 5 and larger): $F_c = 975$ psi and $E = 1{,}600{,}000$ psi.
2. *Determine* F_c^* *and* E'.

$$F_c^* = F_c C_D = (975 \text{ psi})(1.15) = 1121 \text{ psi}$$
$$E' = E = 1{,}600{,}000 \text{ psi}$$

3. *Determine the effective length(s).* In this example $\ell_e = \ell = 5, 10, 15,$ and 20 ft ($K_e = 1.0$ for all four cases).
4. *Determine the slenderness ratio(s) and check against maximum value of 50.* $b = d = 5\frac{1}{2}$ in.

For the 5-ft column: $$\frac{\ell_{e1}}{d_1} = \frac{\ell_{e2}}{d_2} = \frac{(5 \text{ ft})(12 \text{ in./ft})}{5.5 \text{ in.}} = 10.9$$

For the 10-ft column: $$\frac{\ell_{e1}}{d_1} = \frac{\ell_{e2}}{d_2} = \frac{(10 \text{ ft})(12 \text{ in./ft})}{5.5 \text{ in.}} = 21.8$$

For the 15-ft column: $$\frac{\ell_{e1}}{d_1} = \frac{\ell_{e2}}{d_2} = \frac{(15 \text{ ft})(12 \text{ in./ft})}{5.5 \text{ in.}} = 32.7$$

For the 20-ft column: $$\frac{\ell_{e1}}{d_1} = \frac{\ell_{e2}}{d_2} = \frac{(20 \text{ ft})(12 \text{ in./ft})}{5.5 \text{ in.}} = 43.6$$

None of the four slenderness ratios exceeds 50; good.

5. *Determine F_{cE} and c.* From above, $K_{cE} = 0.3$. For $\ell_e = 5$ ft:

$$F_{cE} = \frac{K_{cE}E'}{(l_e/d)^2} = \frac{(0.30)[(1.6)(10^6 \text{ psi})]}{(10.91)^2} = 4033 \text{ psi}$$

For $\ell_e = 10$ ft, $F_{cE} = 1008$ psi;

For $\ell_e = 15$ ft, $F_{cE} = 448$ psi;

For $\ell_e = 20$ ft, $F_{cE} = 252$ psi; and $c = 0.8$ for sawn timber.

6. *Determine C_P:*

$$C_P = \frac{1 + F_{cE}/F_c^*}{2c} - \sqrt{\left(\frac{1 + F_{cE}/F_c^*}{2c}\right)^2 - \frac{F_{cE}/F_c^*}{c}}$$

For the 5-ft column,

$$C_P = \frac{1 + 4033 \text{ psi}/1121 \text{ psi}}{(2)(0.8)} - \sqrt{\left(\frac{1 + 4033 \text{ psi}/1121 \text{ psi}}{(2)(0.8)}\right)^2 - \frac{4033 \text{ psi}/1121 \text{ psi}}{0.8}} = 0.934$$

Similarly, for the 10-ft column, $C_P = 0.653$; for the 15-ft column, $C_P = 0.359$; and for the 20-ft column, $C_P = 0.213$.

7. *Determine F'_c.*

$$F'_c = F_c^* C_P$$

For the 5-ft column, $F'_c = 1121 \text{ psi} \times 0.934 = 1047$ psi; for the 10-ft column, $F'_c = 732$ psi; for the 15-ft column, $F'_c = 403$ psi; and for the 20-ft column, $F'_c = 239$ psi.

8. *Determine the allowable axial load, P.* Let $f_c = P/A = F'_c$, where $A = (5.5 \text{ in.})(5.5 \text{ in.}) = 30.25 \text{ in}^2$, and determine P. For the 5-ft column, $P_{(5\text{-ft})} = (1047 \text{ psi})(30.25 \text{ in}^2) = 31{,}700$ lb; for the 10-ft column, $P_{(10\text{-ft})} = (732 \text{ psi})(30.25 \text{ in}^2) = 22{,}100$ lb; for the 15-ft column, $P_{(15\text{-ft})} = (403 \text{ psi})(30.25 \text{ in}^2) = 12{,}200$ lb; and for the 20-ft column, $P_{(20\text{-ft})} = (239 \text{ psi})(30.25 \text{ in}^2) = 7200$ lb.

Answer: The allowable load for the four SP No. 1 dense 6 × 6's of different lengths are:

$$P_{(5\text{-ft})} = 31{,}7000 \text{ lb}$$

$$P_{(10\text{-ft})} = 22{,}100 \text{ lb}$$

$$P_{(15\text{-ft})} = 12{,}200 \text{ lb}$$

$$P_{(20\text{-ft})} = 7200 \text{ lb}$$

Discussion: This example illustrates the effect of unbraced length on the carrying capacity of members loaded in axial compression.

Example 4.10 Structural Glued Laminated Timber Column

Given: An 18-ft length of $6\frac{3}{4}$ in. × $8\frac{1}{4}$ in. 16F-V2 southern pine is to be used as a column. The top and bottom of the column are held to prevent translation, and lateral support is provided at midheight to resist buckling about the y–y axis (weak direction). The column is used in a dry location.

Find: The allowable concentric load capacity of the column (snow load duration).

Approach: From AITC 117 [1], the 16F-V2 SP grade is normally used as a simple bending member. It will be used in this application as a compression member, taking into consideration the axial loading properties provided in AITC 117-Design.

Solution:

1. The design values from AITC 117-Design for the SP 16F-V2 are as follows: F_c = 1350 psi, E_x = 1,400,000 psi, E_y = 1,400,000 psi, and E_{axial} = 1,500,000 psi.
2. *Determine* F_c^* *and* E'.

$$F_c^* = F_c \text{ multiplied by all applicable factors except } C_P.$$

$$F_c^* = F_c C_D C_M C_t = (1350 \text{ psi})(1.15)(1.00)(1.00) = 1553 \text{ psi}$$

In this case the value for E is the same for both weak axis and strong axis bending (buckling).

$$E'_x = E_x C_M C_t = 1{,}400{,}000 \text{ psi}$$

3. *Determine the effective lengths.* With reference to Figure 4.10, d_1 = 6.75 in., ℓ_1 = 18 ft, d_2 = 8.25 in., and ℓ_1 = 9 ft. The effective lengths are equal to the unbraced lengths for this example; thus,

$$\ell_{e1} = \ell_1 = 9 \text{ ft} = 108 \text{ in.}$$

$$\ell_{e2} = \ell_2 = 18 \text{ ft} = 216 \text{ in.}$$

4. *Determine the slenderness ratio(s).*

$$\frac{\ell_{e1}}{d_1} = \frac{108 \text{ in.}}{6.75 \text{ in.}} = 16 \leq 50 \qquad \text{good}$$

$$\frac{\ell_{e2}}{d_2} = \frac{216 \text{ in.}}{8.25 \text{ in.}} = 26.2 \leq 50 \qquad \text{good}$$

Bending (buckling) in the strong direction governs, with $\ell_{e2}/d_2 = 26.2 \leq 50$.
5. *Determine F_{cE} and c.* From above, $K_{cE} = 0.418$,

$$F_{cE} = \frac{K_{cE}E'}{(\ell_e/d)^2} = \frac{(0.418)[(1.4)(10^6 \text{ psi})]}{(26.2)^2} = 854 \text{ psi}$$

and $c = 0.9$ for glulam.
6. *Determine C_P.*

$$C_P = \frac{1 + F_{cE}/F_c^*}{2c} - \sqrt{\left(\frac{1 + F_{cE}/F_c^*}{2c}\right)^2 - \frac{F_{cE}/F_c^*}{c}}$$

$$= \frac{1 + 854 \text{ psi}/1553 \text{ psi}}{(2)(0.9)}$$

$$- \sqrt{\left(\frac{1 + 854 \text{ psi}/1553 \text{ psi}}{2(0.9)}\right)^2 - \frac{854 \text{ psi}/1553 \text{ psi}}{0.9}} = 0.50$$

7. *Determine F'_c.*

$$F'_c = F_c C_M C_t C_P = F_c^* C_P = (1553 \text{ psi})(0.50) = 776 \text{ psi}$$

8. *Determine the allowable P.* Let $f_c = P/A = F'_c$, where $A = (6.75$ in.)(8.25 in.) $= 55.7$ in^2 and determine P.

$$P = F'_c A = (776 \text{ psi})(55.7 \text{ in}^2) = 43{,}200 \text{ lb}$$

Answer: The allowable snow load capacity of the column is 43,000 lb (snow load duration).

Discussion: In this example, the slenderness ratios for the x–x and y–y axes of the member were significantly different. In cases where they are the same, or close numerically, and values of E_x and E_y differ, C_P must be calculated with respect to each direction, and the lesser value used to adjust F_c. If this member is braced at midheight in both directions, it can be shown that weak axis buckling would govern, giving $C_P = 0.932$ and an allowable axial compression load of 80,000 lb.

4.4.4 Round Columns

Round columns are solid wood members of circular cross section loaded axially as in the preceding section. For design purposes, a square section member is used with equal cross-section area as the round. The equivalent square-section dimensions are thus $d_{eq} \times d_{eq}$, where $d_{eq} = \sqrt{\pi/4}\,D = 0.886D$, D being the diameter of the round section. Once the equivalent square section dimension is determined, the design procedure of the preceding section is used. Design values for round columns used in log construction are to be provided by the manufacturer as established under the guidelines of ASTM D3957 [7]. Depending on application, the designer may wish to require that the individual member grade or design values be certified by an approved grading agency. Requirements and design values for round members used in utility construction are established by ANSI 05.1 [8]. Round members used in pole or post–frame construction and round timber piles are covered in Chapter 6.

4.4.5 Tapered Columns

Columns may be tapered from a larger cross section toward a smaller cross section at one end or both ends. Tapered columns are designed similarly to other columns, as described in the preceding sections, except that calculations are based on representative dimensions for each face. The representative dimension d for each face of the column is derived as follows:

$$d = d_{min} + (d_{max} - d_{min})\left[a - 0.15\left(1 - \frac{d_{min}}{d_{max}}\right)\right] \tag{4.15a}$$

where

d_{min} = the minimum dimension for that face of the column,
d_{max} = the maximum dimension for that face of the column, and
a = depends on the following support conditions:

With the large end fixed and the small end unsupported or simply supported, $a = 0.70$. With the small end fixed and the large end unsupported or simply supported, $a = 0.30$. With both ends simply supported and tapered toward one end, $a = 0.50$; when tapered toward both ends, $a = 0.70$. For all other support conditions,

$$d = d_{min} + (d_{max} - d_{min})(\tfrac{1}{3}) \tag{4.15b}$$

Tapered round columns are designed on the basis of a square column of the same cross-sectional area and having the same degree of taper.

4.4.6 Spaced Columns

Spaced columns (Figure 4.11) are made of individual column pieces connected to one another by shear plates or split rings through end and spacer blocks. The capacity of the spaced column is the sum of capacities of the longitudinal pieces wherein the fasteners and blocks effectively increase the load-carrying capacity of the full-length pieces in their weak direction (buckling perpendicular to the wide faces of the longitudinal pieces).

The limitations for individual longitudinal pieces and block spacing of a spaced column are as follows (Figure 4.11):

1. ℓ_1/d_1 is not to exceed 80, where ℓ_1 is the distance between lateral supports that provide restraint perpendicular to the wide faces of the individual members.

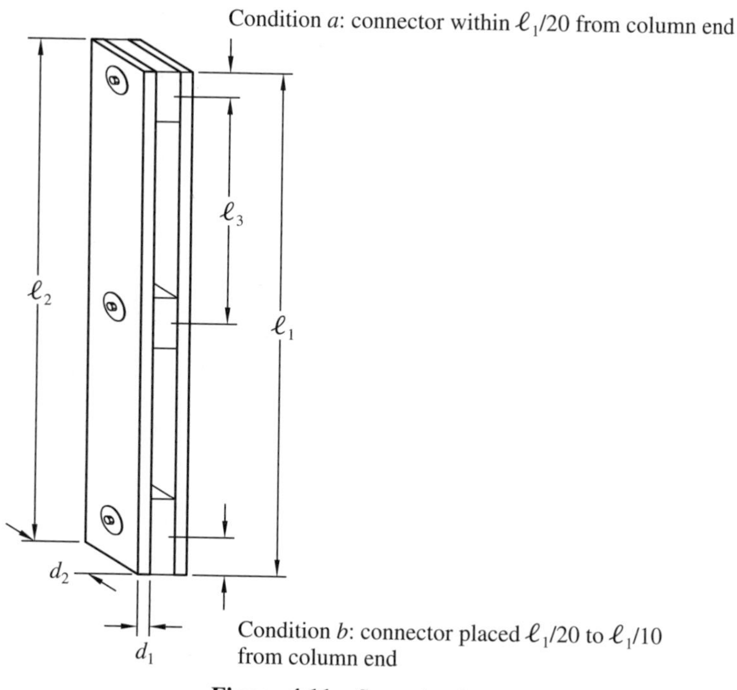

Figure 4.11 Spaced column.

2. ℓ_2/d_2 is not to exceed 50, where ℓ_2 is the distance between lateral supports that provide restraint parallel to the wide faces of the individual members.
3. ℓ_3/d_2 is limited to 40, where ℓ_3 is the distance between the centroid of connectors in an end block and the center of the spacer block.

Requirements for the spacer and end blocks are as follows:

1. For condition a, the centroid of the split-ring or shear plate connector, or group of connectors, in the end blocks, should be within $\ell_1/20$ from the end of the column.
2. For condition b, the centroid of the split-ring or shear plate connector, or group of connectors, in the end blocks, should be between $\ell_1/10$ and $\ell_1/20$ from the end of the column.
3. Where a single spacer block is located in the middle tenth of the column length, ℓ_1, split-ring or shear plate connectors are not required.
4. If there are two or more spacer blocks, split-ring or shear plate connectors are required and the distance between any two adjacent blocks must not exceed one-half the distance between the centers of the split-ring or shear plate connectors in the end blocks.
5. For spaced columns used as compression members of a truss, a panel point that is stayed laterally should be considered as the end of the spaced column, and the portion of the web members, between individual pieces making up a spaced column, should be permitted to be considered as the end blocks.
6. The thickness of spacer and end blocks should not be less than that of the individual longitudinal pieces of the spaced column, nor should the thickness, width, and length of spacer and end blocks be less than required for the split-ring or shear plate connectors of size and number capable of carrying the loads required by the following section.
7. The split-ring or shear plate connectors in each mutually contacting surface of end block and individual member at each end of a spaced column should be of size and number to provide a load capacity equal to the required cross-sectional area in inches of one of the individual members times the end spacer block constant, K_s (Table 4.6).

Split-ring and shear plate connector design is covered in Chapter 5. The species groups listed in Table 4.6 are defined in the *National Design Specification*® and may also be found in Chapter 5.

The capacity of a spaced column, P, is the sum of the capacities of the individual members. For a spaced column with two identical longitudinal pieces, the capacity is given by

TABLE 4.6 End Spacer Block Constant, K_s

Species Group	Constant
A	$K_s = (9.55)(\ell_1/d_1 - 11) \leq 468$
B	$K_s = (8.14)(\ell_1/d_1 - 11) \leq 399$
C	$K_s = (6.73)(\ell_1/d_1 - 11) \leq 330$
D	$K_s = (5.32)(\ell_1/d_1 - 11) \leq 261$

Source: National Design Specification® [2].

$$P = 2F'_c d_1 d_2 \tag{4.16}$$

where $F'_c = F^*_c C_P$. F^*_c is the compression design value parallel to the grain for the longitudinal pieces multiplied by all applicable adjustment factors except C_P. C_P is calculated using equation (4.13) except that F_{cE} is calculated using

$$F_{cE} = \frac{K_{cE} K_x E'}{(\ell_e/d)^2} \tag{4.17}$$

where

$$\begin{aligned} K_x &= 2.5 \text{ for fixity condition a} \\ &= 3.0 \text{ for fixity condition b} \end{aligned}$$

and

$$\begin{aligned} c &= 0.8 \text{ for sawn lumber} \\ &= 0.9 \text{ for glued laminated timber} \end{aligned}$$

The effective length, ℓ_e, of the spaced column must be established by calculation or engineering judgment with the aid of Table 4.5, but in no case may it be taken to be less than the actual column length. The allowable compression value parallel to the grain, F'_c, may not exceed the allowable compression parallel to the grain for the individual pieces treated as simple sawn columns using the slenderness ratio ℓ_2/d_2 (Figure 4.11). Where different grades, species, or thicknesses of members are used, the lesser value of F'_c determined for either member is applied to both. Spaced columns may be loaded eccentrically or loaded to produce combined axial and flexural stresses in the strong direction of the pieces (parallel to d_2). There is no provision, however, to produce combined axial load and flexure in the direction perpendicular to the wide faces of the pieces.

4.4.7 Built-up Columns

Built-up columns are made of two or more pieces of wood with their wide faces fastened together mechanically as illustrated in Figure 4.12. The capacity of the built-up column with respect to potential buckling in the strong direction of the individual pieces is equal to the sum of the capacities of the individual members. The capacity of the column with respect to potential buckling in the weak direction of the pieces is greater than the sum of the individual pieces, although not as great as a solid piece with the same overall cross-sectional area. The strength of the column with respect to the weak axes of the pieces, where fastened properly, is given as a fraction of the capacity of a similar solid section.

Specific requirements of built-up columns are as follows:

1. Columns may be built up of from two to five pieces.
2. Each piece is to be of rectangular cross section and must be at least $1\frac{1}{2}$ in. thick (2 in. nominal).
3. All pieces must have the same wide-face (depth) dimension.
4. All pieces must be in full contact for the full length of the column.

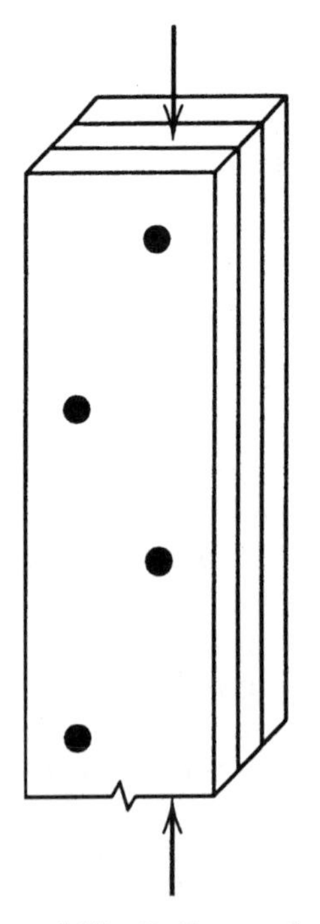

Figure 4.12 Built-up column.

5. The pieces must be fastened together by nails and bolts according to the requirements of this section.

The effective length for the column may be determined using Table 4.5. For built-up columns, a value of C_P for each direction of potential buckling is calculated using the following equation and the smaller of the two is used to calculate F_c':

$$C_P = K_f \left[\frac{1 + F_{cE}/F_c^*}{2c} - \sqrt{\left(\frac{1 + F_{cE}/F_c^*}{2c} \right)^2 - \frac{F_{cE}/F_c^*}{c}} \right] \tag{4.18}$$

where F_c^*, F_{cE}, and c are as defined previously for sawn columns, and

K_f = 0.6 for built-up columns where ℓ_{e2}/d_2 is used to calculate F_{cE} and the built-up column is nailed according to the rules below
= 0.75 for built-up columns where ℓ_{e2}/d_2 is used to calculate F_{cE} and the built-up column is bolted according to the rules below
= 1.0 for built-up columns where ℓ_{e1}/d_1 is used to calculate F_{cE} and the built-up column is either nailed or bolted according to the rules below

In the case of short, builtup columns, C_P as calculated by equation (4.18) may be less than that calculated by assuming each lamination of the column acts independently. C_P need not be taken as less than the value calculated by assuming each lamination acts independently.

4.4.7.1 Nailed Built-up Columns

The requirements for mechanically fastening the pieces of built-up columns using nails are as follows:

1. Adjacent nails are driven from opposite sides of the column.
2. All nails penetrate at least three-fourths of the thickness of the last piece.
3. The end distance for the nail group must be between $15D$ and $18D$, inclusive, where D is the nail diameter.
4. The spacing of nails in a row must be greater than or equal to $20D$ but not greater than six times the thickness of the thinnest lamination, where the row direction is defined as the longitudinal direction of the column.
5. The spacing between rows of nails is to be between $10D$ and $20D$, inclusive.
6. The edge distance for the group must be between $5D$ and $20D$, inclusive.
7. Two or more rows must be provided where the depth of the individual pieces are greater than three times the thickness of the thinnest piece.

8. Where only one row of nails is required, the nails should be staggered across the width of the pieces.
9. Where three or more rows of nails are used, nails in adjacent rows shall be staggered.

4.4.7.2 Bolted Built-up Columns The requirements for mechanically fastening the pieces of built-up columns using bolts are as follows:

1. A metal plate or washer must be provided between the wood and the bolt heads and between the wood and the nuts.
2. Nuts must be tightened to ensure that the faces of the adjacent pieces are in contact.
3. For softwood, the end distance for the bolt group must be between $7D$ and $8.4D$, inclusive; and for hardwoods, between $5D$ and $6D$, inclusive.
4. The spacing between adjacent bolts in a row must be between $4D$ and six times the thickness or the thinnest lamination, inclusive.
5. The spacing between rows of bolts must be between $1.5D$ and $10D$, inclusive.
6. The edge distance for the group must be between $1.5D$ and $10D$, inclusive.
7. Two or more rows of bolts must be provided where the depth of the pieces is greater than three times the thickness of the thinnest piece.

4.4.8 Columns with Flanges

Glued laminated and built-up columns are usually square or rectangular but can also be made with flanges, as shown in Figure 4.13. Because of fabrication and handling difficulties, these shapes are not common. The capacities of flanged columns may be limited by the buckling potential of outstanding flange pieces themselves or the difficulty in transferring shear forces from piece to piece making up the column. The design of a column with flanges should include, at a minimum, the following steps.

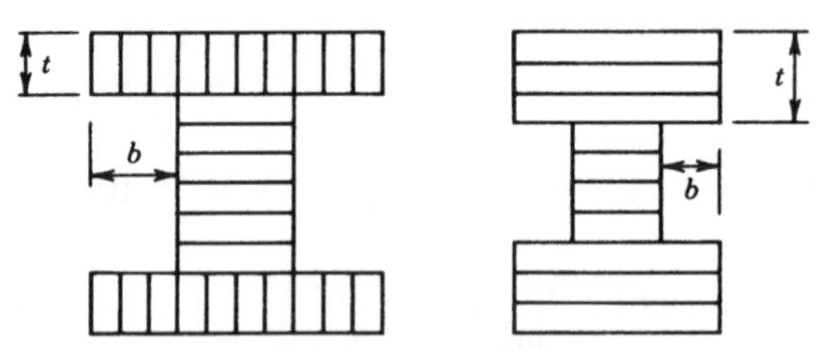

Figure 4.13 Columns with flanges.

1. Determination of the allowable compression parallel to the grain, F'_c, is based on the overall buckling potential of the column and appropriate column stability factor C_P. Effective lengths and slenderness ratios must be determined for each direction of potential buckling. The slenderness ratio, ℓ_e/r, must not exceed 175 in either direction, where r is the radius of gyration, $r = \sqrt{I/A}$ and I is the respective area moment of inertia.
2. The allowable compression parallel to the grain for the outstanding flange piece(s) must be determined by

$$F'_c = 0.016\left(\frac{t}{b}\right)^2 \tag{4.19}$$

   where t and b are defined in Figure 4.13, and the lesser of this stress and the allowable compressive stress for the whole section used.
3. Approved fasteners or adhesives must be used to achieve the necessary shear transfer from piece to piece based on an assumed or known eccentricity or applied flexural load. The designer should require quality assurance in the form of special inspection and/or load capacity verification for such columns.

4.4.9 Tension Members

Tension members in timber construction include struts and chords in trusses and drag struts and ties. Tension members must be proportioned such that the tension stresses under load do not exceed the allowable tension stresses:

$$f_t = \frac{T}{A_n} \tag{4.20}$$

where

T = the tension force
A_n = the net or effective area considering section loss due to holes and notches.

Since tension members are often loaded uniformly (equal tension along the entire length of the member), particular attention must be given to any losses in wood section such as due to notches and holes. Members subject to combined axial tension and flexure are considered in the following section.

4.5 COMBINED AXIAL AND FLEXURAL LOADING

Members subject to the combined action of flexural and axial loading are commonly encountered in timber construction. Such members include top and bottom chords of trusses, columns subject to side loads, and any axial members loaded eccentrically. Combined axial and bending loads arise in any of the following conditions:

1. Axial compression and applied end moments
2. Centric axial load and side loads
3. Eccentric axial end loads
4. Combined end loads, side loads, and eccentricity
5. Loads applied at brackets on columns

4.5.1 General Equations

In this section general equations are presented that are used in all of the cases listed above. Examples of these cases are given in the following sections. The following inequality should be satisfied for any combination of axial compression and flexural loading, including eccentric axial loading:

$$\left(\frac{f_c}{F'_c}\right)^2 + \frac{f_{b1} + f_c(6e_1/d_1[1 + 0.234(f_c/F_{cE1})]}{F'_{b1}[1 - (f_c/F_{cE1})]}$$
$$+ \frac{f_{b2} + f_c(6e_1/d_2)\ \{1 + 0.234(f_c/F_{cE2}) + 0.234\ [f_{b1} + f_c(6e_1/d_1)/F_{bE}]^2\}}{F'_{b2}\ \{1 - (f_c/F_{cE2}) - [f_{b1} + f_c(6e_1/d_1)/F_{bE}]^2\}}$$
$$\leq 1.0 \qquad (4.21)$$

where f_c, F'_c, f_{b1}, f_{b2}, d_2, F'_{b1}, F'_{b2}, F_{bE}, F_{cE1}, and F_{cE2} as defined previously and the subscripts 1 and 2 relate to the wide and narrow face dimensions, respectively. Thus,

d_1 = the wide face dimension,
d_2 = the narrow face dimension, and thus,
f_{b1} = the bending stress due to loads applied to the narrow face (bending in the strong direction), as before;
f_{b2} = the bending stress due to loads applied to the wide face (bending in the weak direction), as before;
e_1 = eccentricity measured parallel to the wide face of the member (from centerline of axial load to centerline of member);
e_2 = eccentricity measured parallel to the narrow face of the member (from centerline of axial load to centerline of member).

The Euler buckling stresses for the strong and weak direction are, thus,

$$F_{cE1} = \frac{K_{cE}E'}{(l_{e1}/d_1)^2}$$

$$F_{cE2} = \frac{K_{cE}E'}{(l_{e2}/d_2)^2}$$

$$F_{bE} = \frac{K_{bE}E'}{R_B^2}$$

where E' in each case is the modulus of elasticity associated with the direction of potential buckling. In all cases, the compressive axial stress and compressive flexural stress must also be less than the corresponding Euler buckling stresses,

$$f_c < F_{cE1} \tag{4.21a}$$

$$f_c < F_{cE2} \tag{4.21b}$$

$$f_{b1} < F_{bE} \tag{4.21c}$$

In the case of concentric axial compression plus bending, equations (4.21) reduce to

$$\left(\frac{f_c}{F'_c}\right)^2 + \frac{f_{b1}}{F'_{b1}(1 - f_c/F_{cE1})} + \frac{f_{b2}}{F'_{b2}[1 - (f_c/F_{cE2}) - (f_{b1}/F_{bE})^2]} \leq 1 \tag{4.22}$$

In the case of concentric axial compression and bending about the strong axis only, equation (4.22) reduces to

$$\left(\frac{f_c}{F'_c}\right)^2 + \frac{f_{b1}}{F'_{b1}(1 - f_c/F_{cE1})} \leq 1.0 \tag{4.23}$$

In the case of eccentric axial compression without bending due to side loads, equation (4.21) becomes

$$\left(\frac{f_c}{F'_c}\right)^2 + \frac{f_c(6e_1/d_1)[1 + 0.234(f_c/F_{cE1})]}{F'_{b1}(1 - (f_c/F_{cE1})}$$
$$+ \frac{f_c(6e_2/d_2)\{1 + 0.234(f_c/F_{cE2}) + 0.234[f_c(6e_1/d_1)/F_{bE}]^2\}}{F'_{b2}\{1 - (f_c/F_{cE2}) - [f_c(6e_1/d_1)/F_{bE}]^2\}}$$
$$\leq 1.0 \tag{4.24}$$

If the axial compression load is eccentric only in the strong direction and no side load bending occurs, equation (4.24) reduces to

$$\left(\frac{f_c}{F'_c}\right)^2 + \frac{f_c(6e_1/d_1)[1 + 0.234(f_c/F_{cE1})]}{F'_{b1}(1 - f_c/F_{cE1})} \leq 1.0 \tag{4.25}$$

The following equations may be used for any combination of axial tension and flexure. For axial tension and flexural tension,

$$\frac{f_t}{F'_t} + \frac{f_b}{F^*_b} \leq 1 \tag{4.26}$$

and for axial tension and flexural compression,

$$\frac{f_b - f_t}{F^{**}_b} \leq 1 \tag{4.27}$$

where

f_t, F'_t, and f_b are as defined previously,

F^*_b is the bending design valued multiplied by all applicable adjustment factors except C_L, and

F^{**}_b is the bending design value multiplied by all applicable adjustment factors except C_V.

The volume factor, C_V, applies to members subjected to combined tension and flexural tension about the x–x axis [equation (4.26)]. The beam stability factor C_L relates to flexural compression and thus is not used as an adjustment factor in equation (4.26). When bending is about the y–y axis, F_{b2} should be used and is adjusted by the flat use factor.

The beam stability factor, C_L, is applicable for the flexural compression side of the member where flexural compression is offset by axial tension [equation (4.27)]. In cases where flexural compression is more than offset by axial tension, equation (4.27) is no longer applicable and equation (4.26) alone governs. Members manufactured for primary use as axial members should be used in such cases. Where glued laminated timbers intended primarily for use as bending members are used, equation (4.26) need only be applied to the face with axial tension plus flexural tension for balanced layup members. For unbalanced layups, equation (4.26) must be satisfied for both faces and their respective design values. In no case should timbers with camber be used unless the eccentricity introduced by the camber is considered in the design and the camber is deemed acceptable from a serviceability standpoint.

In the case of axial tension plus bending, equations (4.26) and (4.27) are used. The following example illustrates the use of an axial member to resist tensile and flexural loading.

4.5.2 Combined Axial Tension and Bending

The following example illustrates the use of the foregoing equations where axial tension and strong axis bending are combined.

Example 4.11 Combined Axial Tension and Bending

Given: A 10-ft-long glued laminated timber tension member made of western species to resist a 30,000-lb axial tension load and a uniform moment of 75,000 lb-in. about the x–x axis. No lateral support is provided between the ends of the member. Design for a normal duration of load and a dry condition of use. Ends of member are supported laterally to prevent lateral rotation.

Determine: Suitability of $5\frac{1}{8}$ in. × 10.5 in. western species member.

Approach: Due to the large axial load, a member intended to be loaded primarily in axial compression or tension will be selected from AITC 117, Table 8.2 [1]. For the visually graded western species layup 3-DF, the following design values are obtained. (It is assumed that the member will have four or more laminations and the member will be specified without special tension laminations.) F_{by} = 2100 psi, F_{bx} = 2000 psi, F_t = 1450 psi, and E_y = 1,900,000 psi.

Solution: The load duration factor, for normal duration is unity, C_D = 1.00.

$$F_t' = F_t C_D = (1450 \text{ psi})(1.00) = 1450 \text{ psi}$$

$$F_{bx}^* = F_{bx} C_D = 2000 \text{ psi}(1.00) = (2000 \text{ psi})$$

$$E' = E = 1{,}900{,}000 \text{ psi}$$

For the $5\frac{1}{8}$ in. × $10\frac{1}{2}$ in. member,

$$A = 53.8 \text{ in}^2 \qquad S_x = 94.17 \text{ in}^3$$

$$f_t = \frac{30{,}000 \text{ lb}}{53.8 \text{ in}^2} = 558 \text{ psi}$$

$$f_{bx} = \frac{75{,}000 \text{ lb-in.}}{94.17 \text{ in}^3} = 796 \text{ psi}$$

To determine the beam stability factor, $\ell_e = 1.84\ell_u$ from Table 4.3 (equal end moments),

$$\ell_e = 1.84\ell_u = (1.84)(10\text{ ft})(12\text{ in./ft}) = 220.8\text{ in.}$$

$$R_B = \sqrt{\frac{\ell_e d}{b^2}} = \sqrt{\frac{(220.8\text{ in.})(10.5\text{ in.})}{(5.125\text{ in.})^2}} = 9.395 \leq 50 \qquad \text{good}$$

$$K_{bE} = 0.610$$

$$F_{bE} = \frac{K_{bE}E'}{R_B^2} = \frac{(0.610)(1{,}900{,}000\text{ psi})}{(9.395)^2} = 13{,}131\text{ psi}$$

$$C_L = \frac{1 + F_{bE}/F_{bx}}{1.9} - \sqrt{\left(\frac{1 + F_{bE}/F_{bx}}{1.9}\right)^2 - \frac{F_{bE}/F_{bx}}{0.95}}$$

$$C_L = \frac{1 + 13{,}131\text{ psi}/2000\text{ psi}}{1.9}$$

$$- \sqrt{\left(\frac{1 + 13{,}131\text{ psi}/2000\text{ psi}}{1.9}\right)^2 - \frac{13{,}131\text{ psi}/2000\text{ psi}}{0.95}} = 0.991$$

The volume factor, C_V, will be unity since the dimensions of the trial member are less than the standard-size member for which the volume effect is calculated ($C_V = 1.00$).

From Equation (4.26),

$$F_{bx}^* = 2000\text{ psi multiplied by all factors except } C_L\text{; thus,}$$

$$F_{bx}^* = F_{bx} = 2000\text{ psi}$$

$$\frac{f_t}{F_t'} + \frac{f_b}{F_{bx}^*} = \frac{558}{1450} + \frac{796}{2000} = 0.783 \leq 1.00$$

From equation (4.27),

$$F_{bx}^{**} = 2000\text{ psi multiplied by all factors except } C_V\text{; thus,}$$

$$F_{bx}^{**} = (2000\text{ psi})(0.991) = 1982\text{ psi and}$$

$$\frac{f_{bx} - f_1}{F_{bx}^{**}} = \frac{796\text{ psi} - 557\text{ psi}}{1982\text{ psi}} = \frac{239\text{ psi}}{1982\text{ psi}} = 0.121 \leq 1.00 \qquad \text{good}$$

Answer: The $5\frac{1}{8}$ in. × 10.5 in. western species layup 3 axial member is satisfactory.

Discussion: If the length of the member had been 30 ft rather than 10 ft,

$$\ell_e = (1.84)(30 \text{ ft})(12 \text{ in./ft}) = 662.4 \text{ in.}$$

$$R_B = \sqrt{\frac{(662.4 \text{ in.})(10.5 \text{ in.})}{(5.125 \text{ in.})^2}} = 16.27$$

$$F_{bE} = \frac{K_{bE}E'}{R_B^2} = \frac{(0.610)(1{,}900{,}000 \text{ psi})}{(16.27)^2} = 4377 \text{ psi}$$

The beam stability and volume factors will be recalculated to reflect the 30-ft length:

$$C_L = \frac{1 + 4377 \text{ psi}/2000 \text{ psi}}{1.9} - \sqrt{\left(\frac{1 + 4377 \text{ psi}/2000 \text{ psi}}{1.9}\right)^2 - \frac{4377 \text{ psi}/2000 \text{ psi}}{0.95}} = 0.962$$

$$C_V = \left(\frac{5.125 \text{ in.}}{5.125 \text{ in.}}\right)^{1/10} \left(\frac{12 \text{ in.}}{10.5 \text{ in.}}\right)^{1/10} \left(\frac{21 \text{ ft}}{30 \text{ ft}}\right)^{1/10} = 0.978$$

From equation (4.26),

$$\frac{557 \text{ psi}}{1450 \text{ psi}} + \frac{796 \text{ psi}}{(2000 \text{ psi})(0.978)} = 0.791 \leq 1.0 \qquad \text{good}$$

From equation (4.27),

$$\frac{796 \text{ psi} - 557 \text{ psi}}{(2000 \text{ psi})(0.962)} = \frac{239 \text{ psi}}{1924 \text{ psi}} = 0.124 \leq 1.00 \qquad \text{good}$$

The selected section would also be adequate for the stated conditions and a length of 30 ft. An unsupported member length has a much smaller effect on members loaded in axial tension compared to those loaded in compression.

4.5.3 Axial Compression and Side Load Bending with Respect to One Axis Only

When a member is subjected to both bending and axial compression, the combination of the compressive stresses and the flexural tensile stresses results in a smaller absolute value of compression stress on the flexural tension side and greater absolute compression on the flexural compression side. This affects the manner in which the volume factor, C_V, and the beam stability factor, C_L, are applied. The volume factor, C_V, reflects the effect that the size of member has on the flexural tension strength, while the beam stability factor, C_L, reflects the reduction in bending strength needed to resist lateral buckling. If no net tension exists in the section, the volume effect need not be considered. Further, if the beam is braced to resist all lateral buckling, beam stability need not be considered. If the member is not braced to resist buckling and if net tension occurs on the section, both effects must be considered for combined loading. In this consideration, since axial compression lessens the net tension on the member, the allowable bending stress, if the tension side governs, may be increased by the amount of the compressive stress, f_c. Thus, the allowable bending stress, F'_b, is the smaller of the values $F^*_b C_L$ (compression side governs) or $F^*_b C_V + f_c$ (tension side governs). In no case may $F^*_b C_V + f_c$ exceed F^*_b:

$$F'_b = \text{smaller of } \begin{Bmatrix} F^*_b C_V + f_c \leq F^*_b \\ F^*_b C_L \end{Bmatrix} \tag{4.28}$$

When $f_c \geq F^*_b\ (1 - C_V)$, the term $F^*_b C_L$ will control.

Example 4.12 Combined Axial Compression and Bending

Given: A 20-ft-long glued laminated timber simple-span member is to resist a 240-lb/ft uniform total load acting perpendicular to the wide faces of the laminations and an axial compressive load of 20,000 lb (Figure 4.14). The member is braced perpendicular to the plane of the y–y axis at the ends and throughout its length at the top. Assume a normal duration of load and dry conditions of use.

Determine: A suitable 20F (F_b = 2000 psi) southern pine bending member section and other minimum design properties.

Approach: A 20F trial-size bending member will be selected and size-adjusted as needed.

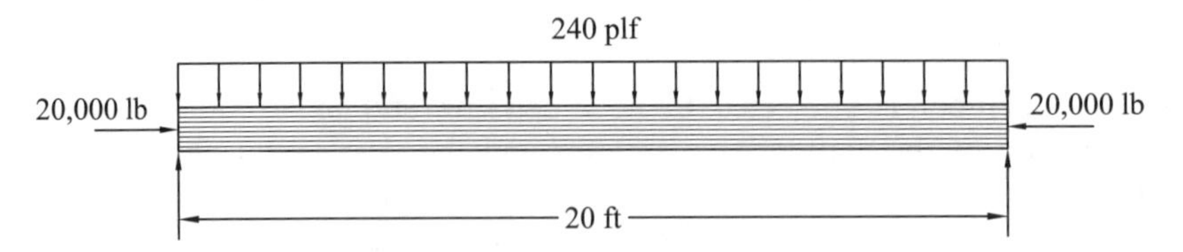

Figure 4.14 Combined axial compression and bending loads.

Solution: For a trial design, use the 20F-V3 SP combination with $F_{bx} = 2000$ psi, $F_c = 1400$ psi, $F_{vx} = 300$ psi, and $E_x = 1{,}500{,}000$ psi, and $E_y = 1{,}400{,}000$ psi. Try a 5 in. × $12\frac{3}{8}$ in. member: $A = 61.88$ in^2, $S_x = 127.6$ in^3,

$$
\begin{aligned}
C_D &= C_M = C_t = 1.00 \\
F_{bx}^* &= F_{bx} = 2000 \text{ psi} \\
E_x' &= E_x = 1{,}500{,}000 \text{ psi} \\
E_y' &= 1{,}400{,}000 \text{ psi} \\
F_c^* &= F_c = 1400 \text{ psi} \\
F_{vx} &= 300 \text{ psi}
\end{aligned}
$$

Determine f_c:

$$
f_c = \frac{P}{A} = \frac{20{,}000 \text{ lb}}{61.88 \text{ in}^2} = 323 \text{ psi}
$$

The member is braced from buckling about the y–y axis; therefore, only strong direction buckling is applicable for C_P. Since the bracing is provided on the top of the member, $C_L = 1.00$ as well.

For C_P, the member is simply supported; thus, $K_e = 1.00$,

$$
\begin{aligned}
\ell_e &= K_e \ell = (1.0)(20 \text{ ft})(12 \text{ in./ft}) = 240 \text{ in.} \\
\frac{\ell_e}{d} &= \frac{240 \text{ in.}}{12.375 \text{ in.}} = 19.39 \leq 50 \\
K_{cE} &= 0.418 \\
F_{cEx} &= \frac{K_{cE} E_x}{(l_e/d)^2} = \frac{(0.418)(1{,}500{,}000 \text{ psi})}{(19.39)^2} = 1668 \text{ psi} \\
c &= 0.9 \\
C_P &= \frac{1 + F_{cE}/F_c^*}{2c} - \sqrt{\left(\frac{1 + F_{cE}/F_c^*}{2c}\right)^2 - \frac{F_{cE}/F_c^*}{c}} \\
&= \frac{1 + 1668 \text{ psi}/1400 \text{ psi}}{(2)(0.9)} \\
&\quad - \sqrt{\left[\frac{1 + 1668 \text{ psi}/1400 \text{ psi}}{(2)(0.9)}\right]^2 - \frac{1668 \text{ psi}/1400 \text{ psi}}{0.9}} = 0.820 \\
F_c' &= F_c^* C_P = (1400 \text{ psi})(0.820) = 1148 \text{ psi}
\end{aligned}
$$

Determine f_{bx} and F'_{bx}:

$$M = \frac{\omega\ell^2}{8} = \frac{(240)\ \text{plf})(20\ \text{ft})^2}{8} = 12{,}000\ \text{lb-ft} = 144{,}000\ \text{lb-in.}$$

$$f_b = \frac{M}{S} = \frac{144{,}000\ \text{lb-in.}}{127.6\ \text{in}^3} = 1128\ \text{psi}$$

$$C_V = \left(\frac{5.125\ \text{in.}}{5\ \text{in.}}\right)^{1/20}\left(\frac{12\ \text{in.}}{12.375\ \text{in.}}\right)^{1/20}\left(\frac{21\ \text{ft}}{20\ \text{ft}}\right)^{1/20} = 1.002 \leq 1.00$$

$$= 1.00$$

From Equation (4.28),

$$F'_b = \text{smaller of} \begin{cases} F_b^* C_V + f_c \leq F_b^*;\ 2000\ \text{psi} + 323\ \text{psi} = 2323\ \text{psi} \leq 2000\ \text{psi} \\ F_b^* C_L = (2000\ \text{psi})(1.00) = 2000\ \text{psi} \end{cases}$$

Thus, $F'_b = 2000$ psi.

From Equation (4.23),

$$\left(\frac{f_c}{F'_c}\right)^2 + \frac{f_{b1}}{F'_{b1}(1 - f_c/F_{cE1})} \leq 1.0$$

From before, $F_{cEx} = F_{cE1} = 1668$ psi, so

$$\left(\frac{323\ \text{psi}}{1148\ \text{psi}}\right)^2 + \frac{1128\ \text{psi}}{(2000\ \text{psi})(1 - 323\ \text{psi}/1668\ \text{psi})} = 0.779 \leq 1.00 \qquad \text{good}$$

Check shear. V at d from the end is

$$V = (240\ \text{lb/ft})\left(\frac{20\ \text{ft}}{2}\right) - (240\ \text{lb/ft})\left(\frac{12.375\ \text{in.}}{12\ \text{in./ft}}\right) = 2153\ \text{lb}$$

$$f_v = \frac{(3)(2153\ \text{lb})}{(2)(61.88\ \text{in}^2)} = 52\ \text{psi} \leq F'_v = 300\ \text{psi} \qquad \text{good}$$

Answer: A 20F-V3 southern pine combination is suitable for the conditions.

Example 4.13 Unbraced Combined Axial Compression and Bending

Given: The conditions stated in Example 4.12 except that no support to resist buckling is provided for either direction.

Determine: A suitable 20F-SP beam section.

Approach: Since no support is given to resist buckling in the weak direction, a wider section is anticipated.

Solution: Try a $6\frac{3}{4}$ in. × $12\frac{3}{8}$ in. beam with the same properties as in Example 4.12. Determine f_c and f_{bx}:

$$A = 83.53 \text{ in}^2 \qquad S = 172.3 \text{ in}^3$$

$$f_c = \frac{P}{A} = \frac{20{,}000 \text{ lb}}{83.53 \text{ in}^2} = 239 \text{ psi}$$

$$f_{bx} = \frac{M_x}{S_x} = \frac{[(240 \text{ lb/ft})(1 \text{ ft/12 in.})][(20 \text{ ft})(12 \text{ in./ft})]^2}{(8)(172.3 \text{ in}^3)} = 836 \text{ psi}$$

Determine C_V:

$$C_V = \left(\frac{5.125 \text{ in.}}{6.75 \text{ in.}}\right)^{1/20} \left(\frac{12 \text{ in.}}{12.375 \text{ in.}}\right)^{1/20} \left(\frac{21 \text{ ft}}{20 \text{ ft}}\right)^{1/20} = 0.987$$

Determine C_L. The member is unbraced but the ends are held in place.

$$\ell_u = (20 \text{ ft})(12 \text{ in./ft}) = 240 \text{ in.} \qquad \frac{\ell_u}{d} = \frac{240 \text{ in.}}{12.375 \text{ in.}} = 19.4$$

$$\ell_e = 1.63\,\ell_u + 3d \quad \text{(from Table 4.3)}$$

$$= 1.63\,(240 \text{ in.}) + (3)(12.375 \text{ in.}) = 428 \text{ in.}$$

$$R_B = \sqrt{\frac{\ell_e d}{b^2}} = \sqrt{\frac{(428 \text{ in.})(12.375 \text{ in.})}{(6.75 \text{ in.})^2}} = 10.78 \leq 50 \qquad \text{good}$$

$$F_{bE} = \frac{K_{bE}E'_y}{R_B^2} = \frac{(0.610)(1{,}400{,}000 \text{ psi})}{(10.78)^2} = 7346 \text{ psi}$$

$$F_b^* = F_b C_D = (2000 \text{ psi})(1.00) = 2000 \text{ psi}$$

$$C_L = \frac{1 + F_{bE}/F_{bx}^*}{1.9} - \sqrt{\left(\frac{1 + F_{bE}/F_{bx}^*}{1.9}\right)^2 - \frac{F_{bE}/F_{bx}^*}{0.95}}$$

$$= \frac{1 + 7346 \text{ psi}/2000 \text{ psi}}{1.9}$$

$$- \sqrt{\left(\frac{1 + 7346 \text{ psi}/2000 \text{ psi}}{1.9}\right)^2 - \frac{7346 \text{ psi}/2000 \text{ psi}}{0.95}} = 0.982$$

F'_b = smaller of

$$\begin{cases} F'_b C_V + f_c \le F^*_b; \ (2000 \text{ psi})(0.987) + 239 \text{ psi} = 2213 \text{ psi} \le 2000 \text{ psi} \\ F^*_b C_L = (2000 \text{ psi})(0.982) = 1964 \text{ psi} \end{cases}$$

$F'_b = 1964$ psi.

Determine C_P and F'_c. The member is unbraced between ends in both directions. Because the section is much deeper than it is wide and $E_x \ge E_y$, compression buckling about y–y axis will be critical and will determine a smaller value for C_P.

$$\ell_e = K_e \ell = 1.0\ell = 20 \text{ ft} = 240 \text{ in.}$$

$$\frac{\ell_e}{d} = \frac{240 \text{ in.}}{6.75 \text{ in.}} = 35.6 \le 50 \qquad \text{good}$$

$$F_{cE} = \frac{K_{cE} E'_y}{(\ell_e/d)^2} = \frac{(0.418)(1{,}400{,}000 \text{ psi})}{(35.6)^2} = 463 \text{ psi}$$

$$F^*_c = 1400 \text{ psi}$$

$$c = 0.9$$

$$C_P = \frac{1 + F_{cE}/F^*_c}{2c} - \sqrt{\left(\frac{1 + F_{cE}/F^*_c}{2c}\right)^2 - \frac{F_{cE}/F^*_c}{c}}$$

$$= \frac{1 + 463 \text{ psi}/1400 \text{ psi}}{(2)(0.9)}$$

$$- \sqrt{\left[\frac{1 + 463 \text{ psi}/1400 \text{ psi}}{(2)(0.9)}\right]^2 - \frac{463 \text{ psi}/1400 \text{ psi}}{0.9}} = 0.316$$

$$F'_c = F^*_c C_P = (1400 \text{ psi})(0.316) = 442 \text{ psi}$$

From Equation (4.23),

$$\left(\frac{f_c}{F'_c}\right)^2 + \frac{f_{b1}}{F'_{b1}(1 - f_c/F_{cE1})} \le 1.0$$

$$\left(\frac{239 \text{ psi}}{442 \text{ psi}}\right)^2 + \frac{836 \text{ psi}}{(1964 \text{ psi})(1 - 239 \text{ psi}/1556 \text{ psi})} = 0.795 \le 1.00 \qquad \text{good}$$

Answer: The SP 20F-V3 combination size $6\frac{3}{4}$ in. × $12\frac{3}{8}$ in. is adequate.

Discussion: Under the new stress class system, the SP 20F class (20F-1.5E) has significantly lower design values for axial compression and E_y. As such, specifying the member by a combination symbol is necessary unless the member is reanalyzed and found suitable using the stress class design values. In some cases, specifying by stress class is more efficient and allows the manufacturer more flexibility. In other cases, where a particular set of design values is critical, it may be beneficial to specify by combination number or by minimum design values necessary for the particular application.

4.5.4 Axial Compression and Side Load Bending with Respect to Both Axes

The following example illustrates the use of equation (4.22) for the case of concentric axial compression and side load bending about both axes.

Example 4.14 Axial Compression and Side Load Bending about Both Axes

Given: A 30-ft-long glued laminated timber beam used in a dry location is loaded about the x–x axis with a uniform load consisting of a dead load of 200 lb/ft and a snow load of 400 lb/ft. A side load P_1 consisting of a 2500-lb wind load is applied at the midpoint. In addition, the member is loaded in compression with a 30,000-lb load P consisting of a 10,000-lb dead load and a 20,000-lb snow load (Figure 4.15). The member is supported at the ends to prevent rotation but is not braced along its length.

Determine: A suitable size 24F-1.8E western species glued laminated beam for the conditions stated.

Approach: Design values for the 24F-1.8E stress class will be used with a trial-size member. Stresses and adjustment factors will be calculated and equation (4.22) will be used to check the suitability of the trial section. In ac-

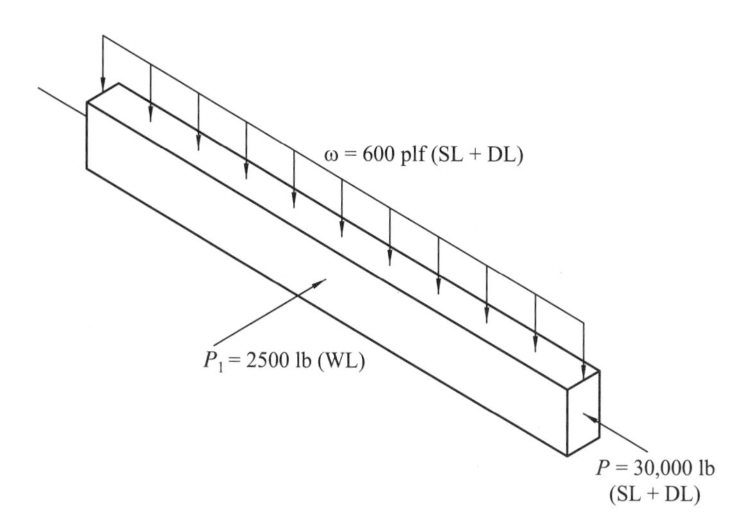

Figure 4.15 Combined axial compression and bending about both axes.

cordance with the load combinations in the IBC [3], the member will be proportioned to sustain the DL, DL + SL, and DL + 0.75(SL + WL) combinations.

Solution: The design values associated with the 24F-1.8E western species stress class are F_{bx} = 2400 psi, F_{by} = 1450 psi, F_c = 1600 psi, E_x = 1,800,000 psi, E_y = 1,600,000 psi, and E_{axial} = 1,700,000 psi. Trial size: $8\frac{3}{4}$ in. × 24 in. From Section A.1.2, A = 210 in^2, S_x = 840 in^3, and S_y = 306 in^3. For load combination DL + 0.75(SL + WL), C_D = 1.60.

$$F^*_{bx} = (2400)(1.60) = 3840 \text{ psi} \qquad F^*_{by} = (1450)(1.60) = 2320 \text{ psi}$$

$$F^*_c = (1600)(1.60) = 2560 \text{ psi}$$

$$E'_x = E_x = 1{,}800{,}000 \text{ psi} \qquad E'_y = E_y = 1{,}600{,}000 \text{ psi}$$

Determine f_c, f_{bx}, and f_{by}:

$$f_c = \frac{P}{A} = \frac{10{,}000 \text{ lb} + (0.75)(20{,}000 \text{ lb})}{210 \text{ in}^2} = 119 \text{ psi}$$

$$f_{bx} = \frac{M_x}{S_x} = \frac{\omega\ell^2}{8S_x} = \frac{[200 \text{ plf} + (0.75)(400 \text{ plf})](30 \text{ ft})^2(12 \text{ in./ft})}{(8)(840 \text{ in}^3)} = 803 \text{ psi}$$

$$f_{by} = \frac{M_y}{S_y} = \frac{PL}{4S_y} = \frac{(0.75 \times 2500 \text{ lb})(30 \text{ ft})(12 \text{ in./ft})}{(4)(306 \text{ in}^3)} = 551 \text{ psi}$$

Determine F'_{bx}. From Equation (4.28),

$$F'_{bx} = \text{smaller of} \begin{cases} F^*_{bx}C_V + f_c \leq F^*_{bx} \\ F^*_{bx}C_L \end{cases}$$

First determine C_V. Bending about the x–x axis for western species:

$$C_V = \left(\frac{5.125 \text{ in.}}{b}\right)^{1/10} \left(\frac{12 \text{ in.}}{d}\right)^{1/10} \left(\frac{21 \text{ ft}}{L}\right)^{1/10}$$

$$= \left(\frac{5.125 \text{ in.}}{8.75 \text{ in.}}\right)^{1/10} \left(\frac{12 \text{ in.}}{24 \text{ in.}}\right)^{1/10} \left(\frac{21 \text{ ft}}{30 \text{ ft}}\right)^{1/10} = 0.853$$

$$F^*_b C_V + f_c = (3840 \text{ psi})(0.853) + 119 \text{ psi} = 3394 \text{ psi}$$

Determine C_L for bending about the x–x axis.

$$F_{bx}^* = 3840 \text{ psi}$$

$$K_{bE} = 0.610$$

$$\ell_u = 30 \text{ ft}(12 \text{ in./ft}) = 360 \text{ in.}$$

$$\frac{\ell_u}{d} = \frac{360 \text{ in.}}{24 \text{ in.}} = 15$$

$$\ell_e = 1.63\ell_u + 3d = (1.63)(360 \text{ in.}) + (3)(24 \text{ in.}) = 659 \text{ in.}$$

$$R_B = \sqrt{\frac{\ell_e d}{b^2}} = \sqrt{\frac{(659 \text{ in.})(24 \text{ in.})}{(8.75 \text{ in})^2}} = 14.37$$

$$F_{bE} = \frac{(0.610)(1{,}600{,}000 \text{ psi})}{(14.37)^2} = 4726 \text{ psi}$$

$$C_L = \frac{1 + F_{bE}/F_{bx}^*}{1.9} - \sqrt{\left(\frac{1 + F_{bE}/F_{bx}^*}{1.9}\right)^2 - \frac{F_{bE}/F_{bx}^*}{0.95}}$$

$$= \frac{1 + 4726 \text{ psi}/3840 \text{ psi}}{1.9}$$

$$- \sqrt{\left(\frac{1 + 4726 \text{ psi}/3840 \text{ psi}}{1.9}\right)^2 - \frac{4726 \text{ psi}/3840 \text{ psi}}{0.95}} = 0.885$$

$$F_b^* C_L = (3840 \text{ psi})(0.885) = 3398 \text{ psi}$$

So F'_{bx} = smaller of 3398 psi and 3416 psi = 3398 psi.

Determine F'_{by}. C_L for bending about the y–y axis is 1.00, since the member is wider than it is deep. C_V will not apply, but C_{fu} will.

$$C_{fu} = \left(\frac{12 \text{ in.}}{d}\right)^{\frac{1}{9}} = \left(\frac{12 \text{ in.}}{8.75 \text{ in.}}\right)^{\frac{1}{9}} = 1.036$$

$$F'_{by} = F_{by}C_D C_{fu} = (1450 \text{ psi})(1.60)(1.036) = 2404 \text{ psi}$$

Determine F'_c. First determine C_p.

$$K_e = 1.0 \qquad \ell_e = (1)(30 \text{ ft})(12 \text{ in./ft}) = 360 \text{ in.}$$

$$K_{cE} = 0.418 \qquad c = 0.9 \text{ for glulam}$$

$$\frac{\ell_e}{d} = \frac{360 \text{ in.}}{24 \text{ in.}} = 15.0 \leq 50 \qquad \text{okay}$$

and

$$\frac{\ell_e}{d} = \frac{360 \text{ in.}}{8.75 \text{ in.}} = 41.1 \leq 50 \qquad \text{good}$$

Potential buckling about the y–y axis will govern; $\ell_e/d = 41.1$.

$$F_{cEy} = \frac{K_{cE}E'_y}{(\ell_e/d)^2} = \frac{(0.418)(1{,}600{,}000 \text{ psi})}{(360 \text{ in.}/8.75 \text{ in.})^2} = 395 \text{ psi}$$

$$F_c^* = 2560 \text{ psi (from above)}$$

$$C_P = \frac{1 + F_{CE}/F_c^*}{2c} - \sqrt{\left(\frac{1 + (F_{cE}/F_c^*)}{2c}\right)^2 - \frac{F_{cE}/F_c^*}{c}}$$

$$= \frac{1 + 395 \text{ psi}/2560 \text{ psi}}{2(0.9)}$$

$$- \sqrt{\left[\frac{1 + 395 \text{ psi}/2560 \text{ psi}}{(2)(0.9)}\right]^2 - \frac{395 \text{ psi}/2560 \text{ psi}}{0.9}} = 0.152$$

Therefore,

$$F'_c = F_c C_D C_P = (1600 \text{ psi})(1.60)(0.152) = 389 \text{ psi}$$

Check the adequacy of the member using equation (4.22).

$$\left(\frac{f_c}{F'_c}\right)^2 + \frac{f_{b1}}{F'_{b1}[1 - f_c/F_{cE1})} + \frac{f_{b2}}{F'_{b2}[1 - (f_c/F_{cE2}) - (f_{b1}/F_{bE})^2]} \leq 1$$

$$F_{cE1} = F_{cEx} = \frac{(0.418)(1{,}800{,}000 \text{ psi})}{(15.0)^2} = 3344 \text{ psi}$$

$$F_{cE2} = F_{cEy} = 395 \text{ psi (from above)}$$

$$\left(\frac{119 \text{ psi}}{389 \text{ psi}}\right)^2 + \frac{803 \text{ psi}}{(3398 \text{ psi})(1 - 119 \text{ psi}/3344 \text{ psi})}$$

$$+ \frac{551 \text{ psi}}{2404 \text{ psi}[1 - (119 \text{ psi}/395 \text{ psi}) - (803 \text{ psi}/4726 \text{ psi})^2]}$$

$$= 0.654 \leq 1.00 \qquad \text{good}$$

Check the load combination with the wind load omitted but a full snow load. When the wind load is not acting, the full dead and snow loads are considered. The appropriate combined stress equation is equation (4.23):

$$\left(\frac{f_c}{F'_c}\right)^2 + \frac{f_{bx}}{F'_{bx}(1 - f_c/F_{cE1})} \leq 1.00$$

In this case,

$$f_c = \frac{P}{A} = \frac{30{,}000 \text{ lb}}{210 \text{ in}^2} = 143 \text{ psi}$$

$$f_{bx} = \frac{M_x}{S_x} = \frac{\omega\ell^2}{8S_x} = \frac{(600 \text{ plf})(30 \text{ ft})^2(12 \text{ in./ft})}{(8)(840 \text{ in}^3)} = 964 \text{ psi}$$

$$C_V = 0.853 \text{ (calculated previously)}$$

$$F^*_{bx} = F_{bx}C_D = (2400 \text{ psi})(1.15) = 2760 \text{ psi}$$

$$F_{bE} = 4726 \text{ psi (calculated previously)}$$

$$C_L = \frac{1 + 4726 \text{ psi}/2760 \text{ psi}}{1.90} - \sqrt{\left(\frac{1 + 4726 \text{ psi}/2760 \text{ psi}}{1.90}\right)^2 - \frac{4726 \text{ psi}/2760 \text{ psi}}{0.95}} = 0.942$$

From Equation (4.28),

F'_b = smaller of

$$\begin{cases} F^*_bC_V + f_c \leq F^*_b \text{ or } (2760 \text{ psi})(0.853) + 143 \text{ psi} = 2497 \text{ psi} \leq 2760 \text{ psi} \\ F^*_bC_L \text{ or } (2760 \text{ psi})(0.942) = 2600 \text{ psi} \end{cases}$$

therefore, $F'_b = 2497$ psi. For axial compression and calculation of C_P, use $F_{cE} = F_{cEy} = 395$ psi.

$$F^*_c = (1600 \text{ psi})(1.15) = 1840 \text{ psi}$$

$$C_P = \frac{1 + F_{cE}/F^*_c}{2c} - \sqrt{\left(\frac{1 + F_{cE}/F^*_c}{2c}\right)^2 - \frac{F_{cE}/F^*_c}{c}}$$

$$= \frac{1 + 395 \text{ psi}/1840 \text{ psi}}{(2)(0.9)} - \sqrt{\left[\frac{1 + 395 \text{ psi}/1840 \text{ psi}}{(2)(0.9)}\right]^2 - \frac{395 \text{ psi}/1840 \text{ psi}}{0.9}} = 0.209$$

$$F'_c = F^*_cC_P = (1840 \text{ psi})(0.209) = 386 \text{ psi}$$

Now for combined one-way bending and axial compression, again, from equation (4.23),

$$\left(\frac{f_c}{F'_c}\right)^2 + \frac{f_{b1}}{F'_{b1}(1 - f_c/F_{cE1})} \leq 1.0$$

$$\left(\frac{143 \text{ psi}}{386 \text{ psi}}\right)^2 + \frac{964 \text{ psi}}{2497 \text{ psi}(1 - 143 \text{ psi}/3344 \text{ psi})} = 0.541 \leq 1.00$$

Check the case of DL + full WL (no SL). The stresses are

$$f_c = \frac{P}{A} = \frac{10{,}000 \text{ lb}}{210 \text{ in}^2} = 48 \text{ psi}$$

$$f_{bx} = \frac{M_x}{S_x} = \frac{\omega \ell^2}{8S_x} = \frac{(200 \text{ plf})(30 \text{ ft})^2(12 \text{ in./ft})}{(8)(840 \text{ in}^3)} = 321 \text{ psi}$$

$$f_{by} = \frac{M_y}{S_y} = \frac{PL}{4S_y} = \frac{(2500 \text{ lb})(30 \text{ ft})(12 \text{ in./ft})}{(4)(306 \text{ in}^3)} = 735 \text{ psi}$$

For this case the load duration factor is 1.60.

$$F^*_{bx} = (2400 \text{ psi})(1.60) = 3840 \text{ psi} \qquad F^*_{by} = (1450 \text{ psi})(1.60) = 2320 \text{ psi}$$

$$F^*_c = (1600 \text{ psi})(1.60) = 2560 \text{ psi}$$

$$E'_x = E_x = 1{,}800{,}000 \text{ psi} \qquad E'_y = E_y = 1{,}600{,}000 \text{ psi}$$

Determine F'_{bx} using equation (4.28).

$$F'_{bx} = \text{smaller of} \begin{cases} F^*_{bx}C_V + f_c \leq F^*_{bx} \\ F^*_{bx}C_L \end{cases}$$

$$C_V = 0.853 \text{ (from before).}$$

$$F^*_b C_V + f_c = (3840 \text{ psi})(0.853) + 48 \text{ psi} = 3324 \text{ psi}$$

Determine C_L for bending about the x–x axis.

$$F^*_b = 3840 \text{ psi}$$

$$K_{bE} = 0.610$$

$$R_B = 14.37 \text{ (from before)}$$

$$F_{bE} = \frac{(0.610)(1{,}600{,}000 \text{ psi})}{(14.37)^2} = 4726 \text{ psi (as before)}$$

$$C_L = \frac{1 + F_{bE}/F_{bx}^*}{1.9} - \sqrt{\left(\frac{1 + F_{bE}/F_{bx}^*}{1.9}\right)^2 - \frac{F_{bE}/F_{bx}^*}{0.95}}$$

$$= \frac{1 + 4726 \text{ psi}/3840 \text{ psi}}{1.9}$$

$$- \sqrt{\left(\frac{1 + 4726 \text{ psi}/3840 \text{ psi}}{1.9}\right)^2 - \frac{4726 \text{ psi}/3840 \text{ psi}}{0.95}} = 0.886$$

which is the same as in the first case.

$$F_b^* C_L = (3840 \text{ psi})(0.886) = 3402 \text{ psi}$$

so F'_{bx} = smaller of 3324 psi and 3402 psi = 3324 psi.

Now for F'_{by}. C_L for bending about the y–y axis is 1.00 since the member is wider than it is deep. C_V will not apply but C_{fu} will. $C_{fu} = 1.036$ (from before).

$$F'_{by} = F_{by} C_D C_{fu} = (1450 \text{ psi})(1.60)(1.036) = 2404 \text{ psi}$$

Determine F'_c. First determine C_P.

$$K_e = 1.0 \qquad \ell_e = (1.0)(30 \text{ ft})(12 \text{ in./ft}) = 360 \text{ in.}$$

$$K_{cE} = 0.418 \qquad c = 0.9 \text{ for glulam}$$

Potential buckling about the y–y axis will govern (greater slenderness).

$$\frac{\ell_e}{d} = \frac{360 \text{ in.}}{8.75 \text{ in.}} = 41.1 \leq 50 \qquad \text{good}$$

$$F_{cEy} = \frac{K_{cE} E'_y}{(\ell_e/d)^2} = \frac{(0.418)(1{,}600{,}000 \text{ psi})}{(360 \text{ in.}/8.75 \text{ in.})^2} = 395 \text{ psi}$$

$$F_c^* = 2560 \text{ psi [from before (first case)]}$$

$$C_P = \frac{1 + F_{cE}/F_c^*}{2c} - \sqrt{\left(\frac{1 + F_{cE}/F_c^*}{2c}\right)^2 - \frac{F_{cE}/F_c^*}{c}}$$

$$= \frac{1 + 395 \text{ psi}/2560 \text{ psi}}{(2)(0.9)}$$

$$- \sqrt{\left[\frac{1 + 395 \text{ psi}/2560 \text{ psi}}{(2)(0.9)}\right]^2 - \frac{395 \text{ psi}/2560 \text{ psi}}{0.9}} = 0.152$$

also as in the first case. Therefore, again,

$$F'_c = F_c C_D C_P = (1600 \text{ psi})(1.60)(0.152) = 389 \text{ psi}$$

Check the adequacy using equation (4.22):

$$F_{cE1} = F_{cEx} = 3344 \text{ psi (from before)}$$

$$F_{cE2} = F_{cEy} = 395 \text{ psi (from before)}$$

$$F_{bE} = 4726 \text{ psi (from before)}$$

$$\left(\frac{f_c}{F'_c}\right)^2 + \frac{f_{b1}}{F'_{b1}(1 - f_c/F_{cE1})} + \frac{f_{b2}}{F'_{b2}[1 - (f_c/F_{cE2}) - (f_{b1}/F_{bE})^2]} \leq 1$$

$$\left(\frac{48 \text{ psi}}{389 \text{ psi}}\right)^2 + \frac{327 \text{ psi}}{3324 \text{ psi}(1 - 48 \text{ psi}/3344 \text{ psi})}$$

$$+ \frac{735 \text{ psi}}{2404 \text{ psi } [1 - (48 \text{ psi}/395 \text{ psi}) - (327 \text{ psi}/4726 \text{ psi})^2]}$$

$$= 0.015 + 0.100 + 0.350 = 0.464 \leq 1.00 \qquad \text{good}$$

The section is also adequate under DL + WL.

Answer: The 24F-1.8E western species member is adequate to carry the stated loads.

Discussion: It is not anticipated that the DL only combination will govern, since DL + SL has already been investigated and the DL only is 33% of the DL + SL. Since the stresses f from DL + SL = TL were found to be less than the design values F multiplied by the load duration factor of 1.15, the stresses from DL only, which are one-third of the total load, should also be less than the design values F multiplied by the load duration factor of 0.90 (i.e., since TL/1.15 = 0.87TL $\leq F$, then 0.33TL/0.90 = 0.37TL is also $\leq F$).

4.5.5 Columns with Side Loads and Eccentric Axial Loads

Equation (4.21) covers the most general case of axial loads with eccentricity about both axes as well as side load bending about both axes. Equation (4.24) covers the case of axial loads with eccentricity about both axes but no side loads. There is some question about using the reduced value of F'_b based on lateral buckling in bending when column buckling is about a different axis. Most references use the more conservative approach of modifying F_b by the beam stability factor C_L, regardless of the direction of column buckling. This procedure is used in the examples in this book. When members are designed primarily as compression members, the depth/breadth ratio is usually close to 1, and the effect of using the more conservative approach is small. When the depth/breadth ratio is larger, in the range 3 to 7, as is the case with members designed primarily as beams, the effect of including C_L in the equation is more significant, but the beam buckling and the column buckling are more likely to be about the same axis.

Example 4.15 Eccentric Axial Compression and Side Load Bending

Given: A $6\frac{3}{4}$ in. × 10.5 in. western species glued laminated timber column is planned for the loading and conditions illustrated in Figure 4.16. The column is braced in the x and y directions at both the top and bottom. Use Douglas fir and assume wet conditions of use. P = 17,000 lb (including 7000 lb of dead load and 10,000 lb of snow load), e_1 = 2 in., and ω = 200 lb/ft (wind load only).

Determine: The adequacy of DF column combination 2 from AITC 117 with respect to DL + 0.75(SL + WL).

Approach: Design values from AITC 117 will be used with equation (4.21) (with $f_{b2} = 0$ and $e_2 = 0$).

Solution: From AITC 117, Table 8.2 [1], assuming four or more laminations and no special tension laminations,

$$F_c = 1950 \text{ psi}$$

$$F_{bx} = 1700 \text{ psi}$$

$$E = 1{,}600{,}000 \text{ psi (all loadings)}$$

The wet service factor for glulam from Table 3.5 is 0.800 for F_b, 0.73 for F_c, and 0.833 for E. From Section A.1.2, the section properties are A = 70.88 in^2 and S_x = 124.0 in^3.

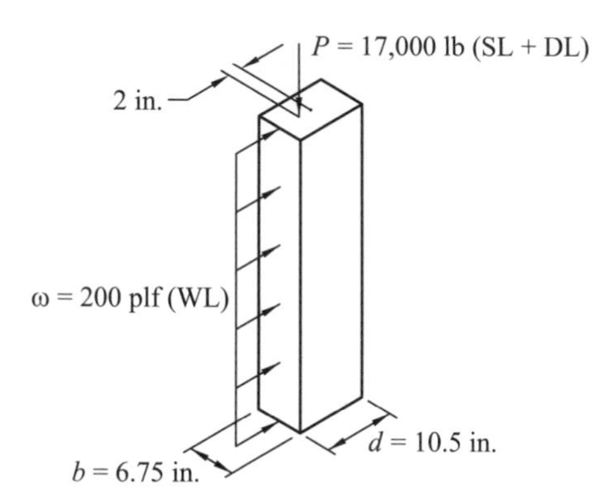

Figure 4.16 Eccentric axial compression and side load bending.

Check DL + 0.75(SL + WL). The load duration factor will be 1.60.

$$F_b^* = F_b C_D C_M = (1700 \text{ psi})(1.60)(0.80) = 2176 \text{ psi}$$

$$F_c^* = F_c C_D C_M = (1950 \text{ psi})(1.60)(0.73) = 2278 \text{ psi}$$

$$E' = EC_M = (1{,}600{,}000 \text{ psi})(0.833) = 1{,}333{,}000 \text{ psi}$$

Determine the stresses.

$$M = \frac{\omega\ell^2}{8} = \frac{(0.75)(200 \text{ lb/ft})(15 \text{ ft})^2}{8} = 4219 \text{ lb-ft} = 50{,}625 \text{ lb-in.}$$

$$f_b = \frac{M}{S} = \frac{50{,}625 \text{ lb-in.}}{124 \text{ in}^3} = 408 \text{ psi}$$

$$f_c = \frac{P}{A} = \frac{[7000 + (0.75)(10{,}000)] \text{ lb}}{70.88 \text{ in}^2} = 205 \text{ psi}$$

Determine F_b'. The column is unbraced along its length with regard to bending; thus, $\ell_u = \ell = 15$ ft. $\ell_u/d_1 = 17$; thus, $\ell_e = 1.63\ell_u + 3d = (1.63)(15 \text{ ft})(12 \text{ in./ft}) + 3(10.5 \text{ in.}) = 325$ in., so

$$R_B = \sqrt{\frac{\ell_e d}{b^2}} = \sqrt{\frac{(325 \text{ in.})(10.5 \text{ in.})}{(6.75 \text{ in.})^2}} = 8.65 \leq 50 \qquad \text{good}$$

$$K_{bE} = 0.610 \text{ (glulam)}$$

$$F_{bE} = \frac{(0.610)(1{,}333{,}000 \text{ psi})}{(8.65)^2} = 10{,}867 \text{ psi}$$

$$F_b^* = 1809 \text{ psi (from above)}$$

$$C_L = \frac{1 + F_{bE}/F_b^*}{1.9} - \sqrt{\left(\frac{1 + F_{bE}/F_b^*}{1.9}\right)^2 - \frac{F_{bE}/F_b^*}{0.95}}$$

$$= \frac{1 + 10{,}867 \text{ psi}/2176 \text{ psi}}{1.9}$$

$$- \sqrt{\left(\frac{1 + 10{,}867 \text{ psi}/2176 \text{ psi}}{1.9}\right)^2 - \frac{10{,}867 \text{ psi}/2176 \text{ psi}}{0.95}} = 0.988$$

$$F_b^* C_L = (2176 \text{ psi})(0.988) = 2150 \text{ psi}$$

Check the volume effect, C_V.

$$C_V = \left(\frac{5.125 \text{ in.}}{b}\right)^{1/10} \left(\frac{12 \text{ in.}}{d}\right)^{1/10} \left(\frac{21 \text{ ft}}{L}\right)^{1/10}$$

$$= \left(\frac{5.125 \text{ in.}}{6.75 \text{ in.}}\right)^{1/10} \left(\frac{12 \text{ in.}}{10.5 \text{ in.}}\right)^{1/10} \left(\frac{21 \text{ ft}}{15 \text{ ft}}\right)^{1/10} = 1.02 \leq 1.00$$

$$= 1.00$$

$$F_b^* C_V + f_c = (2176 \text{ psi})(1.00) + 205 \text{ psi} = 2381 \text{ psi} \leq F_b^* = 2176 \text{ psi}$$

$$F_b' = \text{smaller of} \begin{cases} F_b^* C_V + f_c \leq F_b^* \\ F_b^* C_L \end{cases}$$

which gives $F_b' = 2150$ psi.

Determine F_c'. Assuming pinned ends $K_e = 1.0$.

$$\ell_e = K_e \ell = (1.0)(15 \text{ ft})(12 \text{ in./ft}) = 180 \text{ in.}$$

$$\frac{\ell_e}{d} = \frac{180 \text{ in.}}{6.75 \text{ in.}} = 26.7 \quad \text{or} \quad \frac{180 \text{ in.}}{10.5 \text{ in.}} = 17.1 \qquad 26.7 \text{ governs}$$

$$K_{cE} = 0.418$$

$$F_{cE} = \frac{(0.418)(1{,}333{,}000 \text{ psi})}{(26.7)^2} = 783 \text{ psi}$$

$$c = 0.9 \text{ (glulam)}$$

$$F_c^* = 2278 \text{ psi (from above)}$$

$$C_P = \frac{1 + F_{cE}/F_c^*}{2c} - \sqrt{\left(\frac{1 + F_{cE}/F_c^*}{2c}\right)^2 - \frac{F_{cE}/F_c^*}{c}}$$

$$= \frac{1 + 783 \text{ psi}/2278 \text{ psi}}{1.8}$$

$$- \sqrt{\left(\frac{1 + 783 \text{ psi}/2278 \text{ psi}}{1.8}\right)^2 - \frac{783 \text{ psi}/2278 \text{ psi}}{0.9}} = 0.328$$

$$F_c' = (2278 \text{ psi})(0.328) = 747 \text{ psi}$$

Check the adequacy using equation (4.19), not showing the last term, as it vanishes with $f_{b2} = 0$ and $e_2 = 0$. $e_1 = 2$ in.

$$F_{cE1} = \frac{K_{cE}E'}{(\ell_e/d)^2} = \frac{(0.418)(1{,}330{,}000 \text{ psi})}{(17.1)^2} = 1906 \text{ psi}$$

$$\left(\frac{f_c}{F'_c}\right)^2 + \frac{f_{b1} + f_c(6e_1/d_1)[1 + 0.234(f_c/F_{cE1})]}{F'_{b1}(1 - f_c/F_{cE1})} \leq 1.0$$

$$\left(\frac{205 \text{ psi}}{747 \text{ psi}}\right)^2 + \frac{408 \text{ psi} + (205 \text{ psi})\left[\frac{(6)(2 \text{ in.})}{(10.5 \text{ in.})}\right]\left[1 + (0.234)\left(\frac{205 \text{ psi}}{1906 \text{ psi}}\right)\right]}{(2150 \text{ psi})\left[1 - \left(\frac{205 \text{ psi}}{1906 \text{ psi}}\right)\right]}$$

$$= 0.414$$

$$0.414 \leq 1.00 \qquad \text{good}$$

Answer: The section is adequate for the combination DL + 0.75(SL + WL).

Discussion: The member should also be checked for adequacy for the load combinations DL + SL and DL + WL. Since the equation above produced a value significantly less than 1.00, a smaller section could be investigated.

4.5.6 End Eccentricity and End Fixity

In actual practice, the loads transmitted to columns from beams may be eccentric, especially when a column supports the end of a beam. The eccentricity of axial loads should be determined from the bearing or other loading conditions and accounted for with the preceding equations. Where the eccentricity is unknown or uncontrolled, a minimum eccentricity of one-sixth the column dimension parallel to the supported member is recommended. Framing should be detailed to produce concentric loading of columns as much as possible.

Column design calculations are customarily based on pinned end conditions. In practice, most columns are square cut and bear on significant areas of connection hardware or other members, or may be attached with significant bolt groups, tending to produce some end fixity. Although such bearing conditions may theoretically increase the capacity of the columns due to the increased fixity at the ends (and corresponding smaller effective lengths), the increased fixity is generally not considered in design calculations.

4.5.7 Columns with Side Brackets

Exact solutions of columns loaded by side brackets are complicated and laborious. For these reasons, special simplified design procedures that are safe are recommended when eccentric load is applied through a bracket within the upper quarter of the column length, as shown in Figure 4.17. In the simplified procedure the bracket load P applied at distance a from the center of the column is replaced by the same load P centrally applied at the top of the columns plus a side load P_s applied at midheight, $\ell/2$ (Figure 4.17).

The load P_s is determined by the empirical equation

$$P_s = \frac{3Pa\ell_P}{\ell^2} \tag{4.29}$$

where

P = assumed horizontal side load placed midheight of the column,
P = actual load on the bracket,
a = horizontal distance from the bracket load to the center of the column,
ℓ_P = distance from point of application of load on the bracket to the farthest end of the column, and
ℓ = length of the column.

The assumed axially applied load P should be added to other concentric column loads, and the calculated horizontal side load should be used to determine the flexural stress. The solution is obtained by satisfying the appro-

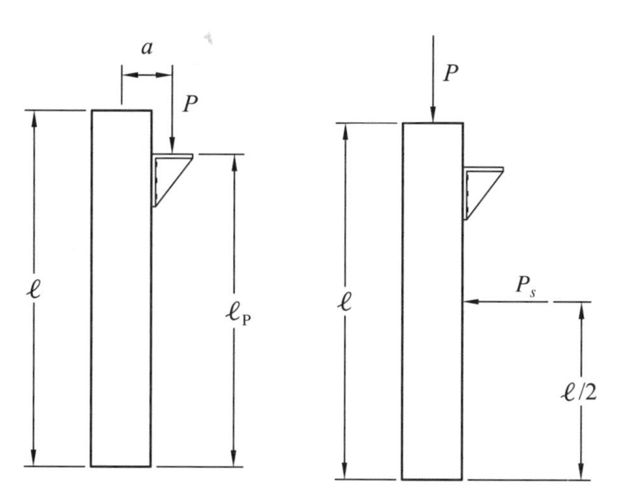

Figure 4.17 Column with load applied at side bracket.

priate stress interaction equation [(4.22) or (4.23)]. The method described above assumes that the column is of such a size that the combination of loads is critical. The empirical method is to be used only in checking the combined axial and bending stress; it should never be used to determine shear or horizontal reaction.

4.6 TAPERED STRAIGHT BEAMS

Glued laminated beams are often tapered to meet architectural requirements, to provide pitched roofs, to facilitate drainage, and to lower wall height requirements at the end supports. Figure 4.18 illustrates the most common forms of these beams. Section 4.2.8 covered tapers at the ends of beams. Generally, tapers at the ends are field-cut into prismatic beams, resulting in the modified design stresses of Table 4.4 for the tapered portions. In this section, straight beams with tapers along their entire lengths are discussed. These tapers may be provided by field-cutting straight prismatic members, in which case the stresses of Table 4.4 are to be used, or the tapers may be manufactured into the beams, with no required reduction in design values. In cases where the tapers are manufactured into the beam, higher-grade laminations are maintained in the zones of high bending stress, allowing the use of the design values published for the particular class or grade of beam in question. In all cases it is recommended that taper cuts be made only on the compression face and that the laminations be parallel to the tension face of the members.

Camber may be provided in beams that are taper cut on the compression face by providing camber on the tension face similar to that in nontapered

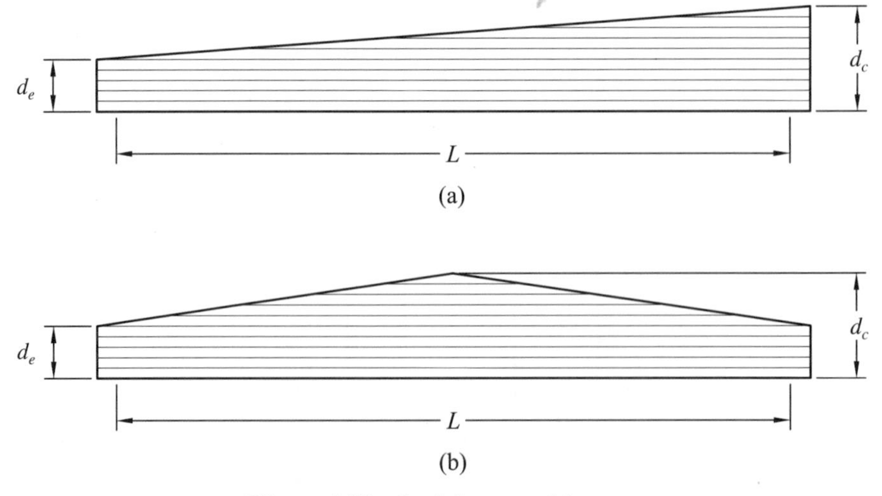

Figure 4.18 Straight tapered beams.

beams and by sawing camber into the compression face. For single-tapered straight beams, the minimum camber based on dead load deflection may be built into the tension face as well as sawn into the tapered face. For double-tapered straight beams, the minimum camber is usually built into the tension face, and if this camber is over 2 in., one-fourth the centerline camber is sawn into the tapered face. The designer should specify the required camber for both the tension face and the sawn compression face. Camber is usually not provided by the fabricator for custom glued laminated timber unless specified by the designer.

The design methods for tapered beams are based on the procedures contained in *Deflection and Stresses of Tapered Wood Beams* [9]. The procedures presented are based on the Bernoulli–Euler theory of bending and beams of isotropic material. This results in an approximate solution for wood beams, which is considered satisfactory for design purposes. The design procedure presented herein is for beams with constant width and varying depth. The example provided in this section assumes that the taper is manufactured into the beam, thus maintaining the published design values for the grade or stress class being considered.

When the top and bottom of a beam are not parallel to each other, consideration must be given to the combined effects of bending, compression, tension, and shear parallel to the grain, and also to compression or tension perpendicular to the grain. An illustration of the distribution of the shear stresses existing in a tapered beam is shown in Figure 4.19. In a straight prismatic beam, the locations of maximum shear and bending stress do not occur at the same point. However, as indicated in Figure 4.19, with a tapered section, the locations of maximum shear and bending stresses at a given section may coincide. At such locations, the following interaction equation is used:

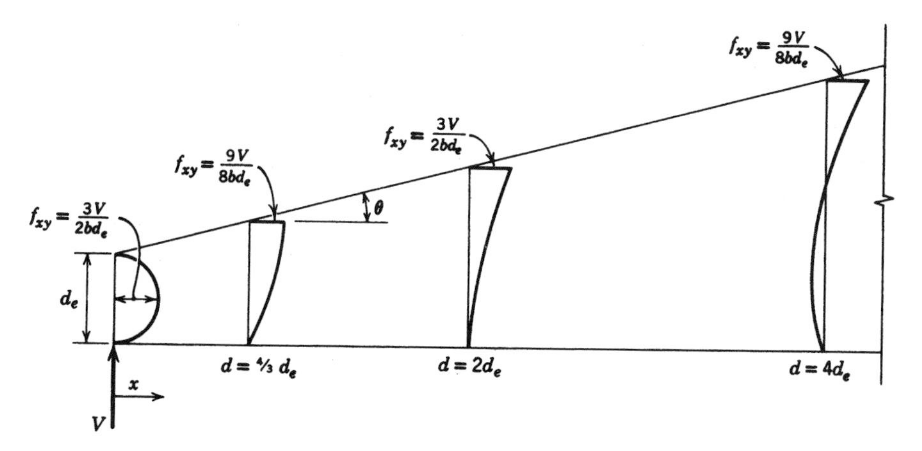

Figure 4.19 Shear stresses in a straight tapered beam.

$$\frac{f_x^2}{F_x^2} + \frac{f_y^2}{F_y^2} + \frac{f_{xy}^2}{F_{xy}^2} = \left(\frac{f_x}{F_x}\right)^2 + \left(\frac{f_y}{F_y}\right)^2 + \left(\frac{f_{xy}}{F_{xy}}\right)^2 \leq 1.00 \qquad (4.30)$$

where

f_x = actual bending stress,
f_y = actual compression or tension stress perpendicular to the grain (compression if the taper cut is on the compression face and tension if the taper cut is on the tension face);
f_{xy} = actual shear stress,
F_x = design value in bending, F_b, multiplied by all applicable adjustment factors except C_V, C_I, and C_L;
F_y = design value perpendicular to the grain ($F_{c\perp}$ where the taper cut is on the compression face or F_{rt} where the taper is on the tension face), multiplied by all applicable adjustment factors;
F_{xy} = design value in shear parallel to the grain (F_v for nonprismatic members), multiplied by all applicable adjustment factors.

Tapering is not recommended on the tension faces of beams, because it causes tension stresses to develop perpendicular to the grain. Where tapering on the tension face is unavoidable, the design value for radial tension, F_{rt}, is used for F_y, as shown above. For tapered beams, the actual stresses at the tapered edge can be expressed as a function of the bending stress, f_x, as follows:

$$f_{xy} = f_x \tan\theta$$

$$f_y = f_x \tan^2\theta = f_{xy} \tan\theta$$

where

$f_x = f_b = M/S$, the extreme fiber bending stress, and
θ = slope of the tapered face.

For a taper on the compression face,

$$\frac{f_x^2}{(F_b^*)^2} + \frac{f_x^2}{F_v^2}\tan^2\theta + \frac{f_x^2}{F_{c\perp}^2}\tan^4\theta \leq 1.00$$

Solving this equation for f_x yields

$$f_x \leq F_b^* C_I$$

where

C_I = the stress interaction factor, first presented in Section 4.2.8:

$$C_I = \left[\frac{1}{1 + (F_b \tan \theta / F_v)^2 + (F_b \tan^2 \theta / F_c)^2} \right]^{1/2} \tag{4.10}$$

To account for stress interactions in design, the design value in bending, F_b, is multiplied by C_I:

$$F'_b = F^*_b C_I$$

where F^*_b = the tabular design value in bending (psi) multiplied by all applicable adjustment factors except C_L and C_V.

For a taper on the compression side of a beam, C_I is cumulative with adjustments for lateral stability C_L but is not cumulative with C_V. Although taper on the tension side of beams is not recommended, C_I in such cases is cumulative with C_V but not with C_L. Where the taper is on the tension side, the design value obtained by the use of C_I and other applicable adjustment factors is then compared to the actual calculated bending stress, f_b, to determine if the section is adequate.

Unless otherwise specified, glued laminated timber beams manufactured with taper will have the same grades on the tapered side as required for the compression side of the combination being used. The tabular design values required for calculating $C_I(F_{bx}$, F_{vx}, and $F_{c\perp}$ *or* $F_{rt})$ can be obtained from AITC 117 [1] or the *National Design Specification*® [2]. As already discussed, tapered beams can also be manufactured by cutting through the compression zone as for a straight beam with end cuts. If this is done, the tabular design values in Table 4.4 for F_{bx}, F_{vx}, and $F_{c\perp}$ should be used to calculate C_I. Other design values remain unchanged. There must be a clear understanding between the designer and the manufacturer regarding how the taper is to be made.

In the calculation of stresses and the interaction of stresses, it is necessary to know where the maximum stress conditions will occur. The shear stress, f_{xy}, and the stress, f_y, perpendicular to the grain are functions of the bending stress, f_x. Determination of the location of the highest bending stress also gives the location of the highest values of f_{xy} and f_y. The location of the point of maximum stress can be obtained by simple calculation or by inspection in some members. In other members, it may be necessary to calculate the maximum bending stress at several cross sections to determine the location of maximum stress.

The reduced design values of Table 4.4 are applicable for tapers cut in beams, with the depth of taper at the end up to two-thirds of the original beam depth. The reduced compression value perpendicular to the grain is applicable to bearing on the cut face. For tapers cut in the compression face and bearing on the tension face, the full design values for compression perpendicular to the grain from AITC 117 [1] or the *National Design Specification®* [2] are applicable.

The design values of Table 4.4 are intended for beams with relatively short taper cuts at the ends. Where taper cuts extend for significant portions, or all, of the beam lengths, the design values of at any section of interest may be obtained by straight-line interpolation between the values at the end (from Table 4.4) and the full design values associated with no taper cut. Where the design values are so determined, the stress interaction factor should also be calculated based on the interpolated design values.

4.6.1 Uniformly Loaded, Single- or Symmetrical Double-Tapered Beams

In addition to the considerations of taper geometry and field cutting of tapers on design values, tapered beams differ from prismatic beams in that, under uniform load, the point of maximum stress is not at midspan. The depth d of tapered beams where maximum bending stress occurs for uniformly loaded single-tapered straight beams or symmetrical double-tapered beams can be determined by

$$d = 2d_e \frac{d_e + \ell \tan \theta}{2d_e + \ell \tan \theta} \tag{4.31}$$

where

d_e = the depth at the end of the beam (smaller end for single-tapered beams)

ℓ = span.

For both single- and symmetric double-tapered beams, the distance x from the small end to the point of maximum stress is given by

$$x = \frac{\ell d_e}{2d_e + \ell \tan \theta} \tag{4.32}$$

Equations (4.31) and (4.32) can be converted to the following forms for easier computation. For single-tapered straight beams under uniform load,

$$d = 2d_e \frac{d_c}{d_e + d_c} \tag{4.31a}$$

$$x = \frac{\ell d_e}{d_e + d_c} \tag{4.32a}$$

where d_c is the depth at the deeper end. For symmetrical double-tapered straight beams under uniform load,

$$d = \frac{d_e}{d_c}(2d_c - d_e) \tag{4.31b}$$

$$x = \frac{\ell d_e}{2d_c} \tag{4.32b}$$

where d_c is the depth at the center (midspan).

The bending stress, f_x, at the point of maximum stress for both types of tapered beams can be determined by

$$f_x = \frac{3\omega\ell^2}{4bd_e(d_e + \ell \tan \theta)} \tag{4.33}$$

where ω = total uniform load. For single-tapered straight beams, equation (4.33) reduces to

$$f_x = \frac{3\omega\ell^2}{4bd_e d_c} \tag{4.33a}$$

For symmetrical double-tapered straight beams under uniform load, equation (4.33) becomes

$$f_x = \frac{3\omega\ell^2}{4bd_e(2d_c - d_e)} \tag{4.33b}$$

4.6.2 Single- or Symmetrical Double-Tapered Beams under Concentrated Load

For cases in which the concentrated load is applied at a section with depth greater than twice the end depth, the location in which the bending stress is maximum will occur at the location in which the depth is twice the end depth (at the location $x = d_e/\tan \theta$). For cases in which the concentrated load is applied at a section where the depth is less than or equal to twice the end depth, the location of maximum bending stress occurs at the same section as the concentrated load.

The shear stress for tapered beams must always be checked at the ends of the beam using equation (4.6). The shear stress at other locations is taken into account by use of the stress interaction factor discussed previously.

4.6.3 Single- or Double-Tapered Members with Other Loads

When a tapered member supports loads or combinations of loads other than a uniform load or a single concentrated load or if a nonsymmetrical double-tapered straight member is used, the simplified equations for determining the depth at which the maximum stress occurs do not apply. In these cases it is generally necessary to determine the bending stress at various sections based on the depth at those sections and check it against the bending stress design value adjusted by all applicable factors, including the stress interaction factor. The shear stress at the end must also be checked using equation (4.6).

4.6.4 Deflection of Tapered Beams

The shear deflection in tapered members is larger than in prismatic members, and the resulting total deflection, including both bending deflection and shear deflection, is slightly larger than that obtained by the customary methods of calculating deflection in prismatic members. Acceptable accuracy in determining the deflection of tapered members can be obtained by determining the deflection of an equivalent prismatic member. The depth of an equivalent member of constant cross section of the same width that will have the same deflection as a tapered beam can be determined by the equation

$$d = C_{dt} d_e \tag{4.34}$$

where

C_{dt} = empirical constant derived from the relationship of equations for deflection of tapered beams and straight prismatic beams.

For a uniformly loaded symmetrical double-tapered beam,

$$C_{dt} = \begin{cases} 1 + 0.66\, C_y & \text{where } 0 < C_y \leq 1 \\ 1 + 0.62 C_y & \text{where } 1 < C_y \leq 3 \end{cases}$$

where

$$C_y = \frac{d_c - d_e}{d_e}$$

For a uniformly loaded single-tapered beam,

$$C_{dt} = \begin{cases} 1 + 0.46\, C_y & \text{where } 0 < C_y \leq 1.1 \\ 1 + 0.43\, C_y & \text{where } 1.1 < C_y \leq 2 \end{cases}$$

Theoretically, both the shear deflection, Δ_s, and the bending deflection, Δ_b, should be determined. The total deflection Δ is the summation of the two ($\Delta = \Delta_s + \Delta_b$). The design values for modulus of elasticity effectively contain a small amount of shear deflection because they are based on the deflection of actual members with a span/depth (ℓ/d) ratio of 21. In most cases, sufficient accuracy in deflection can be obtained by using only the modulus of elasticity in deflection calculations for bending deflection, Δ_b, and ignoring the shear deflection, Δ_s, because it has been largely compensated for by the lower design values for E.

If it is desired to calculate both Δ_b and Δ_s, the modulus of elasticity, E, in the equation for bending deflection Δ_b should be 5% larger than the design value of E. The shear deflection for uniformly load beams can be approximated by

$$\Delta_s = \frac{3\omega\ell^2}{20Gbd_e} \qquad (4.35)$$

where

G = shear modulus (approximately equal to $E/16$).

Example 4.16 Symmetrical Double-Tapered Beam

Given: The double-tapered straight glued laminated timber roof beams in Figure 4.20 to meet the following requirements:

Design conditions:
- Length = 60 ft
- Spacing (tributary width) = 16 ft
- Roof slope: 1:12
- SL = 30 psf
- DL = 15 psf

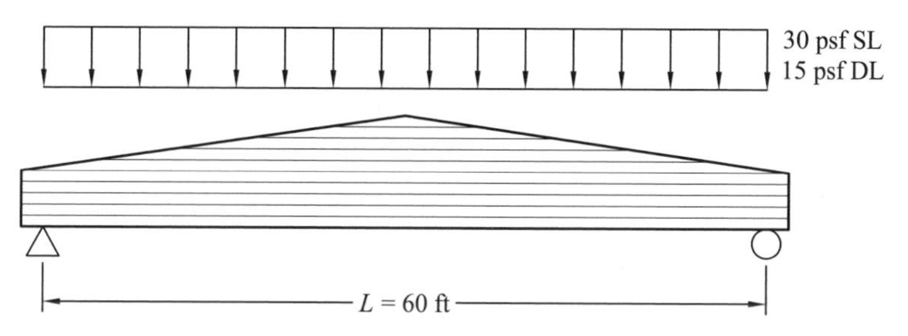

Figure 4.20 Example of double-tapered beam under uniform load.

Tops of beams braced by roof decking
Deflection under total load not to exceed $\ell/180$
Camber to be 1.5 DL deflection

Manufacturing conditions:

Manufactured taper: western species 24F-1.7E

$$F_b = 2400 \text{ psi}$$

$$F_v = (210 \text{ psi})(0.72) = 151 \text{ psi}$$

$$F_{c\perp} = 650 \text{ psi (top and bottom)}$$

$$E_x = 1{,}700{,}000 \text{ psi}$$

Determine: A suitably sized tapered beam.

Approach: The beam will be manufactured with higher-grade material at the tapered surface (taper manufactured into the beam) so that the design values for the 24F-1.7E stress class may be used in design considering the taper.

Solution:

1. *Determine the end depth based on shear.* Assume a western species width of $5\frac{1}{8}$ in. and end depth of 30 in.

$$V = \omega\ell/2$$

$$\omega = (16 \text{ ft})(30 + 15) \text{ psf} = 720 \text{ plf}$$

$$V = (720 \text{ plf})(60 \text{ ft}/2) = 21{,}600 \text{ lb}$$

From Equation (4.6),

$$f_v = \frac{3}{2}\frac{V}{A} = \frac{3}{2}\left[\frac{21{,}600 \text{ lb}}{(5.125 \text{ in.})d_e}\right]$$

and letting $f_v = F'_v$, where $F'_v = F_v C_D = (151 \text{ psi})(1.15) = 174$ psi. The minimum value of d_e becomes

$$\text{min. } d_e = \frac{3}{2}\left[\frac{21{,}600 \text{ lb}}{(5.125 \text{ in.})(174 \text{ psi})}\right] = 36.3 \text{ in.}$$

Use $d_e = 37$ in.

2. *Determine the midspan depth.*

$$d_c = d_e + \frac{\ell}{2}\tan\theta = 37 \text{ in.} + \frac{(60 \text{ ft})(12 \text{ in./ft})}{2}\left(\frac{1}{12}\right) = 67 \text{ in.}$$

3. *Check the deflection.* Calculate the depth d of an equivalent prismatic beam that has the same deflection as the tapered beam:

$$d_{\text{equivalent}} = d_e C_{dt}$$

$$C_y = \frac{d_c - d_e}{d_e} = \frac{67 \text{ in.} - 37 \text{ in.}}{37 \text{ in.}} = 0.81$$

Thus,

$$C_{dt} = 1 + 0.66C_y = 1 + (0.66)(0.81) = 1.54$$

$$d_{\text{equivalent}} = d_e C_{dt} = (37 \text{ in.})(1.54) = 56.8 \text{ in.}$$

$$I_{\text{equivalent}} = \frac{(5.125 \text{ in.})(56.8 \text{ in.})^3}{12} = 78{,}300 \text{ in}^4$$

$$\Delta_{TL} = \frac{5\omega\ell^4}{384\, EI_{\text{equivalent}}} = \frac{(5)(720 \text{ plf})(60 \text{ ft})^4(12 \text{ in./ft})^3}{(384)(1{,}700{,}000 \text{ psi})(78{,}300 \text{ in}^4)} = 1.58 \text{ in.}$$

$$\frac{\ell}{180} = \frac{(60 \text{ ft})(12 \text{ in./ft})}{180} = 4.0 \text{ in.}$$

Since 1.58 in. $\leq$ 4.0 in.; good.

4. *Determine the depth at which maximum stresses occur.*

$$d = \frac{d_e}{d_C}(2d_C - d_e) = \frac{37 \text{ in.}}{67 \text{ in.}}[(2)(67 \text{ in.}) - 37 \text{ in.}] = 53.6 \text{ in.}$$

5. *Check the bending stress.* $C_L = 1.00$. Since beam is braced on top for the full length,

$$F_b^* = F_b C_D = (2400 \text{ psi})(1.15) = 2760 \text{ psi}$$

Therefore, either C_V or C_I calculated at the critical section will control:

$$C_V = \left(\frac{5.125 \text{ in.}}{5.125 \text{ in.}}\right)^{1/10} \left(\frac{12 \text{ in.}}{53.6 \text{ in.}}\right)^{1/10} \left(\frac{21 \text{ ft}}{60 \text{ ft}}\right)^{1/10} = 0.780$$

Calculate C_I using equation (4.10):

$$C_I = \left[\frac{1}{1 + (F_b \tan\theta/F_v)^2 + (F_b \tan^2\theta/F_{c\perp})^2}\right]^{1/2}$$

$$= \left\{\frac{1}{1 + [2400 \text{ psi}(1/12)/151 \text{ psi}]^2 + [2400 \text{ psi}(1/12)^2/650 \text{ psi}]^2}\right\}^{1/2}$$

$$= 0.602$$

Therefore, C_I controls.

$$F'_b = F_b C_I = (2760 \text{ psi})(0.602) = 1662 \text{ psi}$$

$$f_x = \frac{2\omega\ell^2}{4bd_e(2d_C - d_e)} = \frac{(3)(720 \text{ plf})(60 \text{ ft})^2(12 \text{ in./ft})}{(4)(5.125 \text{ in.})(37 \text{ in.})(134 \text{ in.} - 37 \text{ in.})}$$

$$= 1268 \text{ psi}$$

Since $f_x = f_b = 1268$ psi ≤ 1662 psi; good.

6. *Check the shear stress.* The shear stress at the end is satisfactory because of step 1. The shear stress at other points is also satisfactory through the stress interaction factor.

7. *Check the compression perpendicular to the grain.* The compression perpendicular to the grain, f_y, at various sections is satisfactory through the stress interaction factor. Compression perpendicular to the grain at the bearing condition would be checked as a part of the connection design. The trial size is satisfactory.

8. *Determine the camber.*

$$\text{Dead load deflection} = \Delta_{DL} = \Delta_{TL}\frac{W_{DL}}{W_{TL}}$$

$$\Delta_{DL} = 1.58 \text{ in.}\left(\frac{15 \text{ psf}}{30 \text{ psf} + 15 \text{ psf}}\right) = 0.53 \text{ in.}$$

$$\text{Midspan camber} = 1.5\Delta_{DL} = (1.5)(0.53 \text{ in.}) = 0.79 \text{ in.; use } \tfrac{7}{8} \text{ in.}$$

No compression face camber is required because the camber is less than 2 in.

Answer: The following tapered beam is satisfactory:

24F-1.7E with 1:12 manufactured taper

$$b = 5\tfrac{1}{8} \text{ in.}$$

$$d_e = 37 \text{ in.}$$

$$d_c = 67 \text{ in.}$$

camber = $\frac{7}{8}$ in. at midspan (camber tension face)

Discussion: Some manufacturers may supply members with lower design values at lower cost. In such cases, the designer may wish to investigate the suitability of the section cited above at lower design values. The design check above assumed adequate attachment of roof decking to provide lateral support of the compression face ($C_L = 1.0$). Such attachment should be verified by the designer and specified in the plans.

4.7 CURVED MEMBERS

Horizontally glued laminated timbers are unique in that they can be manufactured in curved, pitched, and tapered forms to accommodate architectural and aesthetic purposes. Common curved, pitched, and tapered shape beams are shown in Figure 4.21. Laminations of straight wood can be bent into curved shapes and the interlamination adhesive cured to produce a permanently curved shape. Tapers can be manufactured into the members as discussed previously. The design procedures for these members tend to be more complex than for straight beams.

4.7.1 Minimum Radii of Curvature

Due to the stresses induced into the laminations during the process of manufacturing curved members, minimum radii of curvature are recommended

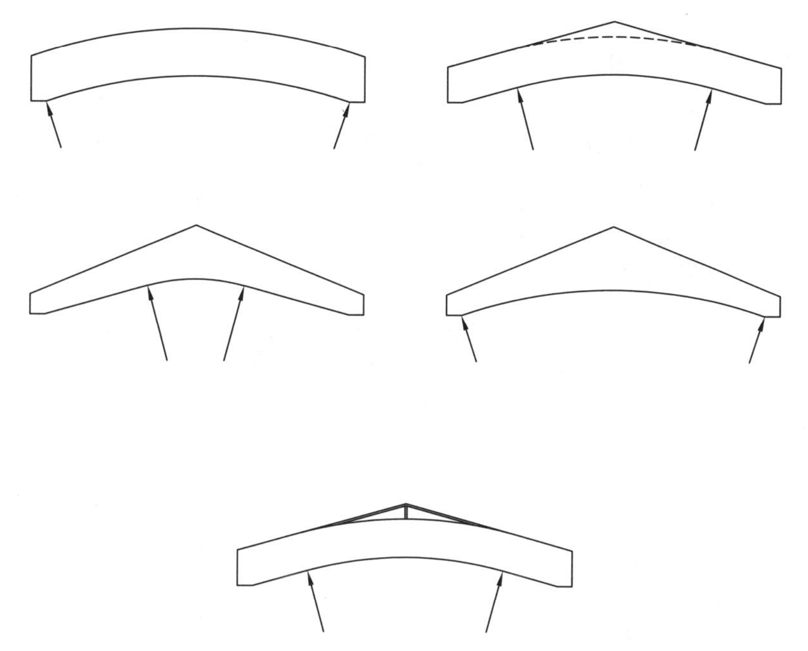

Figure 4.21 Examples of curved beams.

for curved structural glued laminated timbers. The minimum radii for $\frac{3}{4}$-in. laminations are 9.33 ft for western species and 7.0 ft for southern pine. For a lamination thickness of 1.5 in., the recommended minimum radius is 27.5 ft for all species. If required for architectural or other design considerations, other radius-thickness combinations may be used provided that the lamination thickness/radius ratio, t/R, does not exceed $\frac{1}{100}$ for hardwoods and southern pine, or $\frac{1}{125}$ for other softwoods.

4.7.2 Curvature Factor

The prestress induced in the manufacturing of curved timbers effectively reduces the flexural capacity of the member. The reduced capacity is accounted for by multiplying the bending design value with the curvature factor, C_c:

$$C_c = 1 - 2000\left(\frac{t}{R}\right)^2 \tag{4.36}$$

where

t = lamination thickness and
R = radius of curvature (bending radius) of the lamination.

Timbers may be manufactured with curved portions and straight portions. In such cases, the curvature factor C_c is applied for the curved portions only.

4.7.3 Radial Tension and Compression

Where curved members are subjected to a bending moment, radial stresses are set up in a direction parallel to the radius of curvature R of the centerline of the member (perpendicular to the grain). If the moment increases the radius of curvature (causes the member to become straighter), the stress is tension; if it decreases the radius (causes the member to become more sharply curved), the stress is compression.

4.7.3.1 Constant Cross-Section Curved Members The equation for computing the radial stress f_r in members of constant cross section is given by

$$f_r = \frac{3M}{2R_m bd} \tag{4.37}$$

where

M = bending moment,
b = width of the rectangular member,

d = the depth of the rectangular member, and
R_m = radius of curvature of the centerline of the member.

Where the loads tend to straighten the member, the radial stress is in *tension,* $f_r = f_{rt}$. Design values for radial tension, F_{rt}, are provided in AITC 117 [1] and are subject to the adjustment factors of Table 3.8. Where the calculated radial tension stress exceeds the allowable radial tension stress (design value multiplied by all applicable adjustment factors), the design should be re-evaluated. The radial tension stress may be reduced by increasing the section size or by changing the geometry of the member to increase the radius of curvature or depth.

For western species glued laminated timber subject to dead, live, snow, and other loads for which the design values for radial tension are relatively low, mechanical reinforcement may be used to resist the radial tension load in the member, as shown in the next section. In no case may the calculated radial tension stress exceed one-third of the allowable shear stress for non-prismatic members. Where radial reinforcement is used, it is to be designed to resist the full radial tension load. Radial reinforcement is not generally used for southern pine members.

Where radial reinforcement is used, it resists shrinkage that may occur between the time of installation of the reinforcement and the time the member reaches equilibrium. If the shrinkage is excessive, this restraint may cause cracking in the member. To minimize this condition, reinforced members for use in dry conditions are required to be manufactured from laminations with a maximum moisture content of 12%.

In cases where the loads tend to increase the curvature of the member, the radial stress is in compression, $f_r = f_{rc}$. This stress should be limited to the adjusted design value in compression perpendicular to the grain, $F_{c\perp y}$, for the stress class or combination published in AITC 117 [1]. Use of the y value accommodates the compression stress acting in the typically weaker inner laminations.

4.7.3.2 Variable-Depth Curved Members When designing a curved bending member of variable cross section such as a pitched and tapered curved beam, the radial stress f_r is computed by

$$f_r \text{ or } f_{rt} = K_r C_r \frac{6M}{bd_c^2} \tag{4.38}$$

where

K_r = radial stress factor obtained from equation (4.39) and Table 4.7;
C_r = reduction factor based on the shape of the member, obtained from equation (4.40) and Table 4.8;
M = bending moment at midspan;

b = width of the cross section;
d_c = depth of the cross section at midspan.

The radial stress factor is given by

$$K_r = A + B\frac{d_c}{R_m} + C\left(\frac{d_c}{R_m}\right)^2 \qquad (4.39)$$

where the constants A, B, and C are obtained from Figure 4.22 or Table 4.7.

Equation (4.38) may be written as

$$f_r \text{ or } f_{rt} = K_r C_r f_o \qquad (4.38a)$$

where

$$f_o = \frac{6M}{bd_c^2} \qquad (4.38b)$$

The reduction factor for shape C_r is obtained from the following equation and Table 4.8:

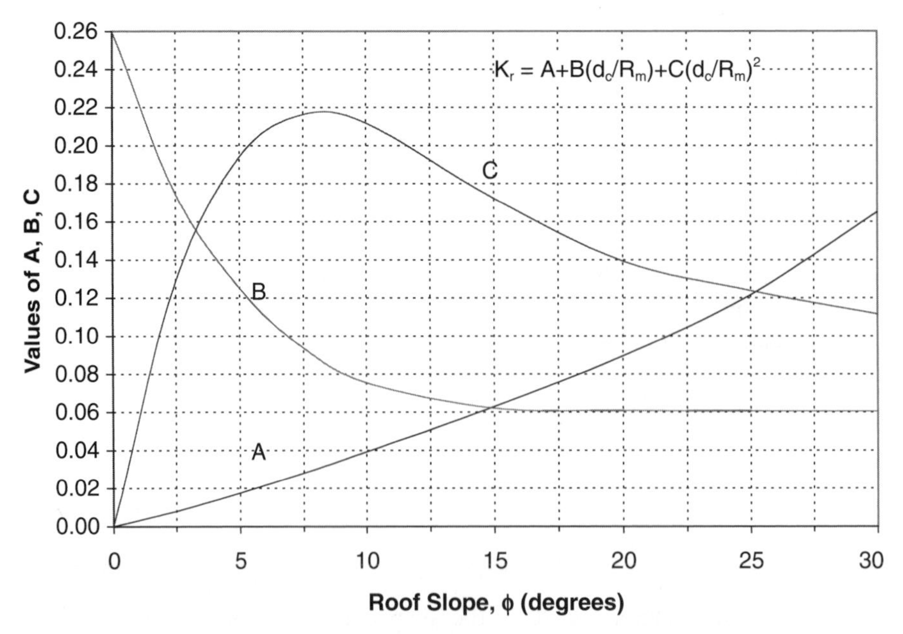

Figure 4.22 Radial tension stress coefficient K_r.

TABLE 4.7 Coefficients for the Determination of K_r[a]

Angle of Upper Tapered Surface, ϕ_T (deg)	Factor		
	A	B	C
2.5	0.0079	0.1747	0.1284
$4.67 = \tan^{-1}(\frac{1}{12})$	0.0165	0.1298	0.1877
5.0	0.0174	0.1251	0.1939
7.5	0.0279	0.0937	0.2162
$9.46 = \tan^{-1}(\frac{2}{12})$	0.0367	0.0793	0.2128
10.0	0.0391	0.0754	0.2119
$14.0 = \tan^{-1}(\frac{3}{12})$	0.0583	0.0645	0.1798
15.0	0.0629	0.0619	0.1722
$18.4 = \tan^{-1}(\frac{4}{12})$	0.0810	0.0611	0.1496
20.0	0.0893	0.0608	0.1393
$22.6 = \tan^{-1}(\frac{5}{12})$	0.1061	0.0606	0.1312
25.0	0.1214	0.0605	0.1238
$26.6 = \tan^{-1}(\frac{6}{12})$	0.1350	0.0604	0.1200
30.0	0.1649	0.0603	0.1115

[a] Straight-line interpolation may be used for intermediate values of ϕ_T.

$$C_r = \alpha + \beta \frac{d_c}{R_m} \tag{4.40}$$

The design values in radial tension, F_{rt}, for pitched and tapered curved beams are the same as the design values given for members of constant cross section. The radial stress factor K_r varies with the ratio of the slope of the bottom face, ϕ_B, to the slope of the top face, ϕ_T. To reduce the design complexity, K_r has been selected for one ϕ_B/ϕ_T ratio ($\phi_B/\phi_T = 1.0$). This results in a slightly conservative design for other ϕ_B/ϕ_T ratios.

In addition, generation of the K_r curves was based on the assumption that the member was subjected to pure bending. It can be shown from equilibrium equations that for a given bending moment, the case of pure bending will result in higher values of radial stresses compared to other more common loading conditions, such as a uniformly distributed load. Since most structural roof members such as pitched and tapered curved beams are seldom designed for the case of pure bending, the C_r factor was introduced to allow the designer to reduce the K_r value to account for uniform loading conditions.

For equal concentrated loads positioned at third points of the span and for single concentrated loads at midspan, the value of C_r as determined for the uniformly distributed load case is to be adjusted by the values in Table 4.9. The adjustment factors for loading conditions other than uniformly distributed loads are approximate values based on a ratio of the area of the moment diagrams in the curved portion of the pitched and tapered curved member for equal maximum moment conditions. For other, more complex conditions, the

TABLE 4.8 Shape Reduction Factor Constants[a]

Roof Slope	ℓ/ℓ_c	α	β
2:12	1	0.44	−0.55
	2	0.68	−0.65
	3	0.82	−0.70
	4	0.89	−0.68
	≥8	1.00	0.00
3:12	1	0.62	−0.85
	2	0.82	−0.87
	3	0.94	−0.83
	4	0.98	−0.63
	≥8	1.00	0.00
4:12	1	0.71	−0.87
	2	0.88	−0.82
	3	0.97	−0.82
	4	1.00	−0.23
	≥8	1.00	0.00
5:12	1	0.79	−0.88
	2	0.95	−0.78
	3	0.98	−0.68
	4	1.00	0.00
	≥8	1.00	0.00
6:12	1	0.85	−0.88
	2	1.00	−0.73
	3	1.00	−0.43
	4	1.00	0.00
	≥8	1.00	0.00

[a] For other ℓ/ℓ_c ratios, coefficients α and β may be obtained through linear interpolation. It is conservative to use the coefficients for the higher ℓ/ℓ_c ratio.

TABLE 4.9 Adjustment to C_r for Curved Beams with Concentrated Loads

ℓ/ℓ_c	Multiply C_r by
Equal concentrated loads at third points	
All ℓ/ℓ_c	1.05
Concentrated load at midspan	
$\ell/\ell_c = 1.0$	0.75
$\ell/\ell_c = 2.0$	0.80
$\ell/\ell_c = 3.0$	0.85
$\ell/\ell_c = 4.0$	0.90

designer may wish to determine these adjustment factors by comparing the area of the moment diagram in the curved portion of the pitched and tapered curved beam with the area for an equivalent uniformly distributed loading condition.

Pitched and tapered curved glued laminated timber beams are among the most popular types of structural roof members where a sloping roof and maximum interior clearance are desired. In these beams, as shown in Figure 4.23, the top edge slopes from the apex at the centerline toward the supports at an angle to the horizontal ϕ_T and the lower edge is curved between the points of tangency (P.T.) and slopes at an angle ϕ_B between the tangent point and the supports. The end portion of the member is usually tapered but may be of constant cross section between the end and the tangent point. The tangent points are usually located near the quarter points, but the location can vary from a point located as close to the center as the minimum radius of curvature permits to a point at the end support. The procedure illustrated applies only to symmetrical pitched and tapered curved beams that are laterally supported continuously along the top edge. The bending and radial stresses in pitched and tapered curved beams are affected by the variable shape of the section, and their exact determination is complex. The procedures presented in this book are based on a simplification of those contained in *Behavior and Design of Double-Tapered Pitched and Curved Glulam Beams* [10]. Because the factors for determining bending and radial tension stresses are a function of the geometric configuration of the member, they cannot be determined accurately until the final size and shape are known. Therefore, the procedure presented in this book is a trial-and-error method whereby a trial size is determined and adjusted until the correct size is obtained. Other methods may be used provided that the design criteria are satisfied.

4.7.4 Radial Reinforcement

The reinforcement used to resist the radial tension stresses in western species glued laminated timbers must be designed on the basis of sound engineering principles. In no case may the calculated radial tension stress exceed one-third the allowable shear stress for nonprismatic members. Where reinforcement is used, it is designed to carry the full radial tension stress (no sharing of radial tension between reinforcement and wood). Any type of reinforcement, such as lag screws mechanically attached to wood or deformed steel bars bonded by adhesive, that will effectively transfer the radial tension stresses between the wood and the reinforcement throughout the entire depth of embedment may be used. The method of bonding or attaching the reinforcement to the wood should be of a durable quality and capable of developing the required tensile strength of the reinforcement. When radial reinforcement is used, it tends to restrain shrinkage due to moisture loss.

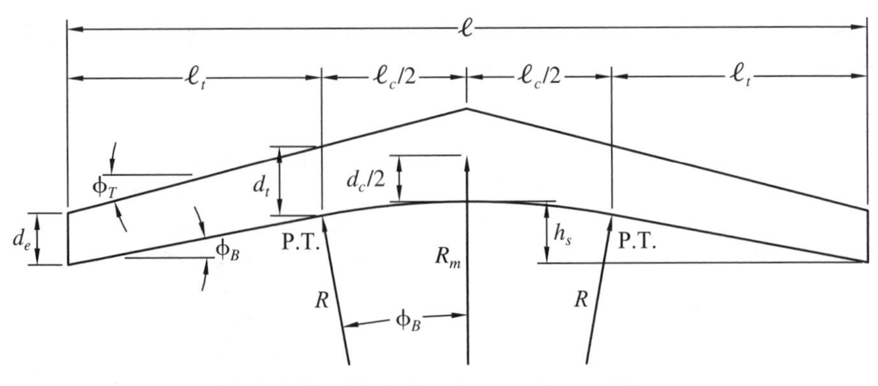

Figure 4.23 Pitched and tapered curved beam.

Because of this, the moisture content of the laminations prior to manufacture must not exceed 12% for dry conditions of use.

A typical design is to utilize fully threaded lag screws as radial tension reinforcement. The following comments provide general installation recommendations and design guidelines for the use of lag screw reinforcement.

1. Lag screws threaded for their full length are to be used for radial reinforcement. These screws should be shop-installed in prebored holes from the top of the member on the width centerline of the member. They should be used only in the curved portion of the member. The screws should be installed normal to the axis of the laminations (90° to the direction of the glue lines). No washer should be used under the head of the screw. It is desirable, for structural reasons, not to countersink the head of the screw into the top of the beam. If it is necessary to install the screw in such a way that the top surface is smooth following installation, the head of the screw can be sawn off flush with the member or it can be countersunk, but the countersunk hole should only be large enough to install the head flush with the top of the beam.
2. The prebored lead hole for the threaded portion of the lag screw should have a diameter not greater than 85% of the shank diameter for species with a specific gravity of 0.50 or greater, and 80% for species with a specific gravity of less than 0.50, and a depth equal to the length of the threaded portion. The lead holes for radial reinforcement are necessarily larger than for lag screws in ordinary application, due to the longer screws necessary for radial reinforcement. Design values for lag screws used as radial reinforcement are provided for Douglas fir–larch in Table 4.10. These withdrawal values in pounds per inch of wood embedment are 85% of the design values for lag screws in withdrawal from the *National Design Specification®* [2]; the reduction is due to the larger-

TABLE 4.10 Lag Screw Reinforcement Data for Douglas Fir–Larch

	Steel		Wood		
			Allowable Withdrawal Load (lb/in.) of Threaded Portion[b]		
Lag Screw Shank Diameter, D (in.)	Net Area at Root of Thread (in^2)	Allowable Tension Load (lb) for a Unit Stress of 20,000[a] (psi)	Normal 1.00[c]	Snow 1.15[c]	7-Day 1.25[c]
$\frac{1}{4}$	0.0235	470	191	220	239
$\frac{5}{16}$	0.0405	810	226	260	283
$\frac{3}{8}$	0.0552	1,105	259	298	324
$\frac{7}{16}$	0.0845	1,690	291	334	363
$\frac{1}{2}$	0.108	2,160	321	369	402
$\frac{9}{16}$	0.149	2,980	349	401	436
$\frac{5}{8}$	0.174	3,485	380	437	475
$\frac{3}{4}$	0.263	5,265	436	501	545
$\frac{7}{8}$	0.366	7,330	490	563	612
1	0.478	9,555	541	622	676
$1\frac{1}{8}$	0.618	12,360	591	679	736
$1\frac{1}{4}$	0.804	16,090	639	735	799

[a] Rounded to nearest 5 lb.
[b] Based on Douglas fir–larch; specific gravity of 0.50 based on weight and volume when oven dry. For other species, see *National Design Specification for Wood Construction* [2]. Values are 85% of the values for lag screws in ordinary withdrawal, due to the larger lead hole used for the longer lag screws necessary for radial reinforcement.
[c] Load duration factors.

diameter lead holes. For other species, lag screw withdrawal values must similarly be multiplied by 85% for use as radial reinforcement.

3. The length of the full thread of the lag screw should extend from the top (head end of the lag screw) of the beam to between 2 and 3 in. from the bottom face (soffit).
4. In general, lag screws from $\frac{5}{8}$ to 1 in. in diameter should be limited to a maximum length of 60 in. and screws less than $\frac{5}{8}$ in. diameter to a maximum length of 30 in. If greater lengths of reinforcement are required, larger-diameter screws should be specified.
5. The fully threaded lag screws should be designed to take the entire radial tension stress developed in the member with no radial tension stress carried by the wood. The screws should be so spaced along the width centerline throughout the curved portion of the member as to carry the entire radial tension force.
6. The magnitude of the radial tension force to be carried by each lag screw should not exceed either of the following:

a. The allowable tension load for the screw based on the allowable tensile stress in lag screws acting on the net area at the root of the thread. Table 4.10 provides the allowable tension load for each size screw (third column) based on an allowable stress in the metal of 20,000 psi.
b. The allowable load on the screw based on withdrawal in pounds per inch (the fourth, fifth, and sixth columns) multiplied by the effective embedment. The effective embedded thread length to be used in this determination is the embedded thread length from the neutral axis of the member to the point end of the lag screw or the thread length from the neutral axis of the member to the head end of the lag screw, whichever is less.

7. The section modulus is reduced somewhat by the hole bored for the radial reinforcement. In many cases, the effect is small, but the reduction should be considered in design. The reduced section modulus is determined by subtracting only that portion of the lag screw hole below the neutral axis, because the lag screw fills the hole completely on the compression side. The section modulus should be reduced by an amount due to the area of the depth of embedment below the neutral axis multiplied by the shank diameter.

4.7.5 Bending Stress in Curved Beams

The bending stress in curved beams of constant cross section are calculated from the flexure formula for extreme fiber stress in bending, equation (4.1). The bending stress, f_b, for curved beams of variable cross section, such as pitched and tapered curved beams, is greater than for straight prismatic members and is increased by the bending stress factor, K_θ, due to the shape of the member:

$$f_b = K_\theta \frac{6M}{bd_c^2} = K_\theta f_o \tag{4.41}$$

where

M = moment at midspan,
b = width,
d_c = depth at midspan,

$$K_\theta = D + E\frac{d_c}{R_m} + F\left(\frac{d_c}{R_m}\right)^2 \tag{4.42}$$

D, E, and F are coefficients from Table 4.11 and the other terms are as defined previously.

TABLE 4.11 Coefficients for Determining Bending Stress Coefficient, K_θ[a]

ϕ_T (deg)	D	E	F
2.5	1.042	4.247	−6.201
4.76 = $\tan^{-1}(\frac{1}{12})$	1.139	2.245	−2.239
5.0	1.149	2.036	−1.825
9.46 = $\tan^{-1}(\frac{2}{12})$	1.131	0.219	0.631
10.0	1.330	0.0	0.927
14.0 = $\tan^{-1}(\frac{3}{12})$	1.659	0.0	0.179
15.0	1.738	0.0	0.0
18.4 = $\tan^{-1}(\frac{4}{12})$	1.891	0.0	0.0
20.0	1.961	0.0	0.0
22.6 = $\tan^{-1}(\frac{5}{12})$	2.309	−1.482	1.854
25.0	2.625	−2.829	3.538
26.6 = $\tan^{-1}(\frac{6}{12})$	2.762	−2.755	3.194
30.0	3.062	−2.594	2.440

[a] Straight-line interpolation may be used for intermediate values of ϕ_T.

4.7.6 Deflection of Curved Beams

The deflection is also affected by the shape of the member. The following equation provides a close approximation of midspan deflection based on test data:

$$\Delta_c = \frac{5\omega\ell^4}{32E'b(d_{\text{equiv}})^3} \tag{4.43}$$

where

Δ_c = deflection at midspan,
ω = uniform distributed load,
ℓ = span,
E' = modulus of elasticity multiplied by appropriate adjustment factors,
b = width,

$$d_{\text{equiv}} = (d_e + d_c)(0.5 + 0.735 \tan \phi_T) - 1.41 d_c \tan \phi_B \tag{4.44}$$

where

ϕ_B = slope of the bottom (soffit), at the ends;
ϕ_T = slope of the top,
d_c = midspan depth,
d_e = depth at the end of the member.

The geometric features in the preceding equations are illustrated in Figure 4.23. Other methods of determining deflection such as by finite elements or virtual work may also be used.

4.7.7 Design Procedure for Pitched and Tapered Curved Beams

The procedure for the design of symmetrically pitched and tapered curved (PTC) glued laminated timber beams under uniform load is described in this section. Other procedures may be used as long as the appropriate design stress and serviceability checks are met and the appropriate adjustments are made for various loading conditions. The procedure that follows is based on the selection of a trial geometry and subsequent design checks. Trial geometry may be determined based on a desired curved portion length or by a minimum curved portion radius. The trial or final geometry must also be checked for aesthetic appeal.

4.7.7.1 Trial Geometry The geometries of pitched and tapered beams are governed by architectural considerations such as top (roof) slope, bottom (soffit) slope, and overall appearance (length and radius of curved portion), as well as the section requirements (depths) to satisfy stress checks and serviceability criteria. The information necessary for determining the trial geometry is summarized below.

1. Overall geometry conditions, including top (roof) slope and span
2. Desired or trial bottom (soffit) slope
3. Desired or trial ratio of curved portion to total length (if known)
4. Applied load, load duration, and serviceability criteria (deflection limitations)
5. Beam class or design values
6. Minimum radius of curved portion (based on species and lamination thickness)
7. Beam width

Once the information above is obtained, the trial beam geometry is determined as follows.

1. Determination of minimum end depth based on desired beam width and horizontal shear
2. Determination of bottom (soffit) geometry based on bottom slope and ratio of curved portion to total length or based on radius of curvature

Once the end depth has been determined by calculation, the bottom geometry of the beam may be determined by calculation with the equations that follow or by any suitable (mechanical or computer-aided graphical) means. Beam width is often governed by support conditions and availability per species. Experience has indicated that successful trial geometries are those for which the end depth based on shear is somewhat conservative, and for which the difference in the angles of the top and bottom slopes of the tapered ends is not excessive. Geometries for which end shear stress is at or near the allowable stress level may be unsatisfactory with regard to the other design checks. Geometries with large differences between the top and bottom slopes tend to be less efficient due to the relatively large reduction in allowable stresses due to the stress interaction factor.

The shear parallel to the grain for the ends of a pitched and tapered curved beam, f_v, may be calculated using

$$f_v = \frac{3}{2}\frac{V}{bd_e} \tag{4.45}$$

where

V = vertical end shear or the vertical end reaction,
b = desired beam width,
d_e = vertical end depth.

Equation (4.45) may be rearranged as follows, where the end shear is set to the allowable shear parallel to the grain, $f_v = F'_v$, and where the design value for shear, F_v, is taken to be the shear value for nonprismatic members provided in AITC-117 [1] or the *NDS®* [2]:

$$d_e(\text{min.}) = \frac{3}{2}\frac{V}{bF'_v} \tag{4.45a}$$

With this information the geometry may be calculated as follows using either the desired ratio of total span to curved portion length, or the soffit radius, as a function of the bottom slope. Where the ratio of total length to curved portion length is used, the corresponding radius may be calculated from

$$R = \frac{\ell}{2\ell/\ell_c \sin \phi_B} \tag{4.46}$$

where

ℓ = span,
ℓ_c = curved portion of the span,
R = bottom (soffit) radius or curved portion of the beam,
ϕ_B = bottom (soffit) slope of the tapered ends.

The depth at midspan, d_c, may be calculated as the difference between the apex height, h_a, and soffit height at midspan, h_s, both measured with respect to the bottom of the ends of the beam:

$$d_c = h_a - h_s \tag{4.47}$$

where

$$h_a = \frac{\ell}{2} \tan \phi_T + d_e \tag{4.48}$$

where ϕ_T the top (roof) slope and

$$h_s = \frac{\ell}{2} \tan \phi_B - R(\sec \phi_B - 1) \tag{4.49}$$

If the soffit radius is given, the length of the curved portion of the beam, ℓ_c, may be calculated using

$$\ell_c = 2R \sin \phi_B \text{ (plan length)} \tag{4.46a}$$

The length of each tapered end, ℓ_t, is given by

$$\ell_t = \frac{\ell - \ell_c}{2} \text{ (plan length)} \tag{4.50}$$

4.7.7.2 Radial Tension Stress Check The radial tension stress is calculated using equation (4.38) (Section 4.7.3), or

$$f_{rt} = K_r C_r f_o \tag{4.38a}$$

where

$$f_o = \frac{6M}{bd_c^2} \tag{4.38b}$$

The radius of the curved portion of the beam at midspan may be calculated using

$$R_m = R + \frac{d_c}{2} \tag{4.51}$$

The radial tension stress is calculated at midspan using equation (4.38) or (4.38a) and (4.38b). The radial tension design check is given by

$$f_{rt} \leq F'_{rt}$$

where F'_{rt} is the design value for radial tension as given in AITC 117 [1] or the *NDS*® [2] adjusted by the appropriate factors of Chapter 3:

$$F'_{rt} = F_{rt}C_D C_M C_t$$

Where the radial tension stress design check is not satisfied, the geometry must be adjusted to reduce the radial tension stress, or, for western species glued laminated timber, radial reinforcement may be used to carry the radial tension stress, except that in no case may the calculated radial tension stress exceed one-third of the allowable shear stress for nonprismatic members. Design of radial reinforcement is discussed in greater detail in Section 4.7.4 and shown by example later in this section. Increasing the radius of curvature or increasing the beam depth are effective ways to reduce the radial tension stress.

4.7.7.3 Bending Stress Check Bending stress in the beam at midspan is calculated using equation (4.41) from Section 4.7.5 with the bending stress factor, K_θ, from equation (4.42) and Table 4.11.

$$f_b = K_\theta \frac{6M}{bd_c^2} \qquad (4.41)$$

The allowable stress in bending, F'_b, is obtained from the appropriate design value in bending, F_b, and the adjustment factors of

$$F'_b(\text{midspan}) = F_b C_D (C_V \text{ or } C_L) C_c C_M C_t \qquad (4.52)$$

The bending stress must also be checked along the tapered portions of the beam, including the tangent point. In the tapered section the stress interaction factor must be calculated and the lesser of it or the beam stability factor or the volume factor applied. At the tangent point, both the stress interaction factor and the curvature factor are applicable. As such, the tangent point may be the critical design point of a pitched and tapered curved beam. For many pitched and tapered beam applications, the roof may be attached to the top of the beam in such a way as to provide lateral bracing, in which case the beam stability factor, C_L, may be taken to be 1.00. The allowable stress equations above are

$$F'_b(\text{tapered section}) = F_b C_D (C_V \text{ or lesser of } C_I \text{ or } C_L) C_M C_t \qquad (4.53)$$

$$F'_b(\text{tangent point}) = F_b C_D (C_V \text{ or lesser of } C_I \text{ or } C_L) C_c C_M C_t \qquad (4.54)$$

Customarily the bending stress is checked at midspan, the tangent point, and at several representative points between the tangent point and the ends. If the bending stress check, $f_b \leq F'_b$, is not satisfied at all locations, the beam geometry must be adjusted to either reduce the bending stress or to increase the allowable stress. Increasing the beam depth at the location of overstress will reduce the bending stress. Decreasing the angle of taper in the tapered leg will increase the allowable bending stress.

The depth of the beam along the straight tapered portion, d'_x, may be determined using

$$d'_x = [d_e + x(\tan \phi_T - \tan \phi_B)][\cos \phi_B - \sin \phi_B \tan(\phi_T - \phi_B)] \quad (4.55)$$

where

d'_x = the depth of the beam at x, measured perpendicular to the soffit
x = the distance from the end toward the tangent point.

4.7.7.4 *Deflection Check* The deflection of a pitched and tapered curved beam under uniform load may be calculated using

$$\Delta_c = \frac{5\omega\ell^4}{32E'b(d_{\text{equiv}})^3} \quad (4.43)$$

where

Δ_c = the midspan deflection (typically, due to either live load or total load),
ω = uniform load,
ℓ = span,
E' = adjusted modulus of elasticity = EC_MC_t,
b = width, and
d_{equiv} = depth of a straight prismatic beam having a deflection equivalent to the pitched and tapered curved beam, given by

$$d_{\text{equiv}} = (d_e + d_c)(0.5 + 0.735 \tan \phi_T) - 1.41d_c \tan \phi_B \quad (4.44)$$

The deflection criteria must be satisfactory to the owner and building official and as appropriate include the time-dependent deflection (creep). Where deflection criteria are not satisfied, the beam geometry must be adjusted (typically, by increased depth).

4.7.7.5 *Horizontal Deflection at Supports* The horizontal movement (push-out) at each support may be estimated using

$$\Delta_H = \frac{2\Delta_c}{\ell} \tag{4.56}$$

where

Δ_H = calculated horizontal movement at each support
h = the rise in middepth of the beam from the end to midspan, which may be calculated using

$$h = h_a - \frac{d_c}{2} - \frac{d_e}{2} \tag{4.57}$$

Δ_c = deflection at midspan.

The designer must detail the connections to the support structure to either accommodate the above push-out at each end or the combined effect at one end only. AITC 104 [11] provides typical connection details that may be of assistance to the designer in this regard. If the calculated horizontal displacement is excessive, a stiffer section may need to be designed. The designer must also ensure that the beam ends have adequate bearing and are also properly anchored against lateral forces parallel and perpendicular to the beam axis as well as any applicable uplift forces.

4.7.7.6 Radial Reinforcement (If Applicable) The application of and design procedures for radial reinforcement are discussed in Section 4.7.4 and are shown by example in the following section.

4.7.8 Design of Pitched and Tapered Curved Beams Made of Douglas Fir with Radial Reinforcement

The design of a pitched and tapered curved beam made of Douglas fir laminations utilizing radial reinforcement is illustrated by the following example. This example utilizes hand calculations for the beam geometry.

Example 4.17 Douglas Fir Pitched and Tapered Curved Beam (with Radial Reinforcement)

Given: The following design conditions for a pitched and tapered curved glued laminated timber roof beam made of Douglas fir.

Stress class 24F-1.7E
ℓ = 60 ft (720 in.)
Spacing: 16 ft

Roof slope: 3:12 (14.0°)
Snow load: 30 psf
Dead load: 15 psf
Deflection limitation: $\ell/240$ for total load
Beam is supported laterally along the top by roof decking
Desired ratio of total length to curved portion length: 3.0

Design values for the 24F-1.7E stress class from AITC 117 [1]:

$$F_b = 2400 \text{ psi}$$

$$F_v = 150 \text{ psi (nonprismatic member)}$$

$$E = 1{,}700{,}000 \text{ psi}$$

$$F_{rt} = 15 \text{ psi}$$

$$F_{c\perp} = 500 \text{ psi}$$

$$C_D = 1.15$$

The minimum soffit radius (1.5-in. laminations) is 27.5 ft (330 in.).

Determine: A suitable Douglas fir pitched and tapered curved beam. Radial reinforcement may be used if necessary.

Approach: Follow the design procedure of the preceding section. The beam will be assumed to have a self-weight of 100 plf in addition to the area loads stated above.

Solution:

1. *Trial geometry.* From the given information of a 14° top slope and a span of 60 ft, a trial bottom slope of 10° will be used.

Applied loads:
$\omega_{SL} = (30 \text{ psf})(16 \text{ ft}) = 480 \text{ plf}$
$\omega_{DL} = (15 \text{ psf})(16 \text{ ft}) = 240 \text{ plf}$
Assumed self-weight: 100 plf
Total load ω: 480 plf + 240 plf + 100 plf = 820 plf
Load duration: two months ($C_D = 1.15$)
Deflection limitation: $\ell/240 = [(60 \text{ ft})(12 \text{ in./ft})]/180 = 3.0$ in.
Beam width: a trial beam width of 6.75 in.

The minimum beam end depth may be calculated from the preceding information using the allowable stress for shear:

$$F'_v = (150 \text{ psi})(1.15) = 172.5 \text{ psi}$$

and the end shear

$$V = \frac{\omega\ell}{2} = \frac{(820 \text{ plf})(60 \text{ ft})}{2} = 24{,}600 \text{ lb}$$

From equation (4.45a),

$$d_e \text{ (min.)} = \frac{3}{2}\frac{V}{bF'_v} = \frac{3}{2}\left[\frac{24{,}600 \text{ lb}}{(6.75 \text{ in.})(172.5 \text{ psi})}\right] = 31.7 \text{ in.}$$

Use $d_e = 32$ in. From equation (4.46),

$$R = \frac{\ell}{2(\ell/\ell_c)\sin\phi_B} = \frac{720 \text{ in.}}{(2)(3)\sin 10°} = 691 \text{ in.}$$

The radius desired (691 in.) is greater than the minimum radius (330 in.); good.

From the information above, a complete trial geometry may be calculated as follows. The apex height is

$$h_a = \frac{\ell}{2}\tan\phi_T + d_e = \frac{720 \text{ in.}}{2}\tan 14° + 32 \text{ in.} = 122 \text{ in.}$$

The soffit height at midspan is

$$h_s = \frac{\ell}{2}\tan\phi_B - R(\sec\phi_B - 1) = \frac{720 \text{ in.}}{2}\tan 10° - (691 \text{ in.})(\sec 10° - 1) = 52.8 \text{ in.}$$

The depth of beam at midspan is

$$d_c = h_a - h_s = 122 \text{ in.} - 52.8 \text{ in.} = 69.2 \text{ in.}$$

The length of the curved portion of the beam in this example is

$$\ell_c = \frac{720 \text{ in.}}{3} = 240 \text{ in.}$$

The length of each tapered end is, from equation (4.50),

$$\ell_t = \frac{\ell - \ell_c}{2} = \frac{720 \text{ in.} - 240 \text{ in.}}{2} = 240 \text{ in.}$$

In this case it is also one-third of the total length.

2. *Radial tension stress check.* Radial tension is checked at midspan as follows:

$$R_m = R + \frac{d_c}{2} = 691 \text{ in.} + \frac{69.2 \text{ in.}}{2} = 725.6 \text{ in.}$$

$$\frac{d_c}{R_m} = \frac{69.2.6 \text{ in.}}{725.6 \text{ in.}} = 0.0954$$

Interpolating values of A, B, and C from Table 4.7, $A = 0.0581$, $B = 0.0646$, and $C = 0.1801$.

$$K_r = A + B\frac{d_c}{R_m} + C\left(\frac{d_c}{R_m}\right)^2$$

$$= 0.0581 + (0.0646)(0.095) + (0.1801)(0.095)^2 = 0.066$$

From Table 4.8, $\alpha = 0.94$ and $\beta = -0.83$.

$$C_r = \alpha + \beta\frac{d_c}{R_m} = 0.94 + (-0.83)(0.095) = 0.86$$

$$M = \frac{\omega\ell^2}{8} = \frac{(820 \text{ lb/ft})(60 \text{ ft})^2}{8}$$

$$= 369{,}000 \text{ lb-ft} = 4{,}428{,}000 \text{ lb-in.}$$

$$f_o = \frac{6M}{bd_c^2} = \frac{(6)(4{,}428{,}00 \text{ lb-in.})}{(6.75 \text{ in.})(69.2 \text{ in.})^2} = 822 \text{ psi}$$

Therefore,

$$f_{rt} = K_r C_r f_o = (0.066)(0.86)(822 \text{ psi}) = 46.7 \text{ psi}$$

$$F'_{rt} = (15 \text{ psi})(1.15) = 17.25 \text{ psi}$$

$$\tfrac{1}{3} F'_v = \tfrac{1}{3}(150 \text{ psi})(1.15) = 57.5 \text{ psi}$$

$$F'_{rt} = 17.25 \text{ psi} < f_{rt} = 46.7 \text{ psi} \leq \tfrac{1}{3} F'_v = 57.5 \text{ psi}$$

Since the radial stress is greater than the allowable stress but does not exceed one-third the allowable shear stress, the trial geometry is acceptable with respect to radial tension, but radial reinforcement will be required.

3. *Bending stress check.* From Table 4.11:

$$D = 1.656$$

$$E = 0.000$$

$$F = 0.185$$

From equation (4.42),

$$\begin{aligned} K_\theta &= D + E\frac{d_c}{R_m} + F\left(\frac{d_c}{R_m}\right)^2 \\ &= 1.656 + (0.0)(0.095) + (0.185)(0.095)^2 \\ &= 1.658 \end{aligned}$$

$$f_b = K_\theta \frac{6M}{bd_c^2} = K_\theta f_o = (1.658)(822 \text{ psi}) = 1363 \text{ psi}$$

For the allowable bending stress at midspan,

$$C_L = 1.00 \text{ (full lateral support at top of beam)}$$

$$C_V = \left(\frac{5.125 \text{ in.}}{6.75 \text{ in.}}\right)^{1/10} \left(\frac{12 \text{ in.}}{69.2 \text{ in.}}\right)^{1/10} \left(\frac{21 \text{ ft}}{60 \text{ ft}}\right)^{1/10} = 0.735$$

$$C_c = 1 - 2000\left(\frac{1.5 \text{ in.}}{691 \text{ in.}}\right)^2 = 0.991$$

Therefore, at midspan,

$$F_b' = (2400 \text{ psi})(1.15)(0.735)(0.991) = 2010 \text{ psi}$$

Since $f_b = 1363 \text{ psi} \leq F_b' = 2010 \text{ psi}$, the trial beam geometry checks at midspan with respect to bending.

The bending stress is also calculated and checked at four sections along the pitched and tapered segments of the beam in Table 4.12 using equations (4.53) and (4.54) as illustrated in Figure 4.24. Along the tapered portion, the depth at any particular section is calculated using equation (4.55). The table shows that the bending stresses check at the selected locations along the pitched and tapered portions of the beam.

4. *Deflection check.* From equation (4.46) the equivalent depth for deflection is calculated to be

$$\begin{aligned} d_{\text{equiv}} &= (32 \text{ in.} + 69.2 \text{ in.})(0.5 + 0.735 \tan 14.04°) \\ &\quad - (1.41)(69.2 \text{ in.}) \tan 10° = 51.99 \text{ in.} \end{aligned}$$

$$\omega = 820 \text{ plf} = (820 \text{ lb/ft})(1 \text{ ft}/12 \text{ in.}) = 68.33 \text{ lb/in.}$$

From equation (4.43) the deflection at midspan is

TABLE 4.12 Bending Stress Check at Four Locations in Pitched and Tapered Portion of Pitched and Tapered Curved Beam

Section	x (in.)	d'_x (in.)	M (lb-in.)	I (in^4)	S (in^3)	f_b (psi)	C_v	C_I	C_c	F'_b (psi)	f_b/F'_b
A	60	35.42	1,353,000	24,996	1411	959	0.786	0.663	1.0	1830	0.524
B	120	39.72	2,460,000	35,249	1775	1386	0.777	0.663	1.0	1830	0.757
C	180	44.02	3,321,000	47,981	2180	1523	0.769	0.663	1.0	1830	0.833
P.T.	240	48.32	3,936,000	63,460	2627	1498	0.762	0.663	0.991	1813	0.826

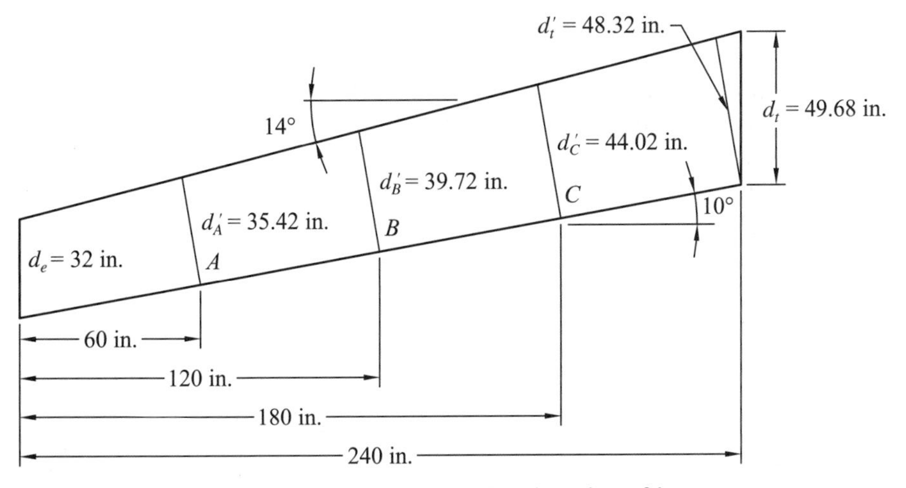

Figure 4.24 Straight tapered end section of beam.

$$\Delta_c = \frac{5\omega\ell^4}{32E'b(d_{\text{equiv}})^3} = \frac{(5)(68.33 \text{ lb/in.})(720 \text{ in.})^4}{(32)(1{,}700{,}000 \text{ psi})(6.75 \text{ in.})(51.99 \text{ in.})^3} = 1.78 \text{ in.}$$

The deflection limitation in this example is $\ell/240$ for the total load. Since 1.78 in. $\leq$ 720 in./240 = 3.00 in., the trial geometry is satisfactory with regard to deflection.

5. *Horizontal deflection.* The horizontal deflection (push-out) at each support is estimated using equation (4.56) using h from equation (4.57).

$$h = h_a - \frac{d_c}{2} - \frac{d_e}{2} = 122 \text{ in.} - \frac{69.2 \text{ in.}}{2} - \frac{32 \text{ in.}}{2} = 71.4 \text{ in.}$$

$$\Delta_H = \frac{2h\Delta_c}{\ell} = \frac{(2)(71.4 \text{ in.})(1.78 \text{ in.})}{720 \text{ in.}} = 0.353 \text{ in.}$$

6. *Radial reinforcement.* Since the radial tension stress exceeded the allowable value (but did not exceed one-third the corresponding allowable shear stress perpendicular to the grain), radial reinforcement is necessary. The radial force to be carried by the reinforcement is equal to the radial tension stress multiplied by the beam width, (46.7 psi) (6.75 in.) = 315 pli. Assuming $\frac{3}{4}$-in. lag screws, from Table 4.10, the lag screws will carry 501 lb per inch of penetration (snow load duration) but not greater than the allowable steel tension load = 5265 lb. All screws will be installed to penetrate to between 2 and 3 in. from the soffit face. For the design of lag screws themselves, assume the lesser penetration, or 3 in. distance. Thus, the effective thread penetration is $(d_c/2) - 3$ in. = (69.2 in./2) − 3 in. = 31.6 in. The allowable withdrawal load on each screw is (501 lb/in.)(31.6 in.) = 15,832 lb. The allowable load based on steel governs: 5265 lb each screw. The maximum lag screw spacing

is therefore (5265 lb/315 lb/in.) = 16.7 in.; use 16-in. spacing. Install the lag screws symmetric about midspan and straddle midspan so that a screw does not coincide with the point of maximum bending moment.

To determine the number of lag screws needed, the length of the curve must be calculated based on the middepth radius at midspan, R_m. This length can be calculated by the formula

$$S_c = 2R_m \left(\phi_B \frac{\pi}{180°} \right) \tag{4.58}$$

where

S_c = length along the curve.

For this example, S_c = (2)(725.6 in.)(10°)((π/180°) = 253.3 in. The curved portion on each side of midspan is 253.3 in./2 = 126.6 in. Use 127 in./16 in. per lag = 7.94 lags. Use eight lags on each side of midspan, starting at 16 in./2 = 8 in. each side of midspan. Each lag screw is to be installed from the top to within 2 to 3 in. from the soffit face. The lag screws are to be installed perpendicular to the glue lines of the laminated timber.

Check: 16 screws total: allowable radial tension carried by the lag screws is (16 × 5265 lb)/(253 in. × 6.75 in.) = 49.3 psi. The calculated radial tension stress of 46.7 psi ≤ 49.3 psi; good (enough lag screws).

The lag screws will effectively reduce the section modulus due to the removal of wood for the screws. The loss of wood is considered for the tension side only, as the screws installed may still resist compression. The distance from the soffit face to the neutral axis (Figure 4.25) of the modified section may be determined by

$$c = \frac{bd^2 - b'c^2 + b'a^2}{2(bd - b'c + b'a)} \tag{4.59}$$

where

c = distance from the soffit to the neutral axis of the modified section,
b = member width,
d = member depth,
b' = effective width of the lag screw hole (lag screw diameter),
a = distance from the lag screw tip to the soffit surface (taken to be 2 in. for the case of maximum penetration).

The unmodified section modulus at the midspan is

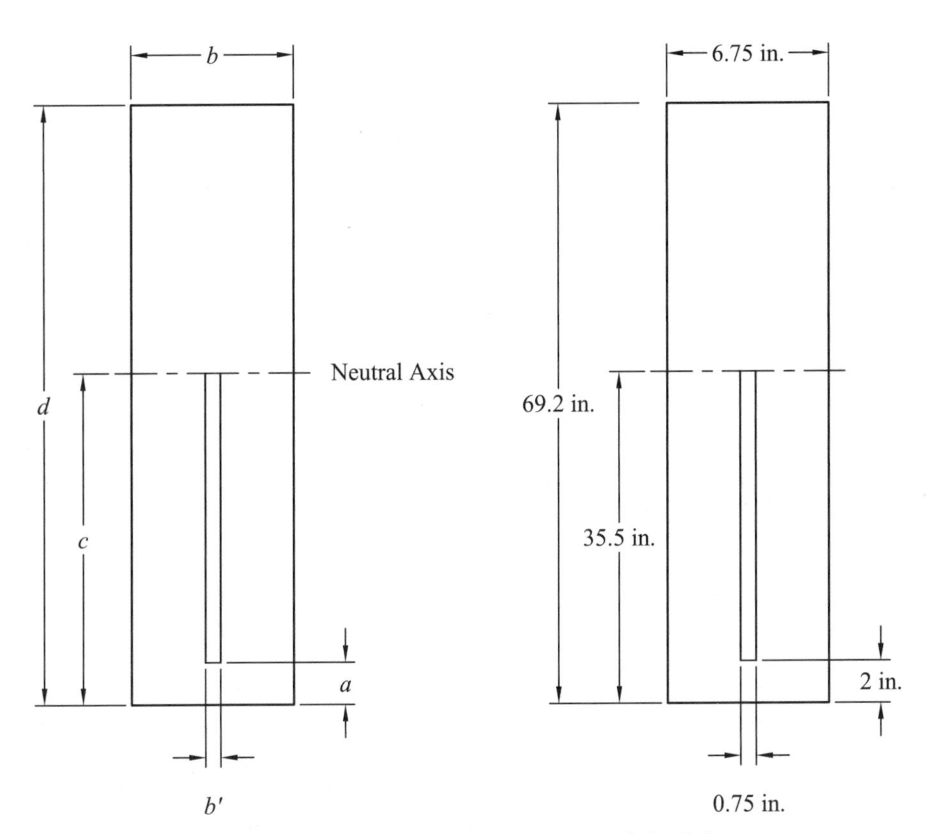

Figure 4.25 Section loss at midspan due to radial reinforcement.

$$S = \frac{bd^2}{6} = \frac{(6.75\text{ in.})(69.2\text{ in.})^2}{6} = 5387\text{ in}^3$$

With the presence of the lag, c = 35.5 in. from equation (4.59), giving a modified section modulus S of 4996 in^3, or 4996/5387 = 0.93 of the original section modulus. The bending stress at (or near) midspan corresponding to the reduced section calculated above is 1363 psi/0.93 = 1466 psi. Since f_b = 1466 psi $\leq F'_b$ = 2010 psi, the beam is still adequate in bending at midspan with consideration of loss of section due to the radial reinforcement.

Similarly, at the tangent point, the modified section modulus becomes 2462 in^3, which is 94% of 2627 in^3 (unmodified section modulus). Since the ratio of the stress applied to allowable stress at the tangent point for the unmodified section was found to be 0.826, the section is still adequate considering section loss due to reinforcement (0.826/0.94 = 0.88 $\leq$ 1.00).

Finally, the assumed self-weight should be checked. A beam with a prismatic section of 6.75 in. $\times$ 69.2 in. will have a self-weight of

$$\omega_{\text{s.w.}} = \left(\frac{6.75 \text{ in.}}{12 \text{ in./ft}}\right)\left(\frac{69.2 \text{ in.}}{12 \text{ in./ft}}\right)(33 \text{ pcf}) = 107 \text{ plf}$$

Since most of the beam will actually have significantly less section, the use of 100 plf for self-weight should be suitable, indeed conservative.

Answer: The 24F-1.7E Douglas fir pitched and tapered curved beam is satisfactory as follows (Figure 4.26):

End depth: 32 in.
Midspan depth: 69.2 in.
Top slope: 3:12 (14.04°)
Bottom slope: 10°
Distance from end to point of tangency: 240 in.
Overall length/length of curved section: 3.0
Inside (soffit) radius: 691 in.
Reinforcement: sixteen $\frac{3}{4}$-in. lag screws at 16 in. o.c. (eight each side) symmetric about midspan

Lag screws should be threaded full length and installed to within 2 to 3 in. of the soffit face of the beam. Lag screws are to be installed perpendicular to the soffit face and centered across the width. The lag screw lead hole diameter should be $0.85 \times \frac{3}{4}$ in. $= \frac{5}{8}$ in. and should be for the full threaded length of the screw. Where lag screws must be flush with the top of the beam, cut the heads off the screws (do not countersink).

4.7.9 Design of Pitched and Tapered Curved Beams Made of Southern Pine

The design of a pitched and tapered curved beam made of southern pine is illustrated in the following example.

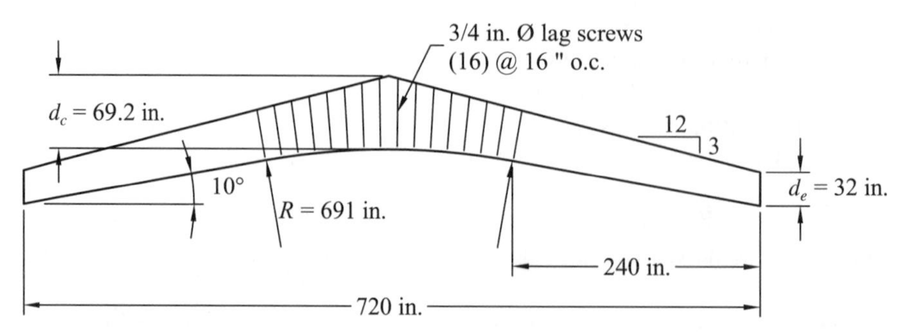

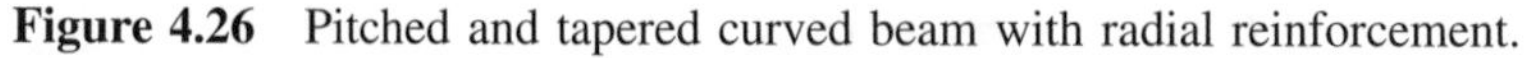
Figure 4.26 Pitched and tapered curved beam with radial reinforcement.

Example 4.18 Pitched and Tapered Curved Beam of Southern Pine

Given: A southern pine pitched and tapered curved glued laminated timber roof beam to meet the following requirements:

Beam length: 45 ft (540 in.)
Spacing: 14 ft
Roof slope: 4:12 (18.43°)
Snow load: 30 psf
C_D: 1.15 (snow)
Dead load: 15 psf
Δ_{max}: ℓ /180 for total load

Assume: Decking is applied to the top of the beam, providing adequate lateral support. The preferred beam width is 5 in. SP design values available from the manufacturer are as follows:

$$F_b = 2400 \text{ psi}$$

$$F_v = 215 \text{ psi (nonprismatic member)}$$

$$F_{rt} = 72 \text{ psi (one-third of the nonprismatic member shear design value)}$$

$$E = 1.8 \times 10^6 \text{ psi}$$

$$F_{c\perp} = 650 \text{ psi}$$

The minimum radius of curvature is $R = 18$ ft $= 216$ in.; the lamination thickness $t = 1\frac{3}{8}$ in. The beam will be manufactured as a tapered curved member; thus, full design values will be used when considering the stress interaction.

Determine: Suitable section and geometric properties.

Approach: The design procedure for pitched tapered glued laminated beams will be used. An assumed weight of 50 plf will be used for the beam itself in addition to the stated loads.

Solution:

1. *Trial geometry.* The assumed trial bottom slope is 14°. The end depth computed from the total load and design value for shear is determined as follows. The total uniform load, including self-weight, is

$$\omega = (15 \text{ psf} + 30 \text{ psf})(14 \text{ ft}) + 50 \text{ plf} = (680 \text{ plf})\ (56.67 \text{ lb/in.})$$

The end reaction, R_v, is

$$R_v = \frac{\omega\ell}{2} = \frac{(680 \text{ plf})(45 \text{ ft})}{2} = 15{,}300 \text{ lb}$$

$$F'_v = F_v C_D = (215 \text{ psi})(1.15) = 247 \text{ psi}$$

The minimum end depth is

$$d_e(\text{min.}) = \frac{3R_v}{2bF'_v} = \frac{(3)(15{,}300 \text{ lb})}{(2)(5 \text{ in.})(247 \text{ psi})} = 18.6 \text{ in.}$$

Using $d_e = 19$ in., the remaining trial geometry, based on minimum radius, is established.

From equation (4.46a),

$$\ell_c = 2R \sin \phi_B = 216 \text{ in. } (\sin 14°) = 104.5 \text{ in.}$$

The apex height is

$$h_a = \frac{\ell}{2} \tan \phi_T + d_e = \frac{540 \text{ in.}}{2} \tan 18.43° + 19 \text{ in.} = 109.0 \text{ in.}$$

The soffit height at midspan is

$$h_s = \frac{\ell}{2} \tan \phi_B - R(\sec \phi_B - 1) = \frac{540 \text{ in.}}{2} \tan 14° - (216 \text{ in.})(\sec 14° - 1) = 60.7 \text{ in.}$$

The midspan depth is

$$d_c = h_a - h_s = 109.0 \text{ in.} - 60.7 \text{ in.} = 48.3 \text{ in.}$$

and

$$\frac{\ell}{\ell_c} = \frac{540 \text{ in.}}{104.5 \text{ in.}} = 5.17$$

The radius at middepth at midspan is

$$R_m = 216 \text{ in.} + \frac{60.7 \text{ in.}}{2} = 246.35 \text{ in.}$$

$$\frac{d_c}{R_m} = \frac{48.3 \text{ in.}}{246.35 \text{ in.}} = 0.20$$

2. *Radial tension stress check.* For checking radial stress,

$$M = \frac{\omega \ell^2}{8} = \frac{(56.67 \text{ lb/in.})(540 \text{ in.})^2}{8} = 2{,}065{,}500 \text{ lb-in.}$$

$$f_o = \frac{6M}{bd_c^2} = \frac{(6)(2{,}065{,}500 \text{ lb-in.})}{(5 \text{ in.})(48.3 \text{ in.})^2} = 1062 \text{ psi}$$

$$C_r = \alpha + \beta \frac{d_c}{R_m} = 1.00 + (-0.16)(0.20) = 0.97$$

where the values for α and β are from Table 4.8 (interpolated):

$$K_r = A + B\frac{d_c}{R_m} + C\left(\frac{d_c}{R_m}\right)^2 = 0.0810 + (0.0611)(0.20)$$
$$+ (0.1496)(0.20)^2 = 0.099$$

and where the values for A, B, and C are interpolated from Table 4.7:

$$f_{rt} = K_r C_r f_o = (0.099)(0.97)(1062 \text{ psi}) = 102 \text{ psi}$$

$$F'_{rt} = F_{rt} C_D = (72 \text{ psi})(1.15) = 82.8 \text{ psi}$$

Since $f_{rt} = 102$ psi $> F'_{rt} = 82.8$ psi, the trial geometry is not acceptable with regard to radial tension stress.

The radial stress may be decreased by increasing the depth, width, or radius of curvature. Further design checks with the geometry above would reveal that the beam is not acceptable with regard to bending in parts of the pitched and tapered section. Decreasing the bottom slope will also increase the midspan depth and reduce radial tension stress; however, decreased bottom slope will also decrease the allowable bending stress through the stress interaction factor. The increased end depth could have been established initially by being additionally conservative in the analysis of the end shear condition. Since determining a satisfactory geometry may involve a number of iterations and trial geometries, developing computational spreadsheets is recommended.

In this example, a new trial geometry will be used based on increased end depth and larger radius of curvature. The increased end depth will increase the depth and thus reduce the bending stress throughout. The increased radius of curvature will increase the length and depth of the curved portion, reduce radial stress, and increase the allowable bending stress through increasing the curvature factor. New trial end depth = 21 in.; new trial soffit radius = 300 in. The new trial geometry becomes

$$\ell_c = 145 \text{ in.}$$

$$\frac{\ell}{\ell_c} = 3.72$$

$$h_a = 111.0 \text{ in.}$$

$$h_s = 58.1 \text{ in.}$$

$$d_c = 52.8 \text{ in.}$$

$$R_m = 326.4 \text{ in.}$$

$$\frac{d_c}{R_m} = 0.162$$

for which

$$A = 0.081$$

$$B = 0.061$$

$$C = 0.150$$

$$K_r = 0.095$$

$$\alpha = 0.99$$

$$\beta = -0.40$$

$$C_r = 0.925$$

$$f_o = \frac{6M}{bd_c^2} = \frac{(6)(2{,}065{,}500 \text{ lb-in.})}{(5 \text{ in.})(52.8 \text{ in.})^2} = 889 \text{ psi}$$

$$f_{rt} = K_r C_r f_o = (0.095)(0.925)(889 \text{ psi}) = 78 \text{ psi}$$

Since $f_{rt} = 78$ psi $\leq F'_{rt} = 82.8$ psi, the revised geometry is acceptable with regard to radial tension.

3. *Bending stress check*. From Table 4.11, $D = 1.89$, $E = 0.0$, and $F = 0.0$:

$$K_\theta = D + E\frac{d_c}{R_m} + F\left(\frac{d_c}{R_m}\right)^2 = 1.89 + (0.0)(0.162) + (0.0)(0.162)^2 = 1.89$$

$$f_b = K_\theta f_o = (1.89)(889 \text{ psi}) = 1680 \text{ psi}$$

For the allowable bending stress at midspan

$$C_L = 1.00 \text{ ((full lateral support at top of beam)}$$

$$C_V = \left(\frac{5.125 \text{ in.}}{5 \text{ in.}}\right)^{1/20} \left(\frac{12 \text{ in.}}{52.8 \text{ in.}}\right)^{1/20} \left(\frac{21 \text{ ft}}{45 \text{ ft}}\right)^{1/20} = 0.895$$

$$C_c = 1 - 2000 \left(\frac{1.375}{300}\right)^2 = .958$$

Therefore, at midspan,

$$F_b' = (2400 \text{ psi})(1.15)(0.895)(0.958) = 2366 \text{ psi}$$

Since $f_b = 1680$ psi $\leq F_b' = 2366$ psi, the trial beam geometry checks at midspan with respect to bending.

The bending stress is also calculated and checked at four sections along the pitched and tapered segments of the beam in Table 4.13 using equations (4.53) and (4.54). The table shows that the bending stresses check at the locations selected along the pitched and tapered portions of the beam.

4. *Deflection check.* The equivalent depth for deflection is, from equation (4.44), $d_{\text{equiv}} = 36.4$ in. The deflection at midspan is, from equation (4.43),

$$\Delta_c = \frac{(5)(56.67 \text{ lb/in.})(540 \text{ in.})^4}{(32)(1{,}800{,}000 \text{ psi})(5 \text{ in.})(36.4 \text{ in.})^3} = 1.73 \text{ in.}$$

The deflection limitation in this example is $\ell/180$ for total load. Since 1.73 in. $\leq$ 540 in./180 = 3.00 in., the trial geometry is satisfactory with regard to deflection.

5. *Horizontal deflection.* The horizontal deflection (push-out) at each support is estimated using equation (4.56), using h from equation (4.57): h = 74 in.

$$\Delta_H = \frac{2h\Delta_c}{\ell} = \frac{2(74 \text{ in.})(1.73 \text{ in.})}{540 \text{ in.}} = 0.47 \text{ in.}$$

6. *Radial reinforcement.* Radial reinforcement is not used with southern pine.

Answer: The southern pine pitched and tapered curved beam is satisfactory as follows (Figure 4.27):

End depth: 21 in.
Midspan depth: 52.8 in.
Top slope: 4:12 (18.43°)
Bottom slope: 14°

TABLE 4.13 Bending Stress Check at Four Locations in Pitched and Tapered Portion of Southern Pine Pitched and Tapered Curved Beam

Section	x (in.)	d'_x (in.)	M (lb-in.)	I (in^4)	S (in^3)	f_b (psi)	C_v	C_I	C_c	F'_b (psi)	f_b/F'_b
A	49.4	23.90	686,676	5,688	476	1443	0.918	*0.756*		2087	0.691
B	58.7	27.90	800,482	9,049	649	1234	0.911	*0.756*		2087	0.591
C	148	31.80	1,644,478	13,399	843	1951	0.905	*0.756*		2087	0.935
TP	197	35.75	1,916,162	19,038	1065	1799	0.900	*0.756*	*0.958*	1999	0.900

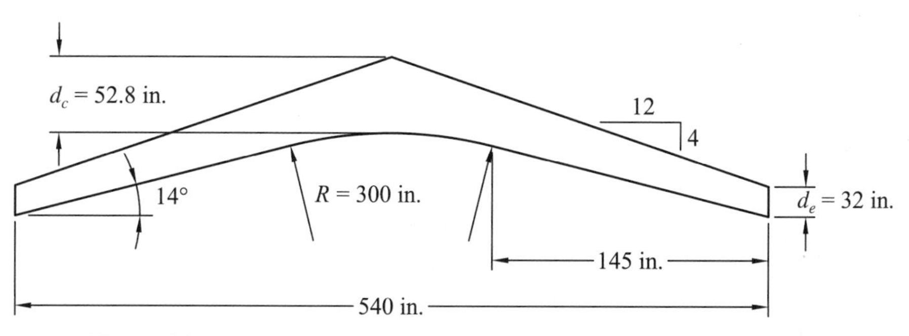

Figure 4.27 Pitched and tapered curved beam made of southern pine.

Distance from end to point of tangency: 197 in.
Overall length/length of curved section: 3.72
Inside (soffit) radius: 300 in.
Reinforcement: none (radial reinforcement is not used in southern pine beams)

4.7.10 Curved Beams of Constant Cross Section with Mechanically Attached Haunch

Various examples of curved beams are shown in Figure 4.21. Beams with detached haunches have the advantage that they may be made of constant cross section, simplifying engineering design. The curved section of such beams may also be designed to be of large enough radius to eliminate the need for radial reinforcement. The haunch may be of nonstructural glued laminated timber attached mechanically, or may be framed. The haunch may or may not be designed to provide lateral stability for the structural beam. The design of such beams is generally accomplished by trial and error to produce a member that is aesthetically pleasing and structurally adequate. Beams with curved portions of approximately one-half to one-third of the beam span tend to be the most appealing aesthetically. The straight section may be prismatic or may be designed with a taper to accommodate the geometric and structural requirements. The following example illustrates the design of a beam with a nonstructural or detached haunch. The beam has constant cross section in the circular portion, straight tapered sections, and tangent points at one-fourth of the beam span.

Example 4.19 Curved Beam with Mechanically Attached Haunch

Given: Design conditions are as follows:

Beam length: 60 ft (720 in.)
Roof slope: 3:12

Ratio of span to curved portion of span: $\ell/\ell_c = 2.0$
Spacing of beams: 16 ft
Roof load: 25 psf (snow)
Roof dead load: 15 psf
Total load deflection limit: $\ell/180$

Determine: The geometry of a curved beam of constant cross section in the curved portion should be established using design values of 24F-1.8E western species from AITC 117 [1] or *NDS*® [2]. The geometry should be such that no radial reinforcement is required.

Approach: It will be assumed that in the straight tapered section, the roof will be attached to the top of the beam to provide lateral support. It will also be assumed that the haunch will be constructed and attached in such a way as to provide lateral support for the curved section. As such, for the tapered section, C_V and C_I will be calculated and the smaller value used. For the curved portion, C_V and C_c will be applied. At the tangent point C_c and the smaller of C_V, C_I will be applied.

Solution: From AITC 117 [1]:

$$F_b = 2400 \text{ psi}$$

$$F_v = 190 \text{ psi (nonprismatic shape)}$$

$$E = 1{,}800{,}000 \text{ psi}$$

$$F_{rt} = 15 \text{ psi}$$

$$F_{c\perp} = 650 \text{ psi}$$

Lamination thickness $t = 1.5$ in. Assume a beam width of $b = 6.75$ in.

1. *Trial geometry*. A trial soffit slope angle of 10° will be used. A minimum end depth may be calculated based on shear. The estimated self-weight of the beam is 80 plf.

$$\text{Total uniform load, } \omega = (25 \text{ psf} + 15 \text{ psf})(16 \text{ ft}) + 80 \text{ plf}$$
$$= (720 \text{ plf})\ (60 \text{ pli})$$

End shear,

$$V = \frac{\omega\ell}{2} = \frac{(720 \text{ plf})(60 \text{ ft})}{2} = 21{,}600 \text{ lb}$$

Calculate the trial end depth by letting $f_v = \frac{3}{2}\, V/A = F_v'$; $A = bd$; $F_v' = (190$ psi$)(1.15) = 218.5$ psi. So

$$d_e = \frac{3}{2}\frac{21{,}600 \text{ lb}}{(6.75 \text{ in.} \times 218.5 \text{ psi})} = 21.97 \text{ in.}$$

Use $d_e = 22.5$ in.

Calculate the soffit radius based on the requirement $\ell/\ell_c = 2.0$.

$$R = \frac{\ell}{2(\ell/\ell_c)\sin\phi_B} = \frac{720 \text{ in.}}{(2)(2.0)\sin 8°} = 1293 \text{ in.}$$

Calculate the length of the curved portion of the span and the length of the straight tapered portion of the span.

$$\ell_c = \frac{\ell}{2.0} = \frac{720 \text{ in.}}{2.0} = 360 \text{ in.}$$

$$\ell_t = \frac{\ell - \ell_c}{2} = \frac{720 \text{ in.} - 360 \text{ in.}}{2} = 180 \text{ in.}$$

Calculate the depth of the structural portion of the beam at the tangent point, midspan, and at A = $\ell_t/4$, B = $\ell_t/2$, and C = $\frac{3}{4}\ell_t$ from the support.

$$d_c = d_t' = [d_e + 1_t\,(\tan\phi_T - \tan\phi_B)][\cos\phi_B - \sin\phi_B\tan(\phi_T - \phi_B)]$$

$$d_c = d_t' = [22.5 \text{ in.} + 180 \text{ in.}\,(\tan 14.04° - \tan 8°)]$$

$$[\cos 8° - \sin 8°\tan(6.04°)] = 41.17 \text{ in.}$$

$$d_A' = [d_e + \ell_A(\tan\phi_T - \tan\phi_B)][\cos\phi_B - \sin\phi_B\tan(\phi_T - \phi_B)]$$

$$d_A' = [22.5 \text{ in.} + 45 \text{ in.}\,(\tan 14.04° - \tan 8°)][\cos 8° - \sin 8°\tan(6.04°)]$$

$$= 26.76 \text{ in.}$$

$$d_B' = [d_e + \ell_B(\tan\phi_T - \tan\phi_B)][\cos\phi_B - \sin\phi_B\tan(\phi_T - \phi_B)]$$

$$d_B' = [22.5 \text{ in.} + 90 \text{ in.}\,(\tan 14.04° - \tan 8°)][\cos 8° - \sin 8°\tan(6.04°)]$$

$$= 31.56 \text{ in.}$$

$$d_C' = [d_e + \ell_A(\tan\phi_T - \tan\phi_B)][\cos\phi_B - \sin\phi_B\tan(\phi_T - \phi_B)]$$

$$d_C' = [22.5 \text{ in.} + 135 \text{ in.}\,(\tan 14.04° - \tan 8°)][\cos 8° - \sin 8°\tan(6.04°)]$$

$$= 36.37 \text{ in.}$$

Calculate the radius at middepth, R_m:

$$R_m = R + \frac{d_c}{2} = 1293 \text{ in.} + \frac{41.17 \text{ in.}}{2} = 1314 \text{ in.}$$

2. *Radial tension stress check.* Calculate the moment at midspan:

$$M = \frac{\omega \ell^2}{8} = \frac{(60 \text{ lb/in.})\,(720 \text{ in.})^2}{8} = 3{,}888{,}000 \text{ lb-in.}$$

Calculate the radial tension stress:

$$f_{rt} = \frac{3M}{2bdR_m} = \frac{3(3{,}888{,}000 \text{ lb-in.})}{2(6.75 \text{ in.})(41.2 \text{ in.})(1314 \text{ in.})} = 16.0 \text{ psi}$$

$$f_{rt} = 16.0 \text{ psi} \leq F'_{rt} = 15 \text{ psi } (1.15) = 17.2 \text{ psi} \qquad \text{good}$$

3. *Bending stress check.* Determine the allowable bending stress at midspan:

$$F'_b = F_b C_D (C_L \text{ or } C_V) C_c$$

$$C_D = 1.15$$

$$C_V = \left(\frac{5.125 \text{ in.}}{6.75 \text{ in.}}\right)^{1/10} \left(\frac{12 \text{ in.}}{41.2 \text{ in.}}\right)^{1/10} \left(\frac{21 \text{ ft}}{60 \text{ ft}}\right)^{1/10} = 0.774$$

$$C_c = 1 - 2000\left(\frac{t}{R}\right)^2 = 1 - 2000\left(\frac{1.5 \text{ in.}}{1293 \text{ in.}}\right)^2 = 0.997$$

So

$$F'_b = (2400 \text{ psi})(1.15)(0.774)(0.997) = 2130 \text{ psi}$$

$$M = 3{,}888{,}000 \text{ lb-in.}$$

Calculate the bending stress at midspan:

$$f_b = \frac{M}{S} = \frac{6M}{bd_c^2} = \frac{(6)(3{,}888{,}000 \text{ lb-in.})}{(6.75 \text{ in.})(41.17 \text{ in.})^2} = 2039 \text{ psi}$$

$$f_b = 2040 \text{ psi} \leq F'_b = 2130 \text{ psi} \qquad \text{good}$$

Determine the allowable bending stress, bending moment, and bending stress at the tangent point:

$$C_V = 0.774$$

$$C_c = 0.997$$

$$C_I = \left[\frac{1}{1 + (F_b \tan\theta/F_v)^2 + (F_b \tan^2\theta/F_{c\perp})^2}\right]^{1/2}$$

$$= \left\{\frac{1}{1 + [(2400\text{ psi})\tan 6.04°/190\text{ psi}]^2 + [(2400\text{ psi})\tan^2 6.04°/650\text{ psi}]^2}\right\}^{1/2}$$

$$= 0.599$$

$$F'_{bt} = F_b C_D C_c\,(C_V \text{ or } C_I) = (2400\text{ psi})(1.15)(0.997)(0.599) = 1650\text{ psi}$$

$$M_t = \frac{\omega \ell_t}{2}(\ell - \ell_t) = \frac{60\text{ lb/in. }(180\text{ in.})}{2}(720\text{ in.} - 180\text{ in.})$$

$$= 2{,}916{,}000\text{ lb-in.}$$

$$f_{bt} = \frac{6M_t}{b(d'_t)^2} = \frac{6(2{,}916{,}000\text{ lb-in.})}{(6.75\text{ in.})(41.2\text{ in.})^2} = 1530\text{ psi}$$

$$f_{bt} = 1530\text{ psi} \le F'_{bt} = 1650\text{ psi} \qquad \text{good}$$

Determine the allowable bending stress, bending moment, and bending stress at quarter points along the tapered section, A, B, and C. $F'_b = F_b C_D\,(C_V$ or $C_I) = (2400\text{ psi})\,(1.15)(0.599) = 1650$ psi for all three sections, because the interaction factor governs.

$$M_A = \frac{\omega \ell_A}{2}(\ell - \ell_A) = \frac{(60\text{ lb/in.})\,(45\text{ in.})}{2}(720\text{ in.} - 45\text{ in.})$$

$$= 911{,}000\text{ lb-in.}$$

$$M_B = \frac{\omega \ell_B}{2}(\ell - \ell_B) = \frac{(60\text{ lb/in.})\,(90\text{ in.})}{2}(720\text{ in.} - 90\text{ in.})$$

$$= 1{,}701{,}000\text{ lb-in.}$$

$$M_C = \frac{\omega \ell_C}{2}(\ell - \ell_C) = \frac{(60\text{ lb/in.})\,(135\text{ in.})}{2}(720\text{ in.} - 135\text{ in.})$$

$$= 2{,}369{,}000\text{ lb-in.}$$

$$f_{bA} = \frac{6M_A}{b(d'_A)^2} = \frac{(6)(911{,}000\text{ lb-in.})}{(6.75\text{ in.})(26.76\text{ in.})^2} = 1130\text{ psi} \le F'_b = 1650\text{ psi} \qquad \text{good}$$

$$f_{bB} = \frac{6M_A}{b(d'_A)^2} = \frac{(6)(1{,}701{,}000\text{ lb-in.})}{(6.75\text{ in.})(31.56\text{ in.})^2} = 1520\text{ psi} \le F'_b = 1650\text{ psi} \qquad \text{good}$$

$$f_{bC} = \frac{6M_A}{b(d'_A)^2} = \frac{(6)(2{,}369{,}000\text{ lb-in.})}{(6.75\text{ in.})(36.37\text{ in.})^2} = 1590\text{ psi} \le F'_b = 1650\text{ psi} \qquad \text{good}$$

4. *Deflection check.* Assume that deflection will be 10% higher than for a beam of constant depth, d_c.

$$\Delta_{TL} = \frac{(1.1)5\omega\ell^4}{32E'bd_c^3} = \frac{(1.1)(5)(60 \text{ lb/in.})(720 \text{ in.})^4}{(32)[(1.8)(10^6)\text{psi}](6.75 \text{ in.})(41.2 \text{ in.})^3}$$

$$= 3.26 \text{ in.} \leq \frac{\ell}{180} = \frac{720 \text{ in.}}{180} = 4.0 \text{ in.} \qquad \text{good}$$

5. *Horizontal deflection at supports.* The horizontal displacement at each support can be estimated using $h = (\ell/2)\tan\phi_B$ with equation (4.56).

$$\Delta_H = \frac{2h\Delta_c}{\ell} = \frac{(2)[360 \text{ in. } (\tan 8°)](3.26 \text{ in.})}{720 \text{ in.}} = 0.46 \text{ in.}$$

Answer: The curved beam of constant curved portion depth and tapered straight section depth is shown in Figure 4.28. The important geometric features follow:

End depth: $d_e = 22.5$ in.
Apex height: $h_a = 112.5$ in.
Midspan depth: $d_c = 41.2$ in. (depth of curved section)
Soffit height: $h_s = 37.9$ in.
Haunch height: 33.4 in.
Soffit radius: $R = 1293$ in.
Length of tangent section: $\ell_t = 180$ in.
Length of curved section: $\ell_c = 360$ in.
Top slope: 14.04° 3:12
Bottom (soffit) slope: 8°

The beam design assumes that the tapered section will be manufactured into the beam such that the design values for 24F-1.8E in AITC 117 are applicable. The design assumed that the haunch was constructed and attached in such a way as to provide lateral support for the curved section of the beam.

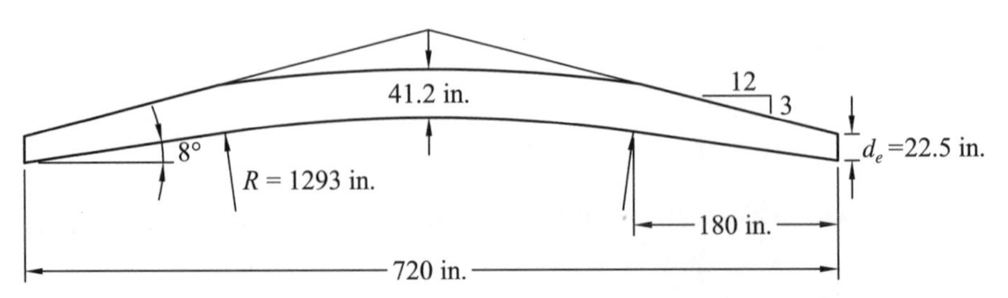

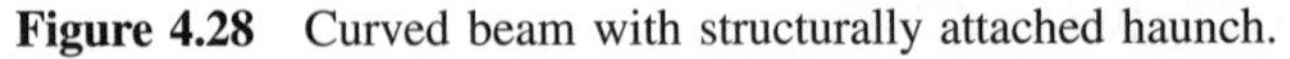

Figure 4.28 Curved beam with structurally attached haunch.

Design of the haunch to provide lateral support must be included in the construction documents. Connections at the supports must be designed to accommodate approximately $\frac{1}{2}$ in. of displacement at each support or 1 in. at one support.

Discussion: Once an acceptable geometry is determined, its aesthetic suitability must be approved by the owner and the constructability of the beam must be verified by the manufacturer.

4.8 ARCHES

An advantage of glued laminated timber is its ability to be manufactured into a variety of long curved and tapered shapes. Glued laminated timber arches have been and continue to be very popular for use in large open structures such as churches and gymnasiums, where they provide structure and are aesthetically appealing. Timber arches are also used for vehicle and pedestrian bridges. Figure 4.29 illustrates a number of common arch configurations used. The three-hinged tudor arch is particularly popular in glued laminated timber. Two-hinged arches may also be designed and be manufactured of glued laminated timber. Arch segment lengths are essentially unlimited from a manufacturing standpoint but are limited by transportation and erection constraints. In such cases, properly designed moment splices may be used.

Detailed design procedures for timber arches may be found in *Mathematical Solution for the Design of a Three-Hinged Arch,* AITC Technical Note 23 [12], in the fourth edition of this book [13] for a graphical solution, and in *Deflection of Glued Laminated Timber Arches,* AITC Technical Note 2 [14]. Manufacturers of glued laminated timber arches and engineering firms specializing in glued laminated timber design may also have propriety computer methods to aid in design. Finite element techniques may also be used to compute member forces and deflections under load.

Many of the arch member shapes used are curved and tapered. The significant difference between the design of curved and tapered beams and the design of arch members is that axial forces developed in the arch are considered in design. In this regard, significant thrust forces are developed at the base of the arch, which must be resisted by the foundation, by tie rods, or by other means. In addition, significant changes in beam geometry may result from moisture content changes in curved portions of the arch. Both of these issues are addressed in the references cited above. The design of the arch members is generally performed by trial and error starting with desired roof slopes, soffit or clearance heights, and desired architectural or aesthetic form.

The basic procedure for the design of arches is as follows.

1. Determine all applicable loads and load combinations.
2. Determine a trial geometry.

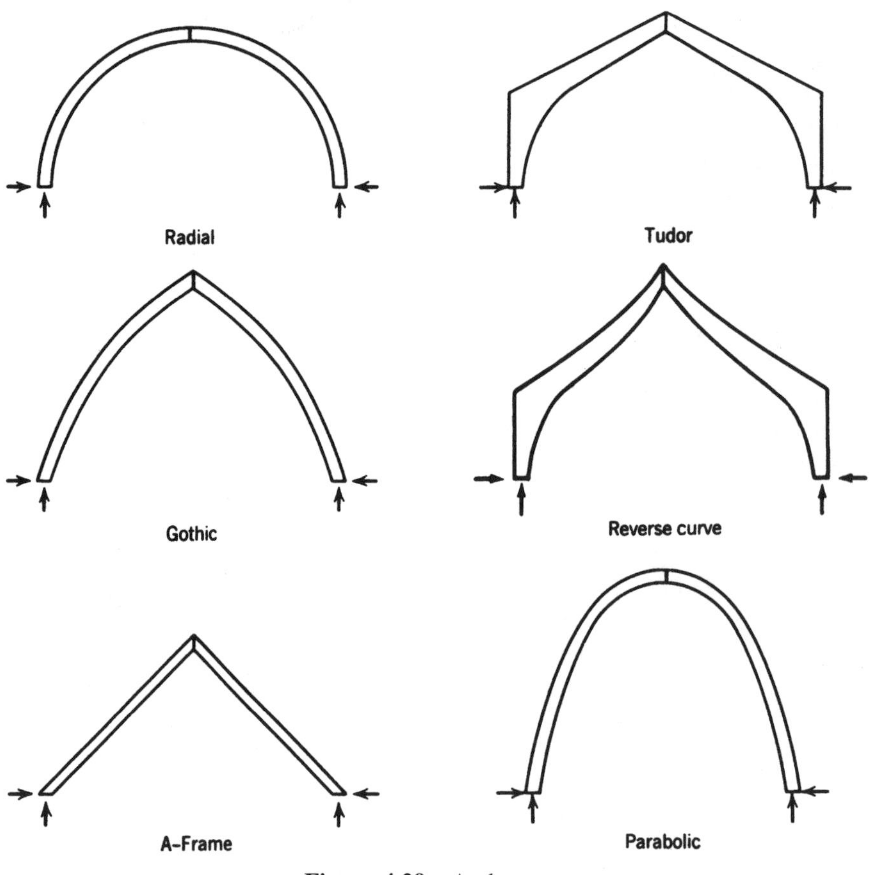

Figure 4.29 Arch types.

3. Determine the base reactions.
4. Assuming that the member is oriented approximately vertically at the base, a trial minimum base size or cross section is determined based on the horizontal reactions

$$bd_b = \frac{3R_H}{2F'_v} \qquad (4.60)$$

where

b = arch width,
d_b = arch depth at the base,
R_H = horizontal reaction at the base,

F'_v = allowable horizontal shear stress.

5. Determine trial depths at the crown and tangent points.
6. Determine trial depths and radii for the curved portions.
7. Check the combined axial and flexural stresses using

$$\frac{f_c}{F'_c} + \frac{f_{b1}}{F'_{b1}} \leq 1.00 \tag{4.61}$$

where f_{b1} *and* F'_{b1} are with regard to loads acting parallel to the depth (strong direction) of the arch. The allowable stress in compression parallel to the grain, F'_c, includes the column stability factor, C_P, taking into consideration resistance to buckling and slenderness. The beam stability factor C_L is not customarily applied to arches. In the determination of F'_b the curvature factor C_c is applied in the curved portions, with the radius of curvature values satisfying the criteria of Section 4.6.1. The volume factor C_V is applied in accordance with

$$\text{When } f_c > F_b^*(1 - C_V)\text{:} \quad F'_b = F_b^* \cdots (C_V \text{ is not applied})$$

$$\text{When } f_c < F_b^*(1 - C_V)\text{:} \quad F'_b = F_b^* C_V + f_c \cdots (C_V \text{ is applied}) \tag{4.62}$$

8. The radial stress, f_{rc} or f_{rt}, must be calculated using equation (4.37) for constant cross-dimension sections or equation (4.38) for pitched and tapered curved sections, and must not exceed the corresponding allowable radial stress value.
9. The deflections must be calculated by any appropriate means and checked against appropriate limitations. Deflection calculations must take into account the following:
    a. Immediate deflection due to applied loads
    b. Creep (permanent deformation due to sustained loads)
    c. Deflection or change in geometry due to changes in moisture content
10. The supports and connections for the arch and any splices must be designed properly.

CHAPTER 5

TIMBER CONNECTIONS AND FASTENERS

5.1 INTRODUCTION

The scope of this chapter is the design of the connections and fasteners between wood members and the connections and fasteners attaching wood members to other structural materials. In general, wood connections fall into several broad categories: bearing-type connections, bolt-type or lag screw connections involving prebored holes in the wood members, nail-type fasteners, split-ring and shear plate connections; and connections with heavy or light steel or other side plates, timber rivets, and moment splices. Connections generally incorporate one or more fasteners, may also incorporate bearing, and may or may not utilize bottom or side plates. With the exception of moment splices, wood connections are generally not expected to transfer moments from one member or element to another. Individual fasteners may transfer load perpendicular to the fastener axis (lateral load) or parallel to the fastener axis. Loads parallel to fastener axis are transferred through friction or threads (withdrawal loading), or by bearing of the bolt head or nut and washers.

For fasteners such as bolts, lag screws, split-ring and shear plate connectors, and timber rivets, the angle of load to the grain must be considered; in others, such as with nails and wood screws, fastener design values are independent of direction of load with respect to the grain. Design values for fasteners loaded perpendicular to the fastener axis and which load the wood parallel to the grain are generally denoted P or $Z_{\|}$. Design values associated with wood loaded perpendicular to the grain are generally denoted by Q or $Z_{\perp}$. Design values associated with loading the fastener in withdrawal are denoted W. Design values adjusted by all appropriate adjustment factors are denoted P', Q', Z', and W'. Design values for nails and wood screws are simply denoted Z and W (irrespective with direction of load to the grain).

In the design of connections, the following general considerations must be taken into account: (1) connections must allow the wood members to shrink and swell with varying moisture content without causing undue stress on the wood; (2) connections must account for varying strength with wood species and angle of the grain; (3) connections and fasteners must not develop undue tension stresses perpendicular to the grain; (4) connections must not trap moisture; (5) design, detailing, and fabrication should not produce undue stress concentrations (e.g., due to notches); (6) eccentric loading should be avoided if possible; and (7) the strength of associated *nonwood* connection components, hardware, and base materials (i.e., steel bolts, base plates, etc.) must also be considered properly.

This chapter contains the procedures and example calculations associated with designing various types of connections. The design approach used in this chapter is allowable stress design (ASD), where, in general, connections will be designed to provide allowable loads equal to or exceeding expected service loads. Load and resistance factor design (LRFD) as it relates to connection design is discussed in Chapter 7. Examples of good and bad connection configurations or details are contained in *Typical Construction Details,* AITC 104 [1], included in Chapter 8 of this book.

In general, the fasteners in any one connection should not be mixed but be of the same size and type. For any single connection, design values for a single fastener (Z or W for bolts, lag screws, wood screws, nails, spikes, and drift pins, and P or Q for split-ring, shear plate connectors, and rivets) may be calculated or are obtained from published tables such as those in the *National Design Specification® for Wood Construction* [2]. The design values are then adjusted for the specific conditions of service and the connection configuration and accumulated to give the theoretical allowable load for the connection *based on the fasteners* (P', Q', Z', and W'). The effect of the fasteners *as a whole* must also be checked so as not to create overall failure of the timber pieces at the connection.

As mentioned above, timber connections should be designed, where possible, to avoid eccentricity of loading. In the design of trusses, for example, all axial loads at a joint or connection should pass through the same point. Where eccentricity cannot be avoided, for example as shown in Figure 5.1, the moment developed by the load at the connection must be accounted for in both the wood and nonwood components of the connection.

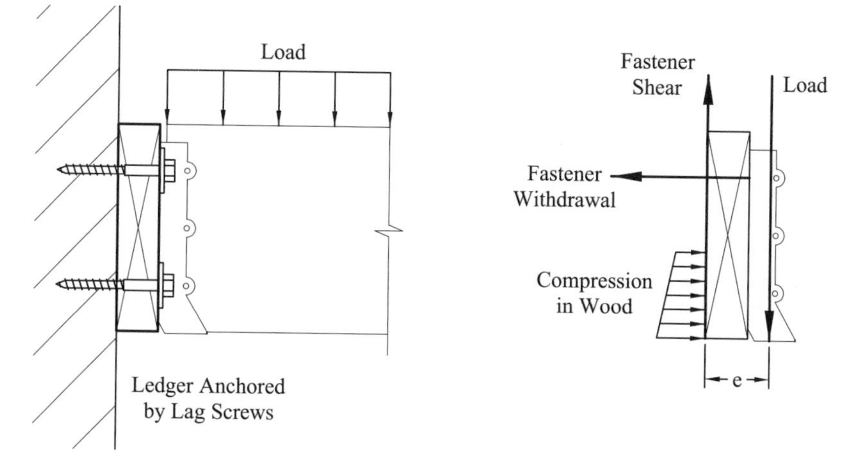

Figure 5.1 Eccentricity and prying action.

Timber fastener design values are based on the oven-dry weight and oven-dry volume specific gravity of the wood members being connected. Species and species groups of wood typically used in timber construction and their accepted specific gravity values are shown in Table 5.1. Where the wood members in a connection are of different species and specific gravity, the lesser of the specific gravity values are generally used in design.

In addition to designing connections "from scratch" (using the procedures of this book), proprietary fasteners and connection hardware are available from a number of manufacturers. The general considerations and design principles described in this book should be applied to the selection of such hardware. Only hardware and fasteners that have been established to meet the appropriate code approvals should be used. Unlike the equations provided in other chapters, the equations for connection design are typically dimensionally dependent.

5.2 ADJUSTMENT FACTORS

Design values for fasteners obtained, for example, from the *National Design Specification®* [2], are generally for a single fastener in wood; are fabricated with the necessary edge distance, end distance, and spacing requirements to provide full design value; and are used under the stated direction of load to grain, moisture condition, and normal duration of load. Where other conditions exist, these design values must be multiplied by appropriate adjustment factors. Where more than one fastener is used, group action must be considered and is treated as an adjustment factor for the design value for the single fastener. The capacity of the fasteners is then taken to be the single fastener

TABLE 5.1 Species and Species Groups and Specific Gravity

Species or Species Combination	Specific Gravity[a] (Oven-Dry Volume)
Aspen	0.39
Balsam Fir	0.36
Beech–Birch–Hickory	0.71
Coast Sitka Spruce	0.39
Cottonwood	0.41
Douglas Fir–Larch	0.50
Douglas Fir–Larch (north)	0.49
Douglas Fir–South	0.46
Eastern Hemlock	0.41
Eastern Hemlock–Balsam Fir	0.36
Eastern Hemlock–Tamarack	0.41
Eastern Hemlock–Tamarack (North)	0.47
Eastern Softwoods	0.36
Eastern Spruce	0.41
Eastern White Pine	0.36
Engelmann Spruce–Lodgepole Pine	0.38
Engelmann Spruce–Lodgepole Pine (higher grades)	0.46
Hem Fir	0.43
Hem Fir (north)	0.46
Mixed Maple	0.55
Mixed Oak	0.68
Mixed Southern Pine	0.51
Mountain Hemlock	0.47
Northern Pine	0.42
Northern Red Oak	0.68
Northern Species	0.35
Northern White Cedar	0.31
Ponderosa Pine	0.43
Red Maple	0.58
Red Oak	0.67
Red Pine	0.44
Redwood, close grain	0.44
Redwood, open grain	0.37
Sitka Spruce	0.43
Southern Pine	0.55
Spruce–Pine–Fir	0.42
Spruce–Pine–Fir (higher grades)	0.50
Spruce–Pine–Fir (south)	0.36
Western Cedars	0.36
Western Cedars (North)	0.35
Western Hemlock	0.47
Western Hemlock (North)	0.46
Western White Pine	0.40
Western Woods	0.36
White Oak	0.73
Yellow Poplar	0.43

[a] Specific gravity values from the *National Design Specification® for Wood Construction* [2] based on oven-dry weight and oven-dry volume.

capacity multiplied by the number of fasteners. The capacity of the fasteners, however, cannot be taken to exceed the capacities of the other components or parts of the connection.

The adjustment factors used in connection design are listed in Table 5.2, and their applicability with various fasteners is shown in Table 5.3. A general description of the adjustment factors for fasteners follows. The adjustment factors for the fasteners are applied generally in the same manner as the adjustment factors for strength properties of wood in that the allowable value based on the fastener is equal to the published (or calculated) design value multiplied by the applicable adjustment factors.

5.2.1 Load Duration Factor, C_D

The duration of load factors applicable to the strength properties of wood (Table 3.6) are generally applicable to fasteners except as noted:

1. The duration of load factor for fasteners is not to exceed 1.6.
2. The duration of load factor is not applicable to bearing stresses perpendicular to grain.
3. The duration of load factor is not applicable to the nonwood parts of the connection, such as metal sideplates.

Building codes may limit the load duration factor for connections to values less than those in Table 3.6. For example, the *Uniform Building Code* [3] limits C_D to 1.33 in allowable stress design for wind and earthquake loads except in some load combinations and failure modes (Section 5.11), where 1.60 may be used.

5.2.2 Wet Service Factor, C_M

The nominal design values for connections in wood are based on wood that is seasoned to a moisture content of 19% or less and used under continuously dry conditions. When the wood is unseasoned or partially seasoned at the time of fastener fabrication, or when the connection is exposed to wet service

TABLE 5.2 Adjustment Factors for Fasteners in Wood Connection Design

C_D, Load duration factor	C_Δ, Geometric factor designated as:
C_M, Wet service factor	$C_{\Delta s}$, Spacing factor
C_t, Temperature factor	$C_{\Delta n}$, End distance factor
C_d, Penetration depth factor	$C_{\Delta e}$, Edge distance factor
C_g, Group action factor	C_{eg}, End grain factor
C_{st}, Metal plate factor	C_{di}, Diaphragm factor
	C_{tn}, Toenail factor

TABLE 5.3 Applicability of Adjustment Factors for Connections

		Load Duration Factor	Wet Service Factor	Temperature Factor	Group Action Factor	Geometry Factor	Penetration Depth Factor	End Grain Factor	Metal Side Plate Factor	Diaphragm Factor	Toe-Nail Factor
Lateral Loads											
Dowel-type Fasteners	$Z' = Z$ x	C_D	C_M	C_t	C_g	C_Δ	-	C_{eg}	-	C_{di}	C_{tn}
Split Ring and Shear Plate Connectors	$P' = P$ x	C_D	C_M	C_t	C_g	C_Δ	C_d	-	C_{st}	-	-
	$Q' = Q$ x	C_D	C_M	C_t	C_g	C_Δ	C_d	-	-	-	-
Timber Rivets	$P' = P$ x	C_D	C_M	C_t	-	-	-	-	C_{st}	-	-
	$Q' = Q$ x	C_D	C_M	C_t	-	C_Δ	-	-	C_{st}	-	-
Metal Plate Connectors	$Z' = Z$ x	C_D	C_M	C_t	-	-	-	-	-	-	-
Spike Grids	$Z' = Z$ x	C_D	C_M	C_t	-	C_Δ	-	-	-	-	-
Withdrawal Loads											
Nails, Spikes, Lag Screws, Wood Screws, and Drift Pins	$W' = W$ x	C_D	C_M	C_t	-	-	-	C_{eg}	-	-	C_{tn}

Reprinted with permission from *National Design Specification® for Wood Construction.* Copyright © 2001 American Forest & Paper Association, Inc.

conditions, the nominal design value must be multiplied by the wet service factors in Table 5.4. These factors apply to both sawn lumber and glued laminated timber. Sawn lumber may be manufactured dry, partially seasoned, or wet and may be in service in dry, wet, partially seasoned, or exposed to weather conditions. Glued laminated timber is manufactured dry and may be used under the dry, wet, or exposed to weather conditions.

5.2.3 Temperature Factor, C_t

In cases where connection capacity is controlled by the wood and the fasteners are subject to sustained elevated temperatures, the temperature factor of Table 5.5 is to be applied.

5.2.4 Group Action Factor, C_g

Research has shown that the load carried by a row of fasteners, such as bolts, lag screws, drift pins and drift bolts, and split-ring and shear plate fasteners, is not divided equally among the fasteners; end fasteners in a row tend to carry a larger portion of the load than do the intermediate fasteners. The distribution of load among fasteners is a function of the relative stiffness of the main member and the side member(s). The unequal load sharing is accounted for by the group action factor, C_g.

With regard to determining the group action factor, C_g, the following conditions apply:

1. A group of fasteners consists of one or more rows of fasteners.
2. A row of fasteners consists of either two or more bolts loaded in single or multiple shear or two or more split rings, shear plates, or lag screws loaded in single shear. The row is aligned with the direction of the load.
3. When fasteners in adjacent rows are staggered and the distance between the rows is less than one-fourth of the spacing between the closest fasteners in an adjacent row (as shown in Figure 5.2), the fasteners in adjacent rows should be considered as one row.
4. The load for each row of fasteners is determined by summing the individual loads for each fastener in the row and then multiplying this value by the adjustment factor C_g. For convenience, C_g may be applied to individual fastener values prior to summation of values. The design value for the group of fasteners is the sum of the design values of the rows in the group.
5. When a member is loaded perpendicular to grain, its equivalent cross-sectional area is the product of the thickness of the member and the overall width of the fastener group for calculating cross-sectional area ratios. When only one row of fasteners is used, the width is equal to the minimum spacing for full load for the type of fastening used. In general, long rows of fasteners perpendicular to grain should be avoided.

TABLE 5.4 Wet Service Factors for Connections

Fastener Type	Moisture Content		C_M
	At Time of Fabrication	In-Service	
Lateral Loads			
Shear Plates & Split Rings[1]	≤ 19%	≤ 19%	1.0
	> 19%	≤ 19%	0.8
	any	> 19%	0.7
Metal Connector Plates	≤ 19%	≤ 19%	1.0
	> 19%	≤ 19%	0.8
	any	> 19%	0.7
Dowel-type Fasteners	≤ 19%	≤ 19%	1.0
	> 19%	≤ 19%	0.4[2]
	any	> 19%	0.7
Timber Rivets	≤ 19%	≤ 19%	1.0
	≤ 19%	> 19%	0.8
Withdrawal Loads			
Lag Screws & Wood Screws	any	≤ 19%	1.0
	any	> 19%	0.7
Nails & Spikes	≤ 19%	≤ 19%	1.0
	> 19%	≤ 19%	0.25
	≤ 19%	> 19%	0.25
	> 19%	> 19%	1.0
Threaded Hardened Nails	any	any	1.0

1. For split ring or shear plate connectors, moisture content limitations apply to a depth of $\frac{3}{4}''$ below the surface of the wood.
2. $C_M = 0.7$ for dowel type fasteners with diameter, D, less than $\frac{1}{4}''$. For dowel type fastener connections with:
   1) one fastener only, or
   2) two or more fasteners placed in a single row parallel to grain, or
   3) fasteners placed in two or more rows parallel to grain with separate splice plates for each row,

   $C_M = 1.0$.

Source: Reprinted with permission from *National Design Specification® for Wood Construction.* Copyright © 2001 American Forest & Paper Association, Inc.

TABLE 5.5 Temperature Factors, C_t, for Connections

In-Service Moisture Conditions[a]	C_t		
	$T \leq 100°F$	$100°F < T \leq 125°F$	$125°F < T \leq 150°F$
Dry	1.0	0.8	0.7
Wet	1.0	0.7	0.5

Source: National Design Specification® [2].

[a] Wet service conditions for connections are specified in Table 5.4.

The group action factor is taken to be unity ($C_g = 1.00$) for dowel-type fasteners with diameters of less than $\frac{1}{4}$ in. (nails and wood screws), as these fasteners tend to deform and more uniformly accept load than the larger, stiffer fasteners. Design values for timber rivets already account for group action. The group action factor for dowel fasteners $\frac{1}{4}$ in. in diameter and larger and for split-ring and shear plate connectors may be calculated by Equation 5-1. Tabulated values for specific fastener size, type, and spacing are provided in the *National Design Specification®* [2].

$$C_g = \frac{m(1 - m^{2n})}{n[(1 + R_{EA}m^n)(1 + m) - 1 + m^{2n}]} \frac{1 + R_{EA}}{1 - m} \tag{5.1}$$

where

n = number of fasteners in a row,
R_{EA} = the lesser of or E_sA_s/E_mA_m or E_mA_m/E_sA_s,
E_m = modulus of elasticity of the main member (psi),
E_s = modulus of elasticity of the side members (psi),
A_m = gross cross-sectional area of main member (in^2),
A_s = sum of the gross cross-sectional areas of the side members (in^2)
$m = u - \sqrt{2u^2 - 1}$

$$u = 1 + \gamma \frac{s}{2}\left(\frac{1}{E_mA_m} + \frac{1}{E_sA_s}\right)$$

s = the center-to-center spacing between adjacent fasteners in a row (in.), and
γ = load/slip modulus for a connection (lb/in.)
= 500,000 lb/in. for 4-in. split-ring or shear plate connectors
= 400,000 lb/in. for $2\frac{1}{2}$-in. split-ring or $2\frac{5}{8}$-in. shear plate connectors
= $(180{,}000)(D^{1.5})$ for bolts or lag screws in wood-to-wood connections
= $(270{,}000)(D^{1.5})$ for bolts or lag screws in wood-to-metal connections
D = the diameter of the bolt or lag screw (in.).

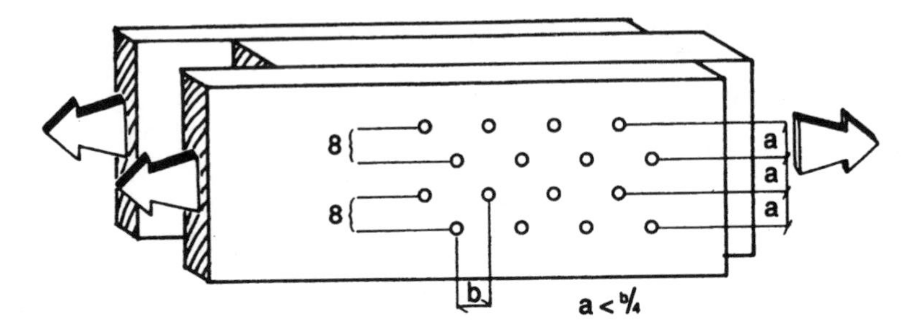

Consider as 2 rows of 8 fasteners

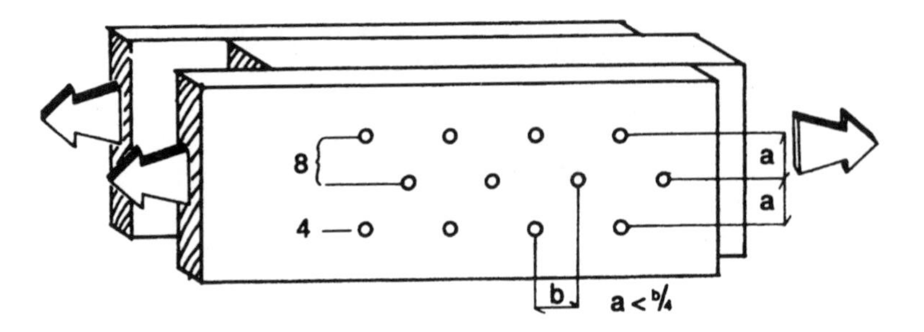

Consider as 1 row of 8 fasteners and 1 row of 4 fasteners

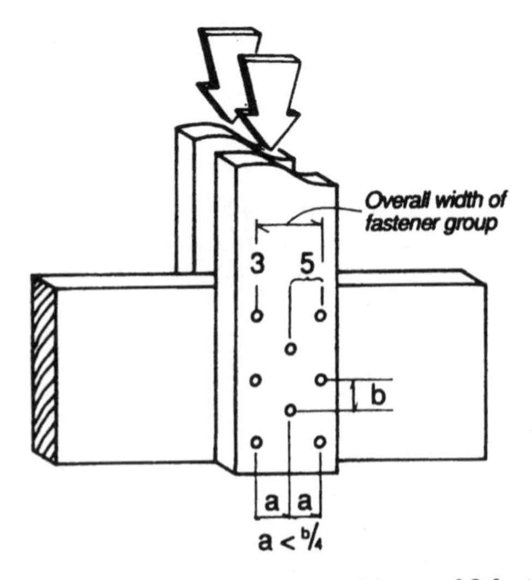

Consider as 1 row of 5 fasteners and 1 row of 3 fasteners

Figure 5.2 Group action for staggered fasteners. Reprinted with permission from *National Design Specification® for Wood Construction.* Copyright © American Forest & Paper Association, Inc.

5.2.5 Geometry Factor, C_Δ

The design values for bolts, split rings, shear plates, lag screws, drift bolts, and drift pins are based on specific end distance, edge distance, and spacing requirements for use of the full design values. Edge, end, and spacing requirements are generally specified in terms of fastener diameter D and are measured from the fastener centerline. Where these distances are less than required for full design values but not less than specified minimums, the design values for the fasteners are reduced by the geometry factor C_Δ.

The geometry factor is applicable to dowel fasteners with diameters equal to or greater than $\frac{1}{4}$ in., to split-ring and shear plate connectors, and to timber rivet connections where the rivet design value is controlled by the capacity of the wood perpendicular to the grain. It is not applicable to nails and screws. However, installation of nails and screws must be such that splitting of the wood members does not occur. Certain specific edge, end, and spacing requirements for nails and screws may be applicable, however, in the construction of structural diaphragms and as required by the manufacturer of proprietary engineered wood products.

5.2.6 Penetration Depth Factor, C_d

The effect of penetration depth of dowel fasteners is taken into direct account with the yield mode equations of Section 5.11 and is used to adjust tabular design values in the *National Design Specification*® [2] where fastener penetration is less than the values used in the *NDS*® tables but greater than stated minimum fastener penetration. For split-ring and shear plate connectors (Section 5.19), a depth penetration factor is used where lag screws are used instead of through bolts and the lag screw penetration is less than the penetration required for full design value but greater than or equal to the minimum penetration permitted.

5.2.7 End Grain Factor, C_{eg}

The nominal design values for fasteners are based on the fasteners being installed in the side grain of wood. Where dowel fasteners are installed in the end grain (fastener axis parallel to the grain), the end grain factor C_{eg} is to be applied.

5.2.8 Metal Side Plate Factor, C_{st}

The design values of 4-in. shear plates loaded parallel to the grain and timber rivets where capacity is controlled by the rivet are adjusted by the metal side plate factor, C_{st}. These factors are discussed in greater detail in the sections on shear plates and timber rivets.

5.2.9 Diaphragm Factor, C_{di}

When nails or spikes are used for diaphragms, the nominal lateral design values are multiplied by the diaphragm factor C_{di}, which is equal to 1.10.

5.2.10 Toenail Factor, C_{tn}

When toenailed connections are used, the nominal design values for nails and spikes are to be multiplied by the toenail factor, C_{tn}, which is equal to 0.83 for lateral loading and 0.67 for withdrawal loading. For toenailed nails and spikes loaded in withdrawal, the wet service factor C_M need not be applied simultaneously with the toenail factor.

5.3 NET AND EFFECTIVE SECTION

The installation of fasteners and other connection hardware in many cases reduces the carrying capacity of the member. For determining flexural capacity, the net section must be considered, where the loss in wood section due to all boring, grooving, dapping, and other means is taken into consideration. Figure 5.3 illustrates various net section conditions. Where bolts, drift bolts or pins, or lag screws are staggered in adjacent rows, but where the parallel to the grain spacing of fasteners in adjacent rows is less than four fastener diameters, the adjacent fasteners are considered to act at the same cross section for determining net section. In determining the shear capacity of members at connections, the locations of the fasteners and directions of loading determine the effective section. Effective depths for various connection conditions are illustrated in Figure 5.4.

In cases where the fasteners are located less than five times the depth of the member from the end of the member, the allowable design shear at the

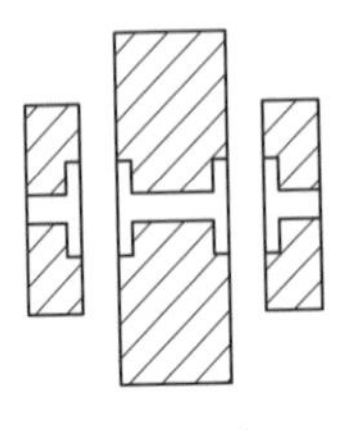

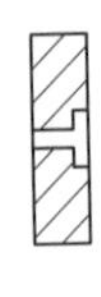

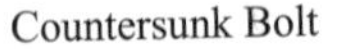

Split Ring or Shear Plate Connectors

Countersunk Bolt

Vertical Hole in Beam (effective hole diameter taken as 1.5 times actual hole diameter)

Figure 5.3 Net section at various connections.

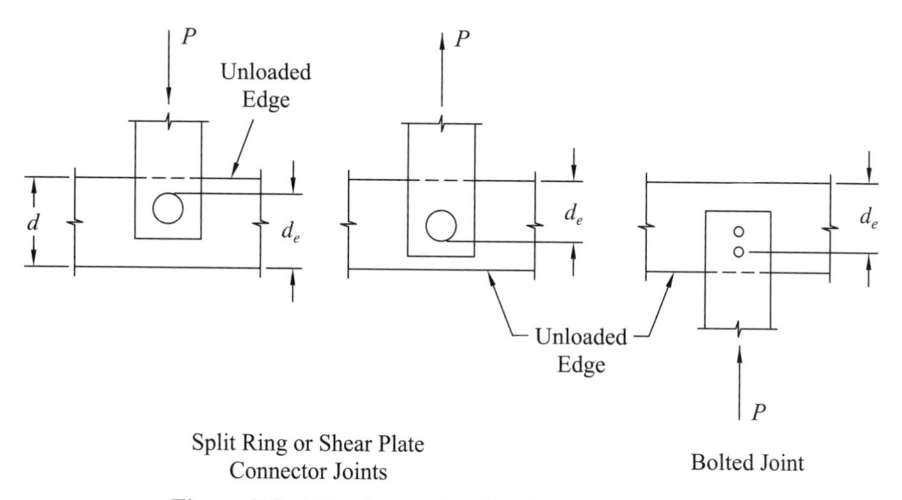

Figure 5.4 Effective section for shear at connections.

connection is determined by Equation 5.2a. Where the connection is at least five times the depth from the end, the allowable design shear for rectangular sections at the connection is given by Equation 5.2b.

$$V'_r = \left(\frac{2}{3} F'_v b d_e\right)\left(\frac{d_e}{d}\right)^2 \tag{5.2a}$$

$$V'_r = \tfrac{2}{3} F'_v b d_e \tag{5.2b}$$

Example 5.1 Effective Depth for Shear

Given: A 6 × 10 Hem Fir, No. 1 timber beam has an applied load P at distance x from the left end as illustrated in Figure 5.5. The span of the beam is 15 ft 6 in. taken center to center of the $5\frac{1}{2}$-in. supporting columns. Two values of x are considered: $x = 3$ ft and $x = 4$ ft. The beam will be in dry use and have loads of normal duration in service.

Determine: The maximum value of P based on shear at the connection.

Approach: The effective depth will be used in conjunction with equation (5.2a) or (5.2b), depending on the relative distance of the connection from the end.

Solution: From the Appendix, the dressed size of 6 × 10 timber is taken to be $5\frac{1}{2}$ in. × $9\frac{1}{2}$ in. For the case of $x = 4$ ft, the joint is located (4 ft)(12 in./ft)/9.5 in. = 5.05d from the end; therefore, the allowable design shear in the connection is calculated using equation (5.2b). From the *National Design Specification*® [2], $F_v = 140$ psi; thus, $F'_v = F_v C_D = (140 \text{ psi})(1.00 = 140$ psi. From the figure, $d_e = 6.5$ in.

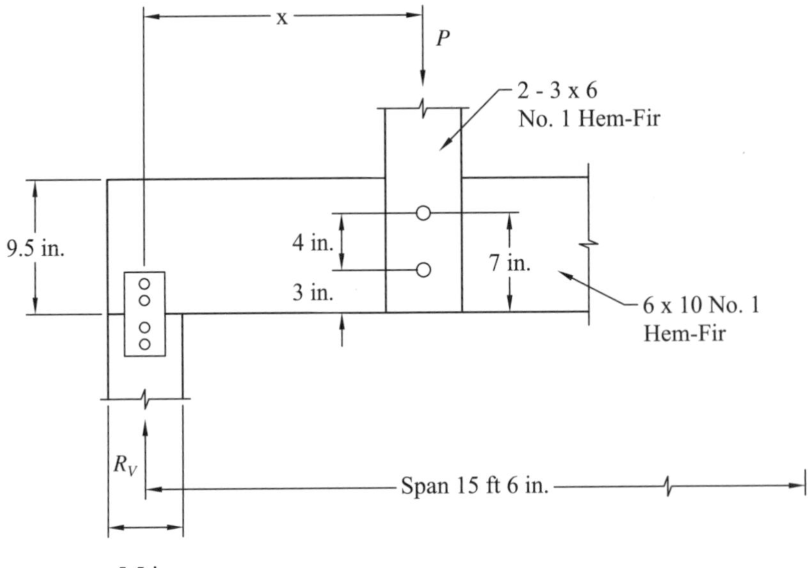

Figure 5.5 Joint detail for Example 5.1.

From equation (5.2b)

$$V'_r = \tfrac{2}{3}F'_v bd_e = \tfrac{2}{3}(140\text{ psi})(5.5\text{ in.})(6.5\text{ in.}) = 3337\text{ lb}$$

The effective span may be taken to be not less than the distance between the inside faces of the supports plus one-half the required bearing distance at each support. For convenience the span will be taken as the center-to-center column spacing, which assumes that half the required bearing distance is less than the distance from column centerline to inside face, which is often the case. Thus, $\ell = 15.5$ ft.

From the Appendix, the reaction due to P (and the applied shear from the support to the connection) may be computed by, $R = V = Pb/\ell$, where b is the distance from the applied load to the other end of the span. From Figure 5.5, $b = 15.5\text{ ft} - [4\text{ ft} - \tfrac{1}{2}(5.5\text{ in.}/(12\text{ in./ft}))] = 11.73$ ft. Since the applied load (and connection) are significantly closer to the left end than the right end, the shear between the connection and the left end will control:

$$P = \frac{R\ell}{b} = \frac{V\ell}{b} = \frac{(3337\text{ lb})(15.5\text{ ft})}{11.73\text{ ft}} = 4410\text{ lb}$$

The allowable load P based on shear at the connection for $x = 4$ ft is 4400 lb.

For the case of $x = 3$ ft, the connection is located (3 ft)(12 in./ft)/9.5 ft $= 3.79d$ from the end; therefore, equation (5.2a) must be used.

$$V_r' = \left(\frac{2}{3}F_v'bd_e\right)\left(\frac{d_e}{d}\right)^2 = \frac{2}{3}(140 \text{ psi})(5.5 \text{ in.})(6.5 \text{ in.})\left(\frac{6.5 \text{ in.}}{9.5 \text{ in.}}\right)^2$$

$$= (3337 \text{ lb})(0.468) = 1562 \text{ lb}$$

From Figure 5.5, $b = 15.5 \text{ ft} - [3 \text{ ft} - \frac{1}{2}(5.5/12) \text{ ft}] = 12.73 \text{ ft}$. Thus, $P = R\ell/b = V\ell/b = (1562 \text{ lb})(15.5 \text{ ft}/12.73 \text{ ft}) = 1902 \text{ lb}$.

Answer: The allowable loads P for the connection shown, based on effective depth for shear in the main member for the cases $x = 4$ ft and $x = 3$ ft, are 4400 lb and 1900 lb, respectively.

Discussion: The allowable load P based on shear at the connection for the case of $x = 3$ ft is significantly less than for $x = 4$ ft. The capacity of the connection based on the bolts must also be checked, taking into consideration edge distance in the main member (timber beam) and end and edge distances in the vertical wood plates. The bearing distance at the end of the beam must also be checked for suitability with regard to available bearing and also with regard to the use of the center-to-center column spacing for the span of the member. While the beam has a greater capacity in shear at the connection for $x = 4$ ft, it can be shown, in the case illustrated, that flexural and/or deflection considerations for the beam will be limiting.

5.4 ANGLE OF LOAD TO GRAIN

For dowel fasteners with diameters equal to and greater than $\frac{1}{4}$ in. (bolts, lag screws, etc.), shear plate and split-ring connectors, and timber rivets, the capacities of the fasteners loaded laterally (perpendicular to the fastener axis) are related directly to the angle of load with respect to the grain. The *National Design Specification®* [2] publishes tabulated nominal design capacities for these connectors loaded both parallel to and perpendicular to the grain. These fasteners in general have greater capacity loaded parallel to the grain than perpendicular to the grain. The capacities of nails and screws ($D < \frac{1}{4}$ in.) are considered to be independent of load direction with respect to the grain. However, the effect of large numbers of such fasteners causing stresses perpendicular to the grain near a loaded edge should be avoided.

The angle of load with respect to the grain, θ, may be different for the various members being connected by the same fastener. For example, in Figure 5.6 the angle of load with respect to the grain for the side members is 0° (parallel to the grain), whereas the angle of load to the grain with respect to the main member is 90° (perpendicular to the grain). Tabulated fastener design values for bolts and lag screws are provided in the *National Design Specification®* [2] for both side and main members loaded parallel to the grain, both loaded perpendicular to the grain, and one member loaded perpendicular to the grain and the other parallel. In cases where the angle of load to the grain

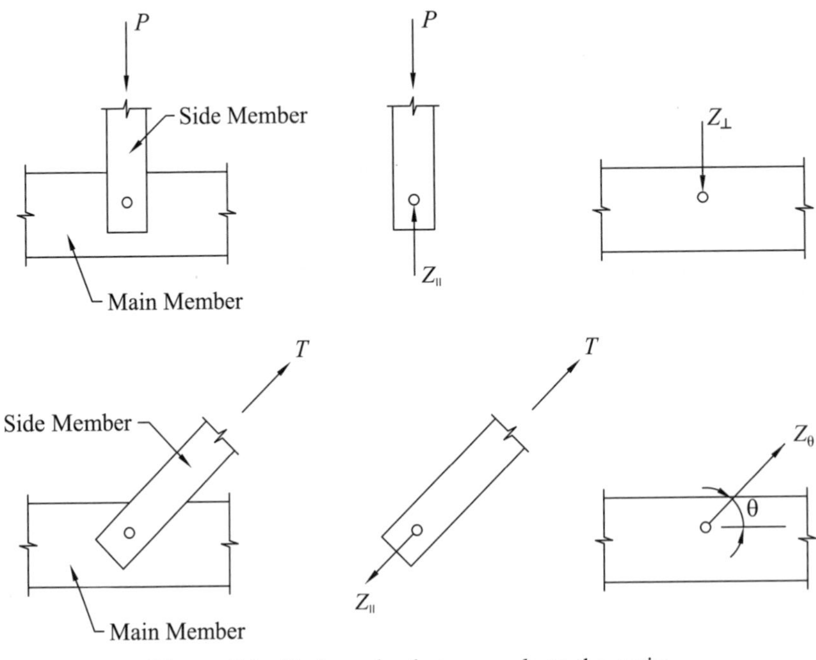

Figure 5.6 Fastener load at an angle to the grain.

in either member is at an angle other than 0° or 90°, the design value must be calculated taking into consideration the dowel bearing strengths in the direction of load of each member by utilizing the bearing strengths parallel and perpendicular to the grain and the Hankinson formula (Section 5.11). For split-ring, shear plate, and timber rivets, the design value for load at an angle between 0° and 90° is determined using the design values for loading parallel and perpendicular to the grain and the Hankinson formula (Section 5.20).

5.5 LOAD AT ANGLE TO FASTENER AXIS

Where the load on a fastener is at angle other than perpendicular to the fastener axis (Figure 5.7), the fastener must be capable of transmitting the load component parallel to the axis, either by the bearing of a bolt head or nut and washer on wood or other surface, or by withdrawal capacity developed by threads or friction. For the load perpendicular to the fastener axis, the length of fastener in the member is used for the (side or main) member thickness. Adequate end distance must be provided such that the equivalent shear area illustrated in Figure 5.7 for parallel members of the same main and side member thickness values satisfies the row and group tear-out provisions of Section 5.8 and the end distance requirements of Section 5.13.

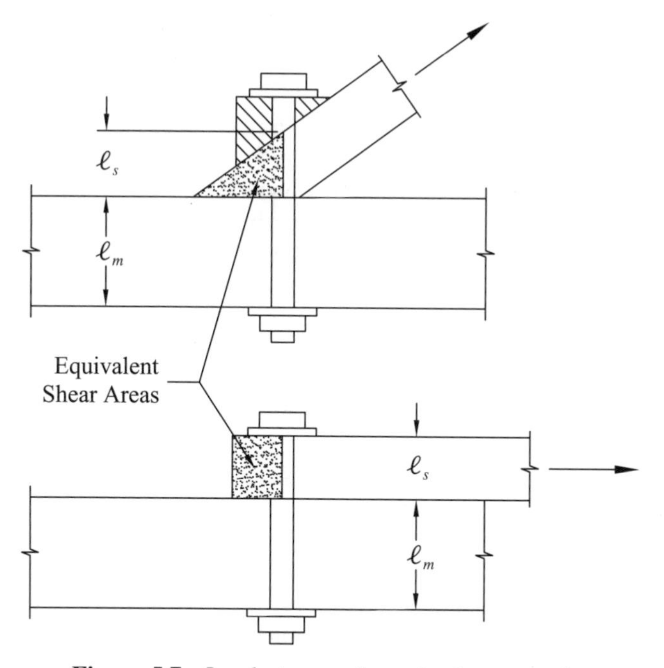

Figure 5.7 Load at an angle to the fastener axis.

5.6 EFFECT OF TREATMENT

No reduction in fastener design values is generally taken for preservative-treated wood except for fire-retardant treatment. Where fire-retardant treatments are applied, appropriate reduction values should be obtained from the company providing the treatment.

5.7 UPLIFT AND ANCHORAGE LOADS

The primary purpose of many wood framing members is to carry gravity loads. The loads carried by such members (beams, girders, and joists) are often transferred to supporting members by direct bearing of the member on supporting members or on connection hardware. In many cases, however, the framing members must also carry uplift and/or horizontal loads resulting from wind, seismic, or construction conditions. Where such loads are not explicitly considered in design, building codes and good design practice also require such members to be anchored against any horizontal or vertical movements or incidental forces. As such, in addition to providing for adequate bearing, anchorage resistance to uplift and lateral movement must be provided in connection design.

To provide anchorage, fasteners may be placed toward the lower edge of the member as shown in Figure 5.8. Placement of the fastener near the lower edge reduces the tendency of shrinkage to lift the member from the bearing condition and thus bear on the fastener in gravity load. However, the fasteners must also be placed high enough on the member to satisfy edge distance or other requirements for the fastener or connection. Where shrinkage effects may be significant, slotted connection hardware may be necessary. Where uplift loads are calculated, the shear capacity of the member under uplift load must be evaluated with the distance d_e applicable to the uplift loading condition (as shown in Figure 5.8).

5.8 FASTENER ROW AND GROUP TEAR-OUT

Groups of closely spaced bolts or split-ring and shear plate connectors near the ends of members loaded parallel to the grain may produce stresses sufficient to tear the fastener row or group out of the member. Two conditions must be considered, as illustrated in Figure 5.9: first, the tearing out of single row of fasteners; and second, the tearing out of a group of several rows of fasteners. These conditions may be examined using Equations 5.3a and 5.3b, respectively. Tear-out of timber rivets does not need to be considered inde-

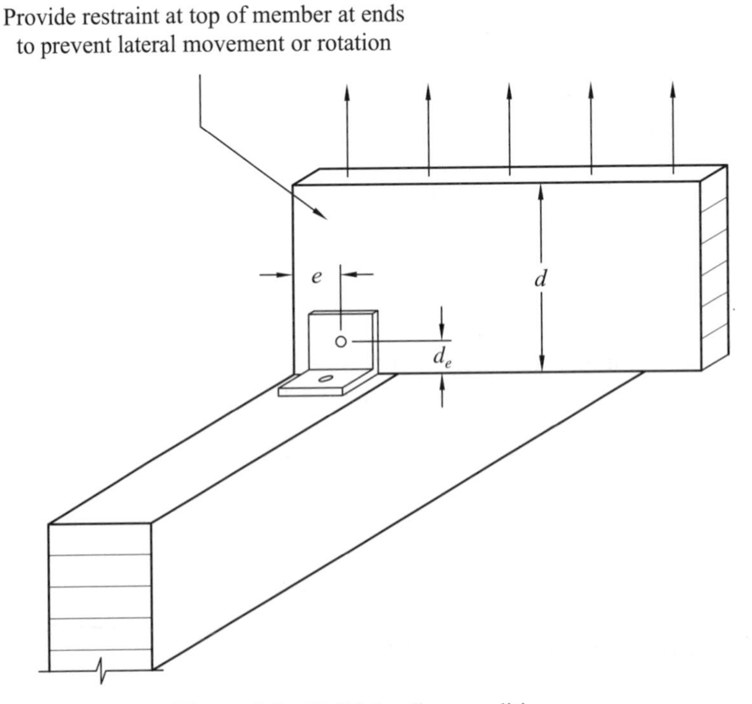

Figure 5.8 Uplift loading conditions.

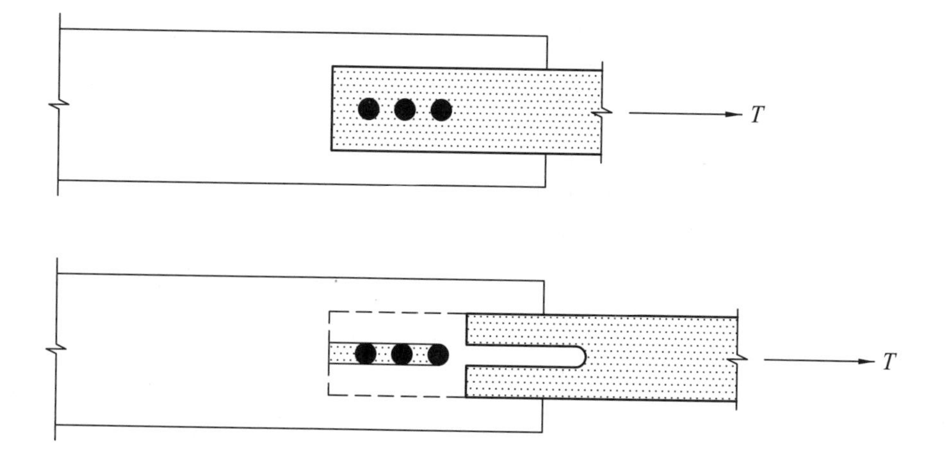

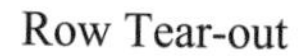

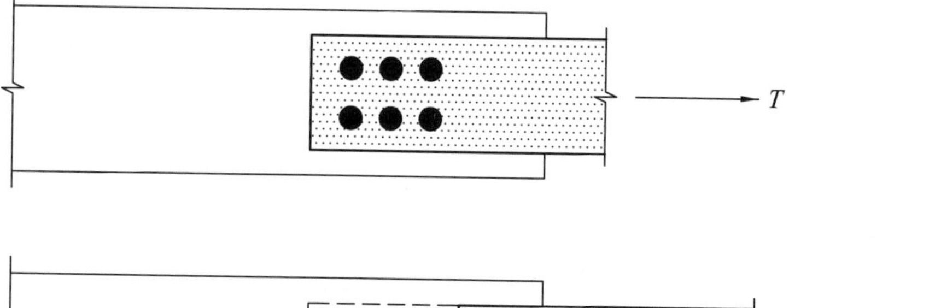

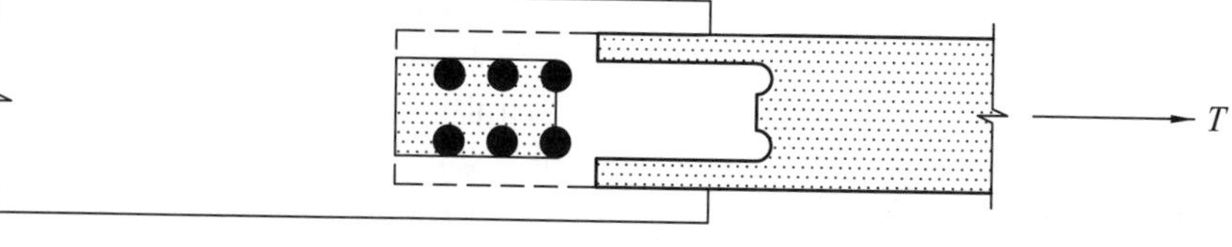

Figure 5.9 Fastener row and group tear-out.

pendently, because the design values for timber rivets already consider group tear out behavior.

$$Z'_{\mathrm{RT}} = n\frac{F'_v}{2} A_{\mathrm{crit\ shear}} \tag{5.3a}$$

$$Z'_{\mathrm{GT}} = \frac{(Z'_{\mathrm{RT}})_{\mathrm{top}}}{2} + \frac{(Z'_{\mathrm{RT}})_{\mathrm{bot}}}{2} + F'_t A_{\mathrm{eff\ tension}} \tag{5.3b}$$

where

Z'_{RT} = allowable load on the row of fasteners based on row tear-out,
n = the number of fasteners in the row,
F'_{v} = the shear parallel to the grain design value multiplied by all applicable adjustment factors,
$A_{\text{crit shear}}$ = the critical shear area described below,
Z'_{GT} = allowable load on the group of fasteners based on group tear-out,
$(Z'_{\text{RT}})_{\text{top}}$, $(Z'_{\text{RT}})_{\text{bot}}$ = the tear-out capacities of the bounding (top and bottom) rows,
F'_{t} = the tension design value parallel to the grain multiplied by all applicable adjustment factors,
$A_{\text{eff tension}}$ = the net section under tension due to group tear-out.

The tear-out capacity of a row of bolts is determined by the allowable shear stress multiplied by the critical shear surface, where the allowable shear stress is taken to be $F'_{v}/2$, times the number of fasteners in the row. The critical shear surface is taken to be the smallest fastener-to-fastener distance or fastener-to-end distance, multiplied by the member width (thickness), multiplied by 2 (effectively, two shear planes per row). The effective shear surface of fasteners that do not go through the member, such as split rings or shear plates (or a group of lag screws, wood screws, or nails), is determined by shortest fastener-to-fastener or fastener-to-end spacing, multiplied by two times the fastener penetration plus the shear area at the back of the connector parallel with the connected face (three shear planes per row).

The tear-out capacity of a fastener group is determined by effectively two shear planes and a net area across which tension stresses are developed for fasteners going through the member. For split-ring, shear plate, and other fasteners that do not penetrate the member fully, the tear-out acts over three shear planes and a net section across which tension stresses are developed. The effective shear area in both cases is determined by the bounding rows for the group, and thus by the smallest of the fastener-to-fastener or fastener-to-end distances. Examples of row and group tear-out capacities are calculated in the following sections.

5.9 FASTENERS INSTALLED IN END GRAIN

Fasteners installed in the ends of members or inserted parallel to the grain require special consideration. The dowel bearing strength for lateral loading of dowel fasteners is taken to be the dowel bearing strength perpendicular to grain. In addition, the end grain factor, $C_{eg} = 0.67$, is applied to the lateral design values. With the exception of lag screws, dowel fasteners must not be loaded in withdrawal when installed in end grain. Similarly, wood screws, nails, and spikes must not be relied upon for anchorage or resisting transient, incidental, or construction loads or movements where such would tend to load the fasteners in withdrawal from end grain. Such loads or movements must be resisted by other means.

Where lag screws are installed in end grain and loaded in withdrawal, the withdrawal design values are multiplied by $C_{eg} = 0.75$. Where lag screws are installed in end grain and are loaded in both withdrawal and laterally, *both* end grain factors apply. Considerations for split rings, shear plates, and timber rivet connectors installed in end grain are discussed in detail in their respective sections later in this chapter.

5.10 BEARING

Framing members generally transfer gravity loads through bearing on supporting members or supporting connection hardware. For horizontally framed wood members, bearing typically occurs perpendicular to the grain. Adequate bearing area must be provided, generally specified in terms of bearing length, so that the applied bearing stresses under design loading conditions do not exceed allowable bearing stresses:

$$f_{c\perp} = \frac{R}{b\ell_b} \leq F'_{c\perp} \tag{5.4}$$

where

R = the applied bearing force, in this case the reaction,
b = member width,
ℓ_b = length of the member in bearing contact. Allowable bearing stresses perpendicular to grain are determined from published design values multiplied by all applicable adjustment factors (Chapter 3). The load duration factor generally is not applied to compression perpendicular to the grain, because compression perpendicular to the grain is typically determined by a deformation limit, not wood rupture.

Where bearing occurs parallel to the grain, for example at the top of a column from a horizontal member, the applied bearing pressure on the column must not exceed the allowable bearing stress parallel to the grain. In this case the allowable bearing stress is determined by the tabulated compression parallel to the grain design value multiplied by all applicable adjustment factors, except C_P. Compression parallel to the grain must also be checked for the member, where C_P is considered. Compression parallel to the grain for the member is generally more critical than bearing stresses parallel to the grain, which generally govern only in conditions where the compressive forces are transmitted across areas less than the cross-sectional areas of the members.

Example 5.2 Bearing Length

Given: The 6 × 10 timber beam of Example 5.1.

Determine: The required bearing distance for the case $x = 3$ ft.

Approach: The bearing distance for $x = 3$ ft will be determined using the published value for compression perpendicular to the grain for Hem–Fir No. 1 timbers and using the reaction determined in Example 5.1 for $x = 3$ ft.

Solution: From example 5.1, the reaction corresponding to $x = 3$ ft is 1562 lb. The allowable compression perpendicular to the grain stress, $F'_{c\perp}$, is obtained by multiplying the published design value for $F'_{c\perp}$ for Hem–Fir No. 1 timbers, 405 psi, by all applicable adjustment factors. From Chapter 3 the applicable factors are C_M, C_t, and C_b. For the conditions stated, C_M and C_t are taken to be 1.00. The bearing area factor, C_b, Chapter 3, is not applicable to bearing at the end of the member. Thus, $F'_{c\perp} = F_{c\perp} = 405$ psi. The required bearing distance, ℓ_b, may be determined by equation (5.4) by letting $f_{c\perp} = F'_{c\perp} = R/b\ell_b$. Thus, the required bearing distance is

$$\ell_b = \frac{R}{F'_{c\perp}b} = \frac{1562 \text{ lb}}{(405 \text{ psi})(5.5 \text{ in.})} = 0.70 \text{ in.}$$

The available bearing distance in Example 5.1 is 5.5 in.; hence the bearing condition is adequate.

The span used in Example 5.1 was taken to be the center-to-center column spacing. The minimum span permitted is the distance between the inside faces of the supports plus one-half the required bearing distance at each support. Assuming each support to be 5.5 in., the minimum span permitted would be 15.5 ft $-$ (2) ($\frac{1}{2} \times 5.5/12$) ft + (2)($\frac{1}{2} \times 0.70/12$) ft = 15.1 ft, which is less than the span used; good.

Answer: The minimum required bearing length for the timber beam of Example 5.1 for $x = 3$ ft is 0.70 in.

Discussion: Although the minimum bearing length is calculated to be 0.70 in., building codes may prescribe a greater minimum distance (1 in. or more). The designer may also specify a greater bearing distance for ease and safety during construction, to accommodate minor framing length discrepancies in the field and/or to accommodate shrinkage of supporting members.

Where bearing stresses occur at an angle to grain, such as in Figure 5.10, the allowable bearing stress is calculated by the Hankinson formula,

$$F'_\theta = \frac{F_c^* F'_{c\perp}}{F_c^* \sin^2\theta + F'_{c\perp} \cos^2\theta} \tag{5.5}$$

where

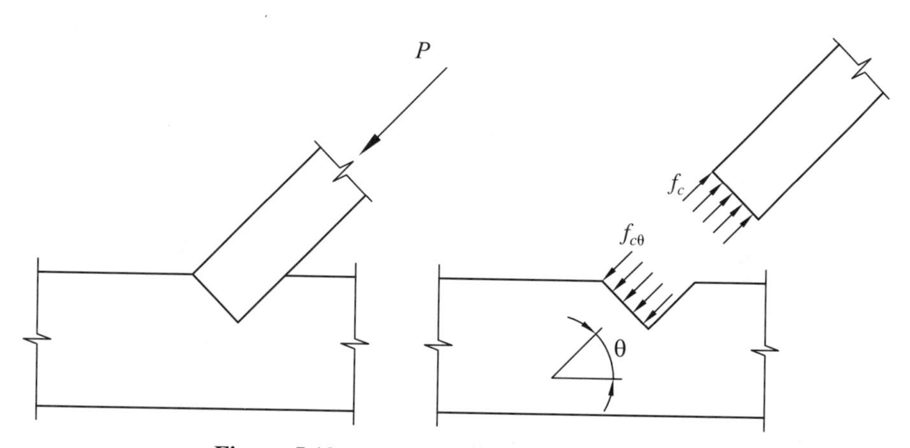

Figure 5.10 Bearing at an angle to the grain.

F'_θ = the allowable bearing stress at an angle to the grain,
θ = angle between the direction of the load (perpendicular to bearing surface) and the direction of the grain,
F^*_c = the design value for compression parallel to the grain multiplied by all applicable adjustment factors except C_P, and
$F'_{c\perp}$ = the design value for compression perpendicular to the grain multiplied by all applicable adjustment factors.

5.11 BOLTS AND LAG SCREWS LOADED LATERALLY

The design values for lateral strength Z for dowel fasteners (bolts, lag screws, wood screws, and nails and spikes) may be calculated directly or may be obtained from tables in the *National Design Specification®* [2] for single- and double-shear connections. Tabulated design values are based on the lowest or critical load considering the various modes of fastener and wood behavior shown in Figure 5.11. The design values, whether calculated or tabulated, assume that the faces of the connected members are in contact; the load acts perpendicular to the fastener axis; the edge distance, end distance, and spacing values are sufficient to develop the full design value; and the penetration of the fastener in the receiving member (main member for single shear or side member for double shear) is greater than or equal to the minimum penetration specified for the type of fastener.

The modes of fastener and member behavior illustrated in Figure 5.11 are described below.

1. Mode I_m represents the bearing-dominated yield of wood bearing on the main member.
2. Mode I_s represents bearing-dominated yield of the side member(s).

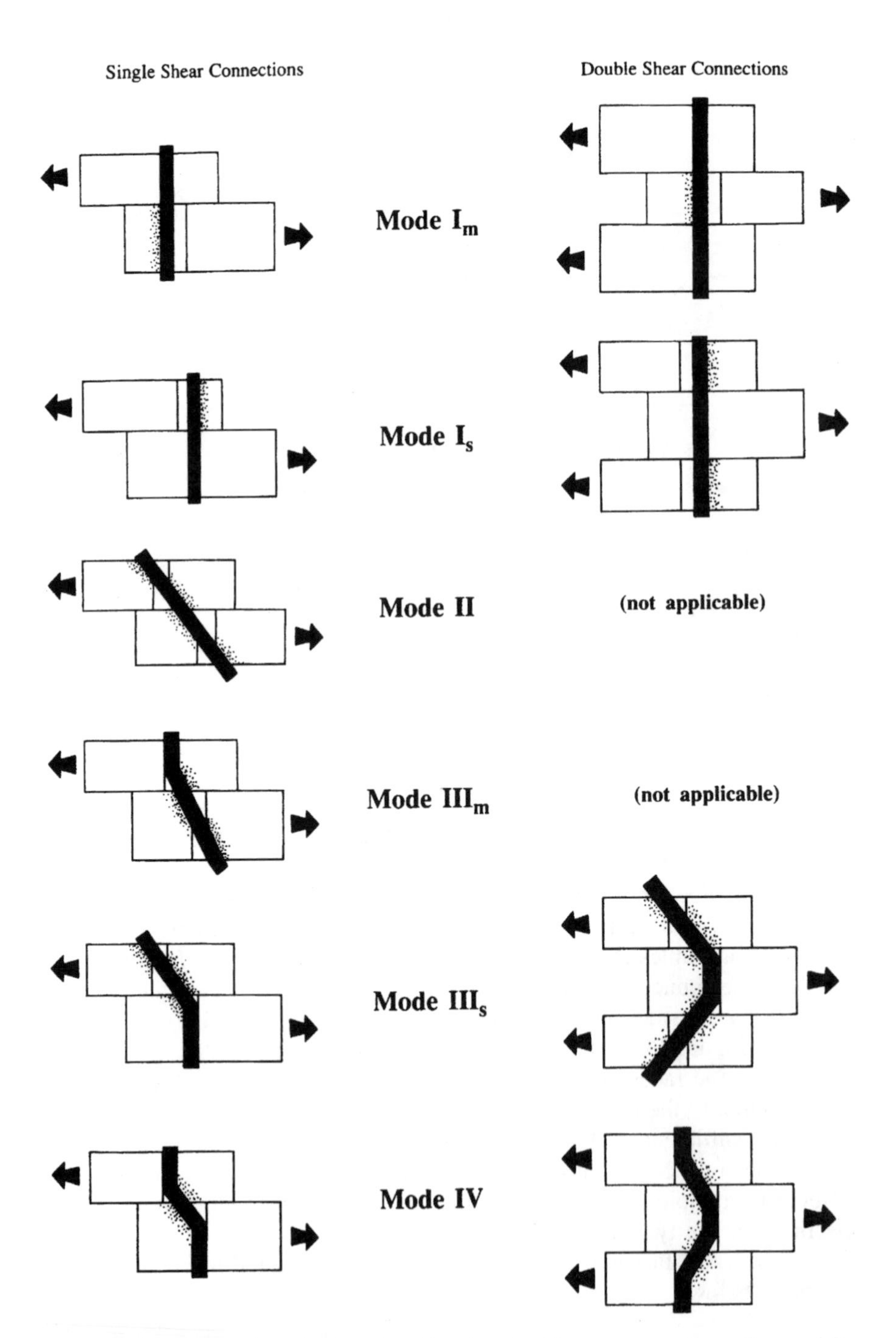

Figure 5.11 Connection yield modes. Reprinted with permission from *National Design Specification® for Wood Construction.* Copyright © American Forest & Paper Association, Inc.

3. Mode II represents fastener pivoting, producing limited localized crushing near the adjacent member faces.
4. Mode III represents fastener yielding one plastic hinge point per shear plane.
5. Mode IV represents fastener yielding with two plastic hinge points per shear plane.

Design values for dowel fasteners are generally denoted by Z. For fasteners with diameters equal to or greater than $\frac{1}{4}$ in., design values are tabulated [2] for the cases of both members loaded parallel to the grain ($Z_{\|}$), main member loaded parallel to the grain and side member(s) loaded perpendicular to the grain ($Z_{s\perp}$), main member loaded perpendicular to the grain and side member(s) loaded parallel to the grain ($Z_{m\perp}$), and both side and main members loaded perpendicular to the grain ($Z_{\perp}$). Design values multiplied by all applicable adjustment factors are denoted Z', $Z'_{s\perp}$, $Z'_{m\perp}$, and $Z'_{\perp}$. Where the load in either member is at an angle other than parallel or perpendicular to grain, the design value must be calculated based on the dowel bearing strength for the direction of load with respect to the grain for each of the members. Design values for dowel fasteners with diameters less than $\frac{1}{4}$ in. are assumed to be independent of load direction with respect to the grain and are denoted simply Z, and the adjusted values Z'.

Certain building codes allow use of a larger load duration factor for mode III and IV failures for wind and earthquake loads under certain load combinations. For this reason it may be advantageous to use the yield mode equations to determine fastener capacity, as tabulated values generally do not indicate the failure mode governing the tabulated design value.

5.11.1 Single-Shear Yield Limit Equations

Design values for dowel-type fasteners in single shear (two-member connections, Figure 5.12) may be calculated using the equations below. Values of Z

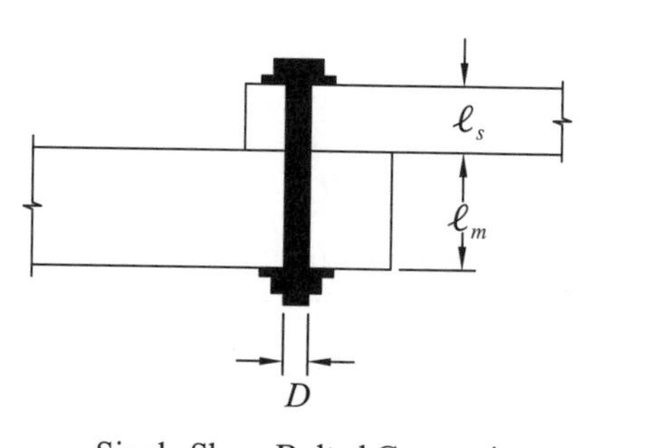

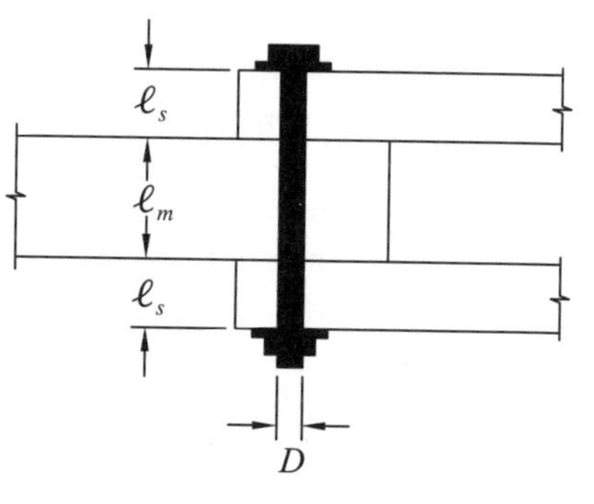

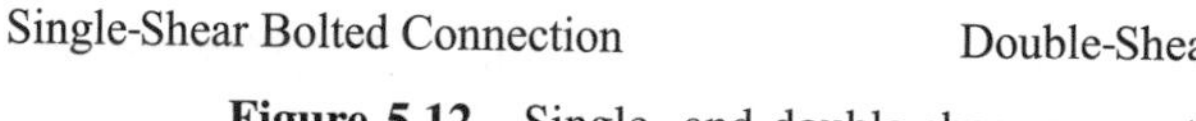

Figure 5.12 Single- and double-shear connections.

are calculated for each mode, with the Z value for the fastener being the lowest value of Z obtained.

$$Z = \frac{D\ell_m F_{em}}{R_d} \qquad \text{Mode I}_\text{m} \tag{5.6a}$$

$$Z = \frac{D\ell_s F_{es}}{R_d} \qquad \text{Mode I}_\text{s} \tag{5.6b}$$

$$Z = \frac{k_1 D\ell_s F_{es}}{R_d} \qquad \text{Mode II} \tag{5.6c}$$

$$Z = \frac{k_2 D\ell_m F_{em}}{(1 + 2R_e)R_d} \qquad \text{Mode III}_\text{m} \tag{5.6d}$$

$$Z = \frac{k_3 D\ell_s F_{em}}{(2 + R_e)R_d} \qquad \text{Mode III}_\text{s} \tag{5.6e}$$

$$Z = \frac{D^2}{R_d}\sqrt{\frac{2F_{em}F_{yb}}{3(1 + R_e)}} \qquad \text{Mode IV} \tag{5.6f}$$

where

$$k_1 = \frac{\sqrt{R_e + 2R_e^2\,(1 + R_t + R_t^2) + R_t^2R_e^3} - R_e\,(1 + R_t)}{1 + R_e} \tag{5.6g}$$

$$k_2 = -1 + \sqrt{2(1 + R_e) + \frac{2F_{yb}(1 + 2R_e)D^2}{3F_{em}\ell_m^2}} \tag{5.6h}$$

$$k_3 = -1 + \sqrt{\frac{2(1 + R_e)}{R_e} + \frac{2F_{yb}(2 + R_e)D^2}{3F_{em}\ell_s^2}} \tag{5.6i}$$

and where

D = fastener diameter (in.),
ℓ_m = main member dowel bearing length (in.),
F_{em} = dowel bearing strength of the main member (psi),
R_d = reduction term described below,
ℓ_s = side member dowel bearing length (in.),
F_{es} = dowel bearing strength of the side member (psi),
F_{yb} = dowel bending yield strength of the fastener (psi),
R_e = $F_{em}/F_{es}F_{em}/F_{es}$, and
R_t = ℓ_m/ℓ.

The fastener diameter to be used in the equations above is the body diameter for full-body fasteners and the reduced or root diameter for threaded fasteners except that the body diameter may be used if the threaded portion does not bear over more than one-fourth of the bearing length of the fastener. Bolts and lag screws must meet the requirements of ANSI/ASME Standard B18.2.1 [6]. Dimensions for lag screw fasteners, including root diameter and typical length of threaded portion, are included in the Appendix. Tabulated design values for fasteners in the *National Design Specification®* [2] assume full body diameter for bolts and root or reduced diameters for lag screws. The reduction term R_d is a function of fastener diameter, yield mode, and direction of load with respect to grain and is determined by the use of Table 5.6.

The dowel bearing strength used in the equations above may be obtained from Table 5.7 for loading parallel and perpendicular to the grain. For loading at an angle to the grain, the dowel bearing strength is determined by use of the Hankinson formula:

$$F_{e\theta} = \frac{F_{e\parallel} F_{e\perp}}{F_{e\parallel} \sin^2\theta + F_{e\perp} \cos^2\theta} \tag{5.7}$$

where

$F_{e\theta}$ = the dowel bearing strength at angle θ to the grain;
$F_{e\parallel}$ = the dowel bearing strength parallel to the grain, from Table 5.7,
$F_{e\perp}$ = the dowel bearing strength perpendicular to the grain as a function of fastener diameter, also from Table 5.7. Dowel bending yield strength values, F_{yb}, for various fasteners are given in the Appendix.

5.11.2 Double-Shear Dowel Yield Limit Equations

Design values for dowel fasteners in double shear may be calculated using the equations below or may be obtained from tables in the *National Design Specification®* [2]. As with single shear, values of Z are calculated for each

TABLE 5.6 Reduction Factor, R_d[a]

Fastener Size	Yield Mode	Reduction Term, R_d
0.25 in. < $D \leq$ 1 in.	I_m, I_s	$4K_\theta$
	II	$3.6K_\theta$
	III_m, III_s, IV	$3.2K_\theta$
$D <$ 0.25 in.	I_m, I_s, II, III_m, III· IV	K_D[b]

Source: National Design Specification® [2].

[a] $K_\theta = 1 + 0.25\ (\theta/90°)$; θ, maximum angle of load to grain ($0° \leq \theta \leq 90°$) for any member in a connection; D, diameter (in.); $K_D = 2.2$ for $D \leq 0.17$ in.; $K_D = 10D + 0.5$ for 0.17 in. < $D \leq 0.25$ in.

[b] For threaded fasteners where nominal diameter is greater than or equal to 0.25 in. and root diameter is less than 0.25 in., $R_d = K_D K_\theta$.

TABLE 5.7 Dowel Bearing Strength for Fasteners

Specific Gravity, G^a	Dowel Bearing Strength (psi)[b]										
	$D < \frac{1}{4}$ in.	$D \geq \frac{1}{4}$ in.									
			$F_{e\perp}$ for Diameter, D (in.)								
	F_e	$F_{e\parallel}$	$\frac{1}{4}$	$\frac{5}{16}$	$\frac{3}{8}$	$\frac{7}{16}$	$\frac{1}{2}$	$\frac{5}{8}$	$\frac{3}{4}$	$\frac{7}{8}$	1
0.73	9300	8200	7750	6900	6300	5850	5450	4900	4450	4150	3850
0.72	9050	8050	7600	6800	6200	5750	5350	4800	4350	4050	3800
0.71	8850	7950	7400	6650	6050	5600	5250	4700	4300	3950	3700
0.70	8600	7850	7250	6500	5950	5500	5150	4600	4200	3900	3650
0.69	8400	7750	7100	6350	5800	5400	5050	4500	4100	3800	3550
0.68	8150	7600	6950	6250	5700	5250	4950	4400	4050	3750	3500
0.67	7950	7500	6850	6100	5550	5150	4850	4300	3950	3650	3400
0.66	7750	7400	6700	5950	5450	5050	4700	4200	3850	3550	3350
0.65	7500	7300	6550	5850	5350	4950	4600	4150	3750	3500	3250
0.64	7300	7150	6400	5700	5200	4850	4500	4050	3700	3400	3200
0.63	7100	7050	6250	5600	5100	4700	4400	3950	3600	3350	3100
0.62	6900	6950	6100	5450	5000	4600	4300	3850	3500	3250	3050
0.61	6700	6850	5950	5350	4850	4500	4200	3750	3450	3200	3000
0.60	6500	6700	5800	5200	4750	4400	4100	3700	3350	3100	2900
0.59	6300	6600	5700	5100	4650	4300	4000	3600	3300	3050	2850
0.58	6100	6500	5550	4950	4500	4200	3900	3500	3200	2950	2750
0.57	5900	6400	5400	4850	4400	4100	3800	3400	3100	2900	2700
0.56	5700	6250	5250	4700	4300	4000	3700	3350	3050	2800	2650
0.55	5550	6150	5150	4600	4200	3900	3650	3250	2950	2750	2550
0.54	5350	6050	5000	4450	4100	3750	3550	3150	2900	2650	2500
0.53	5150	5950	4850	4350	3950	3650	3450	3050	2800	2600	2450
0.52	5000	5800	4750	4250	3850	3550	3350	3000	2750	2550	2350
0.51	4800	5700	4600	4100	3750	3450	3250	2900	2650	2450	2300

0.50	4650	5600	4450	4000	3650	3400	3150	2800	2600	2400	2250
0.49	4450	5500	4350	3900	3550	3300	3050	2750	2500	2300	2150
0.48	4300	5400	4200	3750	3450	3200	3000	2650	2450	2250	2100
0.47	4150	5250	4100	3650	3350	3100	2900	2600	2350	2200	2050
0.46	4000	5150	3950	3550	3250	3000	2800	2500	2300	2100	2000
0.45	3800	5050	3850	3450	3150	2900	2700	2400	2200	2050	1900
0.44	3650	4950	3700	3300	3050	2800	2600	2350	2150	2000	1850
0.43	3500	4800	3600	3200	2950	2700	2550	2250	2050	1900	1800
0.42	3350	4700	3450	3100	2850	2600	2450	2200	2000	1850	1750
0.41	3200	4600	3350	3000	2750	2550	2350	2100	1950	1800	1650
0.40	3100	4500	3250	2900	2650	2450	2300	2050	1850	1750	1600
0.39	2950	4350	3100	2800	2550	2350	2200	1950	1800	1650	1550
0.38	2800	4250	3000	2700	2450	2250	2100	1900	1750	1600	1500
0.37	2650	4150	2900	2600	2350	2200	2050	1850	1650	1550	1450
0.36	2550	4050	2750	2500	2250	2100	1950	1750	1600	1500	1400
0.35	2400	3900	2650	2400	2150	2000	1900	1700	1550	1400	1350
0.34	2300	3800	2550	2300	2100	1950	1800	1600	1450	1350	1300
0.33	2150	3700	2450	2200	2000	1850	1750	1550	1400	1300	1200
0.32	2050	3600	2350	2100	1900	1750	1650	1500	1350	1250	1150
0.31	1900	3450	2250	2000	1800	1700	1600	1400	1300	1200	1100

Source: National Design Specification® [2].

[a] Specific gravity based on weight and volume when oven dry. Different specific gravities (G) are possible for different grades of MSR and MEL lumber.

[b] $F_{e\parallel} = 11{,}200G$; $F_{e\perp} = 6100G^{1.45}/\sqrt{D}$; F_e (for $D < \frac{1}{4}$ in.) $= 16{,}600G^{1.84}$. Values are rounded to the nearest 50 psi.

mode, with the Z value for the fastener being the lowest value of Z obtained. Calculated and tabulated values assume symmetric connections with side members of equal thickness and load orientation with respect to grain. Yield modes II and $\mathrm{III_m}$ are not applicable to double-shear connections. A typical double-shear (three-member) connection in illustrated in Figure 5.12.

$$Z = \frac{D\ell_m F_{em}}{R_d} \qquad \text{Mode } \mathrm{I_m} \qquad (5.8a)$$

$$Z = \frac{2D\ell_s F_{es}}{R_d} \qquad \text{Mode } \mathrm{I_s} \qquad (5.8b)$$

$$Z = \frac{2k_3 D\ell_s F_{em}}{(2 + R_e)R_d} \qquad \text{Mode } \mathrm{III_s} \qquad (5.8c)$$

$$Z = \frac{2D^2}{R_d}\sqrt{\frac{2F_{em}F_{yb}}{3(1 + R_e)}} \qquad \text{Mode IV} \qquad (5.8d)$$

The values of D, ℓ_m, F_{em}, ℓ_s, R_d, F_{es}, F_{yb}, and R_e are as defined previously. It should be noted that ℓ_s is the effective dowel bearing length of one side member (assumed to be equal to the other).

5.11.3 Geometry Factor for Bolts and Lag Screws

The geometry factors for dowel fasteners of diameter greater than or equal to $\frac{1}{4}$ in. are described in greater detail below. The geometry factors applicable to shear plates and timber rivets are described in their appropriate sections later in the chapter. The geometry factor, C_Δ, is generally subdivided into $C_{\Delta s}$, the spacing factor, $C_{\Delta e}$ the edge distance factor, and $C_{\Delta n}$, the end distance factor. These factors are not accumulative and the smallest controls in determining C_Δ. When fasteners are used in a group and any one of the fasteners is adjusted by any of the geometry factors $C_{\Delta e}$, $C_{\Delta n}$, or $C_{\Delta s}$, the entire group must be adjusted by the lowest factor obtained.

5.11.3.1 Geometry Factor for Edge Distance, $C_{\Delta e}$ Edge distance requirements for dowel fasteners in wood members are shown in Table 5.8. Edge distance requirements are based on the direction of load with respect to the edge, direction of load with respect to the grain, and relative length of fastener in the member with respect to fastener diameter. Edge distances (as well as end distances and spacing values) are generally given in terms of fastener diameter, D. Figure 5.13 illustrates various edge, end, and spacing conditions. As such, for dowel fasteners, all of the edge distances of Table 5.8 are to be met in all circumstances, and the geometry factor for edge distance is not applicable ($C_{\Delta e} = 1.00$).

TABLE 5.8 Edge Distance Requirement: Dowel Fasteners

Direction of Loading	Minimum Edge Distance[a]
Parallel to the grain	
When $\ell/D \leq 6$	$1.5D$
When $\ell/D > 6$	$1.5D$ or one-half the spacing between rows, whichever is greater
Perpendicular to the grain	
Loaded edge	$4D$
Unloaded edge	$1.5D$

[a] The ℓ/D ratio used to determine the minimum edge distance is to be the lesser of (a) the length of fastener in the wood main member/$D = \ell_m/D$ or (b) the total length of fastener in wood side member(s)/$D = \ell_s/D$. Heavy or medium concentrated loads should not be suspended below the neutral axis of a single sawn lumber or glued laminated timber beam except where mechanical or equivalent reinforcement is provided to resist tension stresses perpendicular to the grain.

5.11.3.2 Geometry Factor for End Distance, $C_{\Delta n}$ End distance requirements are shown in Table 5.9. End distance requirements are a function of species (softwood or hardwood), direction of load with respect to the grain, and whether the load is directed away from or toward the end of the member. Table 5.9 gives the end distances required for use of the full design value of the fastener, as well as minimum end distance values for reduced fastener design values. In cases where the end distances for the full design values are met, no end distance geometry factor need be applied ($C_{\Delta n} = 1.00$). Where the end distance is less than required for full design value but not less than the minimum distance for a reduced design value, the geometry factor is applied as calculated by

$$C_{\Delta n} = \frac{\text{actual end distance}}{\text{minimum end distance for full design value}} \tag{5.9}$$

In cases of nonparallel main and side members, as illustrated in Figure 5.7, the minimum end distance for the full design value for the fastener for any end is the end distance that produces the same shear area, as would be determined by the end distance of Table 5.9 for parallel members with thicknesses equal to the length of fastener bearing in each member. In cases where the shear area is less than that required by the end distances for full design value in Table 5.9, the geometry factor for end distance is applied as calculated by equation (5.10), except that in no case may a fastener be installed such that the shear area is less than one-half of the shear area associated with full end distance and full design value.

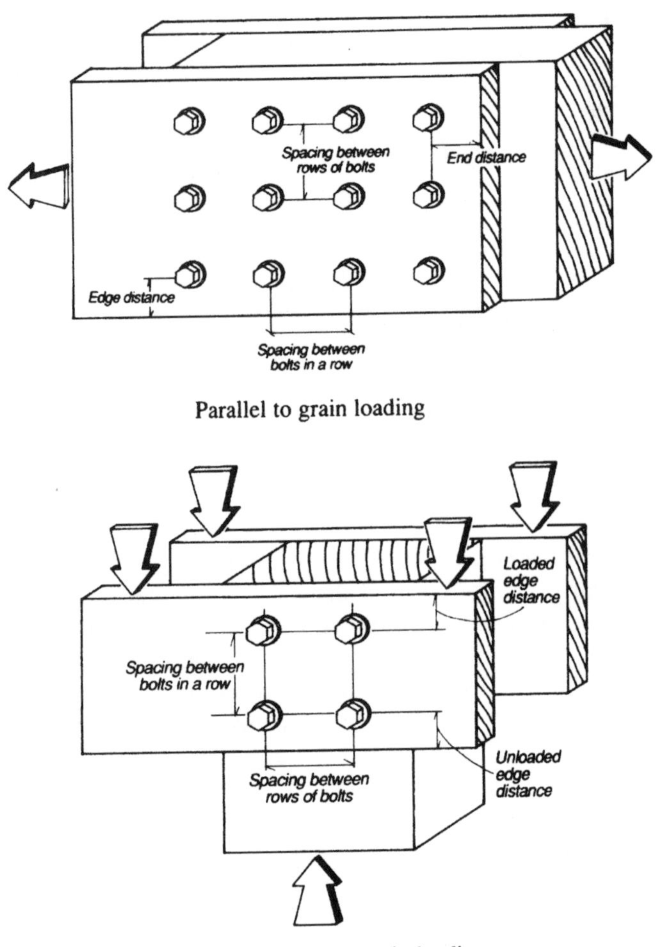

Figure 5.13 Connection geometry. Reprinted with permission from *National Design Specification® for Wood Construction.* Copyright © American Forest & Paper Association, Inc.

$$C_{\Delta n} = \frac{\text{actual shear area}}{\text{minimum shear area for full design value}} \tag{5.10}$$

5.11.3.3 Geometry Factor for Spacing, $C_{\Delta s}$ Spacing requirements for fasteners in a row and for spacing between rows are shown in Table 5.10. Where the spacing requirements for full design value are met, no adjustment is necessary ($C_{\Delta s}$ = 1.00). Where spacing is less than the value required for full design value, but not less than the minimum for reduced design value, the geometry factor is calculated by

TABLE 5.9 End Distance Requirements: Dowel Fasteners

Direction of Loading	Minimum End Distance	
	Reduced Design Value	Full Design Value
Perpendicular to the grain	$2D$	$4D$
Parallel to the grain, compression		
Fastener bearing away from member end	$2D$	$4D$
Parallel to Grain tension		
Fastener bearing toward member end		
For softwoods	$3.5D$	$7D$
For hardwoods	$2.5D$	$5D$

$$C_{\Delta s} = \frac{\text{actual spacing}}{\text{minimum spacing for full design value}} \tag{5.11}$$

Spacing between fasteners in a row and spacing of rows must both be considered, with the lowest geometry factor controlling.

In addition to the required spacing between rows, the *overall* spacing of rows of fasteners parallel to a member must be limited to 5 in. where the fasteners attach to a single side or splice plate (Figure 5.14) unless special detailing is provided to ensure that fabrication is such that accommodation is made for expansion or contraction due to moisture content or other changes. Bearing connections with metal side plates attached to the bearing plates should not have fasteners located more than 5 in. above the bearing surface. Spacing of fasteners in a row and between rows for fasteners of $D < \frac{1}{4}$ in. must be such that splitting or other damage to the wood members is avoided (or as required by the manufacturer for proprietary engineered wood products).

TABLE 5.10 Spacing Requirements Between Rows[a]

Direction of Loading	Minimum Spacing
Parallel to the grain	$1.5D$
Perpendicular to the grain	
When $\ell/D \leq 2$	$2.5D$
When $2 < \ell/D < 6$	$(5\ell + 10D)/8$
When $\ell/D \geq 6$	$5D$

[a] The ℓ/D ratio used to determine the minimum spacing between rows is to be the lesser of (a) the length of fastener in the wood main member$/D = \ell_m/D$ or (b) the total length of fastener in wood side member(s)$/D = \ell_s/D$. The spacing between outer rows of fasteners paralleling the member on a single splice plate should not exceed 5 in. (Figure 5.14).

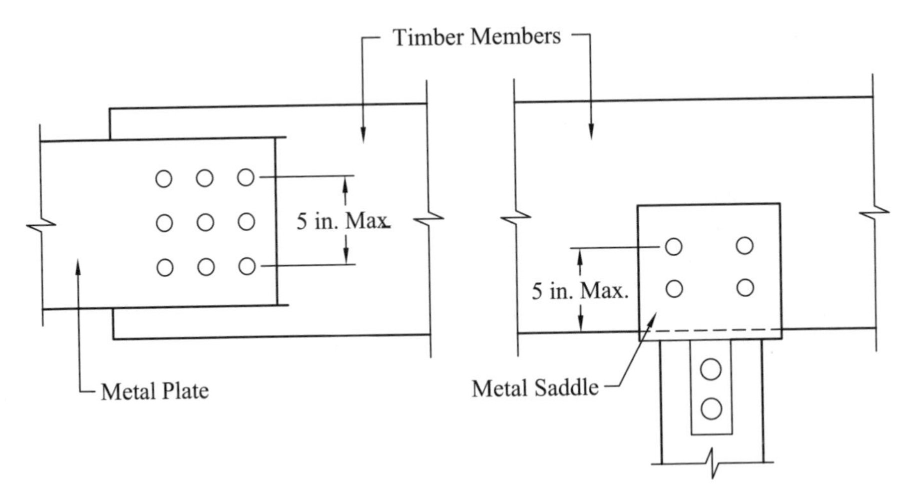

Figure 5.14 Overall spacing of rows of fasteners.

Example 5.3 Single-Shear Connection

Given: A horizontal $5\frac{1}{8}$ in. × 12 in. glued laminated timber member is fastened to an $8\frac{3}{4}$ in. × 9 in. glued laminated timber vertical column with two $\frac{3}{4}$-in. through bolts as shown in Figure 5.15. The species of each member is Douglas fir.

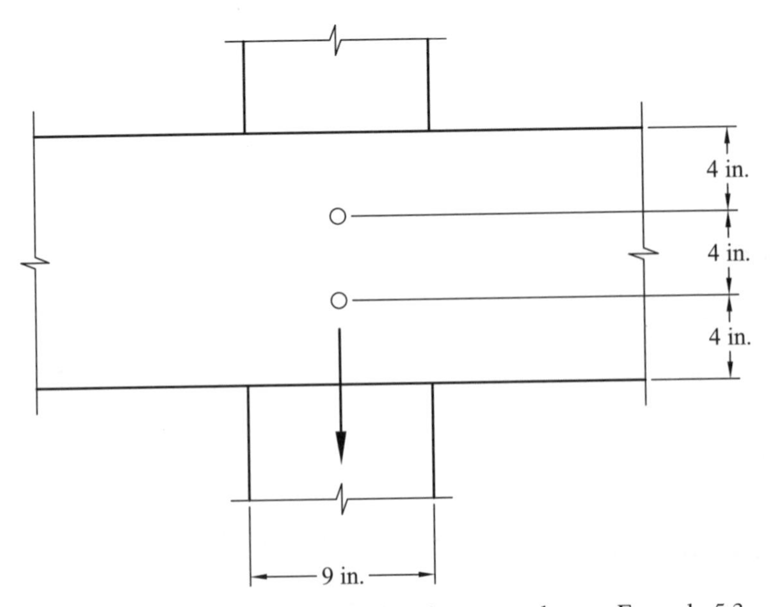

Figure 5.15 Single-shear connection, beam to column—Example 5.3.

Determine: The maximum downward load that can be transferred by the connection from the beam to the column assuming a load duration factor of 1.25.

Approach: Check the edge distance, end distance, and spacing requirements and apply a geometry factor if applicable. Determine the group action factor. Obtain design values using equations (5.6a) through (5.6f) or obtain from tables published in the *National Design Specification*® [2].

Solution: Assume that the bolts are spaced evenly from edge to edge and with respect to one another such that the edge distances and spacing in the horizontal (side) member are 4 in. each. The 9-in. dimension for the column will be taken parallel to the side member; thus, the edge distances in the vertical member are $4\frac{1}{2}$ in. each. The vertical member is assumed to extend upward past the connection; therefore, end distance requirements are not applicable in this case. The load direction is downward; therefore, for both members there is a single row of two bolts.

For determining the minimum required edge distance for the member loaded parallel to the grain (the vertical column), the ratio of the fastener bearing length to the fastener diameter, ℓ/D, is determined with respect to side and main members, and the smaller value is used in Table 5.8. In this example, $\ell/D = 5.125$ in./0.75 in. $= 6.833$ for the side member and 8.75 in./0.75 in. $= 11.67$ for the main member, the smaller of which is 6.833. From Table 5.8 the minimum edge distance is the greater of $1.5D$ or one-half the spacing between rows, in this case $1.5D = 1.5 \times 0.75 = 1.125$ in. (since spacing between rows is not applicable). The required edge distances for the horizontal member, from Table 5.8, are $1.5D = 1.125$ in. for the unloaded edge (below the bottom bolt) and $4.0D = 3$ in. above the top bolt. With the bolts located and spaced as described above, all the edge distance requirements are satisfied.

For group action, $A_s = 5\frac{1}{8}$ in. $\times$ 4 in. $= 20.5$ in^2, $A_m = 8.75$ in. $\times$ 9 in. $= 78.75$ in^2, and $A_s/A_m = 0.26$. It will be assumed that the modulus of elasticity values for both members are 1,600,000 psi for the purpose of calculating group action; thus,

$$R_{EA} = \frac{E_s A_s}{E_m A_m} = \frac{A_s}{A_m} = 0.26$$

$$\gamma = (180{,}000)D^{1.5} = (180{,}000)(0.75^{1.5}) = 116{,}900 \text{ lb/in.}$$

$$s = 4 \text{ in.}$$

$$u = 1 + \gamma \frac{s}{2}\left(\frac{1}{E_m A_m} + \frac{1}{E_s A_s}\right)$$

$$= 1 + 116{,}900\left(\frac{4}{2}\right)\left[\frac{1}{(1{,}600{,}000)(78.75)} + \frac{1}{(1{,}600{,}000)((20.5)}\right]$$

$$= 1.009$$

$$m = u - \sqrt{u^2 - 1} = 1.009 - \sqrt{1.009^2 - 1} = 0.875$$

$$n = 2$$

From equation (5.1)

$$C_g = \frac{m(1 - m^{2n})}{n[(1 + R_{RE}m^n)(1 + m) - 1 + m^{2n}]} \frac{1 + R_{EA}}{1 - m}$$

$$= \frac{(0.875)(1 - 0.875^{(2)(2)})}{2[(1 + (0.26)(0.875^2))(1 + 0.875) - 1 + 0.875^{(2)(2)}]} \left(\frac{1 + 0.26}{1 - 0.875}\right)$$

$$= 0.995$$

Equations (5.6a) through (5.6f) are used with the following inputs to determine the design value, Z, for each bolt in the connection.

$$F_{yb} = 45{,}000 \text{ psi (from the Appendix)}$$

$$G = 0.50 \text{ (from Table 5.1 or Table 2.2)}$$

$$F_{em} = F_{e\parallel} = 5600 \text{ psi (from Table 5.7)}$$

$$F_{es} = F_{e\perp} = 2600 \text{ psi (from Table 5.7)}$$

$$R_e = \frac{F_{em}}{F_{es}} = \frac{5600}{2600} = 2.154$$

$$\ell_m = 8.75 \text{ in.}$$

$$\ell_s = 5.125 \text{ in.}$$

For use in Table 5.6, $K_\theta = 1 + (0.25)(\theta/90) = 1 + (0.25)(90/90) = 1.25$. From Table 5.6, the reduction factor R_d is thus calculated as follows:

$$R_d = 4K_\theta = (4)(1.25) = 5.00 \text{ for yield modes } \mathrm{I_m} \text{ and } \mathrm{I_s}$$

$$= 3.6K_\theta = (3.6)(1.25) = 4.50 \text{ for yield mode II}$$

$$= 3.2K_\theta = (3.2)(1.25) = 4.00 \text{ for yield modes } \mathrm{III_m}, \text{ and } \mathrm{III_s}, \text{ and IV}$$

$$= \frac{\ell_m}{\ell_s} = \frac{8.75}{5.125} = 1.707$$

Thus,

$$k_1 = \frac{\sqrt{R_e + 2R_e^2(1 + R_t + R_t^2) + \mathbf{R}_t^2 R_e^3} - R_e(1 + R_t)}{1 + R_e}$$

$$= \frac{\sqrt{2.154 + (2)(2.154^2)(1 + 1.707 + 1.707^2) + (1.707^2)(2.154^3)} - (2.154)(1 + 1.707)}{1 + 2.154}$$

$$= 1.047$$

$$k_2 = -1 + \sqrt{(2)(1 + R_e) + \frac{2F_{yb}(1 + 2R_e)D^2}{3F_{em}\ell_m^2}}$$

$$= -1 + \sqrt{(2)(1 + 2.154) + \frac{(2)(45{,}000 \text{ psi})[1 + (2)(2.154)](0.75 \text{ in.})^2}{(3)(5600 \text{ psi})(8.75 \text{ in})^2}}$$

$$= 1.553$$

$$k_3 = -1 + \sqrt{\frac{(2)(1 + R_e)}{R_e} + \frac{2F_{yb}(2 + R_e)D^2}{3F_{em}\ell_s^2}}$$

$$= -1 + \sqrt{\frac{(2)(1 + 2.154)}{2.154} + \frac{(2)(45{,}00 \text{ psi})(2 + 2.154)(0.75 \text{ in.})^2}{(3)(5600 \text{ psi})(5.125 \text{ in.})^2}}$$

$$= 0.845$$

Therefore, the following values are calculated from equations (5.6a)–(5.6f):

$$Z = \frac{D\ell_m F_{em}}{R_d} = \frac{(0.75 \text{ in.})(8.75 \text{ in.})(5600 \text{ psi})}{5.00}$$

$= 7350$ lb mode I_m From (5.6a)

$$Z = \frac{D\ell_s F_{es}}{R_d} = \frac{(0.75 \text{ in.})(5.125 \text{ in.})(2600 \text{ psi})}{5.00}$$

$= 1999$ lb mode I_s From (5.6b)

$$Z = \frac{k_1 D\ell_s F_{es}}{R_d} = \frac{(1.047)(0.75 \text{ in.})(5.125 \text{ in.})(2600 \text{ psi})}{4.50}$$

$= 2326$ lb mode II From (5.6c)

$$Z = \frac{k_2 D\ell_m F_{em}}{(1 + 2R_e)R_d} = \frac{(1.553)(0.75 \text{ in.})(8.75 \text{ in.})(5600 \text{ psi})}{[1 + (2)(2.154)](4.00)}$$

$= 2688$ lb mode III_m From (5.6d)

$$Z = \frac{k_3 D \ell_s F_{em}}{(2 + R_e)R_d} = \frac{(0.845)(0.75\text{ in.})(5.125\text{ in.})(56000\text{ psi})}{(2 + 2.154)(4.00)}$$

$$= 1095\text{ lb} \qquad \text{mode III}_s \qquad \text{From (5.6e)}$$

$$Z = \frac{D^2}{R_d}\sqrt{\frac{2F_{em}F_{yb}}{3(1 + R_e)}} = \frac{(0.75\text{ in.})^2}{4.00}\sqrt{\frac{(2)(5600\text{ psi})(45{,}000\text{ psi})}{(3)(1 + 2.154)}}$$

$$= 1026\text{ lb} \qquad \text{mode IV} \qquad \text{From (5.6f)}$$

The smallest of the Z values above is 1026 lb (mode IV). The design values for the $\frac{3}{4}$-in. bolts are thus

$$Z' = ZC_D C_g = (1026\text{ lb/bolt})(1.25)(0.995) = 1276\text{ lb/bolt}$$

For two bolts, the capacity of the connection is (2 bolts)(1276 lb/bolt) = 2552 lb.

Answer: The maximum vertical (downward) load that can be transferred from the horizontal member to the vertical column is 2550 lb (load duration factor 1.25).

Discussion: Since the horizontal member is attached to the side of the main member, prying action should be checked. The downward action of the side member on the bolts will cause the side member to bear on the main member at the bottom corner of the side member and will also tend to pull out the top bolt. Bearing stresses at the bottom should be checked to ensure that crushing does not occur and sufficient washer size for the top bolt to resist the bolt being pulled through or washer crushing wood. If the top bolt in such an arrangement is loaded in withdrawal, such as in the case of a lag screw, the combined effect of lateral and withdrawal load should be considered for the top bolt. Finally, the effect of eccentricity of the load on the vertical member should be considered in analysis of the vertical member capacity.

This solution is based on certain assumptions of bolt location and spacing. Specific bolt locations must be detailed on the design documents to ensure that the connection is fabricated suitably to carry the loads calculated. The group action factor was near unity in this example. Group action generally has a greater effect in connections with greater numbers of fasteners in a row and where side and main members have widely differing stiffness values.

5.11.4 Metal-to-Wood Connections

Lateral design values for metal plate-to-wood connections may be calculated using the yield mode equations of previous sections or may be obtained in tabular form for common combinations of metal plate thickness and wood member size in the *National Design Specification®* [2]. The dowel bearing strength, F_e, of the steel plate(s) in the *NDS®* [2] is taken to be 87,000 psi for ASTM A36 steel. Metal side plates must have adequate edge distances and be of sufficient thickness to suitably develop the capacities of the fasteners.

Example 5.4 Tension Splice Connection

Given: Two $\frac{3}{8}$-in. steel side plates are to be used to splice the ends of two $6\frac{3}{4}$ in. × $8\frac{1}{4}$ in. southern pine glued laminated timbers as illustrated in Figure 5.16. The timbers are to carry a total tensile load (snow plus dead) of 38,000 lb. The glued laminated timbers are combination 48 in AITC 117 [4]. The governing load duration factor is 1.15.

Determine: A suitable double-shear (three-member) connection using 1-in.-diameter through bolts on each side of the splice.

Approach: A preliminary design will be obtained using tabular design values for through bolts in double shear from the *National Design Specification®* [2]. Design values for the through bolts will be calculated from equations (5.8) and adjusted as required, including such factors as group action. The effect of section reduction and overall effect of the bolts on the connected members will also be checked. Edge distance, end distance, and spacing requirements will be checked or determined and noted as part of the solution.

Solution: For a preliminary design, the design value of 1-in.-diameter bolts in double shear for glued laminated timber for a main member width of $6\frac{3}{4}$- and $\frac{1}{4}$-in. steel side plates is obtained from Table 11 I of the 2001 *National Design Specification®* [2]. From the table, a value of 5960 lb is obtained, which assumes the normal duration of load and dry service conditions. The value also assumes that edge, end, and spacing requirements have been met, and it is not adjusted for group action. The design value is based on metal side plate thickness of $\frac{1}{4}$ in. and may be considered conservative for thicker plate values.

To determine a trial number of bolts, a trial group action factor, C_g, of 0.95 is assumed. Thus, the trial Z' value is $Z' = ZC_DC_g = (5960 \text{ lb})(1.15)(0.95) = 6511$ lb. Thus, $n = T/Z' = (38{,}000 \text{ lb})/(6511 \text{ lb/bolt}) = 5.8$, or six through bolts, each side of the splice.

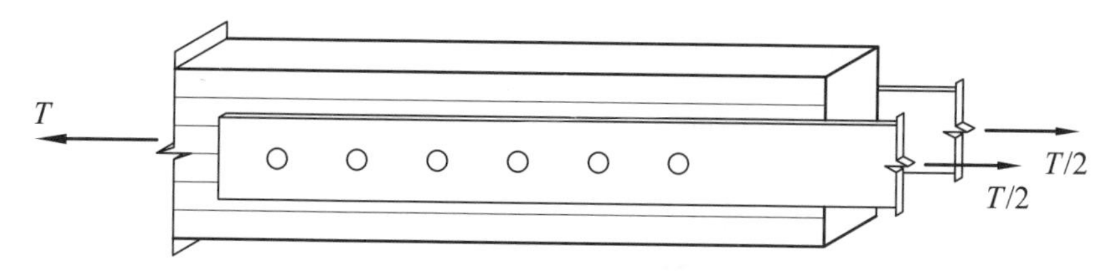

Figure 5.16 Tension splice connection.

The group action factor must be checked. To calculate the group action factor, the modulus of elasticity must be obtained for the timbers. From Table 8.2, AITC 117 [4], combination 48 Southern Pine, a value of $E = 1{,}700{,}000$ psi is obtained for all loading directions. Also, $F_t = 1400$ psi, which will be used to check the capacity of the members, considering the net section resulting from boring for the bolts.

It will be assumed that the steel side plates will be 4 in. wide. All six bolts will be arranged in a single row. The bolt spacing will be at least the minimum for the full design value; thus, $s \geq 4D = 4$ in. Thus,

$$n = 6$$

$$E_s = 29{,}000{,}000 \text{ psi (steel)}$$

$$E_m = 1{,}700{,}000 \text{ psi}$$

Since there are two side plates,

$$A_s = (2)(4 \text{ in.})(0.375 \text{ in.}) = 3.00 \text{ in}^2$$

$$A_m = (6.75 \text{ in.})\ (8.25 \text{ in.}) = 55.69 \text{ in}^2$$

Using $s = 4$ in., a value of $C_g = 0.956$ is obtained from equation (5.1).

This value may be checked using tabulated values of C_g for bolt or lag screw connections with metal side plates in the *National Design Specification®* [2], for which C_g is tabulated as a function of A_s/A_m, A_m, and n for $D = 1$ in.; E for the wood main member is 1,400,000 psi, and E for the metal is taken to be 30,000,000 psi. For this case, $A_s/A_m = 55.69/3.00 = 18.56$. Using $A_s/A_m = 18$, $C_g = 0.95$ for $A_m = 64$ in^2 and $C_g = 0.93$ for $A_m = 40$ in^2, respectively. Interpolating for $A_m = 56$ in^2 gives $C_g = 0.94$. The tabular values are conservative in this case as the metal and side plates values assumed have a greater difference in stiffness.

A group action factor, C_g, of 0.95 was assumed initially, which is slightly conservative. Equations (5.8a) to (5.8d) may be used to determine a more precise lateral design value for the bolts and will also indicate the failure mode that governs the fasteners. For use in the equations,

$$D = 1 \text{ in.}$$

$$\ell_m = 6.75 \text{ in.}$$

$$\ell_s = 0.375 \text{ in.}$$

$$F_{em} = 6150 \text{ psi (from Table 5.7; } G = 0.55 \text{ from Table 5.1)}$$

$$F_{es} = 87{,}000 \text{ psi}$$

$$F_{yb} = 45{,}000 \text{ psi}$$

$$\theta_{max} = 0°$$

which give the following values for Z:

$$Z = 10{,}378 \text{ lb} \qquad \text{mode I}_m$$
$$= 16{,}313 \text{ lb} \qquad \text{mode I}_s$$
$$= 6338 \text{ lb} \qquad \text{mode III}_s$$
$$= 8204 \text{ lb} \qquad \text{mode IV}$$

Mode III_s represents plastic hinges forming in the main member and localized failure in the side members (Figure 5.11), which will be dependent on side member thickness, reflecting the difference between the calculated design value of 6338 lb compared to 5960 lb using the table value assuming $\frac{1}{4}$-in. side plate thickness.

The adjusted design value using the $\frac{3}{8}$-in. thickness and the calculated group action factor is thus

$$Z' = ZC_DC_g = (6338 \text{ lb})(1.15)(0.956) = 6968 \text{ lb}$$

The minimum number of bolts, each side of the splice is thus

$$n = \frac{38{,}000 \text{ lb}}{6968 \text{ lb/bolt}} = 5.45 \text{ bolts, or six bolts}$$

The net section of the timbers is checked as follows. The holes receiving the 1-in. bolts will be bored $\frac{1}{16}$ in. larger than the bolt diameter. Since at any one section there is at most one bolt hole,

$$A_{\text{net}} = (6.75 \text{ in.})(8.25 \text{ in.}) - (1.0625 \text{ in.})(6.75 \text{ in.}) = 48.52 \text{ in}^2$$

The tension stress at the net section will be

$$f_t = \frac{T}{A_{\text{net}}} = \frac{38{,}000 \text{ lb}}{48.52 \text{ in.}^2} = 783 \text{ psi}$$

The allowable stress is

$$F_t' = F_tC_D = (1400 \text{ psi})(1.15) = 1610 \text{ psi}$$

Since $f_t = 783 \text{ psi} \leq F_t' = 1610$ psi, the net section is not overstressed and the timbers specified are satisfactory, with consideration of the section loss due to the bolts in the connection.

The effect of the group of fasteners on the end of each timber is then examined as follows. Equation (5.3a) is used to investigate the action of the bolts tearing out the wood from fastener to fastener in a row, or between the end fastener and the end of the member. For this case, $n = 6$ and $F'_v = (300 \text{ psi})(0.72) = 216$ psi (from AITC 117 [4]); thus,

$$F'_v = (216 \text{ psi})(1.15) = 248 \text{ psi}$$

$$A_{\text{crit}} = (2)(6.75 \text{ in.})(4 \text{ in.}) = 54 \text{ in}^2$$

where the bolt spacing of 4 in. is used, as it is less than the assumed end distance of $7D = 7$ in. Thus,

$$Z'_{\text{RT}} = n \frac{F'_v}{2} A_{\text{crit shear}} = (6)(248 \text{ psi}/2)(54 \text{ in}^2) = 40{,}280 \text{ lb for the row}$$

Tear-out of the row of bolts does not control. Since there is only one row of bolts, the tear-out for the group [equation (5.3b)] need not be considered independently. For this example, edge, end, and spacing values were assumed that would develop the full design values for the fasteners.

The required minimum edge distance $1.5D = 1.5 \times 1.0$ in. $= 1.5$ in. It is assumed that the bolt row will be centered in the timber members; therefore, the actual edge distance will be 8.25 in./2 = 4.125 in., which is greater than 1.5 in. The minimum end distance for use of the full design value for the bolts is $7D = 7 \times 1.0$ in. $= 7.0$ in. The minimum bolt spacing for full design value of the bolts is $4D = 4 \times 1.0$ in. $= 4.0$ in.

Answer: The required double shear splice connection for the glued laminated timbers stated is as follows: six 1-in.-diameter through bolts at 4 in. o.c. spacing, each timber (12 bolts total) centered in two $\frac{3}{8}$ in. $\times$ 4 in. steel plates. Maintain a 7-in. end distance in timbers; bolts and plates are to be centered in the timbers.

Discussion: Since the allowable load for the bolts was greater than the design load, the bolt spacing or end distances could be reduced. However, unless necessary to accommodate other constraints, end and spacing distances should not be detailed less than the minimum values for a full design load. The design checks above consider individual fastener and fastener group behavior in timber members as well as local failure at the fastener in the metal side plates. The side plates should also be examined in accordance with standard practices such as specified in the *Manual of Steel Construction* [5] of the American Institute of Steel Construction.

Example 5.5 Fastener with Load at an Angle to the Grain

Given: A tension connection consisting of two 3 in. by $\frac{1}{4}$ in. steel straps with a $\frac{3}{4}$-in.-diameter through bolt attached to a 5 in. $\times$ $6\frac{7}{8}$ in. glued laminated timber, as shown in Figure 5.17. The timber is combination 48 Southern Pine, and the straps are at an angle of 30° to the main member.

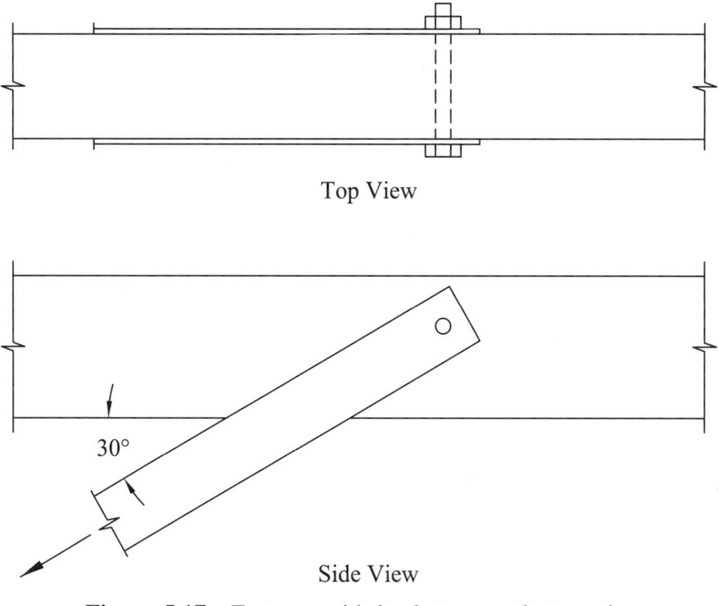

Figure 5.17 Fastener with load at an angle to grain.

Determine: The capacity of the connection for normal load duration.

Approach: Since the load is at an angle to the grain in the main member, the yield mode equations will be used for double shear [equations (5.8a) to (5.8d)]. For the main member, the dowel bearing stress for angle 30° will be calculated using equation (5.10).

Solution: From Table 5.7, the dowel bearing strength values parallel and perpendicular to the grain for Southern Pine are found to be $F_{e\|} = 6150$ psi and $F_{e\perp} = 2950$ psi ($\frac{3}{4}$-in. bolt). From equation (5.7),

$$F_{em} = F_{e\theta} = F_{e30^\circ} = \frac{F_{e\|} F_{e\perp}}{F_{e\|} \sin^2\theta + F_{e\perp} \cos^2\theta}$$

$$= \frac{(6150 \text{ psi})(2950 \text{ psi})}{(6150 \text{ psi})\sin^2(30^\circ) + (2950 \text{ psi})\cos^2(30^\circ)} = 4838 \text{ psi}$$

The other inputs to the double-shear yield mode equations are as follows:

$$\ell_m = 5.0 \text{ in.}$$

$$\ell_s = 0.25 \text{ in.}$$

$$F_{es} = 87{,}000 \text{ psi}$$

$$F_{yb} = 45{,}000 \text{ psi}$$

$$R_e = \frac{F_{em}}{F_{es}} = \frac{F_{e\theta}}{F_{es}} = \frac{4838 \text{ psi}}{87{,}000 \text{ psi}} = 0.05561$$

For use in Table 5.6,

$$K_\theta = 1 + 0.25\left(\frac{\theta}{90}\right) = 1 + 0.25\left(\frac{30}{90}\right) = 1.0833$$

From Table 5.6, the reduction factor, R_d, is calculated as follows:

$$R_d = \begin{cases} 4K_\theta = (4)(1.0833) = 4.33 & \text{for yield modes I}_\text{m} \text{ and I}_s \\ 3.2K_\theta = (3.2)(1.0833) = 3.467 & \text{for yield modes III}_\text{m}\text{, III}_\text{s}\text{, and IV} \end{cases}$$

$$R_t = \frac{\ell_m}{\ell_s} = \frac{5.00}{0.25} = 20.00$$

$$k_3 = 10.009 \text{ from equation (5.6i).}$$

The yield mode equations for double shear [equations (5.8a) to (5.8f)] give

$$\begin{aligned} Z &= 5322 \text{ lb} && \text{mode I}_\text{m} \\ &= 7529 \text{ lb} && \text{mode I}_\text{s} \\ &= 3216 \text{ lb} && \text{mode III}_\text{m} \\ &= 4260 \text{ lb} && \text{mode IV} \end{aligned}$$

Mode III_s governs, giving $Z_{30°} = 3216$ lb.

The $\frac{3}{4}$-in. bolts will be placed at middepth in the timbers; thus, both edge distances will be one-half of 6.875 in., or 3.44 in. from edge. The minimum loaded edge distance is $4D = 4(\frac{3}{4}$ in.$) = 3.0$ in.; the minimum unloaded edge distance is $1.5D$; thus, both edge distances are satisfied. End distance is not applicable in this case; neither is spacing. Hence, the geometry factor is 1.00. Group action is not applicable with one bolt; nor is group or row tear-out. Normal load duration was given for this example; thus, $C_D = 1.00$. The capacity of the connection is thus

$$Z' = ZC_D = (3216)(1.00) = 3216 \text{ lb}$$

Answer: The capacity of the given connection for normal load duration is 3216 lb (mode III_s; behavior governs).

Discussion: The capacity of the side plates themselves must be evaluated. In addition, an effective section for shear as well as the net section for tension in the timber must be evaluated with consideration made for all the loads on the member.

5.11.5 Wood-to-Concrete Connections

The lateral design values for embedded bolts connecting sawn lumber and structural composite lumber wood members to cast-in-place concrete may be found in tabular form in the *National Design Specification*® [2]. These values may be calculated using the yield mode equations of Section 5.11.1 (single shear). In addition to the edge distance, end distance, and spacing requirements for the wood side member, bolts must have adequate embedment and adequate edge and end distances in concrete to prevent failure of concrete in accordance with the *Building Code Requirements for Structural Concrete* [7] by the American Concrete Institute. In general, the wood must also be preservative treated or by some suitable means protected from moisture in the concrete. Proprietary fasteners and connectors for wood to concrete applications are also available commercially.

5.11.6 Lag Screws

Lag screws are sometimes referred to as lag bolts because their size more closely approximates that of bolts than that of wood screws. Sizes range from $\frac{1}{4}$ to $1\frac{1}{4}$ in. in diameter and with lengths up to 12 in. For special purposes, such as reinforcing pitched and tapered curved beams for radial tension, specially made lag screws may be 4 ft or even longer. However, for consistency of terminology in publications on wood fasteners, the term *lag screws* is used in this book. The design values determined herein for lag screws apply to lag screws conforming to ANSI/ASME Standard B18.2.1 [6].

Except as noted below, lag screws must be inserted in prebored lead holes. The lead holes for the shank must be the same diameter as the shank and extend the same depth as the depth of penetration of the unthreaded shank. Dimensions of standard lag bolts, including shank length, threaded portion, and tapered tip, are included in the Appendix. The lead hole for the threaded portion should have a diameter equal to 65 to 85% of the shank diameter in wood with $G > 0.60$, 60 to 75% in wood with $0.50 < G \leq 0.60$, and 40 to 70% in wood with $G \leq 0.50$. The larger percentages in each range apply to the larger-diameter lag screws. Lead holes are not required for $\frac{3}{8}$-in.-diameter and smaller lag screws used in wood with a $G \leq 0.50$, provided that end distance, edge distance, and spacing are such that unusual splitting does not occur.

Lag screws must be installed in a lead hole by turning, not by driving (e.g., with a hammer). Soap or other lubricant should be used to facilitate installation. Washers of proper size or a metal plate or strap should be installed between the wood and the bolt head. Washer or plate or strap area must be of sufficient size to accommodate the axial load of the fastener without damage to wood or metal parts.

Lateral design values for lag screws may be calculated using equations (5.6a) to (5.6f) and (5.8a) to (5.8d) or may be obtained in tabular form in the *National Design Specification®* [2]. When using equations (5.6) and (5.8) the root diameter D_r is used for D and the main member bearing length in single shear connections and side member bearing length in double shear connections is taken to be the penetration depth. The penetration must be not less than $4D$ for lag screws, based on full diameter but excluding the tapered tip length. Tabular values for lag screws loaded laterally in the *National Design Specification®* [2] are based on $8D$ penetration but may be reduced for penetration values not less than $4D$. Dowel bending yield strength values for lag screws are given in the Appendix. The tabular and calculated lag screw lateral design values discussed above assume that minimum edge, end, and spacing requirements have been met as well as normal use with dry conditions. Other installation and service conditions require appropriate adjustment. Group action and row and group tear-out must also be considered for connections utilizing multiple lag screw fasteners.

5.12 FASTENERS LOADED IN WITHDRAWAL

Lag screws, wood screws, nails, and spikes are commonly used to resist withdrawal loading conditions in wood construction. Withdrawal capacity is a function of the tensile capacity of the fastener and the amount of force that can be transferred between fastener and wood either by threads or friction. The amount of load that can be safely transferred between wood and fastener is generally given in terms of W in pounds per inch of the threaded portion in the main member for lag screws and wood screws or in pounds per inch of penetration depth in the main member for nails and spikes. The design value is then multiplied by all applicable adjustment factors (Table 5.3) and then by the effective withdrawal length to give the total safe withdrawal force for the fastener. This force cannot, however, exceed the safe tensile capacity of the fastener itself. Design values (W) for lag screws, wood screws, nails, and spikes are tabulated in the *National Design Specification® for Wood Construction* [2] for various wood specific gravities and fastener size or diameter. Alternatively, safe withdrawal loads may be calculated using

$$P_W = W'p \leq F_t A_{\text{net}} \tag{5.12}$$

where

P_W = allowable withdrawal load in pounds,
W' = withdrawal design value, W, multiplied by all applicable adjustment factors,
p = effective fastener penetration (the threaded portion in the main member minus the tapered tip for lag screws, penetration of threaded portion in main member for wood screws, and fastener penetration in main member for nails and spikes),
$F_t A_{net}$ = allowable tensile capacity of the fastener.

The design value W may be obtained in tabular form as mentioned above or calculated using

$$W = 1800G^{3/2}D^{3/4} \quad \text{for lag screws} \tag{5.13a}$$

$$= 2850G^2D \quad \text{for wood screws} \tag{5.13b}$$

$$= 1380G^{5/2}D \quad \text{for nails and spikes} \tag{5.13c}$$

where

G = the oven-dry weight and oven-dry volume specific gravity (Table 5.1)
D = the fastener diameter (in.).

The foregoing equations for withdrawal design values are for fasteners installed in the side grain of the main member. Where lag screws are installed in the end grain of a member and are loaded in withdrawal (load applied parallel to grain), the end grain factor, $C_{eg} = 0.75$, is to be applied. Wood screws, nails, and spikes are not to be used to resist withdrawal loads where installed in the end grain of members. Typical dimensions of lag screws and wood screws, including the lengths of threaded portions and tapered tips, are included in the Appendix.

5.13 FASTENERS LOADED LATERALLY AND IN WITHDRAWAL

Where dowel fasteners are loaded both laterally and in withdrawal, the following equations may be used:

$$Z'_\alpha = \begin{cases} \dfrac{(W'p)Z'}{(W'\mathbf{p})\cos^2\alpha + Z'\sin^2\alpha} & \text{lag screws and wood screws} \quad (5.14a) \\ \dfrac{(W')Z'}{(W'\mathbf{p})\cos\alpha + Z'\sin\alpha} & \text{nails and spikes} \quad (5.14b) \end{cases}$$

where

Z'_α = combined allowable load at angle α,
α = angle of load with respect to the wood surface,
$W'p$ = adjusted design value in withdrawal, and
Z' = adjusted design value for a lateral load.

Example 5.6 Lag Screw Loaded Laterally and in Withdrawal

Given: A single $\frac{1}{2}$ in. × 5 in. lag screw is loaded by a steel plate at angle $\alpha = 30°$ as shown in Figure 5.18. The wood member is $5\frac{1}{8}$ in. × $6\frac{7}{8}$ in. combination 47 Southern Pine glued laminated timber and the fastener will be installed centered in the top of the member (through the wide faces of the laminations). The metal side plate thickness is $\frac{1}{4}$ in. The timber is preservative treated and may be exposed to wet conditions. The design load will have load duration associated with snow loads. Assume that the fastener is not near either end of the timber. The lateral component of the load acts parallel to the grain in the main timber member.

Determine: The allowable load P_{30} for the fastener.

Approach: The allowable load P_{30} will be calculated using equations (5.6a) to (5.6f) for dowel fasteners (single shear), equations (5.12) and (5.13) for lag screw withdrawal design values, and equation (5.14a) for combined lateral and withdrawal loading. The specific gravity for Southern Pine of 0.55 (Table 5.1) will be used for both lateral and withdrawal design value calculations.

Solution: For the lateral design value, the following values are used in the yield mode equations [equations (5.6a) to (5.6f)].

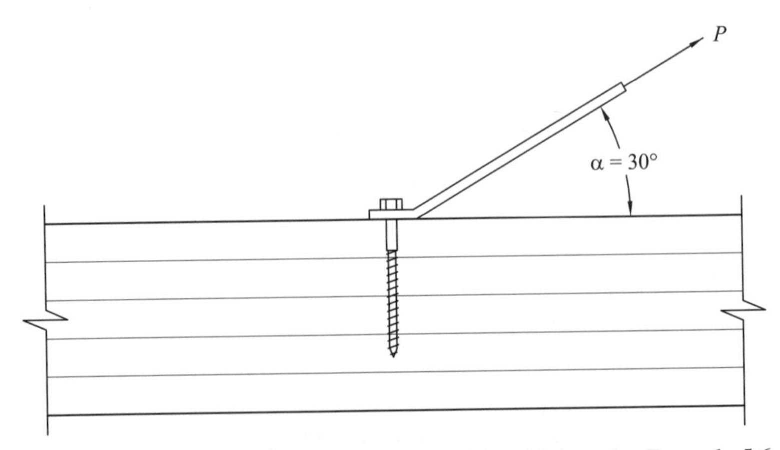

Figure 5.18 Lag screw loaded laterally and in withdrawal—Example 5.6.

$D = D_r$ = root diameter = 0.371 in. (from the Appendix)

ℓ_m = main member dowel bearing length = screw length − assumed $\frac{1}{16}$-in. washer − $\frac{1}{4}$-in. plate = 5.00 in. − 0.06 in. − 0.25 in. = 4.69 in., which must not be less than $4D$ = (4)(0.5 in.) = 2.0 in. good

F_{em} = dowel bearing strength of main member = 6150 psi (Table 5.7)

θ = angle of (lateral) load to the grain = 0°

R_d = reduction term = 4.0 for modes I_m and I_s, 3.6 for mode II, and 3.2 for modes III_m, III_s, and IV

ℓ_m = side member dowel bearing length = 0.25 in.

F_{es} = dowel bearing strength of side member = 87,000 psi

F_{yb} = dowel bending yield strength = 45,000 psi

$$R_e = \frac{F_{em}}{F_{es}} = 0.071$$

$$R_t = \frac{\ell_m}{\ell_s} = 18.76$$

k_1 = 0.542 [from equation (5.6g)]

k_2 = 0.475 [from equation (5.6h)]

k_3 = 6.248 [from equation (5.6i)]

Equations (5.6a) to (5.6f) give

$$\begin{aligned} Z &= 2675 \text{ lb} && \text{mode } I_m \\ &= 2017 \text{ lb} && \text{mode } I_s \\ &= 1215 \text{ lb} && \text{mode II} \\ &= 1392 \text{ lb} && \text{mode } III_m \\ &= 538 \text{ lb} && \text{mode } III_s \\ &= 565 \text{ lb} && \text{mode IV} \end{aligned}$$

Mode III_s governs, with Z = 538 lb.

From Table 5.4, C_M = 0.7. For snow load duration, C_D = 1.15. The fastener is installed distant from the ends; therefore, end distance requirements are met. The edge distances are both one-half of 5.125 in. or 2.56 in., which is greater than the $1.5D$ = (1.5)(0.5 in.) = 0.75 in., required for full design value parallel to the grain. Spacing need not be considered since there is only one fastener; thus, the geometry factor is not applicable ($C_\Delta = 1.00$). Thus,

$$Z' = ZC_DC_M = (538 \text{ lb})(1.15)(0.7) = 433 \text{ lb}$$

For the withdrawal design value, equation (5.13a) is used, where $G = 0.55$ and $D = 0.5$ in. (nominal diameter); thus,

$$W = (1800)(0.55)^{3/2}(0.50)^{3/4} = 437 \text{ lb/in.}$$

The withdrawal design value is in this case multiplied by the same adjustment factors; thus,

$$W' = WC_DC_M = (437 \text{ lb/in.})(1.15)(0.7) = 351 \text{ lb/in.}$$

The penetration, p, of the threaded portion is given by the threaded portion length in the main member minus the tapered tip. From the Appendix, the threaded portion length minus the tapered tip (T − E) is given to be $2\frac{11}{16}$ in. = 2.688 in. Since the fastener penetrates the main member 4.69 in., the threads are fully developed in withdrawal; thus,

$$W'p = (351 \text{ lb/in.})(2.688 \text{ in.}) = 945 \text{ lb}$$

However, the withdrawal load developed by the threads cannot exceed the allowable tension capacity of the lag screw F_tA_{net}, where $F_t = 20{,}000$ psi for grade A307 lag screws, and

$$A_{net} = \frac{\pi D_r^2}{4} = \frac{\pi(0.371 \text{ in.})^2}{4} = 0.108 \text{ in.}^2$$

Thus, $F_tA_{net} = (20{,}000)(0.108) = 2161$ lb. The allowable load developed by the threads does not exceed the tension capacity of the fastener.

For the combined lateral and withdrawal load, equation (5.14a) is used, with $\alpha = 30°$:

$$Z'_\alpha = \frac{(W'p)(Z)}{W'p\cos^2\alpha + Z'\sin^2\alpha}$$

$$Z_{30°} = \frac{(945 \text{ lb})(433 \text{ lb})}{(945 \text{ lb})\cos^2 30° + (433 \text{ lb})\sin^2 30°} = 501 \text{ lb}$$

Answer: The allowable load at angle 30° is 500 lb.

Discussion: If the lateral component of the applied load in the example above had been at angle $\beta = 45°$ with respect to the longitudinal axis of the piece, the lateral resistance would be based on the dowel bearing strength associated with angle β. The edge distance requirements would also need to be reevaluated. The resulting allowable load would be determined as follows.

The dowel bearing strength for the main member is determined using Hankinson's formula [equation (5.7)] with $F_{e\parallel} = 6150$ psi from before and $F_{e\perp} = 3650$ psi based on nominal bolt diameter, giving

$$F_{e45°} = F_{e\theta} = \frac{F_{e\parallel}F_{e\perp}}{F_{e\parallel}\sin^2\theta + F_{e\perp}\cos^2\theta}$$

$$= \frac{(6150 \text{ psi})(3650 \text{ psi})}{(6150 \text{ psi})\sin^2 45° + 3650 \text{ psi} \cos^2 45°} = 4581 \text{ psi}$$

The recalculated Z value becomes 422 lb (mode III_s again governs).

Assuming that the fastener is still centered in the top of the member, the edge distance of 2.56 in. must be evaluated with respect to unloaded and loaded edges perpendicular to the grain, which must be less than $1.5D$ and $4.0D$, respectively. The loaded edge distance must thus not be less than $4.0D = 2.0$ in., which in this case is still satisfied. Thus,

$$Z' = ZC_DC_M = (422 \text{ lb})(1.15)(0.7) = 340 \text{ lb}$$

The allowable withdrawal load remains unchanged. The combined allowable load becomes

$$Z_{30°} = \frac{(945 \text{ lb})(340 \text{ lb})}{(945 \text{ lb})\cos^2 30° + (340 \text{ lb})\sin^2 30°} = 405 \text{ lb}$$

5.14 WOOD SCREWS

Wood screws discussed herein are those conforming to ANSI/ASME Standard B18.6.1 [8]. Wood screw design values for lateral and withdrawal loading may be obtained in tabular form in the *National Design Specification*® [2] or may be calculated using equations (5.6), (5.8), and (5.12) and (5.13). For lateral loading, the screws must penetrate into the main member at least six diameters ($6D$) for reduced design values and 10 diameters ($10D$) for the full design values tabulated in the *National Design Specification*® [2]. The root diameter D_r is used for D in the yield mode equations and to determine the tensile capacity of the screw in withdrawal. Wood screws are not permitted to be used in withdrawal in end grain. Wood screw sizes and dimensions are given in the Appendix. Wood screws are typically available in various lengths, of which approximately two-thirds of the screw length is threaded. Dowel bending yield strength values for wood screws are also provided in the Appendix.

For wood screws loaded laterally, for wood members with $G > 0.60$, the part of the lead hole receiving the shank should have approximately the same diameter as the shank, and the part receiving the threaded portion should have approximately the same diameter as the diameter at the root of the thread. For members with $G \leq 0.60$, the part of the lead hole receiving the shank should be approximately seven-eighths of the shank diameter, and the portion receiving the threaded portion should be about seven-eighths of the diameter of the screw at the root of the thread. Root diameters for wood screws are provided in the Appendix.

For wood screws loaded in withdrawal, wood members with $G > 0.60$ should have a lead hole of approximately 90% of the wood screw root diameter, and for wood members with $0.50 < G \leq 0.60$, the lead hole should be approximately 70% of the wood screw root diameter. For wood members with $G \leq 0.5$, no lead hole is necessary. Edge distance, end distance, and spacing of wood screws must be sufficient to prevent splitting of the wood member.

5.15 NAILS AND SPIKES

The design values for common wire, box, and sinker nails may be calculated using the yield mode equations (5.6a) to (5.6f) and (5.8a) to (5.8d) or may be obtained in tabular form in the *National Design Specification®* [2]. Design values for threaded, hardened nails, and spikes may be calculated where size, length, and dowel bending yield strength values are known and can be specified. Sizes of nails and spikes commonly used in timber construction are shown in Section A.3.4. Because of the variety of nails and spikes available, it is necessary in all engineered construction that the diameter and length of all nail fasteners be clearly specified in the construction documents. Dowel bending yield strengths for nails and spikes may be found in Section A.3.7.

Prebored holes may be used to avoid splitting of the wood. Where prebored holes are used, the holes may not be larger than 90% of the nail or spike diameter for species with a specific gravity greater than 0.6 or 75% of the nail or spike diameter for species with a specific gravity less than or equal to 0.6. Where possible, nails or spikes should not be loaded in withdrawal in general and are *not permitted* to be loaded in withdrawal from end grain. Where withdrawal from side grain must be considered, the nominal withdrawal load in pounds per inch of penetration may be calculated by equation (5.12) and (5.13) or may be obtained in tabular form from the *National Design*

Specification® [2]. Clinched wire nails provide greater resistance to withdrawal than do unclinched nails. However, since proper clinching of nails in the field may be difficult to control and inspect, increased design values considering clinching are not recommended.

For single-shear applications, nails and spikes must penetrate the main member at least six fastener diameters. For double-shear applications, the fastener must penetrate the opposite-side member at least six fastener diameters, except in the case of symmetric double-shear connections with side members at least $\frac{3}{8}$ in., where 12*d* or smaller fasteners extend at least three diameters from the opposite-side member and are clinched.

5.15.1 Toenailed Fasteners

Toenailed fasteners are to be driven at an angle approximately 30° with the face of the side member, as shown in Figure 5.19, and must start approximately one-third the length of the fastener from the end. Care must be taken to avoid splitting the side member for closely spaced toenailed fasteners. Toenail lateral design values are to be multiplied by the toenail factor, $C_{tn} = 0.83$; toenail withdrawal design values are to be multiplied by the factor $C_{tn} = 0.67$. The wet service factor need not be applied simultaneously with the toenail factor for toenails in withdrawal. Some building codes may not allow toenail fasteners to be loaded in withdrawal for some load or load combinations.

For lateral loading of toenailed fasteners, the side member thickness and main member dowel or penetration lengths are taken to be the projection of the fastener lengths perpendicular to the face of the main member [9]. For withdrawal loading, the actual penetration length in the main member is used.

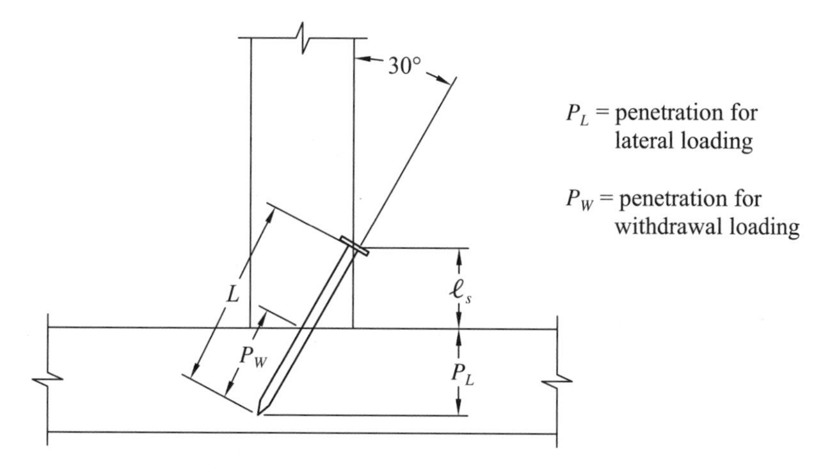

Figure 5.19 Common toenail connection.

5.15.2 Nails Used in Diaphragms

Where nails are used in a structural diaphragm, the lateral design values of the fasteners are permitted to be multiplied by the diaphragm factor, $C_{di} = 1.10$. This adjustment factor applies only to the loads related to the structural diaphragm and not for resistance of other forces that the fasteners may also be dependent upon to resist.

5.16 DRIFT PINS

Drift bolts and pins are long unthreaded bolts or pins that are sometimes used to fasten large timbers. Design values for drift bolts and pins have not been studied as extensively as those for other fasteners and generally are not included in codes. Design values for drift pins or bolts in lateral load should not be taken to be greater than 75% of the design values for common bolts of the same diameter and same main member bearing length. The same considerations and requirements for common bolts may be applied to drift pins and bolts, with the main differences being that drift bolts generally do not fully penetrate the receiving member and may also be loaded in withdrawal.

Design values for withdrawal loading, based on bolts driven in prebored holes $\frac{1}{8}$ in. less in diameter than the bolt and in seasoned wood, may be calculated by [10].

$$W = 1210G^2D \tag{5.15}$$

where

W = design value in withdrawal (pounds per inch of penetration),
G = specific gravity of the receiving member, based on oven-dry weight and oven-dry volume,
D = diameter of drift bolt or pin (in.).

The design value W is then multiplied by the effective depth of penetration and applicable adjustment factors. Suitable means must be provided to transfer any tension load in the bolt to other members or connection parts. Similarly, the tension load developed in the drift pin or bolt may not exceed the allowable tension load in the bolt itself.

5.17 STAPLES

Because staples and nails are similar in nature, the loads for staples may be determined in a manner similar to that for nails. The design value for one staple of a given diameter equals twice the value for a nail of equal diameter, provided that the staple leg spacing (or crown width) is adequate and that the penetration of both legs of the staple into the member receiving the points is approximately two-thirds of their length. In general, nail penetration requirements and other provisions regarding seasoning of members, service conditions, and so on, apply equally to staples.

5.18 LIGHT METAL FRAMING DEVICES

A great many framing anchors and other fastening devices are commercially available for the wood framing industry. Some examples of such devices are illustrated in Figure 5.20. These fasteners are generally made of metal with predrilled holes for special fasteners. Allowable loads for such fasteners are provided by the manufacturer and are generally already adjusted for the com-

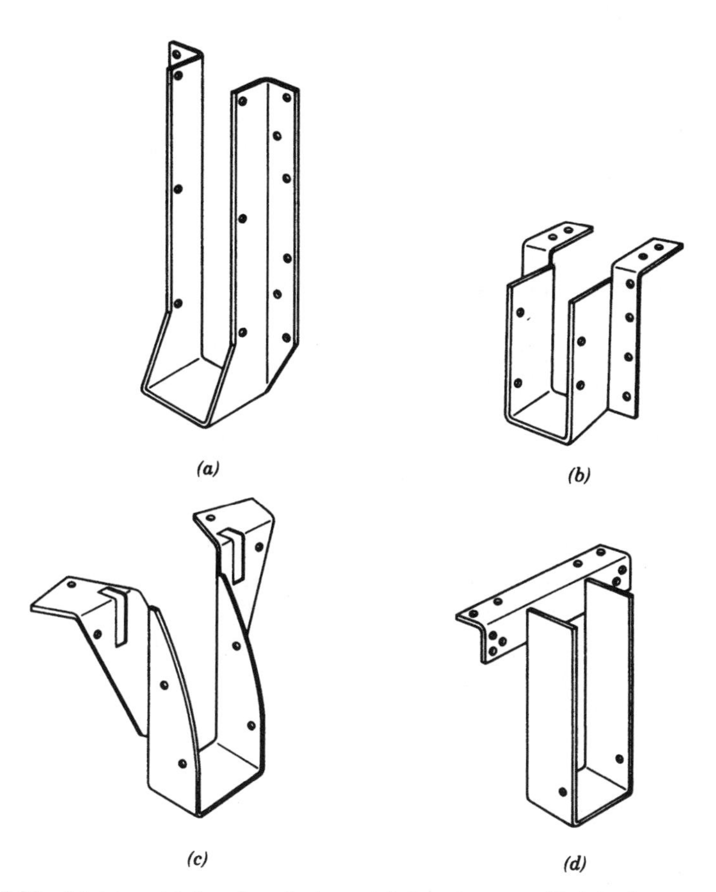

Figure 5.20 Light metal framing devices: (*a*) Joist hanger; (*b*) joist hanger; (*c*) joist and purlin hanger; (*d*) joist and beam hanger.

mon load duration conditions of the expected end use. Such fasteners have become increasingly popular in resisting wind and seismic loads where typical construction nailing is insufficient.

Fasteners should be selected that meet appropriate code requirements. Installation of such fasteners must be in accordance with the manufacturer's requirements. While the capacities of such fasteners may be estimated by the equations of this chapter, code-approved allowable load values provided by the manufacturer should be relied upon in design. Such fasteners should not be modified or used in ways other than as specified by the manufacturer, as modification or such alternative uses will generally void warranty and liability of the manufacturer for such products.

5.19 METAL TRUSS PLATE CONNECTIONS

The fabrication and design of metal plates commonly used to connect dimensional lumber in trusses for residential and light commercial structures are typically proprietary and included in preengineered and premanufactured truss systems. While connections for such trusses may be designed in accordance with the procedures of this book, preengineered/premanufactured systems are typically more economical.

5.20 SPLIT-RING AND SHEAR PLATE CONNECTORS

Split-ring and shear plate connectors used in timber construction are illustrated in Figure 5.21. Connectors discussed in this book have dimensions and specifications shown in Section A.3.5 and are made from SAE 1010 hot-rolled carbon steel (split rings and $2\frac{5}{8}$-in. shear plates) or grade 32510 ASTM A47 malleable iron (4-in. shear plates). Design values for split-ring and shear plate connectors require that the connected members be brought into snug contact with one another and that washers be used between wood members and bolt heads. Through bolts or lag screws are used to hold the connected members together in the connector assembly. Where lag screws are used instead of through bolts, the lead hole for the shank of the lag screw is to be the same diameter as the lag screw shank, and the lead hole for the threaded portion is to be 70% of the shank diameter or as specified in Section 5.11.6 for lag screws in lateral loading. The penetration depth factor, C_d, of Table 5.11 must be applied where lag screws are used instead of through bolts in split-ring and shear plate connectors.

Where wood members are not seasoned to service moisture content conditions, the connector assembly must be tightened periodically until equilibrium is reached. Split-ring and shear plate connectors are not intended to resist loads out of plane with the fastener or connected wood surfaces.

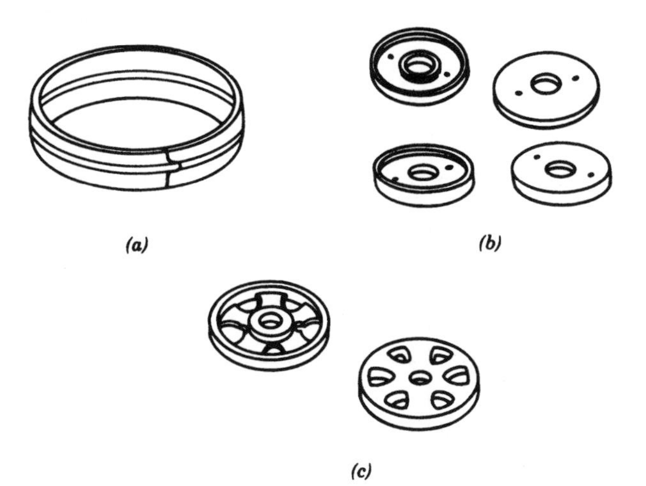

Figure 5.21 Typical split-ring and shear plate connectors. Reprinted with permission from *Design Manual for TECO Connector Construction.* Copyright © 1973 TECO/ Lumberlok. Copyright © 1997 Cleveland Steel Specialty Company.

Design values for split-ring and shear plate connectors are for a single split ring or shear plate. Where more than one connector is used in a connection, the group action factor must be applied and row and group tear-out also considered. Where the end of a connected member is not square, the end distance, measured parallel to the member axis, may be taken from a point $D/4$ from the fastener center as shown in Figure 5.22. The minimum edge distance must also be maintained at the nonsquare end. Spacing of multiple connectors is generally defined as the center-to-center spacing of connectors without respect to orientation of grain or member axis (Figure 5.22).

5.20.1 Design Values for Connectors in Side Grain

Design values for split-ring and shear plate connectors are determined by testing and are provided in tabular form in the *National Design Specification®* [2] for connectors installed in the side grain of the connected members. Design values are based on the wood specific gravity groups shown in Table 5.12. For glued laminated timbers, equivalent specific gravity group designations vary depending on the face of application and are given in AITC 117 [4]. Design values for split-ring and shear plate connectors in the *NDS®* [2] are to be adjusted by all the applicable adjustment factors of Tables 5.2 and 5.3 except that the adjusted values may not exceed published maximum values based on the metal parts also indicated in the tables.

For application of the wet service factor, members are considered seasoned if the moisture content is not greater than 19% for a depth of $\frac{3}{4}$ in. Geometry factors for split-ring and shear plate connectors are published specifically for

TABLE 5.11 Penetration Depth Factors for Connectors Used with Lag Screws

	Side Member	Penetration	Penetration of Lag Screw into Main Member (number of shank diameters) — Species Group (see Table 10A)				Penetration Depth Factor C_d
			Group A	Group B	Group C	Group D	
2-1/2" Split Ring 4" Split Ring 4" Shear Plate	Wood or Metal	Minimum for Full Design Value	7	8	10	11	1.0
		Minimum for Reduced Design Value	3	3-1/2	4	4-1/2	0.75
2-5/8" Shear Plate	Wood	Minimum for Full Design Value	4	5	7	8	1.0
		Minimum for Reduced Design Value	3	3-1/2	4	4-1/2	0.75
	Metal	Minimum for Full Design Value	3	3-1/2	4	4-1/2	1.0

Source: Reprinted with permission from *National Design Specification® for Wood Construction.* Copyright © 2001. American Forest & Paper Association, Inc.

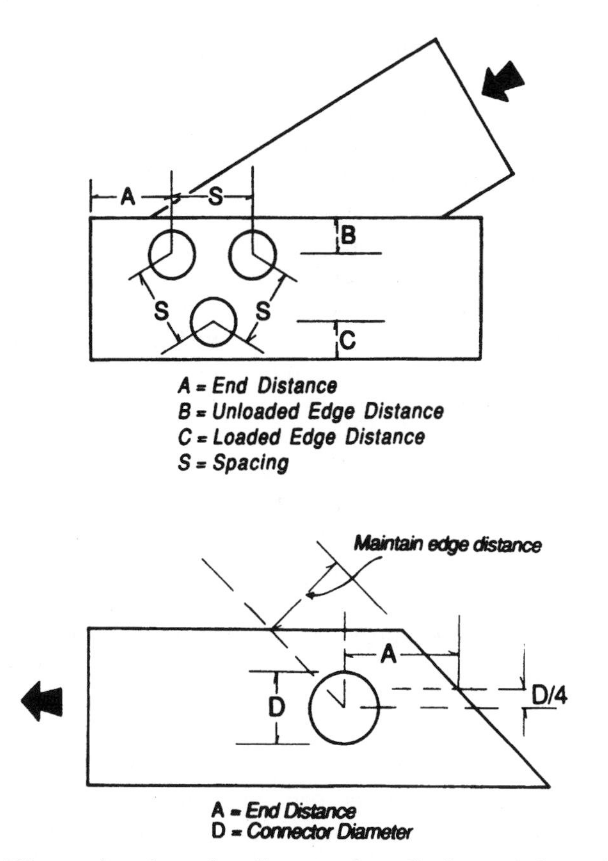

Figure 5.22 Edge, end, and spacing distances for split-ring and shear plate connectors. Reprinted with permission from *National Design Specification® for Wood Construction.* Copyright © American Forest & Paper Association, Inc.

TABLE 5.12 Wood Specific Gravity Groups for Split-Ring and Shear Plate Connectors

Specific Gravity Group	Specific Gravity, G
A	$G \geq 0.60$
B	$0.49 \leq G < 0.60$
C	$0.42 \leq G < 0.49$
D	$G < 0.42$

split-ring and shear plate connectors and are shown in Table 5.13. The geometry factors of Table 5.13 are shown for loading parallel and perpendicular to the grain. For loading at an angle to the grain other than parallel or perpendicular, the geometry factors are determined as follows.

Edge distances are shown in Table 5.13 for unloaded and loaded edges. The unloaded edge distance is a minimum distance and is used both for loading parallel to the grain and for the unloaded edge when the load acts at an angle to the grain. Edge distances are shown for the loaded edge as the distance required for full load and the distance required for reduced load and accompanying geometry factor, C_{Δ}. For intermediate edge distances, straight-line interpolation may be used to determine the corresponding geometry factor. For angles of load to the grain from 45° to 90°, inclusive, the edge distances for perpendicular to grain loading are to be applied. For angles of load to the grain of less than 45°, the required edge distance or corresponding geometry factor may be determined by straight-line interpolation between the value for 45° perpendicular to the grain and the value parallel to the grain.

Edge distances for $2\frac{1}{2}$- and $2\frac{5}{8}$-in. connectors are based on nominal 4-in.-wide pieces (net $3\frac{1}{2}$ in.) and nominal 6-in.-wide pieces (net $5\frac{1}{2}$ in.) for 4-in. connectors. Where timber connectors are centered on the narrow faces of glued laminated timbers of widths smaller than those stated above, the adjustment factors of Table 5.14 may be applied.

End distances for full design load and distances for reduced design load and corresponding geometry factors are given in Table 5.13. Values are given for loading both perpendicular and parallel to the grain. For loading parallel to the grain, values are given for members in both compression (load away from the end) and tension (load toward the end). For end distances intermediate between the minimum and that required for a full load, the corresponding geometry factors may be determined by straight-line interpolation. End distances for members loaded at angles of the grain other than 0° or 90° may also be determined by straight-line interpolation.

Spacing between connectors must be considered in determining connector loads because it controls the shearing area that develops the connector load. Factors that influence spacing include angle of load to the grain and angle of axis of connectors to the grain. The connector axis is formed by a line joining the centers of any two adjacent connectors located in the same face of a member in a joint. The angle of axis of the connectors is the angle formed by the axis line of the connectors and the longitudinal axis of the member as illustrated by angle ϕ in Figure 5.23.

Table 5.13 lists the spacing required for full and reduced load for loading both parallel and perpendicular to the grain. The design values for intermediate spacing values for loading parallel and perpendicular to the grain can be obtained by straight-line interpolation. The spacing in members loaded at an angle to the grain with the connector axis at various angles to the grain can be calculated by

TABLE 5.13 Geometry Factors for Split-Ring and Shear Plate Connectors

		2-1/2" Split Ring Connectors & 2-5/8" Shear Plate Connectors				4" Split Ring Connectors & 4" Shear Plate Connectors			
		Parallel to grain loading		Perpendicular to grain loading		Parallel to grain loading		Perpendicular to grain loading	
		Minimum for Reduced Design Value	Minimum for Full Design Value	Minimum for Reduced Design Value	Minimum for Full Design Value	Minimum for Reduced Design Value	Minimum for Full Design Value	Minimum for Reduced Design Value	Minimum for Full Design Value
Edge Distance	Unloaded Edge	1-3/4"	1-3/4"	1-3/4"	1-3/4"	2-3/4"	2-3/4"	2-3/4"	2-3/4"
	C_Δ	1.0	1.0	1.0	1.0	1.0	1.0	1.0	1.0
	Loaded Edge	1-3/4"	1-3/4"	1-3/4"	2-3/4"	2-3/4"	2-3/4"	2-3/4"	3-3/4"
	C_Δ	1.0	1.0	0.83	1.0	1.0	1.0	0.83	1.0
End Distance	Tension Member	2-3/4"	5-1/2"	2-3/4"	5-1/2"	3-1/2"	7"	3-1/2"	7"
	C_Δ	0.625	1.0	0.625	1.0	0.625	1.0	0.625	1.0
	Compression Member	2-1/2"	4"	2-3/4"	5-1/2"	3-1/4"	5-1/2"	3-1/2"	7"
	C_Δ	0.625	1.0	0.625	1.0	0.625	1.0	0.625	1.0
Spacing	Spacing parallel to grain	3-1/2"	6-3/4"	3-1/2"	3-1/2"	5"	9"	5"	5"
	C_Δ	0.5	1.0	1.0	1.0	0.5	1.0	1.0	1.0
	Spacing perpendicular to grain	3-1/2"	3-1/2"	3-1/2"	4-1/4"	5"	5"	5"	6"
	C_Δ	1.0	1.0	0.5	1.0	1.0	1.0	0.5	1.0

Source: Reprinted with permission from *National Design Specification® for Wood Construction.* Copyright © 2001 American Forest & Paper Association, Inc.

TABLE 5.14 Adjustment Factors for Connectors with Reduced Edge Distances, Connectors Centered on Face

Connector Size		Parallel-to-Grain Loading		Perpendicular-to-Grain Loading	
$2\frac{1}{2}$-in. split ring or $2\frac{5}{8}$-in. shear plate	Width	3 in.	$3\frac{1}{8}$ in.	3 in.	$3\frac{1}{8}$ in.
		0.88	0.91	0.86	0.90
4-in. split ring or shear plate	Width	5 in.	$5\frac{1}{8}$ in.	5 in.	$5\frac{1}{8}$ in.
		0.93	0.95	0.93	0.95

$$R = \frac{AB}{\sqrt{A^2 \sin^2\phi + B^2 \cos^2\phi}} \tag{5.16}$$

where

R = the required spacing along the connector axis for the full design value, A and B and are dependent on the angle of load to the grain from Table 5.15,

ϕ = angle of connector axis to the grain.

Equation (5.16) gives the spacing R required for full design values for various angles of load to the grain and connector axis to the grain. The minimum spacing $R_{\min}$ along the connector axis is also shown in Table 5.15 for reduced load ($C_\Delta = 0.50$). For spacing along the connector axis between $R_{\min}$ and R, straight-line interpolation may be used to determine C_Δ. When three or more connectors are used in one face of a member, the spacing between any two connectors should be checked.

Example 5.7 Spacing at an Angle to the Grain

Given: A connection similar to the one shown in Figure 5.23, with an angle of axis of connectors of 40° and an angle of load to the grain of 30°.

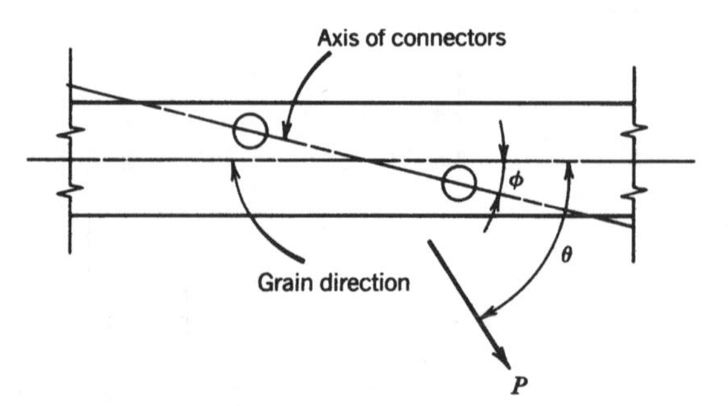

Figure 5.23 Angle of the axis of connectors to the grain.

TABLE 5.15 Factors for Determining Required Spacing along Connector Axis

Connector	Angle[a] of Load to Grain, θ (deg)	A (in.)	B (in.)	R_{min} (in.)
$2\frac{1}{2}$-in. split ring or $2\frac{5}{8}$-in. shear plate	0	6.75	3.50	3.5
	15	6.00	3.75	3.5
	30	5.13	3.88	3.5
	45	4.25	4.13	3.5
	60–90	3.50	4.25	3.5
4-in. split ring or $2\frac{5}{8}$-in. shear plate	0	9.00	5.00	5.0
	15	8.00	5.25	5.0
	30	7.00	5.50	5.0
	45	6.00	5.75	5.0
	60–90	5.00	6.00	5.0

[a] Interpolation is permitted for determining A and B for intermediate angles of loads to the grain.

Determine: The required spacing along the connector axis for full design value and the minimum reduced design value for $2\frac{5}{8}$-in. shear plates.

Approach: The spacing for full design value R will be calculated using equation (5.16) and the values for A and B from Table 5.15. The spacing for the minimum reduced design value, C, will also be obtained from Table 5.15.

Solution: From Table 5.15 for an angle of load to the grain, θ, of 30°, A = 5.125 in., and B = 3.875 in. Using equation (5.16) yields

$$R = \frac{AB}{\sqrt{A^2 \sin^2 \phi + B^2 \cos^2 \phi}} = \frac{(5.125 \text{ in.})(3.875 \text{ in.})}{\sqrt{(5.125 \text{ in. } \sin 40°)^2 + (3.875 \text{ in. } \cos 40°)^2}}$$

$$= 4.48 \text{ in.}$$

The required spacing along the connector axis for full design value is 4.5 in.

The required spacing along the connector axis for the minimum reduced design value is R_{min} from Table 5.15, which in this case gives 3.5 in. The minimum design value associated with this spacing is 50% of the full design values (C_Δ = 0.50).

Answer: For use of the full design value for the case described, the spacing along the connector axis must be 4.5 in. For use of the minimum reduced design value (and corresponding C_Δ = 0.50), the required spacing is 3.5 in.

Geometry factors for edge distance, end distance, and spacing are not cumulative. The smallest geometry factor controls and is applied to all connectors of a connector group. Geometry factors are cumulative with the other applicable adjustment factors listed in Tables 5.2 and 5.3. It is recommended that wherever possible, connector joints be designed with edge distance, end distance, and spacing values equal to or greater than the values required for full design values (C_Δ = 1.00).

The design values for split-ring and shear plate connectors provided in the *National Design Specification*® [2] are for loads parallel and perpendicular to grain with rings or plates installed into the side (side grain) of the members. For load at angles to the grain other than parallel or perpendicular, the design values for parallel and perpendicular loading, P and Q are determined, adjusted by all applicable factors to obtain adjusted design values P' and Q', and then used in the Hankinson formula [equation (5.17)] to obtain the allowable load at an angle to the grain, N':

$$N' = \frac{P'Q'}{P' \sin^2\theta + Q' \cos^2\theta} \tag{5.17}$$

where

N' = allowable load at angle to the grain,
P' = design value for load parallel to the grain adjusted by all applicable factors,
Q' = adjusted design value for load perpendicular to the grain,
θ = angle of load to the grain.

Although split-ring and shear plate connectors are capable of transferring large shear loads, they also result in significant losses in section due to the ring and connecting bolt. This condition is illustrated in Figures 5.3 and 5.24. The net section of members may be calculated using the information in Appendix A3.5 considering the projection of the split ring or shear plate in the member and the section loss due to the bolt hole. For convenience, the sum of the ring or plate projected area and bolt hole section losses are tabulated in Table 5.16 for common-size members.

Example 5.8 Tension Connection with Split-Ring Connectors

Given: A split-ring connection in double shear, illustrated in Figure 5.24, to carry 5000 lb of dead load and 12,000 lb of snow load (C_D = 1.15). Members are specific gravity group (species group) B with design values F_t = 1200 psi, E = 1,800,000 psi, and F_v = 180 psi. The moisture condition of service is dry. The main member is 4 in. × 6 in. and the side members are 2 in. × 6 in. (all nominal dimensions). Connectors are to be $2\frac{1}{2}$-in. split rings.

Determine: A suitable split-ring connection.

Approach: First, the net section will be checked in both main and side members, assuming that the connectors will be placed in a single row. The edge distance, end distance, and spacing values for full design values will be de-

TABLE 5.16 Total Projected Area of Connectors and Bolt Holes for Use in Determining Net Section

Connector		Bolt Diameter (in.)	Placement of Connectors	Member Thickness (in.)									
Type	Size (in.)			$1\frac{1}{2}$	$2\frac{1}{2}$	$3\frac{1}{8}$	$3\frac{1}{2}$	$5\frac{1}{8}$	$5\frac{1}{2}$	$6\frac{3}{4}$	$8\frac{3}{4}$	$10\frac{3}{4}$	$12\frac{1}{4}$
Split Rings													
1	$2\frac{1}{2}$	$\frac{1}{2}$	1 face	1.73	2.29	2.65	2.86	3.77	3.98	4.69	5.81	6.94	7.78
		$\frac{1}{2}$	2 faces	2.62	3.18	3.54	3.75	4.66	4.87	5.58	6.70	7.82	8.67
2	4	$\frac{3}{4}$	1 face	3.05	3.86	4.37	4.68	6.00	6.30	7.32	8.94	10.56	11.79
		$\frac{3}{4}$	2 faces	4.88	5.69	6.20	6.51	7.83	8.13	9.15	10.77	12.40	13.62
Shear Plates													
1	$2\frac{5}{8}$	$\frac{3}{4}$	1 face	2.03	2.85	3.35	3.66	4.98	5.28	6.30	7.92	9.55	10.77
		$\frac{3}{4}$	2 faces	2.84	3.66	4.16	4.47	5.79	6.09	7.11	8.73	10.36	11.58
1-LG[a]	$2\frac{5}{8}$	$\frac{3}{4}$	1 face	1.91	2.72	3.23	3.53	4.85	5.16	6.18	7.80	9.43	10.64
		$\frac{3}{4}$	2 faces	2.60	3.41	3.92	4.22	5.54	5.85	6.87	8.49	10.11	11.33
2	4	$\frac{3}{4}$	1 face	3.26	4.07	4.58	4.89	6.21	6.51	7.53	9.15	10.78	12.00
		$\frac{3}{4}$	2 faces	—	6.11	6.62	6.93	8.25	8.55	9.57	11.19	12.82	14.04
2-A	4	$\frac{7}{8}$	1 face	3.37	4.30	4.89	5.24	6.77	7.12	8.29	10.16	12.04	13.45
		$\frac{7}{8}$	2 faces	—	6.26	6.85	7.20	8.73	9.08	10.25	12.12	14.00	15.41

[a]Light gage.

Source: Design Manual, copyright TECO/Lumberlok.

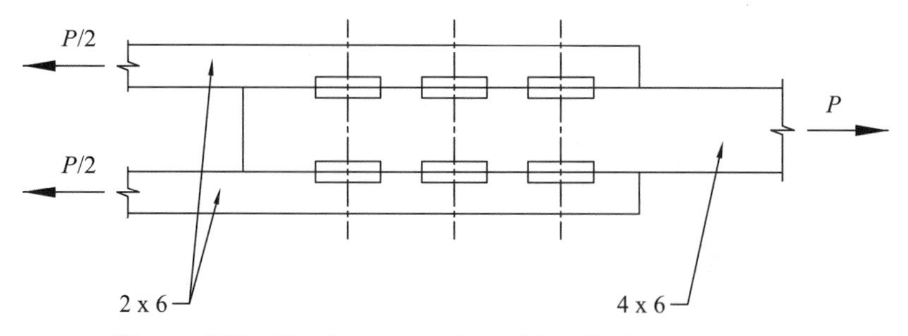

Figure 5.24 Tension connection with split-ring connectors.

termined, and from them an assumed number of connectors needed will be calculated. Connector design values will be determined and adjusted for group action, geometry, and other applicable factors. Finally, the connection will be checked for group or row tear-out.

Solution: Start by determining the cross-sectional area of the main and side members. The cross-sectional area of 4 × 6 dimension lumber is 3.5 in. × 5.5 in. = 19.25 in^2 and the cross-sectional area of each 2 × 6 is 1.5 in. × 5.5 in. = 8.25 in^2.

Next, determine the projected area of the connectors in the side members and subtract it from the cross-sectional area to determine the net sectional area. From Table 5.16, the projected area of a $2\frac{1}{2}$-in. split ring and $\frac{1}{2}$-in. bolt in a $1\frac{1}{2}$-in. member is 1.73 in^2.

For *both* side members the net section available to resist the tension load is

$$A_{\text{net}} = (2 \text{ in.})(8.25 \text{ in.} - 1.73 \text{ in.}) = 13.04 \text{ in}^2$$

Now, check the tensile stress on the net section. The resulting stress is $f_t = T/A_{\text{net}}$, where $T = 5000$ lb + 12,000 lb = 17,000 lb, or $f_t = 17{,}000$ lb/13.04 in^2 = 1304 psi. The allowable stress is $F_t' = F_t C_D C_F$ = (1200 psi)(1.15)(1.3) = 1794 psi (for both side and main members). Since $f_t = 1304 \text{ psi} \leq F_t' = 1794$ psi, the side members are adequate with regard to net section and tension.

The main member must also be checked for tension on the net section. For the main member, the projected area of one connector on each of two faces and the hole for the $\frac{1}{2}$-in. bolt is 3.75 in^2:

$$A_{\text{net}} = 19.25 \text{ in}^2 - 3.75 \text{ in}^2 = 15.50 \text{ in}^2$$

$$f_t = 17{,}000 \text{ lb}/15.50 \text{ in}^2 = 1097 \text{ psi}$$

Since $f_t = 1097 \text{ psi} \leq F_t' = 1794$ psi, the main member is adequate with regard to net section and tension.

The connector edge distance, end distance, and spacing values for the full design value of each connector are obtained from Table 5.13 as follows (parallel to grain loading):

Edge distance (unloaded edge, parallel to grain loading): $1\frac{3}{4}$ in.
End distance (tension member, parallel to grain loading): $5\frac{1}{2}$ in.
Spacing (parallel to grain loading): $6\frac{3}{4}$ in.

For both main and side members, the edge distances, assuming connectors centered in the wide faces, are 5.5 in./2 = 2.75 in. Since 2.75 in. ≥ 1.75 in., the edge distances for full design load are met. It will be assumed that the end and spacing requirements for full design value can be met and that these dimensions will be shown on the construction documents. Hence, $C_\Delta = 1.00$.

To determine the number of connectors, the full design value for a single $2\frac{1}{2}$-in. split-ring connector in specific gravity (species) group B is obtained from the *NDS®* [2] as 2730 lb, both for the side and main members. Thus, $P = 2730$ lb.

The number of connectors required, without considering group action, is

$$n = \frac{(17{,}000 \text{ lb})/2}{(2730 \text{ lb})(1.15)} = 2.7$$

Thus, assume three connectors for each side member. To determine group action, there are two rows of three connectors in each row. The area of the side members, $A_s = (2)(8.25) = 16.50$ in^2 and the area of the main member, $A_m = 19.25$ in^2. From equation (5.1), $C_g = 0.958$. The allowable load per connector is

$$P' = PC_DC_g = (2730 \text{ lb})(1.15)(0.958) = 3008 \text{ lb}$$

The allowable load for both side members is

$$(6 \text{ split rings})(3008 \text{ lb/split ring}) = 18{,}045 \text{ lb}$$

Since the total load of 17,000 lb does not exceed 18,178 lb, the six $2\frac{1}{2}$-in. split-ring fastener connection is adequate with the edge, end, and spacing distances stated above. The connection must also be checked with regard to row tear-out.

The critical area for shear for each split ring consists of the effective shearing surface areas of the sides and the effective shearing area of the back (Figure 5.25). The area of the sides is taken to be twice the split-ring penetration times the smaller of the spacing or end distance. The effective area of the back is taken to be the outside diameter of the rings times the smaller of the spacing or end distance, minus the area of the groove, minus the area of

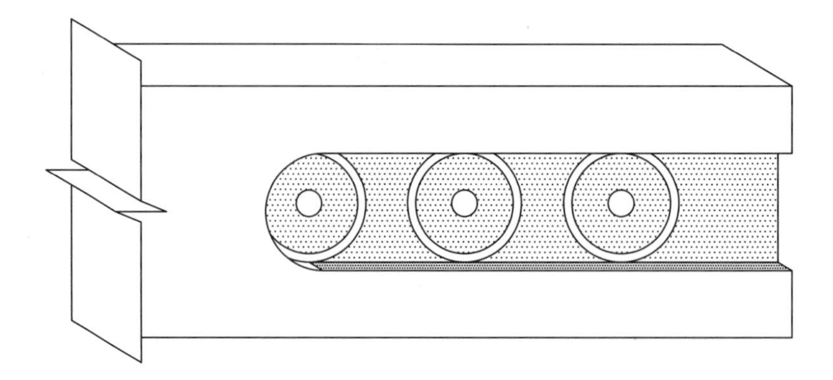

Critical shear area for row tear-out is represented by shading.

Figure 5.25 Effective shearing area for a split-ring connector.

the bolt holes. For the fastener farthest from the end, the shear area also included the area inside the groove (less the bolt hole).

The critical shearing area for split rings may be calculated by

$$A_{\text{crit shear}} = S(2p + D_o) - \frac{\pi}{4}\left(D_o^2 - D_i^2 + D_{bh}^2 - \frac{D_o^2}{2n}\right) \qquad \text{(split rings)} \quad (5.18)$$

where

S = smaller of the spacing or end distance,
p = depth of groove for split ring (one-half of the width of the ring),
D_o = outside diameter of the groove (the inside diameter plus twice the groove width),
D_i = inside groove diameter,
D_{bh} = bolt hole diameter, and
n = number of rings in the group.

Equation (5.18) is conservative in cases where S is governed by end distance rather than spacing. In this example, $A_{\text{crit shear}} = 19.50\ \text{in}^2$. The allowable shear stress is

$$\frac{F_v'}{2} = \frac{(180\ \text{psi})(1.15)}{2} = 103.5\ \text{psi}$$

Using equation (5.3a), the allowable load based on tear-out for *each* side member is therefore

$$Z_{\text{RT}}' = n\frac{F_v'}{2}A_{\text{crit shear}} = (3)(1.03.5\ \text{psi})(19.50\ \text{in}^2) = 6055\ \text{lb}$$

Assuming the double-shear load is equally distributed between the side members, each side member is to carry one-half of 17,000 lb, or 8500 lb each. Since $Z'_{RT} = 6055$ lb is less than $T/2 = 8500$ lb, the connection is not satisfactory with regard to row tear-out.

Since the critical shearing area for row tear-out was determined by the end distance, increasing the end distance, and, if necessary the spacing, may be investigated for providing enough shearing surface. If the end distance is increased to the value of the spacing (6.75 in.), the critical shearing area becomes $A_{\text{crit shear}} = 24.09 \text{ in}^2$:

$$Z'_{RT} = n\,\frac{F'_v}{2}\,A_{\text{crit shear}} = (3)(103.5 \text{ psi})(24.09 \text{ in}^2)$$

$$= 7480 \text{ lb} \qquad \text{still insufficient}$$

Using an end distance *and spacing* of 7.75 in. gives $A_{\text{crit shear}} = 27.76 \text{ in}^2$:

$$Z'_{RT} = n\,\frac{F'_v}{2}\,A_{\text{crit shear}} = 3\ (103.5 \text{ psi})(27.76 \text{ in}^2)$$

$$= 8620 \text{ lb} \qquad \text{good}$$

Since the spacing was changed, the group action factor must be recalculated. For the 7.75-in. spacing, $C_g = 0.953$. The total capacity of the connection based on the split rings is thus (18,045 lb)(0.953/0.958) = 17,950 lb, which is still adequate.

Answer: Six $2\frac{1}{2}$-in. split-ring connectors (three for each side member) are to be used, with connectors placed in single rows, centered in the wide faces of the pieces, with 7.75 in. of minimum end distance and spacing.

Discussion: To the extent that the design load is less than the capacity of the connectors utilizing the full design values, the spacing or end distances could be reduced. Such practice is not advisable unless other requirements in the design make reduced distances necessary. In this example the load was assumed to be shared equally between the two side members. Where conditions exist in which the side members are not of equal stiffness, unequal load sharing must be taken into consideration.

5.20.2 Timber Connectors in End Grain

Where timber connectors are installed in a surface not parallel to the general direction of the grain of a member, such as the square cut or sloping surface at the end of a member, the design values for the connectors must be reduced. Such conditions are illustrated in Figure 5.26. Where the end of a member is square cut, as shown in Figure 5.26*a*, the allowable design value, Q'_{90}, is 60%

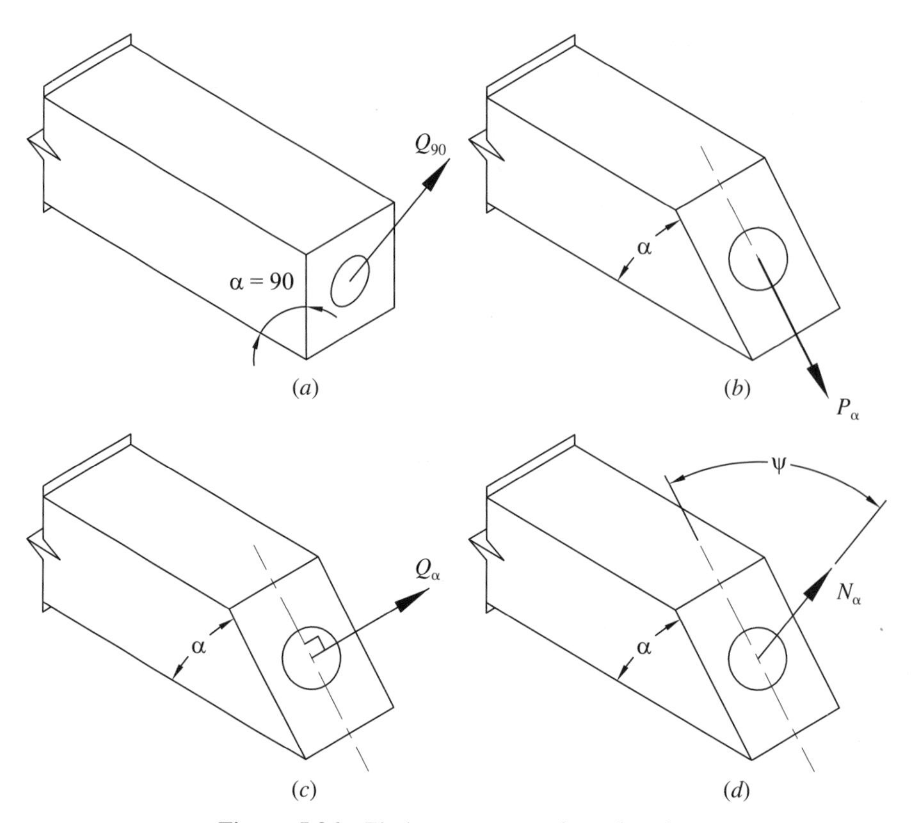

Figure 5.26 Timber connectors in end grain.

of the design value perpendicular to the grain regardless of the direction of the load, Q_{90}, in the plane of the cut:

$$Q'_{90} = 0.60Q' \tag{5.19}$$

where Q' = design value for a connector in a side grain loaded perpendicular to the grain, multiplied by applicable adjustment factors.

For ends with a sloped cut, the axis of the cut is defined as the intersection of a plane that is perpendicular to both the general direction of the grain and the sloped surface with the sloped surface. The angle of slope α of the sloped end is the angle between the sloped surface and the general longitudinal grain direction of the member. Where the load on the connector is parallel with the axis of the cut, as shown in Figure 5.26*b*, the design value $P\alpha$ is determined by

$$P'_\alpha = \frac{P'Q'_{90}}{P' \sin^2\alpha + Q'_{90} \cos^2\alpha} \tag{5.20}$$

where

P'_α = the allowable value for a connector in a sloping surface cut at angle α and loaded parallel to the axis of cut;
P' = design value for a connector in a side grain surface loaded parallel to the grain, multiplied by all applicable adjustment factors;
Q'_{90} = 60% of the design value for a connector in a side grain surface, Q', loaded perpendicular to the grain, multiplied by applicable adjustment factors; and
α = slope of the cut surface, as shown in Figure 5.26.

Where the load on a connector in end grain is perpendicular to the axis of the cut (parallel to the cut surface), as shown in Figure 5.26*c*, the allowable load on the connector is determined by

$$Q'_\alpha = \frac{Q'Q'_{90}}{Q' \sin^2\alpha + Q'_{90} \cos^2\alpha} \tag{5.21}$$

where

Q'_α = allowable load for a connector in a sloping surface, loaded perpendicular to the axis of cut;
Q' = design value for a connector in a side grain surface loaded perpendicular to the grain, multiplied by applicable adjustment factors;
Q'_{90} = 60% of the design value for a connector in a side grain surface loaded perpendicular to the grain, multiplied by applicable adjustment factors; and
α = slope of the cut surface.

Where the load on a connector in end grain is at some angle ψ with respect to the slope axis (Figure 5.26*d*), the allowable load on the connector is determined by

$$N'_\alpha = \frac{P'_\alpha Q'_\alpha}{P'_\alpha \sin^2\psi + Q'_\alpha \cos^2\psi} \tag{5.22}$$

where N'_α = the allowable load for a connector in a sloping surface loaded at an angle ψ to the axis of the cut, and other terms are as defined previously.

The provisions for edge distance, end distance, and spacing shown for connectors in side are applied to the sloping cuts as follows:

1. For square-cut ends, the provisions for loading perpendicular to the grain apply.
2. For sloping surfaces with angle α from 45° to 90° loaded in any direction, the provisions for loading perpendicular to the grain apply.
3. For sloping surfaces with angle α less than 45° loaded parallel to the axis of the cut, the provisions for loading parallel to the grain apply.

4. For sloping surfaces with angle α less than 45° loaded perpendicular to the axes of the cut, the provisions for loading perpendicular to the grain apply.
5. For sloping surfaces with angle α less than 45° loaded at angle ψ to the axis of cut, the provisions for members loaded at angles of grain other than 0° or 90° apply.

5.21 MOMENT SPLICES

The use of glued laminated timber has reduced the need for moment splices in timber beams and girders, as they can be manufactured to long lengths, limited primarily by transportation constraints. However, moment splices in timber members are sometimes used in arches and rigid frames where the transportation of full-sized structural frames is impractical or uneconomical. Moment splices must be designed to resist the bending moments, shear forces, and axial forces for all applicable loading conditions at the splice points. Where possible, moment splices should be located at or near inflection points where bending moments are relatively small.

Resistance to the applicable moments and forces may be achieved by various means. Bending moments may be resisted by tension and compression straps located on the sides or edges, and end grain bearing or compression plates between members to resist bending flexural compression and axial compression. Shear is generally resisted with shear places at the butting faces. A means of holding the two sections in alignment must also be provided. Compression straps may be used to act as tension straps in load reversal conditions. For sawn lumber sections, and for many smaller glued laminated timbers, the simplest form of moment connection is a plywood splice plate (or pair of plates). Such plates can be glued or nailed but are also frequently used in conjunction with bolts and shear connectors.

Close tolerances are required in the fabrication of moment splices. Consideration should be given to inelastic as well as elastic deformation in the joint. Typical moment splice connections for glued laminated timber are illustrated in Figure 5.27 and in AITC 104 (excerpted in Chapter 8). In cases where the combined axial and flexural compressive stress exceeds 75% of the design value in compression parallel to the grain, F'_c, a snug-fitting bearing plate of 20 gauge or thicker metal should be installed between the abutting ends. Flexural tension stress is taken across the splice by means of steel straps and shear plates as illustrated, or by wood splice plates and split rings. Additional side straps and shear plates are generally required to keep the sides and top of the members in position and to take the load reversals from erection loads or wind uplift. Separate side plates should be used for each row of connectors to minimize secondary stresses due to shrinkage effects in the member.

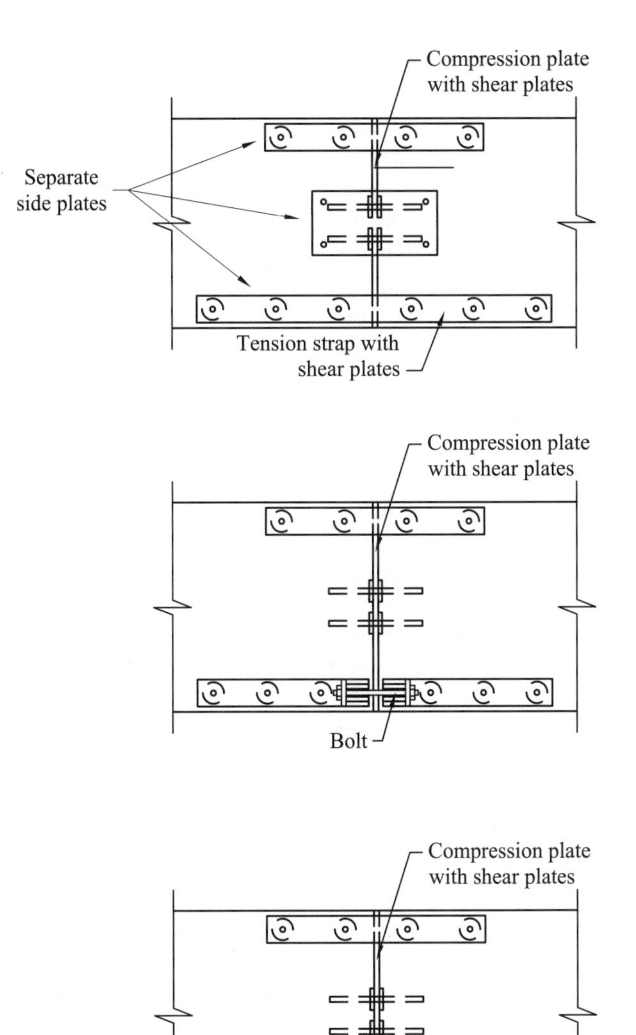

Figure 5.27 Moment splice connections.

Several approaches have been used for the design of moment splice connections. For splices where the axial compressive stress is large relative to the flexural stresses under any and all loading conditions, such that no net tension exists at the section, the moment splice may be designed simply to provide continuity at the section and to resist shear. Such is the case for which

$P > 6M/d$, where P is the applied axial load (compression), M the applied bending moment, and d the member depth. For cases in which the bending moment is not relatively small, $M > Pd/6$, net tension exists across the section. At a minimum, tension straps must be designed to carry the net tension stress at the section. However, due to inelastic slip and fiber crushing in end bearing, accurate determination of the neutral axis for such cases is not generally possible. A conservative design approach for both of the foregoing cases is to design tension straps to carry the full bending moment tension force. The flexural and axial compressive stresses are transferred by end bearing, end bearing plates, or compression straps.

Inelastic slip in the tension splice plates can be offset by using a tension splice with bolts and nuts that can be tightened to eliminate the slip, as shown in Figure 5.27. These also require extra materials and fabrication, but the location of forces to resist the moment can be determined more accurately with this system, and the slippage that occurs in the tension connection can be compensated for by tightening the bolts. Tension plates or straps may also be installed on the tension faces (bottom edges) of the connected members (Figure 5.27). This arrangement places the resultant tension force farther from the resultant compressive force, thereby reducing the required tension force associated with a given applied moment. In addition, the shear plates located in the tension faces of glued laminated timbers may have higher design values, due to a higher split-ring and shear plate connector group value assigned to the tension face material.

Example 5.9 Moment Splice Connection

Given: 5 in. × $17\frac{7}{8}$ in. Southern Pine glued laminated timbers combination 24F-V3 for a two-hinged arch are to be spliced. At the location of the splice, the design positive moment is found to be 255,000 lb-in. caused by DL + SL. The maximum axial compression is found to be 27,000 lb caused by DL + SL, and the maximum shear is found to be 5200 lb caused by DL + SL, (on half the span). The maximum negative moment is 15,000 lb-in. caused by WL (assuming no axial force).

Determine: A suitable moment splice connection.

Approach: In this example the tension strap will be designed to carry the net tension force under the full design load (dead plus snow). Compression parallel to the grain will be checked for the combined axial and flexural compression stresses. The unbalanced load case will be used to design the splice for shear. Finally, straps on the top of the member will be checked with regard to being able to resist the negative moment associated with wind loading.

Solution:

1. *Calculate the axial and flexural stresses.* For the section given, $S = 266$ in^3 and $A = 89.4$ in^2. Thus,

$$f_b = \frac{M}{S} = \frac{255{,}000 \text{ lb-in.}}{266 \text{ in}^3} = 958 \text{ psi}$$

and

$$f_c = \frac{P}{A} = \frac{27{,}000 \text{ lb}}{89.4 \text{ in}^2} = 302 \text{ psi}$$

The combined extreme fiber stresses become

$$f_b + f_c = 958 \text{ psi} + 302 \text{ psi} = 1260 \text{ psi} \qquad \text{compression}$$

$$f_b - f_c = 958 \text{ psi} - 302 \text{ psi} = 656 \text{ psi} \qquad \text{net tension}$$

as illustrated in Figure 5.28.

The location of the neutral axis for the combined flexural and axial load is found as follows. Assuming a linear stress distribution, $x + y = 17.875$ in. and $x/1260 \text{ psi} = y/656 \text{ psi}$; thus, $x = 11.75$ in. and $y = 6.12$ in (y is the distance from the bottom of the section to the neutral axis).

2. *Design the tension strap connection.* The tension stresses to be carried by tension straps result in a force T located two-thirds of 6.12 in. = 4.08 in. below the neutral axis, or 2.04 in. above the bottom face of the members, where

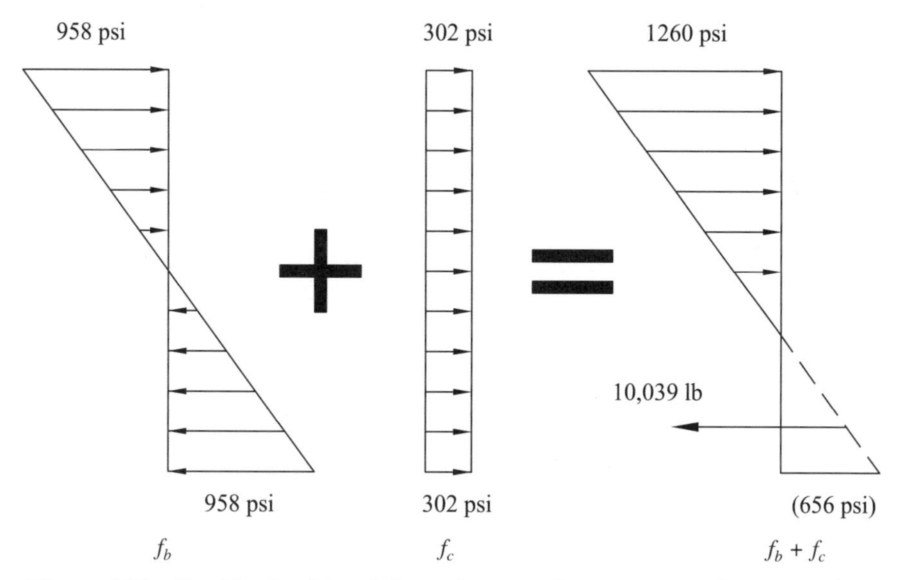

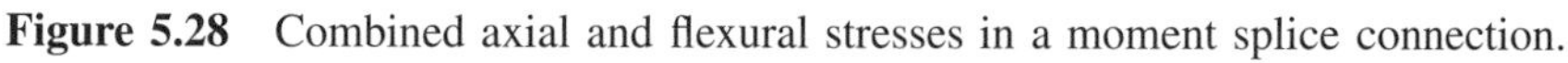

Figure 5.28 Combined axial and flexural stresses in a moment splice connection.

$$T = \tfrac{1}{2}\, by\, (f_b - f_c)$$

$$= \tfrac{1}{2}\,(5\text{ in.})(6.12\text{ in.})(656\text{ psi}) = 10{,}040\text{ lb}$$

Considering $\frac{3}{4}$-in. through bolts, with $F_{e\parallel} = 6150$ psi from Table 5.7, equations (5.8) give $Z = 3480$ lb. Then $Z' = ZC_D = (3480\text{ lb})(1.15) = 4000$ lb assuming a group action factor of unity and that the spacing, edge, and end distance requirements are met for use of the full design values. Thus, the preliminary minimum number of through bolts in tension for each member of the connection is (10,040 lb)/(4000 lb/bolt) = 2.51; use three through bolts for each member.

Assuming $\frac{1}{4}$ in. × 3 in. straps on each side and bolt spacing of 4 in. o.c., the group action factor, C_g [equation (5.1)], becomes 0.980. The minimum end distance for a full design load is $7.0D = 5.25$ in. The minimum edge distance for a full design load is $1.5D = 1.125$ in., which is satisfied with the 2.0-in. distance from the bottom face.

Assuming an end distance of 5.25 in. and a spacing of 4.0 in., row tear-out is investigated using

$$Z'_{\text{RT}} = n\,\frac{F'_v}{2}\,A_{\text{crit shear}} \tag{5.3a}$$

where n = number of fasteners in the row = 3 and

$$F'_v = F_v C_D = (215\text{ psi})(1.15) = 248\text{ psi}$$

where $F_v = 215$ psi from AITC 117 [4] or the *NDS*® [2]. $A_{\text{crit shear}}$, from Chapter 5, is (4 in.)(5 in.)(2) = 40 in^2; thus,

$$Z'_{\text{RT}} = 3\left(\frac{248\text{ psi}}{2}\right)40\text{ in}^2 = 14{,}900\text{ lb}$$

Since the required tension capacity of the connection is 10,040 lb, row tear-out is not critical.

The tension capacity of the straps is checked as follows. The tensile stress in the strap may not exceed $0.60F_y$ based on the gross section or $0.50F_u$ based on net section, from the *Manual of Steel Construction* [5]. Assuming $\frac{1}{4}$ in. × 3 in. straps, A(gross) = 0.25 in. × 3 in. = 0.75 in^2; thus, for each strap,

$$f_t = \frac{\frac{1}{2}(10{,}039\text{ lb})}{0.75\text{ in}^2} = 6693\text{ psi} \qquad \text{(based on gross section)}$$

Assuming $\frac{13}{16}$-in. holes, A(net) = 0.25 in. × [3 in. − ($\frac{13}{16} + \frac{1}{16}$) in.] = 0.0.531 in^2, giving

$$f_t = \frac{\frac{1}{2}(10{,}040 \text{ lb})}{0.531 \text{ in}^2} = 9454 \text{ psi} \qquad \text{(based on net section)}$$

Using an A36 plate with $F_y = 36{,}000$ psi and minimum $F_u = 58{,}000$ psi,

$$f_t = 6693 \text{ psi} \leq (0.50)(36{,}000 \text{ psi}) = 18{,}000 \text{ psi}$$

$$\text{good, based on gross section}$$

$$f_t = 9454 \text{ psi} \leq (0.60)(58{,}000 \text{ psi}) = 34{,}800 \text{ psi,}$$

$$\text{good, based on net section}$$

Checking the strap capacity considering the calculated group action factor yields

$$T = 10{,}039 \text{ lb} \leq (3 \text{ bolts})Z_{\parallel}C_D C_g = (3)(4007 \text{ lb})(1.15)(0.98)$$
$$= 13{,}548 \text{ lb} \qquad \text{good}$$

Each strap is checked with regard to block shear according to the *Manual of Steel Construction* [5], as follows, assuming that the strap extends 2 in. past the end bolt.

$$R_{BS} = 0.3A_V F_u + 0.5A_t F_u$$
$$= (0.3)(2)[(2 \text{ in.} + 4 \text{ in.} + 4 \text{ in.})(0.25 \text{ in})](58{,}000 \text{ psi})$$
$$+ 0.5(0)(58{,}000 \text{ psi})$$
$$= 87{,}000 \text{ lb.}$$

Since the design load on each strap is (10,040 lb)/2 = 5020 lb, which is less than the allowable resistance, $R_{BS} = 87{,}000$ lb, the bolts are also not expected to tear out of the metal strap.

3. *Determine the bearing condition.* The combined flexural and axial compression stress from above is 1260 psi. Since 1260 psi $\leq$ (0.75)(1650 psi)(1.15) = 1423 psi, where 1650 psi is the design value in compression parallel to grain from AITC 117 [4] for the 24F-V3 combination, no metal bearing plate is required.

4. *Design the shear connection.* Shear plates of size $2\frac{5}{8}$ in. will be used. From the *National Design Specification*® [2], the design value for a single shear plate in single shear for loading perpendicular to the grain, Q, is 1990 lb (specific gravity group B). Assuming a group action factor of 0.98 and that the edge and spacing requirements for full design load are met, the allowable load per plate installed in side grain is

$$Q' = QC_DC_gC_\Delta = (1990 \text{ lb})(1.15)(0.98)(1.0) = 2243 \text{ lb}$$

Since the plate is installed in end grain, the allowable value is multiplied by 0.60, giving

$$Q'_{90} = 0.60Q' = (0.60)(2243 \text{ lb}) = 1346 \text{ lb}$$

The required number of shear plates for the design load of 5200 lb is n = (5200 lb)/(1346 lb/shear plate) = 3.8; use four shear plates.

The geometry factor is checked as follows. The shear plates will be arranged for the full design load with regard to spacing, 4.25 in., and should be installed symmetric about the section centerline. The edge distance from plate centers to the wide faces of the timbers (unloaded edges) is 2.5 in., which is greater than the minimum of 1.75 in. for full design load from Table 5.13; good. The required spacing for full design load for $2\frac{5}{8}$-in. shear plates is 4.25 in., also from Table 5.13.

The edge distances from either the top or bottom plates to either the top or bottom of the timbers, assuming the full design value spacing is used, is

$$\frac{17\frac{7}{8} \text{ in.} - (4 - 1 \text{ shear plates})(4.25 \text{ in. spacing})}{2} = 2.563 \text{ in.}$$

Since this value is less than the stated minimum for full design value (loaded edge) but greater than the minimum for reduced design value, a geometry factor is calculated as

$$C_{\Delta e} = 0.83 + \frac{(1.0 - 0.83)(2.56 \text{ in.} - 1.75 \text{ in.})}{2.75 \text{ in.} - 1.75 \text{ in.}} = 0.968$$

The group action factor is calculated using equation (5.1), as follows. The side and main member areas are taken to be the shear plate spacing multiplied by the member thickness. In this case, since the thickness is the member length, the side and main member thickness values are taken to be arbitrarily large, and equal ($A_m = A_x = 50 \text{ in}^2$). The modulus of elasticity of the members is taken to be 1,600,000 psi (weak axis bending) from AITC 117 [4]. Thus, from equation (5.1), $C_g = 0.98$, identical to the assumed value for the preliminary determination of number of shear plates needed. Therefore,

$$Q'_{90} = (0.60)(1990 \text{ lb})(1.15)(0.98)(0.968) = 1303 \text{ lb/shear plate}$$

Since 5200 lb $\leq$ (4)(1303 lb) = 5210 lb, four $2\frac{5}{8}$-in shear plates are adequate.

Since the shear plates are loaded in end grain, steel dowels will be used for the plate bolts. The dowel length in each member should be taken to be at least the required penetration for the full design value of the shear plate with a lag screw fastener, 5 in., from the *National Design Specification*® [2]. Thus, dowels of total length 2(5 in. + 1 in.) = 12 in. should be used.

Shear at the splice should also be checked with respect to effective depth due to the fastener group (Section 5.3 and Figure 5.4). The effective depth is 17.875 in. − 2.563 in. + $\frac{1}{2}$(2.62 in.) = 16.622 in. From equation (5.2a),

$$V'_r = \left(\frac{2}{3} F'_v b d_e\right)\left(\frac{d_e}{d}\right)^2 = \left[\frac{2}{3}(248 \text{ psi})(5 \text{ in.})(16.622)\right]\left(\frac{15.312 \text{ in.}}{17.875 \text{ in.}}\right)^2$$

$$= 11{,}800 \text{ lb}$$

Since V = 5200 lb ≤ 11,880 lb = V'_r = the allowable design shear, the connection is also good with respect to effective section and shear.

5. *Check the load reversal.* The strap required to accommodate the positive design bending moment will be duplicated on the upper part of the splice to accommodate the negative design moment. The connection will be evaluated with respect to flexural stresses only, with the strap at the top of the connection carrying the full tension load. The distance between the resultant compression force and the location of the strap is, from Figure 5.29, 6.94 in. + 5.96 in. = 12.90 in. The resultant tension force is thus T = 15,000 lb-in./12.90 in. = 1163 lb. Since the strap capacity was found to be in excess of 13,000 lb, a duplicate strap at the top of the connection for load reversal is adequate.

Answer: The salient features of the moment splice connection for the stated conditions are shown in Figure 5.30. To simplify detailing, the end distance for bolts in the tension straps with respect to the wood members is increased to 6.0 in., and the end distance with respect to the metal is taken to be 2.0 in., resulting in a total strap length of 32 in.

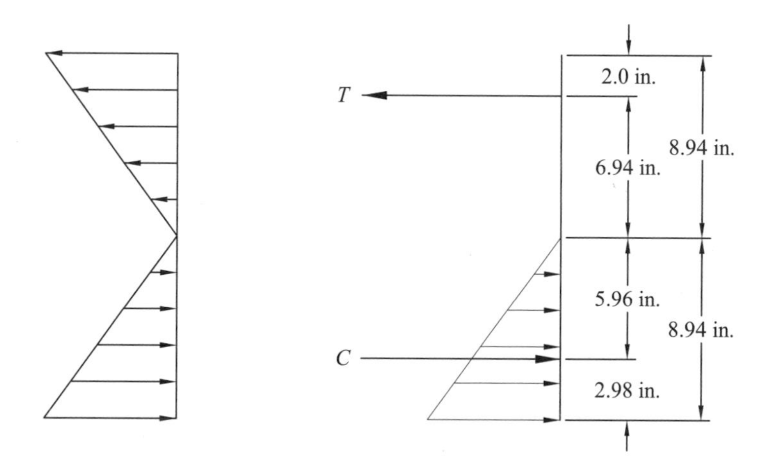

Figure 5.29 Stresses and resultants in load reversal.

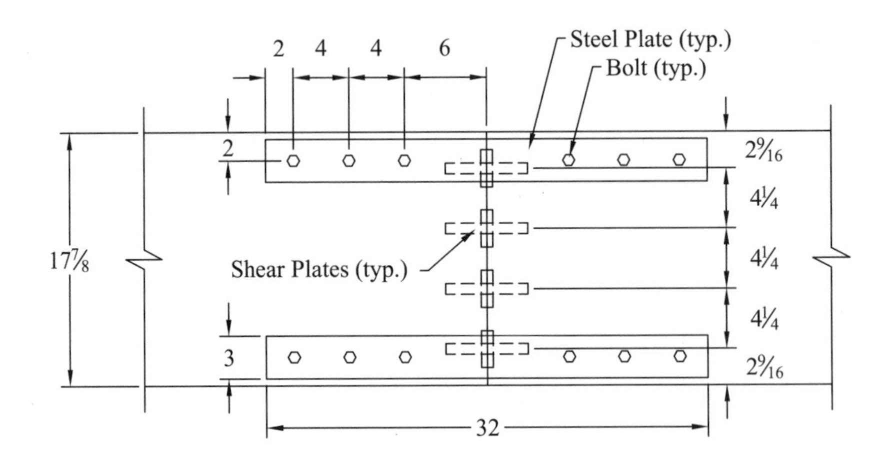

All dimensions are inches. Bolts are $\frac{3}{4}$ in. through bolts. Shear plates are $2\frac{5}{8}$ in. diameter with $\frac{3}{4}$ in. x 6 in. steel dowels. Side plates are $\frac{1}{4}$ in. A36 grade steel.

Figure 5.30 Moment splice connection, Example 5.9.

Discussion: In this example, the strap used to resist tension on the lower part of the connection was based on the net tension associated with combined axial compression and flexural tension stresses. Design to accommodate the full flexural tension force would require a larger strap and larger or more fasteners. For convenience, the strap determined for the bottom part of the connection was also used for the upper part, which in this case was more than adequate. A smaller strap or strap with fewer fasteners could have been used, producing an aesthetically asymmetric splice connection. For aesthetic purposes, it may also be possible to use narrower straps at both top and bottom, provided that the minimum edge distances and the tensile stress requirements for the steel are still satisfied.

5.22 TIMBER RIVETS

Timber rivets, also called *glulam rivets,* may be used to transfer large loads between wood members without section loss, where the rivets are installed with their major cross-sectional axes parallel to the grain. Typical timber rivets and side plates for timber rivet connections are shown in Section A.3.6. Timber rivets are made from AISI 1035 steel with Rockwell C32-39 hardness and minimum ultimate tensile stress of 145,000 psi according to ASTM A370. Side plates are ASTM Standard A36 steel of minimum thickness $\frac{1}{8}$ in.

Design values for timber rivets are determined as being the lesser of the nominal rivet capacity per rivet multiplied by the total number of rivets or the nominal wood capacity based on the number of rows of rivets and rivets in each row. The nominal wood capacity values take into consideration group action and row and group tear-out; thus, the group action factor and row and group tear-out need not be considered independently for timber rivet connections. Rivet and wood capacities are dependent on load orientation with respect to grain.

Timber rivets must be installed such that the maximum penetration of rivet in the wood member is not greater than 70% of the receiving member thickness. In all cases of load, the rivets are to be installed with their long cross-sectional axes parallel to the grain of the wood member. Edge and end distance requirements for rivet connections are shown in Table 5.17 and illustrated in Figure 5.31. Spacing between rivets is not permitted to be less than $\frac{1}{2}$ in. perpendicular to the grain and 1.0 in. parallel to the grain. Where rivets at a connection (joint) are driven into opposite faces of a wood member such that their points overlap, the spacing requirements apply to the rivets at their points and the capacity of the connection is considered to be that of a single-sided connection, with the number of rivets equaling the total number of rivets for the connection.

Nominal rivet capacities based on rivets for loading both parallel and perpendicular to the grain are determined by calculation. Capacities based on wood for loading parallel to the grain are provided in tabular form in the *National Design Specification®* [2] for the total number of rows and rivets per row for $1\frac{1}{2}$-, 2-, $2\frac{1}{2}$-, and $3\frac{1}{2}$-in. rivets at 1- and $1\frac{1}{2}$-in. spacing of rivets in a row. Capacities based on wood for loading perpendicular to the grain are based on tabular data in the *NDS®* [2] and are also subject to a geometry factor particular to rivet connections. Nominal capacities of the rivets are determined based on $\frac{1}{8}$-in. side plate thickness; the side plate factor of Table 5.18 is applied for greater plate thickness values where the value of the rivet controls. Design values are for one plate and the rivets associated with it. Where equal and opposite plates and rivets are used on opposite faces of connected members, the design values are added (the design value for the connection is *twice* the value for either plate). Rivets are taken to be installed in side grain (the sides of the members) except where modified for end grain application.

5.22.1 Loading Parallel to the Grain

For loading parallel to the grain, the nominal design value P for the rivet connection is the lesser of the capacity based on the rivet, P_r, and the value based on the receiving wood, P_w. Design values are provided for connections with rows of an equal number of rivets per row. The capacity based on the rivet is given by

$$P_r = 280p^{0.32}n_R n_C \qquad (5.23)$$

where

TABLE 5.17 Minimum End and Edge Distance Requirements for Timber Rivet Connections

Number of rivet rows, n_R	Minimum end distance, a, in.		Minimum edge distance, e, in.	
	Load parallel to grain, a_p	Load perpendicular to grain, a_q	Unloaded Edge e_p	Loaded edge e_q
1, 2	3	2	1	2
3 to 8	3	3	1	2
9, 10	4	3-1/8	1	2
11, 12	5	4	1	2
13, 14	6	4-3/4	1	2
15, 16	7	5-1/2	1	2
17 and greater	8	6-1/4	1	2

Note: End and edge distance requirements are shown in Figure 5.31.

Source: Reprinted with permission from *National Design Specification® for Wood Construction.* Copyright © 2001. American Forest & Paper Association, Inc.

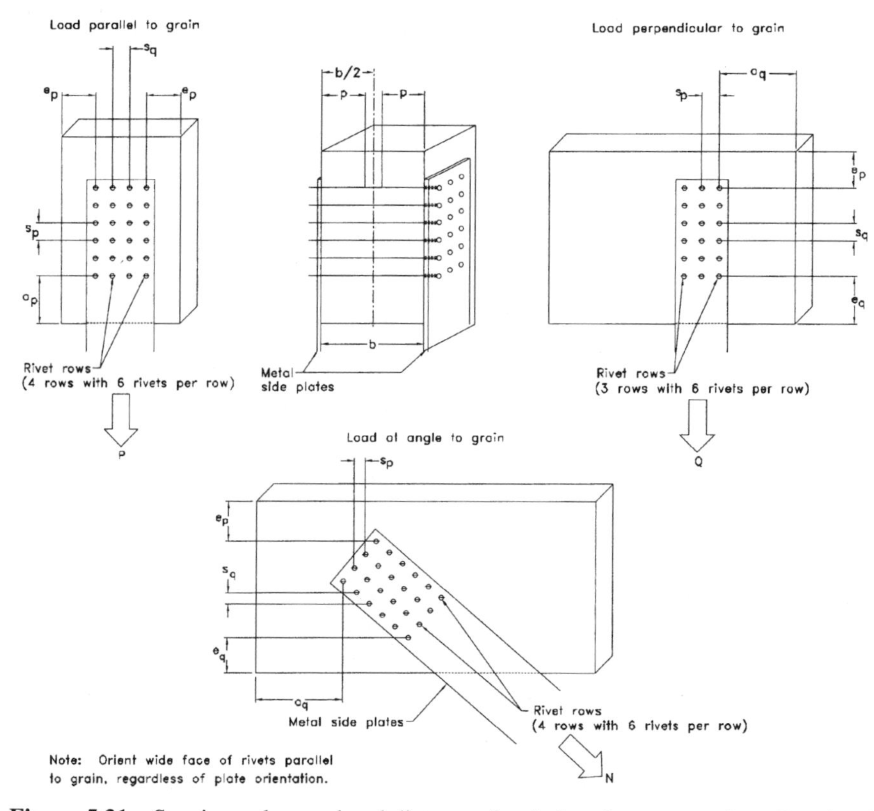

Figure 5.31 Spacing, edge, and end distances for timber rivet connections. Reprinted with permission from *National Design Specification® for Wood Construction.* Copyright © American Forest & Paper Association, Inc.

P_r = nominal capacity parallel to the grain in pounds based on the rivet,
p = depth of penetration of the rivet = rivet length − plate thickness − $\frac{1}{8}$ in.,
n_R = number of rows of rivets parallel to the direction of load,
n_C = number of rivets per row.

The nominal design value based on the receiving wood, P_w, is obtained in tabular form from the *National Design Specification®* [2] as a function of wood member thickness, rivets per row, number of rows per side, rivet length, and spacing. The nominal design value P is obtained as the lesser of P_r and P_w and multiplied by all applicable adjustment factors (Table 5.3) to obtain P'. Where rivet capacity controls, the load duration factor is not used, but the metal side plate factor is applied; where wood controls, the load duration factor is applicable.

TABLE 5.18 Metal Side Plate Factor C_{st}, for Timber Rivet Connections

Metal Side Plate Thickness, t_s	C_{st}
$t_s \geq \frac{1}{4}$ in.	1.00
$\frac{3}{16}$ in. $t_s \leq \frac{1}{4}$ in.	0.90
$\frac{1}{8}$ in. $\leq t_s < \frac{3}{16}$ in.	0.80

5.22.2 Loading Perpendicular to the Grain

Where the rivets are loaded perpendicular to the grain, the nominal design value Q for one plate and rivets is the lesser of Q_r and Q_w as given by

$$Q_r = 160p^{0.32}n_R n_C \tag{5.24}$$

where

Q_r = nominal capacity perpendicular to grain in pounds based on the rivet,
p = depth of penetration of the rivet = rivet length − plate thickness − $\frac{1}{8}$ in.,
n_R = number of rows of rivets parallel to the direction of the load,
n_C = number of rivets per row, and

$$Q_w = q_w p^{0.08} C_\Delta \tag{5.25}$$

where

q_w = tabular value in pounds obtained from the *National Design Specification®* [2] as a function of spacing, rivets per row, and number of rows, and
C_Δ = the geometry factor specific to timber rivet connections, Table 5.19.

The nominal design value Q is obtained as the lesser of Q_r and Q_w and multiplied by all applicable adjustment factors (Table 5.2) to obtain Q'. Where rivet capacity controls, the load duration factor is not used but the metal side plate factor is applied; where wood controls, the load duration factor is applicable.

5.22.3 Loading at an Angle to the Grain

Where the load acts at an angle to the grain, the allowable plate and rivet capacity is determined by

$$N' = \frac{P'Q'}{P' \sin^2\theta + Q' \cos^2\theta} \tag{5.17}$$

TABLE 5.19 Geometry Factor, C_Δ, for Timber Rivet Connections Loaded Perpendicular to the Grain

$\frac{e_p}{(n_{c-1})s_q}$	C_Δ	$\frac{e_p}{n_{c-1}s_q}$	C_Δ
0.1	5.76	3.2	0.79
0.2	3.19	3.6	0.77
0.3	2.36	4.0	0.76
0.4	2.00	5.0	0.72
0.5	1.77	6.0	0.70
0.6	1.61	7.0	0.68
0.7	1.47	8.0	0.66
0.8	1.36	9.0	0.64
0.9	1.28	10.0	0.63
1.0	1.20	12.0	0.61
1.2	1.10	14.0	0.59
1.4	1.02	16.0	0.57
1.6	0.96	18.0	0.56
1.8	0.92	20.0	0.55
2.0	0.89	25.0	0.53
2.4	0.85	30.0	0.51
2.8	0.81		

5.22.4 Timber Rivets in End Grain

For timber rivets in end grain for square-cut members (rivets installed longitudinally parallel to the grain), the allowable lateral resistance is to be taken to be 50% of the value of lateral resistance in side grain. For sloping end cuts, the lateral resistance is determined by interpolation between the 50% for square end cuts and 100% for load (or surface) parallel to the grain.

Example 5.10 Timber Rivet Connection

Given: The tension splice connection shown in Figure 5.32, utilizing ten $1\frac{1}{2}$-in. timber rivets in each row spaced $1\frac{1}{2}$ in. on center in five rows in rows spaced 1 in. row to row on each side of the splice. Rivets and plates are installed on both (opposite) faces of the $3\frac{1}{8}$ in. × 9 in. glued laminated timbers shown. The metal plates are $\frac{1}{8}$-in.-thick ASTM A36 plates. The design load has snow load duration. The glued laminated timbers are 24F-1.8E with F_t = 1100 psi.

Determine: The allowable tension capacity (T) of the connection.

Approach: In this example, the load is parallel to the grain for both members. Proper edge, end, and spacing requirements will be verified and as appropriate, specified for the connection. The capacity of each plate and rivets will be determined by examining both rivet and wood capacity load parallel to the

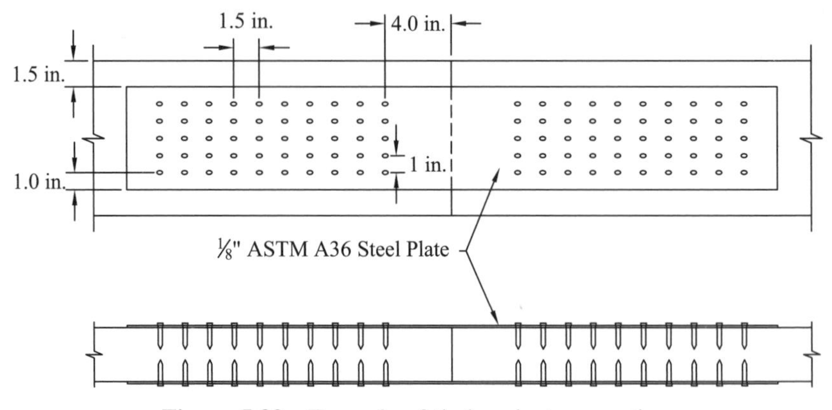

Figure 5.32 Example of timber rivet connection.

grain and adding applicable adjustment factors. Group action and row and group tear-out need not be examined independently; however, the tension capacity of the timber and metal plates themselves must be checked.

Solution: The penetration of each rivet is

$$p = 1.5 \text{ in.} - \tfrac{1}{8} \text{ in. (plate)} - \tfrac{1}{8} \text{ in.} = 1.25 \text{ in.}$$

Rivets will be installed in each face but not overlap ($b/2 = 3.125$ in.$/2 =$ 1.563 in. $> p = 1.25$ in., Section A.3.6). The nominal rivet capacity for the one plate and rivets from equation (5.23) is

$$P_r = 280p^{0.32}n_R n_C = (280)(1.25)^{0.32}\,(5)(10) = 15{,}036 \text{ lb/plate}$$

From the *NDS*® [2], for a rivet length of $1\frac{1}{2}$ in., spacing parallel to the grain of $1\frac{1}{2}$ in., spacing perpendicular to the grain of 1 in., and 10 rivets per row, the following values are obtained for 3- and 5-in. side members and four and six rows per side, from which the value of P_w will be interpolated for the stated example.

Member thickness 3 in.: $P_w = 12{,}600$ lb for four rows and 18,250 lb for six rows; interpolating gives 15,425 lb for five rows.

Member thickness 5 in.: $P_w = 14{,}010$ lb for four rows and 17,090 lb for six rows; interpolating gives 15,550 lb for five rows.

Interpolating for member thickness of 3.125 in. gives 15,433 lb. In this case the rivet capacity governs, and thus the metal side plate factor C_{st} is applied. From Table 5.18, $C_{st} = 0.80$. Thus,

$$P' = P_r C_{st} = (15{,}036 \text{ lb})(0.80) = 12{,}028 \text{ lb (per plate)}$$

From Figure 5.31 and Table 5.17, the edge distance required for an unloaded edge, three to eight rows, is 1 in. In the example given, the edge distance is 2.5 in. (1.0 in. to the edge of the plate and 1.5 in. from the plate edge to the timber edge); since 2.5 in. > 1.0 in.; good. The required end distance for three to eight rows of fasteners is 3 in. for a load parallel to the grain. In the example given, the end distance is 4 in.; good.

The capacity of each plate is the lesser of $0.60F_y$ on the gross (unreduced) area or $0.33F_u$ on the net area. In this case the unreduced area of each plate is 0.125 in. × 6 in. = 0.75 in^2; therefore, 0.75 in^2 × 0.60 × 36,000 psi = 16,200 lb. The reduced area is 0.75 in^2 − 5 × (0.22 in. × 0.125 in.) = 0.6125 in^2; thus, 0.6125 in^2 × 0.33 × 58,000 psi = 11,723 lb (based on F_u = 58,000 psi). Thus, the capacity of the plate is the smaller of the two, or 11,700 lb, which is less than the load capacity of the rivets, 11,700 lb.

The capacity of the connection, considering both plates, is thus T = (2)(11,700 lb) = 23,400 lb. Finally, the capacity of the timbers must be checked. Since no reduction in net section is required with the rivets, the tension capacity of the timbers is

$$F_t'A = (1100 \text{ psi})(1.15)(3.125 \text{ in.})(9 \text{ in.}) = 35{,}578 \text{ lb}$$

Answer: The capacity of the timber rivet connection is 23,400 lb (plate capacity controls). The edge, end, and spacing distances shown in Figure 5.32 must be maintained (except that the end distance may be decreased to 3 in.). The rivet groups should be centered in the plates as shown and plates centered in the side members of the spliced timbers. In this particular example, greater capacity could have been achieved with thicker side plates, increasing the capacities of the plates themselves, as well as allowing for a greater metal side plate factor.

5.23 WOOD JOINERY

A renewed interest in traditional timber framing in recent years is requiring designers in the timber engineering field to provide rational approaches to designing or determining the capacity of all-wood joinery. Design provisions for wood construction are in many cases restrictive or not applicable to such joinery (mortise and tenon joints, dovetail joints, spline connections, wood-peg dowels, and the like). Such joinery often involves large losses of section due to mortises, notching, and housings. Connection design may be further complicated by the relatively large dimension changes involved where unseasoned wood is used.

Whereas bearing-only joints may be designed with the methods and provisions of this manual, other joints involving mortises, tenons, wood pegs, or large slots for splines, in general, may not. For such applications, designers often employ various metal fasteners (hidden, if necessary) to resist the loads and for providing the anchorage requirements of modern building codes. Lateral loads, for example, in many of the newer timber frame structures are resisted by means other than the timber frames themselves: for example, by structural panels or structural insulated panels. Although the recent interest in such construction is motivating new research and the development of code provisions for such joinery, such provisions are not presently available.

5.24 DETAILING

Construction documents for structures having the fasteners and connections described in this chapter must contain enough detail to ensure proper materials and installation. AITC 104 [1] shows typical connection details commonly used in timber construction and may be used as a guide in connection detailing. Construction documents should clearly show bolt locations, including edge distances, end distances, spacing of fasteners in rows and spacing of rows, and any clearance requirements for shrinkage/swelling or for fire protection requirements. Edge, end, and spacing distances must be clearly shown or specified and should, where possible, be at least the minimum for the full design values of the fasteners. Where possible, consistent fastener size and spacing values should be used within individual connections and from connection to connection. Size and grade of metal parts (bolts, split rings, shear plates, rivets, and metal plate), and where used, weld types and lengths, must be clearly identified. Minimum and/or required bearing lengths and anchorage requirements must be shown. Where nails or similar fasteners are used, diameter, length, and bending yield strength must be clearly identified. Wood-to-concrete connections must include minimum concrete compressive strength and the appropriate edge distances, end distances, and penetration (embedment) into concrete to achieve the values used in design. Where required by the designer or by local building code, the connections requiring special inspection or testing must also be identified.

CHAPTER 6

STRUCTURAL SYSTEMS

6.1 INTRODUCTION

The scope of this chapter is the design of structural timber systems. Systems discussed include post and beam framing; pole construction and timber piles; framing systems such as floors, roofs, and decks; timber trusses; and timber bridges. Post–frame, pole, and pile systems are not covered in detail, but reference is made to appropriate design guides. Although roof, floor, and decking systems may be designed and constructed to provide diaphragm action to resist lateral loads, only gravity load resistance is covered in this book. Timber truss and bridge design are covered in detail.

6.2 POST AND BEAM CONSTRUCTION

Post and beam construction is generally comprised of roof and floor panels or decking supported by joists or purlins, which are in turn supported by beams or girders, which are supported by columns. Post and beam construction is illustrated in Figure 6.1. The joists, purlins, beams, and girders are generally framed to carry gravity loads as bending members and the columns

Figure 6.1 Post and beam construction.

are framed as axial members. Individual members are designed in accordance with Chapter 4. Members are typically framed at intervals of 16 in. to 4 ft to accommodate common panel and other material sizes. Spacing of members to make use of repetitive member design values and the use of commonly available size and grade materials generally results in better building economy and reduces design costs.

Although post and beam systems are commonly used for gravity load resistance (roof, snow, and floor loads), they are not inherently suited for lateral load resistance (wind and seismic). For lateral load resistance, bracing is required and is commonly provided by the addition of cross braces (or trussing); let-in bracing, straps, or cables; knee and ankle bracing; and the development of the roof, floor, and wall framing into structural diaphragms and shear walls. Where the lateral forces are resisted by these additional means, it is generally required that the connections in the post and beam system be sufficient to transfer the gravity loads, and provide minimum anchorage against horizontal loads or movements.

Example 6.1 Anchorage in a Post and Beam System

Given: A $5\frac{1}{8} \times 12$ Douglas fir glued laminated timber spans 12 ft as a simple floor beam. The design loads for the beam are 150 plf of dead load and 500 plf of live load, excluding the weight of the beam. The location of the site where the beam will be installed is in northern Idaho, in a location where the governing building code is based on the 1997 *Uniform Building Code* [1]. The building official has established that soil profile type S_E is applicable to the site. The structure is considered standard occupancy.

Determine: The required horizontal anchorage force.

Approach: The 1997 *Uniform Building Code* [1] Section 1633.2.5 requires that all parts of a structure be interconnected, and for beams, girders, and trusses that a positive connection be provided to resist a horizontal force acting parallel to the member of at least $0.3C_aI$ times the dead and live load.

Solution: Northern Idaho is in 1997 UBC Seismic Zone 2B. From 1997 UBC Table 16-I, $Z = 0.2$. From Table 16-Q, soil profile type S_E and $Z = 0.20$, $C_a = 0.34$. From Table 16-K the seismic importance factor, I, is taken to be 1.00. As stated, the given dead load does not include the weight of the beam, which therefore must be calculated and added to the stated dead load. From Table 2.2, assuming 12% moisture content, the specific weight of Douglas fir is 33 pcf. Therefore, the weight of the beam is

$$\omega_{\text{SW}} = \left(\frac{5.125 \text{ in.}}{12 \text{ in./ft}}\right)\left(\frac{12 \text{ in.}}{12 \text{ in./ft}}\right)(33 \text{ pcf}) = 14 \text{ plf}$$

Assuming that equal anchorage is provided by each end, the anchorage force at each end is

$$F = (0.30C_aI)(\tfrac{1}{2})[(\omega_{DL} + \omega_{LL} + \omega_{SW})\ell]$$

$$= [(0.30)(0.34)(1.00)](\tfrac{1}{2})[150 + 500 + 14) \text{ plf}(12 \text{ ft})] = 406 \text{ lb}$$

Answer: A connection must be designed to resist a seismic force of 406 lb parallel to the axis of the beam at each end, or alternatively, code-approved connection hardware must be selected that is rated to resist 406 lb parallel to the axis of the beam in addition to the other loads carried by the connection from the beam.

6.3 POLE CONSTRUCTION, POST–FRAME CONSTRUCTION, AND TIMBER PILES

Pole-type frame structures generally consist of pressure-preservative-treated tapered round timber poles set in the ground as the main, upright supporting members. These poles provide resistance to gravity and lateral loads imposed on a structure. For resistance of gravity loads, the poles (in some cases referred to as *piers*) are generally set to bear on undisturbed native soil, engineered fill, or footings. Resistance of lateral loads is achieved by pole bearing laterally on soil or pole embedment in concrete or other fill that bears laterally on surrounding soil. The quality of surrounding soil, backfill, and backfill placement generally determine the ability of the posts (or piers) to also resist lateral loads.

Post–frame construction typically consists of rectangular section posts supporting trusses and other framing and sheathing or cladding. Whereas the posts in post–frame structures generally resist gravity loads, lateral loads are typically resisted by the embedded posts and diaphragm and shear wall resistance provided by the roof and walls, or a combination thereof. Poles and posts in post–frame are thus both foundational elements and form part of the superstructure.

Timber piles are round tapered timber members generally used as foundation elements that are typically driven into the ground. Piles are generally used where soils near the surface are weak or where the structure must be supported above the surface. Piles may also be used in retaining walls or other structures subject primarily to lateral forces. Piles are often embedded to greater depths than posts or other types of foundations.

6.3.1 Pole Construction

Poles may be used singularly or in system with other poles. Poles and piles shall conform to ASTM Standard D3200 [2] for poles and ASTM Standard D25 [3] for timber piles. Pole sizes and specifications are according to American National Standards Institute (ANSI) Standard 05.1 [4]. Poles must be pressure preservative treated in accordance with the American Wood-

Preservers' Association standards [5]. Sawn and glued laminated timbers may also be used as poles and must also be pressure preservative treated. Poles must be suitably supported on native undisturbed soil or engineered (controlled) fill. For light gravity loads and/or stronger soils, the bearing of a pole on the soil may be adequate. For heavier loads and/or weaker soils, the gravity loads must be distributed through spread footings under the poles. Pole construction relies on the resistance to rotation of the poles provided by pole embedment and backfill. As such, it is critical that the backfill material and placement is suitably specified and that its quality is assured.

General considerations applicable to all pole-type frame structures include the following:

1. Bracing can be provided at the top of a pole in the form of knee braces or cross bracing in order to reduce bending moments at the base of the pole and to distribute loads; otherwise, the poles must be designed as vertical cantilevers. The design of buildings supported by poles without bracing requires good knowledge of soil conditions to eliminate excessive deflection or sidesway.
2. Bearing values under butt ends of poles must be checked with regard to the bearing capacity of the supporting soil. Where the bearing capacity of the soil is not sufficient, a structural concrete footing may first be placed under the pole to spread the load. Backfilling the hole with concrete is common practice. With regard to gravity bearing, the concrete backfill can be used to spread the load if a suitable load transfer mechanism is provided from pole to concrete backfill. Friction between pole and concrete should not be relied upon. Studs or dowels may be used; if metal, they must be galvanized or protected against corrosion by some other means. Installation of the stud or dowel in the post must not compromise the pole's resistance to decay. Boring, notching, or other modifications to poles or posts should be done prior to preservative treatment. If boring or notching must be done in the field, the recommendations of the American Wood-Preservers' Association Standard AWPA M4 [6] must be followed.
3. Where the poles are used to resist lateral loads, it is essential that the backfill material be compacted properly. Sand may be used if placed in shallow, thoroughly compacted (tamped) lifts. Compaction of soil or gravel should be supervised and certified by a geotechnical engineer or other design professional familiar with the site soils conditions and requirements to achieve the needed resistance to lateral movements of the poles. The use of concrete and soil cement backfill will generally result in shallower required embedment depths, as the concrete or soil cement provide greater effective pole diameter with regard to lateral bearing.
4. The use of diaphragms and shear walls in pole structures generally results in smaller poles and shallower pole embedment requirements.

5. Pole structures (as with post–frame structures) may have excessive deflections for some applications, particularly where there is no bracing, diaphragm, or shear wall action. Deflections of structures with gypsum coverings and glazing should be analyzed carefully.
6. Poles require use of adjustment factors common to wood construction, as well as adjustment factors unique to poles. Adjustment factors applicable to timber pole and pile construction are given in the *National Design Specification® for Wood Construction* [7]. Design values for poles graded in accordance with ASTM D3200 are provided in the *NDS®*.
7. The intended use of the structure generally determines such features as height, overall length and width, spacing of poles, height at eaves, type of roof framing, and type of flooring to be used, as well as any special features such as wide bays, unsymmetrical layouts, or the possible suspending of particular loads from the framing.

6.3.2 Post–Frame Construction

Post–frame construction has been used widely for agricultural and utility purposes and has also been used in residential construction. The *Post–Frame Building Design Manual* [8] by the National Frame Builders Association provides guidance and design value information for the design of post–frame structures. The primary distinction between pole–frame and post–frame construction is in nomenclature, where poles are taken to be tapered with round cross section and posts are taken to be prismatic square or rectangular in section. Post–frame construction generally incorporates the use of metal cladding or structural sheathing to develop diaphragm action. The stiffness of the diaphragm and posts must be considered together to determine the distribution of loads to posts and end walls. Methods to determine diaphragm action and the distribution of loads to the posts may be found in the *Post–Frame Design Manual* [8] and *Wood Technology in the Design of Structures* [9], with a simplified approach also available [10]. Once the loads on the posts have been determined, the design checks are similar to those described above for pole construction.

6.3.3 Timber Piles

Piles may be used to perform various functions, including support of lateral and gravity loads from a structure above, to retain earth, water, or other materials, may be used to compact soil and may also be used to resist uplift forces. Pile foundations may be significantly deeper than other foundations and are particularly well suited to conditions of deep, relatively weak soils.

Recommendations for the use of timber piles in foundations may be found in *Pile Foundations Know-How* [11]. Piles may be driven into place or installed in holes prebored by auguring or other means. Tapered timber piles are driven with the small end (tip) down. Piles resist gravity forces by a combination of side or skin friction and end bearing. Short piles or piles driven through weak soils until they bear on stronger soils below tend to carry more of the load in end bearing, with the opposite being true of long piles in relatively homogeneous soils. Piles may be specified by the circumference of the nominal butt or nominal tip in accordance with ASTM D25 [3]. The ASTM classification allows the designer to specify a pile with adequate dimensions at the critical section. For example, a pile depending on frictional forces along the surface of the pile to support the vertical load will generally have a critical section located away from the tip. On the other hand, an end-bearing pile may have the critical section located at the tip.

A number of species are used for piles, with southern pine, Douglas fir, and oak the most commonly used. Piles should be relatively straight and possess the strength to resist driving stress and carry the imposed loads.

Where piles are used entirely below the permanent water table or are completely submerged in fresh water, preservative treatment is not necessary. However, for most permanent structures where these conditions do not exist, preservative treatment is necessary. The preservative treatment should conform to recognized specifications such as Federal Specification TT-W-571 [12] and American Wood-Preservers' Association (AWPA) Standards C1 and C3 [5]. Cutoffs at the tops of piles exposing untreated wood should be field treated in accordance with AWPA Standard M4 [6]. Piles used in salt water are subject to attack by marine borers, and special treatment techniques must be used to minimize degradation. The treatment of piles for use in salt water is covered in Federal Specification TT-W-571 [12] and AWPA Standard C18 [5].

Equipment used for driving timber piles is of special importance. The energy used to drive the piles must be sufficient to drive the pile but must not impart excessive forces. Pile butts and tips may be damaged severely by sharp blows. For this reason it is not desirable to drive timber piles with a drop hammer unless a suitable block is employed to dampen the impact. Air, steam, and diesel hammers are commonly used. Generally, it is desirable to band the butt or driven end of a pile to minimize damage during driving. In cases where tip damage may occur, a special shoe or fitting is used to protect the tip. Pile driving should be monitored by a quality control professional.

Design of pile foundations and other uses of piles should be based on the recommendations of soils investigations of the sites in question. Design values and adjustment factors may be found in the *National Design Specification®* [7]. Such design values typically apply to both wet and dry use. It is generally assumed that piles will be preservative treated and will be used in groups. In cases where no treatment is applied or where piles are used individually, further specific adjustments are applied.

Where the diameter of the pile at the section of critical bending moment exceeds 13.5 in., the bending design value must be multiplied by the size factor [equation (3.3)], based on the depth of a square section of equivalent cross-sectional area. Except in the case of very slender piles or piles in soils providing little or no lateral support, the column stability factor is not applied [13]. The beam stability factor is not applicable to piles.

6.4 SHEATHING AND DECKING

Numerous products are available that provide a structural envelope to buildings and transfer wall, roof, and floor loads to supporting systems. In wood structures, structural sheathing panels such as plywood or other structural panels are commonly used. These panels are used under floor and roof and wall finishes or coverings or are in some cases used alone. Wood lumber sheathing boards of 1 in. nominal thickness have also been used for floors, roofs, and walls. Sawn and glued laminated pieces greater than 1 in. nominal thickness, referred to as *decking,* is commonly used for roofs, floors, and walls. Structural panels have the advantage of being able to develop significant diaphragm and shear wall action. Decking has the advantage of providing both structure and an architectural face where exposed.

6.4.1 Structural Wood Panels

Wood structural panels are commonly used for exterior walls and roof surfaces, for floors, and may be used below grade for wood foundations. Care should be taken to ensure that panels selected meet the appropriate exposure requirements and satisfy the grade and minimum thickness requirements of the local building codes. The design and specification of structural wood panels should be in accordance with the *Plywood Design Specification* [14] and supplements and other appropriate documents available from APA–The Engineered Wood Association (formerly the American Plywood Association). Materials should be specified that meet Product Standard PS 1 [15] or PS 2 [16] product standards.

6.4.2 Lumber Sheathing

Lumber sheathing of boards of 1 in. nominal thickness nailed transversely or diagonally at about 45° to studs or joists is sometimes used in wood frame construction. When subjected to lateral forces such as wind or earthquakes, lumber sheathing and its supporting framework may act as a diaphragm, when properly designed as such, serving to brace the building against the lateral forces and transmitting these forces to the foundations [21].

The edges of lumber sheathing may be square, ship-lapped, splined, or tongue-and-groove. Sheathing runs should be spliced only over supports unless end matched or scarfed and glued end joints or splice blocks are used. Boards used as subflooring in wood frame construction effectively act as shallow beams spanning one or more spaces.

6.4.3 Timber Decking

Tongue-and-groove sawn lumber pieces of 2, 3, and 4 in. nominal thickness are referred to as *heavy timber decking.* Information on species, sizes, patterns, lengths, moisture content, application, specifications, applicable allowable stress, and allowable loads for 2-, 3-, and 4-in.-nominal-thickness tongue-and-groove heavy timber decking used as roof decking may be found in the *Standard for Tongue-and-Groove Heavy Timber Roof Decking,* AITC 112 [17]. Glued laminated timber decking is also available. Glued laminated decking may be manufactured in longer lengths and greater nominal thicknesses and often has higher design values than those of similar sawn decking. Size, length, and design information for laminated decking should be obtained by the individual manufacturer. Edge joints for decking are shown in Figure 6.2.

Decking lengths may be specified and ordered to end over supports, where the pieces are assumed to act as shallow beams spanning one or more joist spaces. In many cases, however, it may be more economical to specify decking of varied or random lengths. Five common layups for timber decking are shown in Figure 6.3. Particular requirements of the layups are given in AITC 112 [17]. The equations in Table 6.1 may be used to compute the allowable uniform load for the decking layups based on flexure and deflection. Additional checks may be necessary for point loads and short spans with heavy loads where shear becomes significant. Design values for heavy timber decking may be found in AITC 112 [17] or may be established by an approved lumber grading agency.

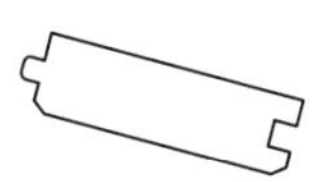

2 in. nominal

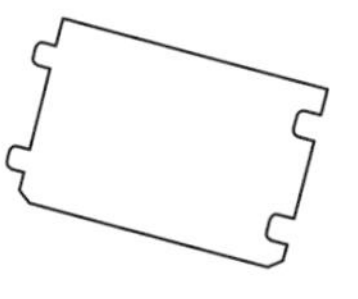

3 in. nominal and
4. in nominal

Sawn Tongue-and-Groove Decking

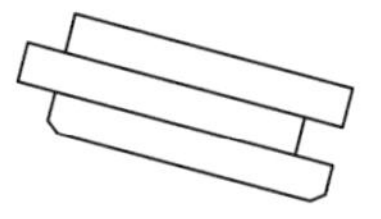

2 in. nominal and
3. in nominal

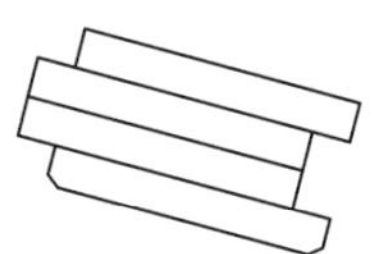

4 in. nominal

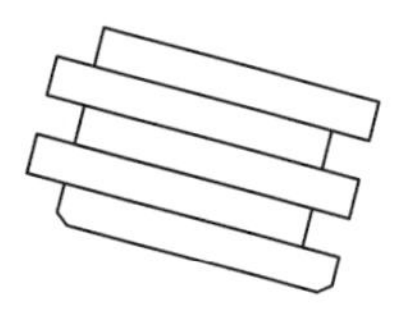

5 in. nominal

Glued Laminated Decking

Figure 6.2 Edge joints of lumber sheathing and decking.

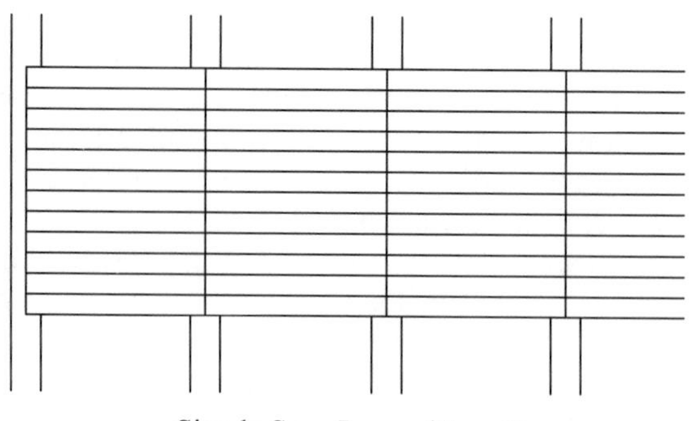

Simple Span Layup (Type 1)

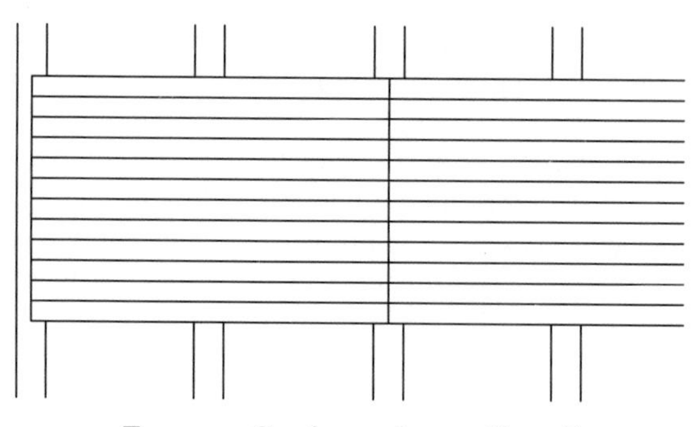

Two-span Continuous Layup (Type 2)

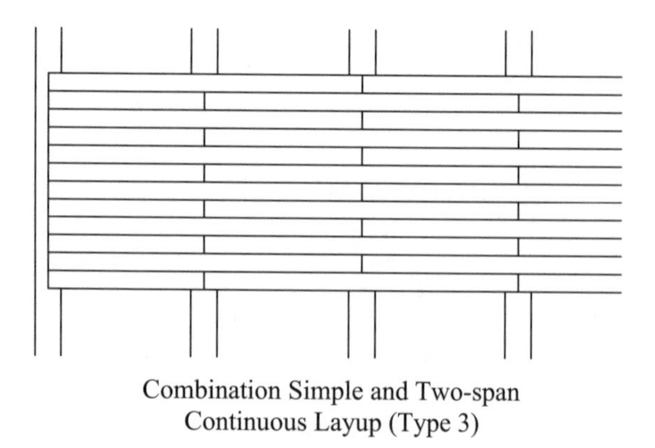

Combination Simple and Two-span
Continuous Layup (Type 3)

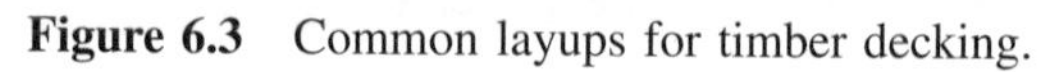

Figure 6.3 Common layups for timber decking.

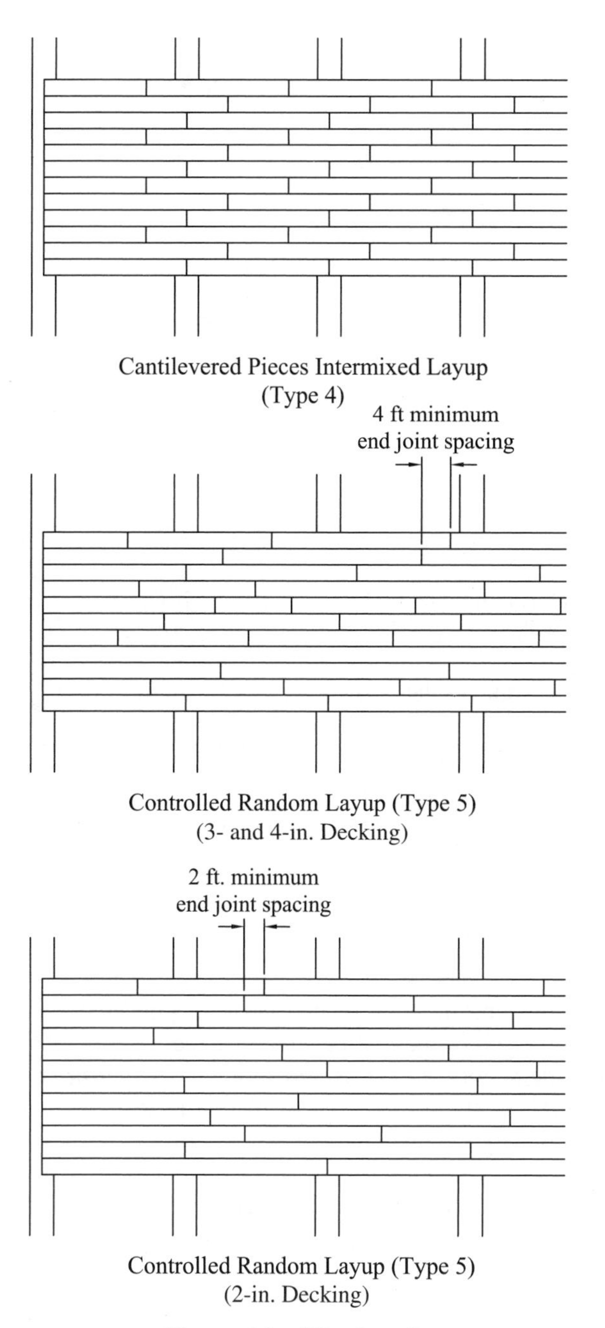

Cantilevered Pieces Intermixed Layup
(Type 4)

Controlled Random Layup (Type 5)
(3- and 4-in. Decking)

Controlled Random Layup (Type 5)
(2-in. Decking)

Figure 6.3 (*Continued*).

TABLE 6.1 Design Formulas for Timber Decking

Layup Type	Allowable Area Load[a] Based On: Bending	Deflection
Simple span (type 1)	$\sigma_b = \dfrac{8F'_b}{\ell^2}\dfrac{d^2}{6}$	$\sigma_\Delta = \dfrac{384\Delta E'}{5\ell^4}\dfrac{d^3}{12}$
Two-span continuous (type 2)	$\sigma_b = \dfrac{8F'_b}{\ell^2}\dfrac{d^2}{6}$	$\sigma_\Delta = \dfrac{185\Delta E'}{\ell^4}\dfrac{d^3}{12}$
Combination simple- and two-span continuous (type 3)	$\sigma_b = \dfrac{8F'_b}{\ell^2}\dfrac{d^2}{6}$	$\sigma_\Delta = \dfrac{131\Delta E'}{\ell^4}\dfrac{d^3}{12}$
Cantilevered pieces intermixed (type 4)	$\sigma_b = \dfrac{20F'_b}{3\ell^2}\dfrac{d^2}{6}$	$\sigma_\Delta = \dfrac{105\Delta E'}{\ell^4}\dfrac{d^3}{12}$
Controlled random layup (type 5)		
Nominal 2-in. decking	$\sigma_b = \dfrac{20F'_b}{3\ell^2}\dfrac{d^2}{6}$	$\sigma_\Delta = \dfrac{100\Delta E'}{\ell^4}\dfrac{d^3}{12}$
Nominal 3- and 4-in. double T/G, horizontally spiked	$\sigma_b = \dfrac{20F'_b}{3\ell^2}\dfrac{d^2}{6}$	$\sigma_\Delta = \dfrac{116\Delta E'}{\ell^4}\dfrac{d^3}{12}$
Glued laminated decking, slant nailed	$\sigma_b{}^b = \dfrac{20F'_b}{3\ell^2}\dfrac{d^2}{6}$	$\sigma_\Delta{}^b = \dfrac{130\Delta E'}{\ell^4}\dfrac{d^3}{12}$

[a] σ_b, allowable total uniform load limited by bending; σ_Δ, allowable total uniform load limited by deflection.; d, actual decking thickness.

[b] Allowable loads may vary by manufacturer. The designer should consult the manufacturer for appropriate layup pattern and equations.

Section modulus and moment of inertia of decking may be determined for individual deck pieces, or may be provided by the manufacturer. Where manufacturers provide section modulus and moment of inertia per unit of width of deck, these values should be used in place of the $d^2/6$ and $d^3/12$ values in Table 6.1, and typically account for section loss from beveled edges, and so on. Where section modulus and moment of inertia are calculated, consideration must be made for the material (section) loss due to the tongues and grooves and any gaps or beveled edges. Care must be taken in the use of design values as to whether the values include or exclude flat use, size, or other applicable factors. Except for layup types 1 and 2, the equations of Table 6.1 may not be used for the loads on individual pieces, as the equations are based on load transfer between pieces of adjacent runs or courses.

Example 6.2 Timber Decking

Given: Glued laminated timber decking of 3 in. nominal thickness is being considered for a roof system with supports at 10 ft o.c. The following design properties are provided by the manufacturer: F_b = 2250 psi (normal load duration); E = 1,800,000 psi; S/b = 9.39 in^3/ft; weight: 6 psf. The design roof load is 45 psf (snow load). The dead load for the roof, including the weight of the decking, is 12 psf.

Determine: Check the suitability of the 3-in. decking for combination simple and two-span layup and for controlled random layup (types 3 and 5, respectively). Deflection criteria to be used in this example are $\ell/240$ for applied load and $\ell/180$ for applied load plus 50% of dead load.

Approach: It will be assumed that the design values provided for bending include consideration of flat use and size; thus, the bending design value will be adjusted for load duration only.

Solution: Case 1: Combination Simple and Two-Span Continuous Layup (Type 3). For a load based on bending, S/b = 9.39 in^3/ft, which is equivalent to $d^2/6$ in the equations in Table 6.1:

$$F'_b = F_b C_D = (2250 \text{ psi})(1.15) = 2585 \text{ psi}$$
$$\ell = 10 \text{ ft}$$

From Table 6.1, for bending, the total allowable load is

$$\sigma_b = \frac{8F'_b}{\ell^2}\frac{d^2}{6} = \frac{(8)(2585 \text{ lb/in}^2)}{(10 \text{ ft})^2}(9.39 \text{ in}^3/\text{ft})\left(\frac{\text{ft}}{12 \text{ in.}}\right) = 162 \text{ psf}$$

Since the applied design load is 45 psf + 12 psf = 57 psf, the layup is satisfactory with regard to bending.

Checking the deflection under applied load yields

$$\frac{\ell}{240} = \frac{(10 \text{ ft})(12 \text{ in./ft})}{240} = 0.50 \text{ in.}$$

and

$$\frac{I}{b} = 10.29 \text{ in}^4/\text{ft}$$

which should be substituted for $d^3/12$ in Table 6.1 equations.

From Table 6.1 for layup type 3, the allowable load based on deflection is

$$\sigma_\Delta = \frac{109\Delta E'}{\ell^4}\frac{d^3}{12}$$

$$= \frac{(109)(0.50 \text{ in.})(1{,}800{,}000 \text{ lb/in}^2)}{(10 \text{ ft})^4}(10.29 \text{ in}^4/\text{ft})\left(\frac{\text{ft}}{12 \text{ in.}}\right)^3 = 58 \text{ psf}$$

Since the applied load is 45 psf, the deflection criterion for live load is satisfied. For applied load plus 50% of dead load, the deflection is limited to $\ell/180$, or (10 ft)(12 in./ft)/180 = 0.667 in.

The corresponding allowable load is (58 psf)(0.667 in./0.50 in.) = 77 psf. Since 45 psf + (0.5)(12 psf) = 51 psf $\leq$ 77 psf, both deflection criteria are satisfied by the 3-in. deck in the type 3 layup.

Case 2: Controlled Random Layup (Type 5) Based on bending, Table 6.1 (glued laminated decking) gives

$$\sigma_b = \frac{20F_b'}{3\ell^2}\frac{d^2}{6} = \frac{(20)(2585 \text{ lb/in}^2)}{(3)(10 \text{ ft})^2}(9.39 \text{ in}^3/\text{ft})\left(\frac{\text{ft}}{12 \text{ in.}}\right) = 135 \text{ psf}$$

Based on applied load deflection,

$$\sigma_\Delta = \frac{130\Delta E'}{\ell^4}\frac{d^3}{12}$$

$$= \frac{(130)(0.50 \text{ in.})(1{,}800{,}000 \text{ lb/in}^2)}{(10 \text{ ft})^4}(10.29 \text{ in}^4/\text{ft})\left(\frac{\text{ft}}{12 \text{ in.}}\right)^3 = 70 \text{ psf}$$

For applied load plus 50% of dead load, the allowable load is (70 psf) (0.667 in./0.50 in.) = 93 psf. Since 51 psf $\leq$ 93 psf, the type 5 layup is also satisfactory.

Answer: The combination simple and two-span continuous layup (type 3) and controlled random layup (type 5) are both satisfactory for the stated loads.

Discussion: In this example a deflection criterion of $\ell/180$ with respect to applied load plus 50% of dead load was used. The deflection value calculated in association with this criterion is *not* the actual deflection that would be expected from the actual applied and dead loads. With regard to deflection, the designer must be sure to satisfy the deflection criteria of the governing building code and to ensure that actual deflections, including creep, are satisfactory to the owner.

The controlled random layup was found to be stiffer than the combination simple and two-span layup, although both were found satisfactory with respect to flexure and the given deflection criteria. The controlled random layup may be more economical, as it allows the manufacturer to provide pieces in varied lengths. This example also illustrates that decking applications of moderate and long span typically are governed by deflection criteria.

The terms $d^2/6$ and $d^3/12$ in Table 6.1 may also be computed directly from actual decking depth values; however, they will generally be slightly nonconservative, as they do not account for section loss due to the beveled edges and gaps between pieces at the tongues and grooves. From the same manufacturer, the actual depth of the 3-in. nominal deck is given to be $2\frac{3}{16}$ in. Hence, $d^2/6 = (2.188 \text{ in.})^2/6 = 0.798 \text{ in}^2$, compared to

$$9.39 \frac{\text{in}^3}{\text{ft}} \frac{\text{ft}}{12 \text{ in.}} = 0.782 \text{ in}^2$$

and

$$\frac{d^3}{12} = \frac{(2.188 \text{ in})^3}{12} = 0.872 \text{ in}^3$$

compared to

$$10.29 \frac{\text{in}^4}{\text{ft}} \frac{\text{ft}}{12 \text{ in.}} = 0.858 \text{ in}^3$$

Mechanically laminated decks consist of square-edged dimension lumber set on edge, wide face to wide face, with the pieces connected by nails or other fasteners. If side nails are used, they should be long enough to penetrate approximately two and one-half lamination thicknesses for load transfer. Where deck supports are 4 ft center to center or less, side nails should be spaced not more than 30 in. on center and staggered one-third of the spacing in adjacent laminations. When supports are spaced more than 4 ft center to center, side nails should be spaced approximately 18 in. on center, alternately near the top and bottom edges and staggered one-third of the spacing in adjacent laminations. Two side nails should be used at each end of the butt-joined pieces. Laminations should be toenailed to supports using 20*d* or larger common nails. When the supports are 4 ft center to center or less, alternate laminations should be toenailed to alternate supports; when supports are spaced more than 4 ft center to center, alternate laminations should be toenailed to every support.

Applications of timber decking for roofs often involve overhangs. In some cases uniform loads on cantilever overhangs may be used to reduce deflections in interior spans. Where appropriate, the unbalanced load condition created by reduced snow load over interior spans due to heating and full snow load over unheated spans should be considered. Ice damming and other snow accumulations must also be considered where appropriate.

6.5 TIMBER TRUSSES

The subject of timber truss design is quite broad. The discussion in this book is limited to basic design procedures and the highlighting of features unique to timber truss construction. The design of metal plate connected wood trusses constructed of dimension lumber is not covered in this book. Metal plate–connected trusses are generally designed by truss manufacturers utilizing proprietary connection and design techniques.

6.5.1 Truss Types

Types of timber trusses commonly used are illustrated in Figure 6.4 as follows: (1) parallel chord, (2) pitched or triangular, (3) bowstring, (4) camelback, and (5) special, such as crescent, scissors, sawtooth, and king or queen post trusses. Architectural considerations generally dictate roof slope and may also dictate truss type. Table 1.2 gives typical economical span ranges for various primary truss framing systems. Span ranges for various secondary framing systems are also given in Table 1.2, which can be used to help determine truss spacing.

Pratt trusses have the advantage that for gravity loads, the longer (diagonal) webs are in tension, whereas the shorter (vertical) webs are in compression. Flat or low-pitched Pratt roof trusses can be used for clear spans up to 120 ft, with the most economy in spans 80 ft and under. Pitched Pratt and Belgian trusses can be used economically for clear spans up to 100 ft. They tend to be more economical than Fink trusses when the pitch is less than $5\frac{1}{2}:12$ and the span is over 50 ft. The roof pitch should be a minimum of 3:12. Fink trusses can be used economically for clear spans up to 80 ft, with roof pitches of more than $5\frac{1}{2}:12$. The optimum span range is 40 to 60 ft for Fink trusses.

Bowstring trusses are most economical for spans from 80 to 150 ft but may range from 50 to 200 ft. Usually, the radius of the top chord is equal to the span. Modified bowstring trusses, where the center portion of the top chord is straight instead of curved, can be used for spans up to 300 ft. The chords and heel connections take the major stresses; the web stresses under uniform loading conditions are negligible, and for unbalanced loading, web stresses are comparatively light. Because of low web stresses, bowstring trusses have light webs and web connections. Chord stresses are nearly equal throughout their length, and chords with constant cross section, have full-length economy. Bowstring trusses usually have the upper chord shaped to the form of a circular arc or parabola. With the normal depth/span ratios used, the variation between the parabola and the circular arc is very slight and is not sufficient to change stresses materially. Circular curves are customarily used. In addition to normal axial and bending stresses, upper chords of bowstring trusses are subject to moments due to the eccentricity of their curved shape.

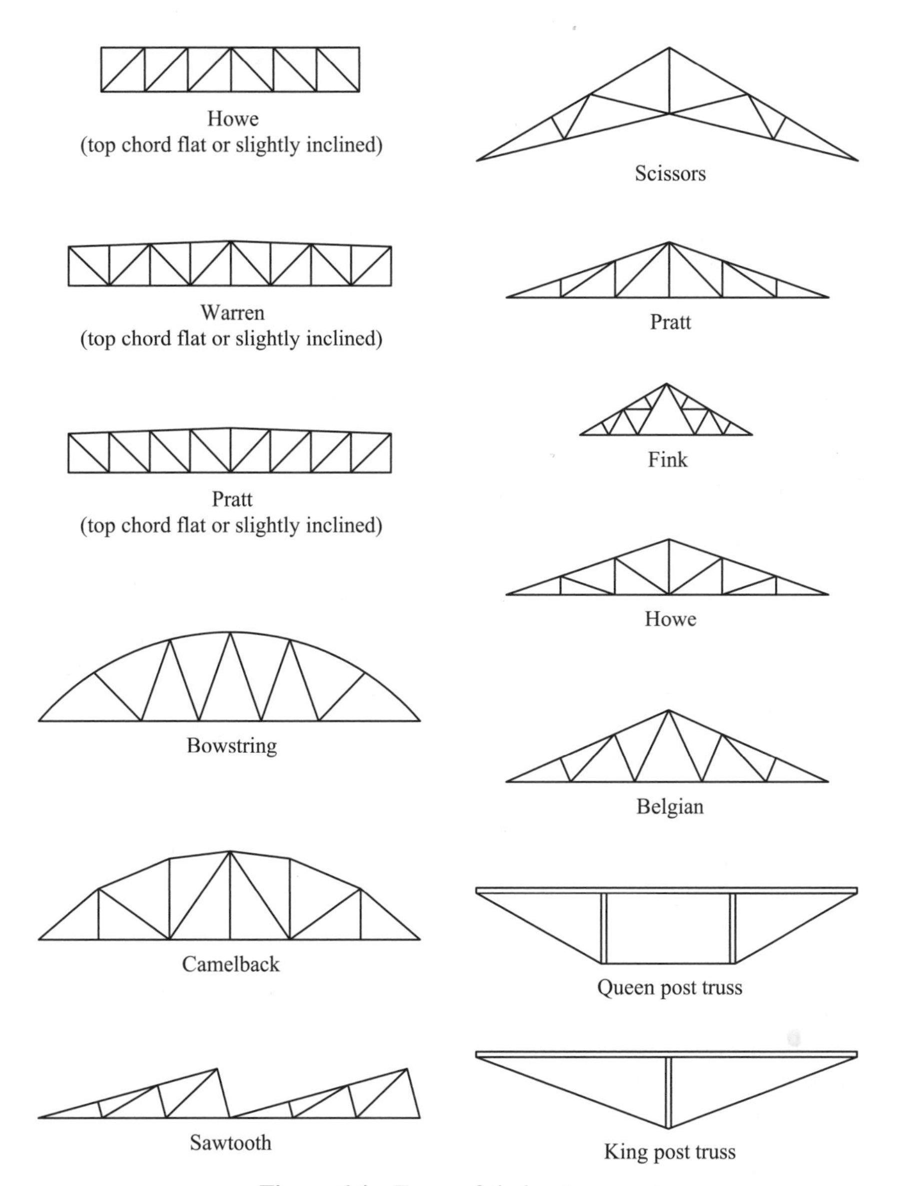

Figure 6.4 Types of timber trusses.

Scissor trusses can be designed efficiently for clear spans up to 80 ft and are used to provide more height clearance toward the midspan. They are commonly used in churches, gymnasiums, various types of assembly halls, and residential construction. Bridge trusses can be designed for economy with clear spans up to 200 ft. Table 6.2 provides recommended depth/span ratios for common truss shapes that may be used for determining preliminary truss

TABLE 6.2 Recommended Timber Truss Depth/Span Ratios[a]

Flat or parallel chord	$\frac{1}{8}$ to $\frac{1}{10}$
Triangular or pitched	$\frac{1}{6}$ or deeper
Bowstring	$\frac{1}{6}$ to $\frac{1}{8}$

[a] Depth measured from centerline (middepth) of top and bottom chord members.

shape dimensions. Smaller depth/span ratios may result in excessive member stresses and deflections; greater ratios may be less economical.

6.5.2 Truss Members

Truss members are designated in three types: *top chord, bottom chord,* and *webs.* Webs in this book refer to all interior vertical or diagonal members between the top and bottom chords. Joints, at which members intersect and connect, are called *panel points.* For flat or nearly flat roof trusses, some pitch must be provided for drainage or the trusses must be designed to resist progressive ponding (Chapter 3). Chords and webs in all truss types may be constructed as single-leaf, double-leaf, or multileaf members. Single-leaf chords are also known as *monochords.* The most common arrangements of trusses are monochord trusses with single-leaf chords and webs and those with double-leaf chords having single-leaf webs located between the chord leaves. Webs may be attached to sides of chords or may be in the same plane. Where single-leaf members are attached to the sides of one another, the effects of twisting (torsion) must either be prevented by bracing or be accounted for in design. Truss members may be sawn lumber or glued laminated timber. The use of steel rods or other steel shapes for members in timber trusses is acceptable if they fulfill all conditions of design and service.

Ideally, timber truss web systems should be selected for convenience of connection and economy. Web locations and panel point spacing may be dictated by selection of secondary purlin framing so as to minimize chord bending stresses. In parallel chord trusses, diagonal webs should be sloped between 45° and 60° from the horizontal for greatest economy. Bowstring trusses using panel lengths from 8 to 14 ft, depending on the truss span, tend to be the most economical.

Monochord trusses may require steel plate connections with through bolts alone or in combination with shear plates. In multileaf trusses, connections can often be made more easily by using through bolts and shear plates or split rings. When split-ring or shear plate connectors are required, it is usually best to have the wood trusses fabricated in a shop. For any truss type, shop fabrication generally affords greater quality control than does field fabrication.

Glued laminated timber provides many features desirable for truss designs. Glued laminated timber can be made in almost any shape, size, or length, and provides higher design values than do sawn timbers. Sawn timbers are limited in maximum length and cross-sectional size and are prone to checking. However, sawn members may provide cost savings. Thus, depending on truss span and loading requirements, trusses can be made of all sawn timber, all glued laminated timber, or a combination of both and may include some steel tension members. If sawn members are used in conjunction with glued laminated timber, care must be taken to match widths at connections and provide for differential shrinkage.

For trusses whose members function essentially as pinned connected axial members, member forces may be found by determinate static analysis and deflections by virtual work or other methods. Many trusses, however, incorporate members that are continuous through some connections, for which static analysis may serve to provide preliminary results, but for which indeterminate analysis should be used for final member forces and deflections.

6.5.3 Truss Deflection and Camber

The effects of truss deflections must be considered in the design of supporting and adjoining columns, structural and nonstructural walls, and other fixtures. Timber trusses should be cambered such that total dead load deflection does not produce sag below a straight line between points of support. Additional camber may be appropriate for sustained or heavy live or other loads where sag may impair serviceability or be aesthetically unappealing. Horizontal deflections of trusses must also be computed and accommodated for in the design of truss support connections. Scissor and crescent trusses may have considerable horizontal deflections. Where horizontal deflection is prevented by the supporting structure, the resulting loads on the truss and support structure must be considered.

6.5.4 General Design Procedure

Given the truss type, span, depth, spacing, and species and grade of timber to be used, the following general procedure may be used for all truss types. Special design features for each particular truss type must also be considered. Dead loads acting on the truss must include weights of all roof and ceiling construction, mechanical and electrical equipment, the estimated truss weight, and other permanently applied loads that will act throughout the service life of the truss. Other loads acting on the truss and their combinations are generally determined from applicable building codes and/or good engineering judgment. Consideration should be made, where appropriate, for increased loads due to drifting and/or sliding of snow and for ponding. Unbalanced loading conditions frequently control truss designs. If net uplift is possible, special considerations for stress reversals and bottom chord buckling will be required.

Timber truss design typically involves selection of trial member sizes and connection types with design iteration until member stress and truss deflection criteria are satisfied. Trial member sizes are generally determined from architectural considerations or engineering judgment, or both. Architectural considerations generally involve total truss depth, whereas structural calculations often involve centerline dimensions. In the absence of other guidance on trial member sizes, preliminary member sizes may be obtained from static analysis considering all joints pinned and the members acting under axial load only. Preliminary deflection information can also be obtained from this idealization.

Where actual trusses are loaded only at joints (panel points), are not continuous through joints, and where the connections do not resist rotation, static analysis for member stresses and truss deflection is sufficient. In most cases, however, timber trusses are to some extent statically indeterminate. Bottom and top chord members are often designed to be continuous through intermediate panel points. Connections may also be modeled and designed to resist some rotation or act as rigid connections. Furthermore, loads to the truss commonly occur at points other than the panel points or joints.

Where all members are discontinuous at joints or where top or bottom chords are continuous and the joints are otherwise free to rotate, a preliminary or approximate analysis involving loads arriving at other than panel points may be obtained by superimposing the axial forces from static truss analysis and the moments due to non-panel-point loads with the chords acting as continuous span beams. Such members are thus subject to combined stresses and subject to the analysis techniques of Chapter 4. Such analysis is limited to relatively short chord lengths where flexure will not affect the loaded truss deflection.

The availability of commercially available and proprietary structural analysis and design software enables designers of timber trusses to account for the statically indeterminate features of truss design, unbalanced loads and load reversals, and check combined stresses and deflection quickly and efficiently. Successive iterations of various design features may also result in successive improvements in the truss economy. However, the final design may not always be the most economical design, as architectural or other features may control. Good engineering judgment and clear communication among engineer, architect, and owner are essential for successful truss design.

Preliminary connection designs should consider first the joint or joints carrying the greatest load. Connections should be concentric and hinged wherever possible or detailed in such a way as to allow relative rotation between members (Figures 6.5 and 6.6). Single through bolts connecting the members of multileaf trusses provide true pin connections. Slotted or overlapping connecting plates may also be used to allow rotation. Consideration should be given to the effect of possible wood shrinkage on the connection. Member sizes may have to be increased to compensate for loss of section at connections. It may be necessary to locate and design splices for chords based on available length of materials.

For final design, the exact geometry of the truss must be determined, including load points, member sizes, and connection geometry. Where connec-

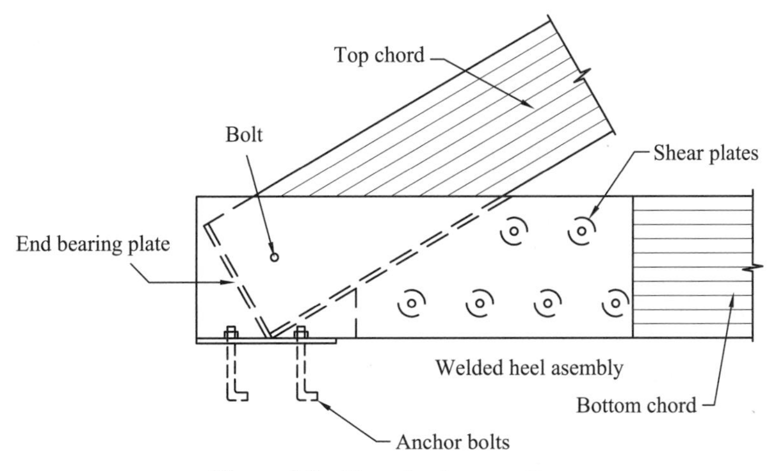

Figure 6.5 Truss heel connection.

tions resist relative rotation between truss members, the connections and fasteners must be designed to transfer the axial and flexural loads. Where there are knee braces or other structural elements attached to the truss that can lend support or influence load distribution, they must be included as integral with the truss in the overall analysis. If chords are curved, moments resulting from eccentricity of the axial forces must be included in the design.

Truss web member designs are governed by axial tension or compression loads for all pinned webs. If web connections are moment resisting, the webs must be sized to resist the resulting combined loading, the same as for chords. Slenderness ratios of compression webs should be checked as columns using the least dimension of the member and the distance between panel points as the length. Unbalanced, wind, wind uplift, or other special loadings can cause increased stresses or reversal of stresses, and thus control portions of a total design. Net member sections must be used where drilling and dapping for connection hardware is required. Where glued laminated timbers are used,

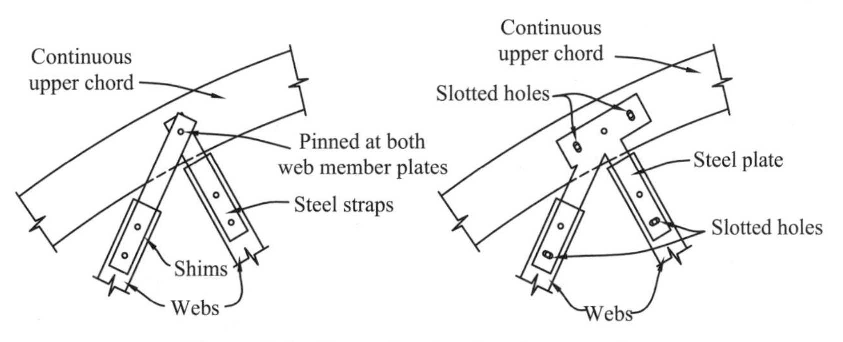

Figure 6.6 Truss chord and web connection.

members loaded primarily in axial tension or axial compression should be specified from Table 8.2, AITC 117 [18] or from Table 5B in the *National Design Specification®* (*NDS®*) [7]. If chord members are highly stressed in bending, members from Table 8.1 in AITC 117 [18] or Table 5A in *NDS®* (members subject primarily to bending) may be more efficient. For hardwood glued laminated timber, members should be specified from AITC 119 [19].

Connections should be designed in accordance with Chapter 5. AITC 104 [20], contains typical construction details that may be of assistance in designing truss connections. Truss connections should be concentric; that is, the centerlines of members at each joint should intersect at a common point so that moments are not induced into the members being connected. Connections should be of the pinned type whenever possible. Truss members tend to rotate in relation to each other as a result of distortion and deflection of the truss under loading. If rotation is restrained by the joints, moments and additional forces in members and connections will result, and the members may tend to split.

Where possible, fastenings of the same size and type should be used throughout the truss. Spacing, end distance, and edge distance requirements for each fastener must be checked. Possible shrinkage of wood members must be considered. For splices in deep members where steel plates are used, it may be necessary to utilize multiple splice plates where the overall cross grain spacing of fasteners for each plate exceeds 5 in.

Design of steel straps in compression must be based on actual support conditions for the plates. For plates connecting webs to chords, generally the web does not provide lateral stability for the plate, but instead, the plate is an extension of the web. Therefore, the end of the plate fastened to the web cannot be considered fixed in determining the plate effective length. When the web plates are pinned at the chord, the spacing between bolts in the web should be large enough to provide stability and prevent splitting of the web. Normal fabrication tolerances must be considered. If steel rods are used as tension elements in a truss, written instruction should be given to the truss assembler regarding tightening of the rods. Otherwise, overtightening could occur, which might overstress truss elements when design loads are applied.

6.5.5 Bracing

In structures employing trusses, a system of bracing is required to provide resistance to lateral forces, to hold the trusses true and plumb, and to hold compression elements in line. Both permanent bracing and temporary erection bracing should be designed according to accepted engineering principles to resist all loads that will normally act on the system. Erection bracing is that which is installed during erection to hold the trusses in a safe position until sufficient permanent construction is in place to provide full stability. Permanent bracing is that which forms an integral part of the completed structure. Part or all of the permanent bracing may also act as erection bracing.

Permanent bracing must be provided to resist both transverse and longitudinal forces and may consist of either of the following systems:

1. A *structural diaphragm in the plane of the top chord.* The diaphragm, acting as a plate girder, transmits forces to end and side walls. This system is the preferred method of providing truss bracing and is usually most economical. It may be necessary to provide additional members acting as the flanges for the diaphragm.
2. *Horizontal bracing between trusses in the plane of the bottom and/or top chords.* In effect, such bracing forms an inclined or horizontal truss. It is a positive method of bracing, but it is more costly and should be used only where the strength of a diaphragm described above is insufficient. If horizontal bracing is used, it should be designed to resist all the transverse and longitudinal loads that will normally act on the system.

Unbraced top chords and web members subject to compressive loads must be adequately braced to satisfy slenderness requirements and resistance to buckling. Similarly, bottom chords and other members subject to compression under wind or other loads may need to be braced. To provide lateral support for truss members subject to compression, the bracing system should be designed to withstand a horizontal force equal to at least 2% of the compressive force in the truss chord if the members are in perfect alignment. If the members are not aligned calculation of the force should be based on the eccentricity of the members due to the misalignment.

Bracing is also required to provide overall building stability and transmit forces from the roof to the ground. This bracing typically consists of shear walls, diagonal bracing, buttresses, cantilevered columns, or knee braces. Use of knee braces between the trusses and columns is a method of bracing particularly adaptable to buildings that have a large length/width ratio. The stresses induced in the trusses and columns can be critical, and members and connections must be designed accordingly. Lateral loads acting in the long direction of the building are usually resisted by means of diagonal bracing and struts in the end bays.

Temporary or erection bracing is not normally the responsibility of the design engineer. Erection truss bracing is that installed to hold trusses true and plumb and in a safe condition until permanent truss bracing and other permanent components, such as joists and sheathing contributing to the rigidity of the complete roof structure, are in place. Erection truss bracing may consist of struts, ties, cables, guys, shores, or similar items. Joists, purlins, and other permanent elements may be used as part of the erection bracing.

6.6 STRUCTURAL DIAPHRAGMS

Structural diaphragms are relatively thin, usually rectangular, structural systems capable of resisting in-plane shear parallel to their edges. Wood-framed roofs, walls, and floors can generally be made into structural diaphragms with proper detailing and in some cases, additional members or connecting hardware. The function of the diaphragm is to brace a structure or parts thereof against lateral forces, such as wind or earthquake loads, and to transmit these forces to the other resisting elements of the structure. Roof and floor systems are often used as horizontal diaphragms. The horizontal diaphragms transfer the lateral forces to vertical diaphragms (shear walls), braced walls of various types, or poles or piles. Common types of diaphragms in wood construction are described below.

1. *Wood structural panels.* Structural panel sheathing used to resist out-of-plane loads (load perpendicular to the panels) can generally also be used to develop diaphragm action by proper detailing of fasteners, boundary members, and attachment to interior support members and blocking. The design of wood structural panel diaphragms is described in the *Plywood Design Specification* [14] and other materials from APA–The Engineered Wood Association.
2. *Transverse (straight or horizontal) sheathing and decking.* Lumber sheathing and decking develop only modest amounts of diaphragm action where placed perpendicular to the support members. Design values and procedure are provided in the *Western Woods Use Book* [21]. Heavy timber and glued laminated timber decking may be covered with wood structural panel sheathing to develop diaphragm action. Design values for this type of system are considered equivalent to those of blocked diaphragms of the same sheathing thickness and with the same nailing schedule. The designer must specify and detail the proper panel and boundary nailing conditions for both sheathing and decking.
3. *Diagonal sheathing.* Diaphragms may be developed by use of 1-in.-nominal-thickness boards or 2-in.-nominal-thickness lumber, nailed at a 45° angle to cross members in a single layer. Considerably greater strength and stiffness result from this placement than where the sheathing is installed perpendicular across supports. Tests verify that although there may be considerable bending in the individual board or sheathing pieces, the primary load resistance of the shear in diaphragm is due to the axial resistance in the pieces (tension or compression). The bending moment of the diaphragm is resisted by the continuous boundary member chords (also acting in tension or compression).
4. *Double diagonal sheathing.* Lumber sheathing may be placed in two layers of diagonal sheathing, one on top of the other, with the sheathing in one layer at a 90° angle with the sheathing in the outer layer. This type is considerably stiffer and stronger than the single-layer diagonal sheathing and lumber sheathing placed perpendicular to supports. Design values and procedures for diagonal and double-diagonal sheathing diaphragms are also provided in the *Western Woods Use Book* [21].

6.7 TIMBER BRIDGES

Sawn and glued laminated timbers have been used successfully for highway, railway, pedestrian and other bridges. Timber bridge design incorporates the methods described in this manual in combination with the loading and other requirements put forth by appropriate organizations such as the American Association of State Highway and Transportation Officials (AASHTO) [22] for highway bridges, and the American Railway Engineering and Maintenance-of-Way Association (AREMA) [23] for railway bridges. Design of nontimber components and support structures (such as steel and concrete) should be done in accordance with the accepted design standards and practices associated with the specific materials. Although not covered in this book, bridge design should also incorporate proper foundation and hydraulic investigations.

6.7.1 Types of Timber Bridges

Timber bridges consist of several basic types, including trestles, girder bridges, truss bridges, and arch bridges. The following sections describe these types in detail.

6.7.1.1 Trestles The trestle is a simple type of timber bridge. Timber trestles consist of stringers supported by pile or frame bents. The bridge deck is applied to the stringers. Pile and frame bents are capped by timbers 12 in. × 12 in. or larger, fastened adequately to the tops of the piles or posts. If pile penetration or height of bent is such that piles longer than those commercially available are required, or if pile bearing values are low and a large number of piles must be driven, posts may be used on top of the pile bents. Frame bents must rest on some type of foundation structure, such as concrete footings or piles. Sway bracing and longitudinal tower bracing, appropriate to the height of the bent, must be provided.

Spacing of bents is determined, in part, by the commercially available lengths of stringers, which are fabricated in even-foot increments. The ends of interior stringers are usually lapped and fastened to the bearing on the caps, whereas exterior stringers are butted to the ends and spliced over the bent caps. Stringers are designed as simple-span beams under the loadings recommended by AASHTO [22] or AREMA [23]. Sizes and spacing are determined by the span and loading conditions. Standard sizes of glued laminated or sawn timbers should be used. Solid blocking should be provided at the ends of stringers to hold them in line and also to serve as a fire-stop. Bridging should also be placed between stringers at midspan and, on long spans, at additional intermediate locations.

6.7.1.2 Glued Laminated Deck and Girder Bridges For short spans, glued laminated deck panels spanning longitudinally from end to end of bridge or between intermediate supports may be used for the bridge structure and driving surface or subsurface. Such bridges may also be especially suited for low-profile applications. For longer spans, girder bridges, consisting of glued laminated or sawn timber girders supporting a transverse bridge deck may be used. Substructures similar to those used for timber trestles can be used for girder bridges, the girder being fastened to the bent caps by means of a fabricated steel girder seat. For intermediate spans, longitudinal stress-laminated girder bridges may be appropriate where longitudinal girders placed side by side are laminated to one another by high-stress tension rods, causing the composite system of girders to act as a large continuous plate. Construction economies will usually result if the standard sizes and readily available grades for glued laminated or sawn timber are used. Timber deck and girders are designed as beams in accordance with the recommendations in Chapter 4 and the requirements and recommendations of AASHTO [22] or other appropriate organization.

6.7.1.3 Truss Bridges Truss bridges may be of either of two types: deck–truss bridges, in which trusses support the bridge deck and roadway; or through-truss bridges, in which the roadway passes between two parallel trusses forming the bridge structure. The deck–truss type is generally more economical, although the use of deck trusses may be limited by underclearance requirements.

Deck trusses may be of the parallel chord type or of the bowstring type, with the truss built up to the level of the floor beams. Through trusses may be of either of these truss types also, but the bowstring is usually more economical. Bowstring trusses are often used as pony truss bridges, which are through trusses with no overhead bracing above the roadway. Through trusses have the disadvantage of potential damage from vehicles.

Substructures for truss bridges may be similar to those for timber trestles; however, because the vertical loads are greater and are concentrated at the ends of the trusses, the bents must be capable of carrying a greater load, and a system of cribbing is required for bent caps. For longer span and heavier loads, timber, stone, or concrete piers may be required. Lateral forces are greater on truss bridges, and a carefully designed substructure sway bracing system is necessary.

The design of trusses for bridges is similar to that for roof trusses, the length of truss panel being determined by economical spacing of support beams, minimum number of joints, and commercially available lengths of timber. As in roof truss design, the joint design is an important consideration. Bridge truss joints should be designed to eliminate or minimize pockets, which may tend to collect moisture.

6.7.1.4 Arch Bridges When site conditions are such that considerable height is required between foundation and roadway or a relatively long clear span is required, an arch bridge may be most economical because there is less need for substructure framing. Arch bridges may be of the two- or three-hinged type, two-hinged designs being used more frequently on short spans and three-hinged designs on long spans. Glued laminated timber arches may be fabricated to the desired shape and the ends built up to the level of the roadway by means of post bents. Post bents may be connected to the arch by means of steel gusset plates, which should be designed for erection loads, possible stress reversals, and lateral forces as well as for the anticipated bridge loads.

6.7.2 Bridge Stringer and Deck Design

The selection of decks for timber bridges is determined by density of traffic and economics. Plank decks may be used for light traffic or for temporary bridges. Laminated decks can be used for heavier traffic conditions. Asphalt wearing surfaces may be applied on the decking, although this is not usually done for plank decks.

Composite timber–concrete decks are sometimes used in timber bridge construction. Composite timber–concrete construction combines timber and concrete in such a manner that the wood is in tension and the concrete is in compression (except at the supports of continuous spans, where negative bending occurs and these stresses are reversed). Composite timber–concrete construction is of two basic types: T-beams and slab decks. T-beams consist of timber stringers, which form the stems, and concrete slabs, which form the flanges of a series of T-shapes. Composite beams of this type are usually simple-span bridges. Slab decks use, as a base for the concrete, a mechanically laminated wooden deck made up of planks set on edge, with alternate planks raised 2 in. to form longitudinal grooves. This grooved surface is usually obtained by using planks of two different widths and alternating them in assembly. This composite type has been used for continuous-span bridges and trestles, but is not very common today.

In both types, a means of shear parallel to grain resistance and a means of preventing separations are needed at the joint between the two materials. In T-beams, resistance to shear parallel to grain is generally provided by a series of notches $\frac{1}{2}$ to $\frac{3}{4}$ in. deep cut into the top of the sawn timber stringer and about $1\frac{1}{2}$ in. deep for a glued laminated timber stringer, while nails and spikes partially driven into the top prevent vertical separation of the concrete and timber. Other adequate methods may be used. In slab decks, shear resistance is accomplished either by means of notches $\frac{1}{2}$-in. deep cut into the tops of all laminations, by triangular steel plate shear developers driven into precut slots in the channels formed by the raised laminations, or by other suitable shear connectors. When the $\frac{1}{2}$-in. notches are used, grooves are milled the full length of both faces of each raised lamination to resist the uplift and separation of the wood and concrete. When the steel shear developers are used, nails or spikes are partially driven into the tops of raised laminations to resist separation.

In T-beam design, secondary shearing stress due to temperature must be considered in designing for shear parallel to grain resistance. These stresses are induced by the thermal expansion or contraction of the concrete, both of which are resisted by the wood, which is assumed to be unaffected by normal temperature changes. Shear connections for temperature change are neglected in slab deck composite construction; however, expansion joints should be provided in the concrete slab. The concrete slab should be reinforced for temperature stress. In continuous spans, steel sufficient to develop negative bending stress is necessary over interior supports. The dead load of the composite structure is considered to be carried entirely by the timber section. The composite structure carries positive bending moment, and over interior supports in continuous spans, steel reinforcing and the wood act to resist negative bending moment.

In designing a composite structure, if it is assumed that the junction between the two materials is without inelastic deformation and has elastic characteristics in keeping with the materials, the structure can be designed by the transformed-area method, that is, by transforming the composite section into an equivalent homogeneous section. This is accomplished by multiplying the concrete width or depth by the ratio of the moduli of elasticity of the materials.

Glued laminated timber bridges are commonly used for highway bridges, and design examples are included herein. Because nail-laminated decks and composite timber–concrete decks are not common, design examples for these decks are not included in this book. For bridges utilizing longitudinal stringers and transverse decks, the following design procedures illustrate the superstructure design with or without interconnecting steel dowels between the panels. Examples herein are for vertical loading from vehicles as described in AASHTO *Standard Specifications for Highway Bridges* [22].

The design examples included in this section do not address camber for the glued laminated stringers because camber may vary with the particular use. Generally, camber equal to dead load deflection plus additional amounts for drainage and appearance is sufficient. Glued laminated timber design values are based on AITC 117 [18]. Wet use design values are used for the design of deck panels because tests in the field indicate that the moisture content is likely to exceed 16% in service. Dry use design values are used for the design of the glued laminated timber stringers, with the exception of compression perpendicular to grain stress, $F_{c\perp}$. The glued laminated timber deck panel, when properly treated and installed, should provide a roof for the stringers, preventing them from exceeding an in-service moisture content of 16%.

Wind-driven moisture can cause a surface wetting on the exposed face of the edge stringers, but this superficial wetting should not affect the design. Localized moisture accumulations can develop, and preservative treatment is recommended for the stringers to protect against possible decay hazards such as at areas of steel connections and at bearing locations.

Example designs of longitudinal stringer with transverse deck and longitudinal deck timber bridges follow, including stringer, deck, and curb, rail, and post design. Additional examples, including design of stress laminated girder bridges, are included in the *Glued Laminated Timber Bridge Systems* [24] by AITC, which also provides tables with optimum stringer size, required deck thickness, and other important features for various spans and typical loading conditions.

6.7.2.1 Longitudinal Stringer Bridge with Doweled Deck A design example of a longitudinal stringer bridge with a doweled deck follows.

Example 6.3 Longitudinal Stringer Bridge with Doweled Deck

Given: A glued laminated timber bridge with a 40 ft 0 in. span, 34 ft 0 in. width, two lanes, with a 3-in. asphalt wearing surface. Transverse decking is to be glued laminated L2 Douglas fir (combination 2) or No. 2 southern pine (combination 47) and will be doweled. Longitudinal stringers are to be Douglas fir or southern pine: 24F-V4 DF/DF or 24F-V3 SP/SP.

Determine: A suitable bridge design with AASHTO HS 15-44 live load. Assume wet conditions of use for the deck only.

Approach: Stringers and deck will be designed in accordance with the following AASHTO [22] specifications and design values for the glued laminated timber from AITC 117 [18].

Dead loads:

Timber	50 pcf
Asphalt pavement	150 pcf

Live loads: combined weight on axles and distance between:

Front		Rear		Trailer
6000 lb	14 ft	24,000 lb	14–30 ft	24,000 lb

Live load moment: for HS 15-44, 337 kip-ft, one lane, simple spans, from AASHTO Appendix A [22].

Distribution Factors:

Glued Laminated Panels on Glued Laminated Stringers	One Traffic Lane	Two or More Lanes
4 in. thick	S/4.5	S/4.0
6 in. or more thick	S/6.0	S/5.0

where S is the average stringer spacing (ft).

Dead load shear: shear occurring at distance d from support.

Live load shear:

$$V_{LL} = 0.50[(0.60V_{LU}) + V_{LD}]$$

where

V_{LL} = distributed live load shear,
V_{LU} = maximum vertical shear at $3d$ or $L/4$ due to undistributed wheel loads,
V_{LD} = maximum vertical shear at $3d$ or $L/4$ due to wheel loads distributed laterally.
d = the stringer depth, L = span.

Load duration factor for live load: 1.15
Overload factor for live load: 2
Percentage of basic unit stress in overload: 150%

Timber bridges need not be designed for impact loading.

Transverse deck (*glued laminated timber*): deck thickness is the greater of

$$t = \sqrt{2\frac{6M_x}{F'_b}} \quad \text{or} \quad t = \frac{3R_x}{2F'_v}$$

where

t = deck thickness.
M_x = primary bending moment (lb-in./in),
= $P[(0.51 \log_{10}s) - K]$
P = design wheel load,
s = effective deck span (in.), and
K = 0.47 for H15 loading.
R_x = primary shear (lb/in.)
= $0.034P$

Dowels:

$$n = \frac{1000}{\sigma_{PL}}\left(\frac{\overline{R}_y}{R_D} + \frac{\overline{M}_y}{M_D}\right)$$

where

$$\overline{R}_y = \frac{P}{2s}(s - 20) \quad \text{for } s > 50 \text{ in.}$$

$$\overline{M}_y = \frac{Ps(s - 30)}{20(s - 10)} \quad \text{for } s > 50 \text{ in.}$$

where

σ_{PL} = proportional limit stress perpendicular to grain (1000 psi for Douglas fir and southern pine),

and R_D and M_D are the shear and moment capacities, respectively, of dowels as specified by AASHTO based on dowel diameter.

Solution: Six girders will be assumed. The average center-to-center stringer spacing, S, is

$$S = \frac{34 \text{ ft}}{6 - 1} = 6.8 \text{ ft}$$

Dead Loads: Assuming $5\frac{1}{8}$-in.-thick deck panels, the dead load on a single stringer is

$$\text{Deck: } (50 \text{ pcf}) \left(\frac{5.125 \text{ in.}}{12 \text{ in./ft}}\right)(6.8 \text{ ft}) = 145 \text{ plf}$$

$$\text{Asphalt wearing surface: } (150 \text{ pcf}) \left(\frac{3.0 \text{ in.}}{12 \text{ in./ft}}\right)(6.8 \text{ ft}) = 255 \text{ plf}$$

The total applied dead load is 145 plf + 255 plf = 400 plf. The assumed stringer self-weight is 100 plf. The total assumed dead load (including self-weight) is 400 plf + 100 plf = 500 plf.

Distribution Factors: The distribution factor for live load design (two lanes) is

$$\frac{S}{5.0} = \frac{6.8}{5} = 1.36$$

The distribution factor for overload (one lane loaded only) is

$$\frac{S}{6.0} = \frac{6.8}{6} = 1.13$$

Stringer Design (Girders)

Bending: The assumed dead load bending moment is

$$\frac{\omega\ell^2}{8} = \frac{(500 \text{ plf})\ (40 \text{ ft})^2}{8} = 100{,}000 \text{ lb-ft} = 1{,}200{,}000 \text{ lb-in.}$$

The live load moment is

$$\frac{(\text{kip-ft/lane})(\text{distribution factor})}{\text{wheel lines/lane}} = \frac{(337{,}000 \text{ lb-ft})(1.36)}{2}$$

$$= 229{,}200 \text{ lb-ft} = 2{,}750{,}000 \text{ lb-in.}$$

The total load moment is

$$1{,}200{,}000 \text{ lb-in.} + 2{,}750{,}000 \text{ lb-in.} = 3{,}950{,}000 \text{ lb-in.}$$

Trying a $6\frac{3}{4}$ in. $\times$ $43\frac{1}{2}$ in. girder, the area is 293.6 in^2, the section modulus is 2129 in^3, and the moment of inertia is 46,300 in^4.

Assuming southern pine, the volume factor is

$$C_V = 1.0\left[\left(\frac{5.125}{6.75}\right)^{1/20}\left(\frac{12}{43.5}\right)^{1/20}\left(\frac{21}{40}\right)^{1/20}\right] = 0.895$$

Assuming Douglas fir, the volume factor is

$$C_V = 1.0\left[\left(\frac{5.125}{6.75}\right)^{1/10}\left(\frac{12}{43.5}\right)^{1/10}\left(\frac{21}{40}\right)^{1/10}\right] = 0.802$$

The bending stress is

$$f_b = \frac{M}{S} = \frac{3{,}950{,}000 \text{ lb-in.}}{2129 \text{ in}^3} = 1855 \text{ psi}$$

The deck is fastened to the stringers to provide lateral support; therefore, C_L = 1.00. For Douglas fir,

$$F'_b = F_b C_D C_V C_M = (2400 \text{ psi})(1.15)(0.802)(1.00) = 2213 \text{ psi}$$

Since 1854 psi $\leq$ 2213 psi, the $6\frac{3}{4}$ in. $\times$ 43.5 in. Douglas fir stringers are adequate in bending.

For southern pine, the allowable stress in bending is

$$F'_b = F_b C_D C_V C_M = (2400 \text{ psi})(1.15)(0.895)(1.00) = 2470 \text{ psi}$$

Since 1854 psi ≤ 2470 psi, the $6\frac{3}{4}$ in. × 43.5 in. southern pine stringers are also adequate in bending.

Shear: For the uniformly distributed dead load, the shear is calculated as

$$V_{\mathrm{DL}} = \frac{\omega\ell}{2} - \omega d = \frac{(500 \text{ plf})(40 \text{ ft})}{2} - (500 \text{ plf})\left(\frac{43.5 \text{ in.}}{12 \text{ in./ft}}\right) = 8190 \text{ lb}$$

For the live load, the wheel loads are taken to be half the axle loads and are placed to produce the maximum shear at the lesser of $3d = 3\ (43.5/12)$ ft = 10.88 ft or $\ell/4 = 40$ ft$/4 = 10$ ft, or 10 ft. The placement of loads to produce maximum shear is shown in Figure 6.7, and the corresponding shear is 13,950 lb.

Thus, the distributed live load vertical shear, from the AASHTO specification, is,

$$V_{\mathrm{LL}} = 0.50[(0.60)(13{,}950 \text{ lb}) + (1.36)(13{,}950 \text{ lb})] = 13{,}670 \text{ lb}$$

The total load V is thus

$$V_{\mathrm{DL}} + V_{\mathrm{LL}} = 8190 \text{ lb} + 13{,}670 \text{ lb} = 21{,}860 \text{ lb}$$

The shear stress parallel to the grain due to the total load is

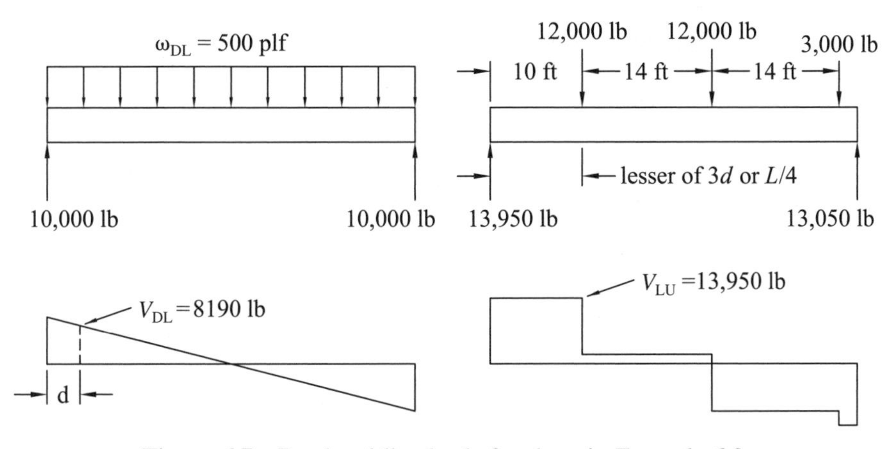

Figure 6.7 Dead and live loads for shear in Example 6.3.

$$f_v = \frac{3V}{2bd} = \frac{(3)(21{,}860 \text{ lb})}{(2)(6.75 \text{ in.})(43.5 \text{ in.})} = 112 \text{ psi}$$

The design value in shear parallel to the grain for bridges is taken from AITC 117 as F_{vx} (for nonprismatic shapes). For Douglas fir the design value is 190 psi, and for southern pine it is 215 psi. The stringers are assumed to be in dry service conditions; therefore, $C_M = 1.00$. Thus,

$$F'_v = F_v C_D C_M = (190 \text{ psi})(1.15)(1.00) = 219 \text{ psi}$$

for the Douglas fir stringer and 247 psi for the southern pine stringer. Since in both cases the actual shear is less than the allowable shear, both the southern pine and Douglas fir sections are adequate with regard to shear.

Overload: In consideration of overload, the AASHTO specifications require the live load to be doubled, and other lanes may be considered not to be loaded concurrently. Thus, a distribution factor of 1.13 is used. The AASHTO specifications also allow the basic unit stress to be increased by 50%. Thus, for bending,

$$M = (2)(2{,}750{,}000 \text{ lb-in.})\left(\frac{1.13}{1.36}\right) + 1{,}200{,}000 \text{ lb-in.} = 5{,}770{,}000 \text{ lb-in.}$$

$$f_b = \frac{M}{S} = \frac{5{,}770{,}000 \text{ lb-in.}}{2129 \text{ in}^3} = 2710 \text{ psi}$$

For Douglas fir the allowable stress in overload is

$$F'_b = (1.5)(2400)(1.5)(0.802) = 3320 \text{ psi}$$

For southern pine, the allowable bending stress in overload is

$$F'_b = (1.5)(2400(1.15)(0.895) = 3705 \text{ psi}$$

Thus, both sections are adequate in overload with regard to bending.

The distributed live load shear in overload is

$$V_{\text{LL}} = (2)(13{,}670 \text{ lb}) = 27{,}340 \text{ lb}$$

The total load V is thus

$$V_{\text{DL}} + V_{\text{LL}} = 8190 \text{ lb} + 27{,}340 \text{ lb} = 35{,}530 \text{ lb}$$

The shear stress parallel to the grain due to the total load in overload is

$$f_v = \frac{3V}{2bd} = \frac{(3)(35{,}530 \text{ lb})}{(2)(6.75 \text{ in.})(43.5 \text{ in.})} = 182 \text{ psi}$$

The allowable shear stresses for Douglas fir and southern pine, increased 50%, are, respectively, 329 and 370 psi. Thus, the stringers are also acceptable with regard to shear in the overload condition.

Deflection: For evaluation of the deflection of the interior stringers, the distributed live load moment may be converted into an equivalent concentrated load placed at midspan and the corresponding deflection used to check the deflection criterion. Such analysis is conservative and further does not account for load distribution to adjacent stringers or consider the composite action that exists between the decking and stringers. Letting $M_{\text{LL}} = 229{,}000$ lb-ft $= P_{\text{equiv}}L/4$, $P_{\text{equiv}} = 22{,}900$ lb. The corresponding live load deflection is

$$\Delta_{\text{LL}} = \frac{PL^3}{48EI} = \frac{(22{,}900 \text{ lb})\ [40 \text{ ft } (12 \text{ in./ft})]^3}{(48)(1{,}800{,}000 \text{ psi})(46{,}300 \text{ in}^4)} = 0.63 \text{ in.}$$

For highway design, the live load deflection should be limited to $L/300$ (Table 3.11), which in this case is $[(40 \text{ ft})(12 \text{ in./ft})]/300 = 1.6$ in. Since 0.63 in. is less than 1.6 in., the stringer deflection is acceptable.

Deck: The design wheel load in this example, P, is 12,000 lb. The effective deck span is taken to be the lesser of the deck clear span plus one-half the stringer width or the deck clear span plus the assumed deck thickness, in this case, 78 in. Thus, the primary bending moment, M_x, is calculated to be

$$M_x = (12{,}000 \text{ lb})[(0.51)(\log 78) - 0.47] = 5940 \text{ lb-in./in.}$$

The decking will be placed with the wide faces parallel to the load; thus, $F_b = F_{by}$. From Table 8.2, AITC 117, $F_{by} = 1800$ psi for combination 2 Douglas fir, and $F_{by} = 1750$ psi for combination 47 southern pine (four or more laminations). For shear parallel to the grain, $F_{vy} = 165$ psi (Douglas fir) and 188 psi (southern pine) for cyclic loading. Since the decking timbers are placed flat, the flat use factor is applicable.

From Chapter 3,

$$C_{fu} = \left(\frac{12 \text{ in.}}{5.125 \text{ in.}}\right)^{1/9} = 1.10$$

Wet use will be considered for the deck, thus, from Chapter 3, $C_M = 0.80$ (bending) and 0.875 (shear parallel to the grain). The allowable bending stresses for the deck are, therefore,

$$F'_b = F_b C_D C_{fu} C_M$$

$$= \begin{cases} (1800)(1.15)(1.10)(0.80) = 1822 \text{ psi} & \text{for Douglas fir} \\ (1750)(1.15)(1.10)(0.80) = 1771 \text{ psi} & \text{for southern pine} \end{cases}$$

Therefore, the minimum deck thickness values, based on bending, are

$$t = \sqrt{6M/F'_b} = \begin{cases} \sqrt{(6)(5940 \text{ lb})/1822 \text{ psi}} = 4.42 \text{ in.} & \text{for Douglas fir} \\ \sqrt{(6)(5940 \text{ lb})/1771 \text{ psi}} = 4.49 \text{ in.} & \text{for southern pine} \end{cases}$$

For deck thickness based on shear,

$$F'_v = F_v C_D C_M = \begin{cases} (165 \text{ psi})(1.15)\ (0.875) = 166 \text{ psi} & \text{for Douglas fir} \\ (188 \text{ psi})(1.15)\ (0.875) = 189 \text{ psi} & \text{for southern pine} \end{cases}$$

$$R_x = (0.034)\ (12{,}000 \text{ lb/in.}) = 408 \text{ lb/in.}$$

Thus,

$$t = \frac{3}{2}\frac{R_x}{F'_v} = \begin{cases} \dfrac{3}{2}\left(\dfrac{408 \text{ lb/in.}}{166 \text{ psi}}\right) = 3.69 \text{ in.} & \text{for Douglas fir} \\ \dfrac{3}{2}\left(\dfrac{408 \text{ lb/in.}}{189 \text{ psi}}\right) = 3.24 \text{ in.} & \text{for southern pine} \end{cases}$$

All of the required thickness values for the deck are satisfied by either species of $5\frac{1}{8}$ in. thickness.

Deck Dowels: Considering $1\frac{1}{2}$-in.-diameter dowels, R_D = 2770 lb and M_D = 8990 lb-in. (AASHTO); thus,

$$\overline{R}_y = \frac{(12{,}000 \text{ lb})(78 \text{ in.} - 20 \text{ in.})}{(2)(78 \text{ in.})} = 4460 \text{ lb}$$

$$\overline{M}_y = \frac{(12{,}000 \text{ lb})(78 \text{ in})(78 \text{ in.} - 30 \text{ in.})}{(20 \text{ in.})(78 \text{ in.} - 10 \text{ in.})} = 33{,}000 \text{ lb-in.}$$

Thus,

$$n = \frac{4460 \text{ lb}}{2770 \text{ lb/dowel}} + \frac{33{,}000 \text{ lb-in.}}{8990 \text{ lb-in./dowel}} = 5.28 \text{ dowels}$$

Use six dowels per span.

AASHTO also requires that the stress in the dowels themselves as given in the following equation does not exceed the minimum yield point of the steel.

$$\sigma = \frac{1}{n}\,(C_R\overline{R}_y + C_M\overline{M}_y)$$

where σ is the combined stress in the dowels, (psi), which may not exceed the minimum yield point specified by AASHTO, and C_R and C_M are steel stress coefficients based on the dowel diameter provided by AASHTO, in this case, 3.11 and 3.02, respectively, for $1\frac{1}{2}$-in.-diameter dowels. In this example,

$$\sigma = \tfrac{1}{6}[(3.11/\text{in}^2)(4460 \text{ lb}) + (3.02/\text{in}^3)(33{,}000 \text{ lb-in.})] = 18{,}900 \text{ psi}$$

Steel dowels with minimum specified yield point of 36 ksi (grade 36) will be adequate. For $1\frac{1}{2}$-in. dowels, AASHTO-specified $19\frac{1}{2}$-in. minimum total lengths are required.

Answer: The longitudinal stringer with-transverse deck bridge with a 40-ft span and 34-ft width with the following specifications is adequate to carry the AASHTO HS 15-44 loading requirements. Stringers: $6\frac{3}{4}$ in. × 43.5 in. DF/DF 24F-V4 or SP/SP 24F-V3, deck (doweled): $5\frac{1}{8}$ in. × 35 ft panels, DF L2 combination 2 or SP No. 2 combination 47. Dowels: six $1\frac{1}{2}$ in. × 19.5 in. per span.

Discussion: Using the AASHTO-specified self-weight for the stringers of 50 pcf, the self-weight line load calculated becomes

$\omega_{\text{s.w.}} = \left(\frac{6.75}{12}\text{ ft}\right)\left(\frac{43.5}{12}\text{ ft}\right)(50 \text{ pcf}) = 102 \text{ plf}$, only slightly greater than the values used in the design checks.

6.7.2.2 Longitudinal Stringer Bridge with Nondoweled Deck

For nondoweled decks carrying HS 25-44, HS 20-44, and HS 15-44 loading, Table 6.3 shows the maximum clear span between stringers and the maximum deck overhang beyond the outside stringer. These values are based on experience in constructing noninterconnected (nondoweled) decks that are fastened to the stringers using the bracket system shown in Figure 6.8 or similar deck connection systems and are used to minimize deflections. The designer must be aware, however, that differential deflections between panels may cause cracking of the wearing surface.

The following example illustrates the procedures to follow in the design of nondoweled decks on longitudinal stringers. Design of the curb, rail, and post for the example is included in the following section.

TABLE 6.3 Maximum Spans for Nondoweled Deck[a]

Deck Thickness (in.)	HS 15 and HS 20[b]		HS 25[c]	
	Main Span	Overhang	Main Span	Overhang
$3\frac{1}{8}$	2 ft 10 in.	1 ft 5 in.	2 ft 6 in.	1 ft 3 in.
$5\frac{1}{8}$	4 ft 9 in.	2 ft $4\frac{1}{2}$ in.	4 ft in.	2 ft 2 in.
$6\frac{3}{4}$	6 ft 2 in.	3 ft 1 in.	5 ft 10 in.	2 ft 11 in.

[a]Douglas fir combination 2; wet use; C_D = 1.15. Live load deflection limited to 0.10 in.
[b]Wheel load for HS 15 and HS 20 is 12,000 lb.
[c]Wheel load for HS 25 is 15,000 lb.

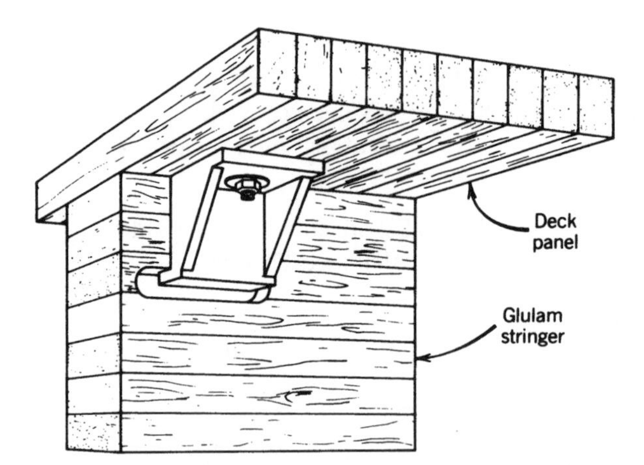

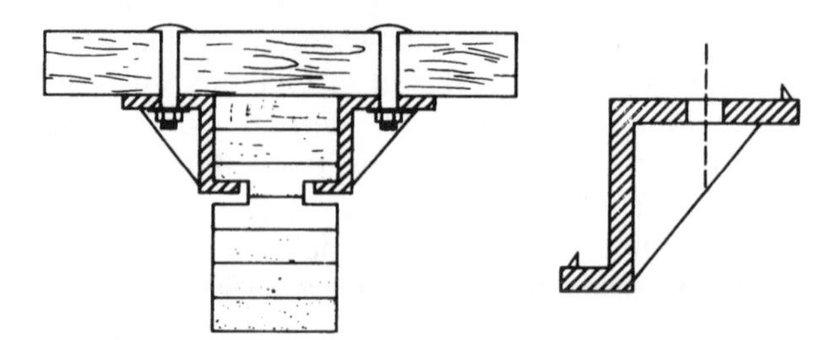

Figure 6.8 Brackets for fastening deck to a glulam stringer.

Example 6.4 Longitudinal Stringer Bridge with Nondoweled Deck

Given: A two-lane timber bridge spanning 50 ft 0 in. center to center of bearings, width of 34 ft 0 in, 3-in. asphalt wearing surface, subject to AASHTO HS 20-44 loading. Stringers are to be $10\frac{3}{4}$ in. × 48 in. 24F-V4 DF. Deck to be $6\frac{3}{4}$-in. combination 1 DF.

Determine: The minimum number of stringers required and a suitable non-doweled timber deck.

Approach: The following AASHTO requirements, along with Table 6.3 and other glulam design properties, will be used to design the glued laminated stringers and noninterconnected glued laminated deck):

Distribution factors:

$S/5.0$ (two or more lanes)
$S/6.0$ (one lane)

Distributed live loads: combined weight on axles and distance between:

Front		Rear		Trailer
8000 lb	14 ft	32,000 lb	14–30 ft	32,000 lb

Maximum live load moment: 627.9 kip-ft, one-lane AASHTO Appendix

Wheel load distribution on deck:

Transverse direction: length such that contact area is $0.01P$ in^2 and is rectangular with ratio of length in direction of traffic to tire width = 1/2.5

Longitudinal direction: 15 in. plus floor thickness, but not to exceed panel length.

Overload: AASHTO does not require consideration of overload for loads of H20 or greater.

Solution: Stringers (Minimum Number) Using $6\frac{3}{4}$-in. deck, the maximum corresponding deck spans and overhangs, from Table 6.3, are, respectively, 6 ft 2 in. and 3 ft 1 in. The minimum number of stringers, n, assuming 3 ft 1 in. overhangs and 6 ft 2 in. interior spans, is therefore determined by

$$3.083 \text{ ft} + (n - 1)6.167 \text{ ft} + 3.083 \text{ ft} = 34 \text{ ft} \quad \text{or} \quad n = 5.5$$

Therefore, six stringers will be used. Letting the overhang distance equal half the interior span center-to-center distance gives 5 ft 8 in. (5.67 ft) center-to-center spacing of stringers and 2 ft 10 in. overhangs.

Design Values: From AITC 117, $F_b = 2400$ psi, $F_v = 190$ psi, $F_{c\perp} = 650$ psi, and $E = 1{,}800{,}000$ psi

Dead Loads:

$$\text{3-in. asphalt: } (3\text{ in.})\left(\frac{1\text{ ft}}{12\text{ in.}}\right)(150\text{ pcf})(5.67\text{ ft}) = 213\text{ plf}$$

$$6\tfrac{3}{4}\text{- in. deck: } (6.75\text{ in.})\left(\frac{1\text{ ft}}{12\text{ in.}}\right)(50\text{ pcf})(5.67\text{ ft}) = 159\text{ plf}$$

Railing, posts, and curb (assumed): 75 plf

$$\text{Stringer: } (10.75\text{ in.})\left(\frac{1\text{ ft}}{12\text{ in.}}\right)(48\text{ in.})\left(\frac{1\text{ ft}}{12\text{ in.}}\right)(50\text{ pcf}) = 179\text{ plf}$$

Total dead load: 626 plf

Bending: The moment due to dead load is

$$M_{\text{DL}} = \frac{\omega\ell^2}{8} = \frac{(626\text{ plf})(50\text{ ft})^2}{8} = 195{,}600\text{ lb-ft} = 2{,}350{,}000\text{ lb-in.}$$

Due to live load, per wheel line,

$$M_{\text{LL}} = \frac{627{,}900\text{ lb-ft per lane}}{\text{two wheel lines per lane}}(\text{distribution factor})$$

$$= \frac{(627{,}900\text{ lb-ft})\ (1.134)}{2} = 356{,}000\text{ lb ft} = 4{,}270{,}000\text{ lb-in.}$$

The total moment is, therefore,

$$M = 2{,}350{,}000\text{ lb-in.} + 4{,}270{,}000\text{ lb-in.} = 6{,}620{,}000\text{ lb-in.}$$

For a $10\tfrac{3}{4}$ in. $\times$ 48 in. stringer, $S = 4128$ in^3 and $I = 99{,}070$ in^4; thus,

$$f_b = \frac{M}{S} = \frac{6{,}620{,}000\text{ lb-in.}}{4128\text{ in}^3} = 1604\text{ psi}$$

With regard to lateral stability, the tops of the stringers will be attached to the deck by brackets on 12-in. centers; thus, C_L will be taken to be unity. The volume factor is

$$C_V = \left(\frac{5.125}{10.75}\right)^{1/10} \left(\frac{12}{48}\right)^{1/10} \left(\frac{21}{50}\right)^{1/10} = 0.741$$

Thus,

$$f_b = 1604 \text{ psi} \leq F'_b = (2400)(1.15)(0.741) = 2045 \text{ psi} \qquad \text{good}$$

Shear: Shear is checked by placing the wheel loads at the closer of one-quarter the span length or three times the stringer depth from the supports. In this case, 50 ft/4 = 12.5 ft or (3)(48 in.) (1 ft /12 in.) = 12 ft; thus 12.0 ft controls. The corresponding maximum shear, per wheel line, becomes V = 20,640 lb. The design live load shear, following Example 6.3, is thus,

$$V_{\text{LL}} = (0.5)[(0.6)(20{,}640 \text{ lb}) + ((1.134)(20{,}640 \text{ lb})] = 17{,}900 \text{ lb}$$

The design dead load shear is taken at distance d from the end; thus,

$$V_{\text{DL}} = \frac{\omega\ell}{2} - \omega d = \frac{(626 \text{ plf})(50 \text{ ft})}{2} - (626 \text{ plf})(4 \text{ ft}) = 13{,}150 \text{ lb}$$

The total design shear is thus

$$V_{\text{TL}} = 17{,}900 \text{ lb} + 13{,}150 \text{ lb} = 31{,}050 \text{ lb}$$

$$f_v = \frac{3}{2} \frac{31{,}050 \text{ lb}}{(10.75 \text{ in.})(48 \text{ in.})} = 92 \text{ psi}$$

$$F'_v = (190 \text{ psi})(1.15)$$

$$= 219 \text{ psi} \qquad \text{good}$$

Deck: Using DF combination 1, from AITC 117 [18], we obtain F_{by} = 1450 psi and F_{vy} = 165 psi. From the AASHTO specification, the tire contact area is taken to be $0.01P$ in^2. For the HS 20-44 loading, P = 16,000 lb; thus the area is 160 in^2. The contact area is taken to be rectangular with the width in the direction of traffic to be 1/2.5 times the transverse width, b_t; thus, A = 160 in^2 = $(b_t/2.5)b_t$, giving b_t = 20 in. The transverse deck span is taken to be the smaller of

$$5 \text{ ft } 8 \text{ in.} - 10.75 \text{ in.} + 10.75/2 \text{ in.} = 62.63 \text{ in.}$$

or

$$5 \text{ ft } 8 \text{ in.} - 10.75 \text{ in.} + 6.75 \text{ in.} = 64 \text{ in.}$$

or in this case, s = 62.63 in.

The live load bending moment for the transverse deck, assuming simple span between stringers, is obtained by placing the wheel load at midspan, as shown in Figure 6.9. From the load tables in Section A2.2 (case 4), where the distributed load is centered in the span, the maximum moment becomes

$$M_{\text{max}} = P\left(\frac{\ell}{4} - \frac{b_t}{8}\right) \qquad \text{where } \ell = s = 62.63 \text{ in.}$$

Thus,

$$M_{\text{LL}} = 16{,}000 \text{ lb}\left(\frac{62.63 \text{ in.}}{4} - \frac{20 \text{ in.}}{8}\right) = 211{,}000 \text{ lb-in.}$$

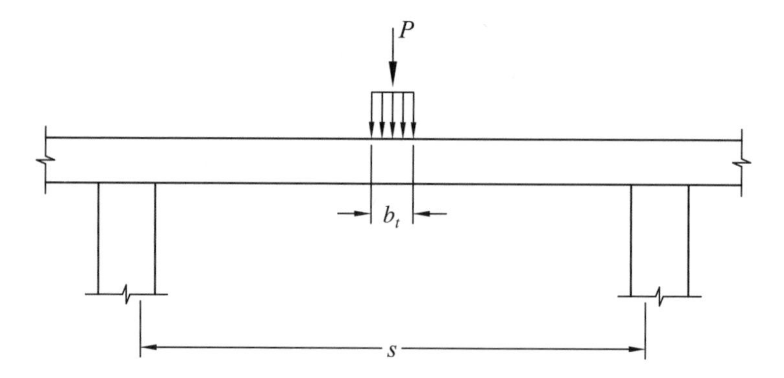

Wheel load distribution for moment

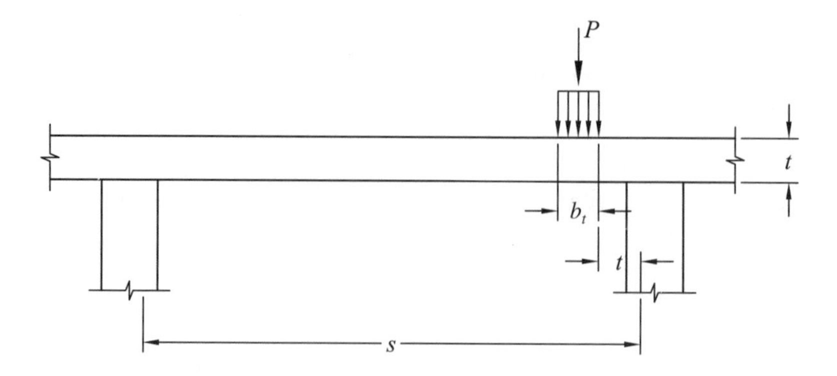

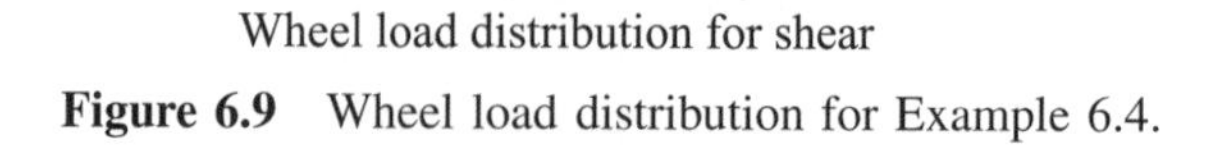

Wheel load distribution for shear

Figure 6.9 Wheel load distribution for Example 6.4.

For determination of the dead load moment, the distribution length of the wheel load in the longitudinal direction is used. For noninterconnected timber flooring, from the AASHTO specifications above, the length is 15 in. + 6.75 in. = 21.75 in. The applied dead load is, therefore,

$$\omega_{\text{DL}} = \left[\left(\frac{3}{12}\text{ ft}\right)(150\text{ pcf}) + \left(\frac{6.75}{12}\text{ ft}\right)(50\text{ pcf})\right]\left(\frac{21.75}{12}\text{ ft}\right) = 119\text{ plf}$$

Assuming a simple-span deck, the dead load bending moment is

$$M_{\text{DL}} = \frac{\omega_{\text{DL}}s^2}{8} = \frac{(119\text{ lb/ft})(62.63/12\text{ ft})^2}{8} = 405\text{ lb-ft} \quad \text{or} \quad 4900\text{ lb-in.}$$

The total bending moment, assuming a simple-span deck, is thus

$$M = 211{,}000\text{ lb-in.} + 4900\text{ lb-in.} = 215{,}900\text{ lb-in.}$$

Because the deck is continuous over two or more stringers, the moment is taken to be 80% of the simple-span moment; thus,

$$M = (0.80)(215{,}900\text{ lb-in.}) = 173{,}000\text{ lb-in.}$$

The section modulus for the 21.75 in. of deck is

$$S = \frac{(21.75\text{ in.})(6.75\text{ in.})^2}{6} = 165\text{ in}^3$$

Thus,

$$f_b = \frac{M}{S} = \frac{173{,}000\text{ lb-in.}}{165\text{ in}^3} = 1050\text{ psi}$$

The deck will be assumed to be in wet use; thus, $C_M = 0.80$ (for bending). The flat use factor for the $6\frac{3}{4}$-in. deck is

$$C_{fu} = \left(\frac{12\text{ in.}}{6.75\text{ in.}}\right)^{1/9} = 1.07$$

Therefore,

$$F'_b = F_bC_DC_{fu}C_M = (1450\text{ psi})(1.15)(1.07(0.8) = 1430\text{ psi}$$

Since, $f_b = 1050$ psi $\leq F'_b = 1430$ psi, the decking is satisfactory with regard to bending.

For investigating the deck with respect to shear, the dead load shear at deck thickness t from the end of the effective span is used. Thus,

$$V_{\mathrm{DL}} = \omega_{\mathrm{DL}}\left(\frac{s}{2} - t\right)$$

$$= (119\text{ plf})\left[\frac{(62.63\text{ in.})(1\text{ ft}/12\text{ in.})}{2} - 6.75\text{ in. }(1\text{ ft}/12\text{ in.})\right]$$

$$= 240\text{ lb}$$

The live load shear is obtained by placing the edge of distributed tire load at deck thickness t from the end of the effective span. The corresponding live load shear is

$$V_{\mathrm{LL}} = \frac{P(2s - 2t - b_t)}{2s}$$

$$= (16{,}000\text{ lb})\left[\frac{(2)(62.63\text{ in.}) - 2(6.75\text{ in.}) - 20\text{ in.}}{(2)(62.63\text{ in.})}\right]$$

$$= 11{,}720\text{ lb}$$

Therefore, the total shear load is $V = 240\text{ lb} + 11{,}720\text{ lb} = 11{,}960\text{ lb}$.

The distribution of stress is determined by the wheel load fraction per panel, specified by AASHTO as $W_p/4.0$ but not less than 1.0, where W_p is the panel width. If panels of minimum width 4.0 ft (48 in.) are specified, the shear stress is

$$f_v = \frac{3}{2}\left[\frac{11{,}960\text{ lb}}{(48\text{ in.})(6.75\text{ in.})}\right] = 55\text{ psi}$$

Since $f_v = 55\text{ psi} \leq F'_v = F_v C_D C_M = (165\text{ psi})(1.15)(0.875) = 166\text{ psi}$, the deck panels are also adequate with respect to shear. An example design of rail, post, and curb for a transverse deck timber bridge is shown in the next section.

Answer: Six $10\frac{3}{4}$ in. × 48 in. DF 24F-V4 stringers should be used with combination 1 DF $6\frac{3}{4}$ in. × 48 in. deck panels.

Discussion: In this example overload considerations were not required by the AASHTO specifications. For any particular project, however, overload checks may be required by the permitting agency.

6.7.2.3 Longitudinal Deck with Transverse Stiffeners A design example of a longitudinal timber deck bridge with transverse stiffeners follows.

Example 6.5 Longitudinal Deck with Transverse Stiffeners

Given: Physical features: 20-ft simple span, width 32 ft, two lanes; loading criteria, HS 15-44; wearing surface, 3-in. asphalt; longitudinal deck, Douglas fir combination 2, grade L2, 4-ft panel widths (eight total).

Determine: The size of the deck members and the requirements for stiffeners.

Approach: The longitudinal deck bridge will be designed in accordance with Chapters 3 and 13 of the AASHTO standards [22] and this book.

Solution: Design values for Douglas fir combination 2 are obtained from AITC 117 [18] as follows (properties with respect to the y–y axis, flat bending, four or more lams): $E = 1{,}600{,}000$ psi, $F_{c\perp} = 560$ psi, $F_{by} = 1800$ psi, and $F_{vy} = 165$ psi (no multiple-piece lams, cyclic loading per bridges). Assuming a deck $8\frac{3}{4}$ in. thick × 48 in. wide, the dead loads are

$$\text{Asphalt: } 150 \text{ pcf (per AASHTO)} \times \tfrac{3}{12} \text{ ft} = 37.5 \text{ psf}$$

$$\text{Deck: } 50 \text{ pcf (per AASHTO)} \times 8\tfrac{3}{4}/12 \text{ ft} = 36.5 \text{ psf}$$

$$\text{Total: } 37.5 \text{ psf} + 36.5 \text{ psf} = 74 \text{ psf}$$

The dead weight line load is thus $\omega_{\text{DL}} = 74 \text{ psf} \times 48/12 \text{ ft} = 296$ plf.

Bending Moment: The corresponding dead load bending moment is

$$M_{\text{DL}} = \frac{\omega_{\text{DL}}\ell^2}{8} = \frac{(296 \text{ plf})(20 \text{ ft})^2}{8} = 14{,}800 \text{ lb-ft}$$

The maximum live load bending moment will occur when the heavy axle is at midspan. For HS 15-44 loading, the corresponding wheel load is $0.4W$, from Figure 3.7.7A of AASHTO [22], where W is the truck weight in tons. Thus, the wheel load is 0.4 (15 tons × 2000 lb/ton) = 12,000 lb.

The wheel load fraction to the panel for is determined by AASHTO [22] as follows for two or more traffic lanes:

$$\text{load fraction} = \text{greater of } \frac{W_p}{3.75 + L/28} \text{ or } \frac{W_p}{500}$$

where

W_p = width of panel (ft)
L = length of span for simple span bridges or length of shortest span for continuous bridges (ft).

Therefore, in this example,

$$\text{load fraction} = \text{greater of } \frac{4}{3.75 + 20/28} = 0.896 \text{ or } \frac{4}{5.00} = 0.80$$

so the load fraction = 0.896. The live load moment is thus

$$M_{\text{LL}} = \frac{(\text{load fraction})PL}{4} = \frac{(0.896)(12{,}000 \text{ lb})(20 \text{ ft})}{4} = 53{,}760 \text{ lb-ft}$$

The total load bending moment is

$$\begin{aligned} M_{\text{TL}} &= M_{\text{DL}} + M_{\text{LL}} = 14{,}800 \text{ lb-ft} + 53{,}760 \text{ lb-ft} \\ &= 68{,}560 \text{ lb-ft, or } 822{,}700 \text{ lb-in} \end{aligned}$$

The corresponding bending stress is

$$f_b = \frac{M}{S} \qquad \text{where } S = \frac{bd^2}{6} = \frac{(48 \text{ in.})(8.75 \text{ in.})^2}{6} = 612.5 \text{ in}^3$$

Thus, f_b = 822,700 lb-in./612.5 in^3 = 1343 psi.

The design value in bending for this case is

$$F'_b = F_{by} C_D C_M C_{fu}$$

where

$$\begin{aligned} C_D &= 1.15 \text{ (AASHTO)}, \\ C_M &= 0.80 \text{ (Table 3.7, or from AITC 117 [18])}, \\ C_{fu} &= \left(\frac{12 \text{ in.}}{\text{d}}\right)^{1/9} = \left(\frac{12 \text{ in.}}{8.75 \text{ in.}}\right)^{1/9} = 1.036 \text{ [from Equation (3.5)]}. \end{aligned}$$

Thus,

$$F'_b = (1800 \text{ psi})(1.15)(0.80)(1.036) = 1715 \text{ psi}$$

Since f_b = 1334 psi $\leq F'_b$ = 1715 psi, the $8\frac{3}{4}$-in.-deep deck panels are satisfactory with regard to bending.

Shear: The design dead load shear is taken at distance d from the end; thus,

$$V_{\text{DL}} = \omega_{\text{DL}} \left(\frac{\ell}{2 - d}\right) = 296 \text{ plf} \left[\frac{20 \text{ ft}}{2} - (8.75/12)\text{ft}\right] = 2740 \text{ lb}$$

The design live load shear obtained by placing the maximum wheel load at the lesser of $3d$ or $\ell/4$ from the support; in this example, $3d = 3[(8.75 \text{ in.})(1 \text{ ft}/12 \text{ in.})] = 2.19$ ft and $\ell/4 = 5$ ft, the lesser of which is 2.19 ft. In the case of the HS loading, two equal loads of 12,000 lb may act as close as 14 ft apart (AASHTO Figure 3.7.7A), as shown in Figure 6.10. The shear at a distance 2.19 ft from one end, with one 12,000-lb force acting at that location and another 12,000-lb force acting 14 ft from the first, is $V = 12{,}970$ lb.

For a longitudinal deck, for shear the load is distributed by the load fraction $W_p/4 \leq 1.00$, in this case, $(48/12)/4 = 1.0$.

The design live load shear becomes

$$V_{\text{LL}} = (\text{load fraction})V = (1.0)(12{,}970) \text{ lb} = 12{,}970 \text{ lb}$$

The total design shear load is

$$V_{\text{TL}} = V_{\text{DL}} + V_{\text{LL}} = 2740 \text{ lb} + 12{,}970 \text{ lb} = 15{,}710 \text{ lb}$$

The shear stress is

$$f_v = \frac{3}{2}\left(\frac{V}{A}\right) = \frac{3}{2}\left[\frac{15{,}710 \text{ lb}}{(48 \text{ in.})(8.75 \text{ in.})}\right] = 56 \text{ psi}$$

The allowable shear stress is

$$F'_v = F_{vy}C_D C_M$$

where

$C_M = 0.875$.

Thus,

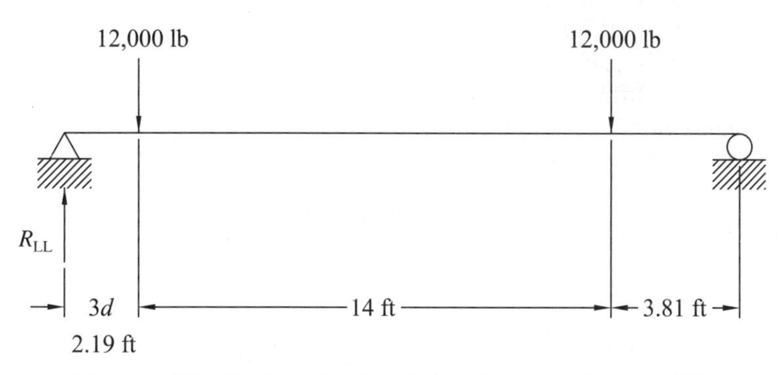

Figure 6.10 Design wheel loads for shear for Example 6.5.

$$F'_v = (165 \text{ psi})(1.15)(0.875) = 166 \text{ psi}$$

Since $f_v = 56$ psi $\leq F'_v = 166$ psi, the 48 in. × $8\frac{3}{4}$ in. deck is satisfactory with respect to shear.

Deflection: Deflection is evaluated based on simple beam action with the wheel load placed and distributed as for maximum moment. The deflection limit criterion is taken to be ℓ as for stringers, from Table 3.11. In this example,

$$\Delta_{\text{LL}} = \frac{(\text{moment load fraction})P\ell^3}{48E'I}$$

where the moment load fraction is 0.896 from above,

$$P = 12{,}000 \text{ lb},$$
$$\ell = (20 \text{ ft})(12 \text{ in./ft}) = 240 \text{ in., and}$$
$$E' = E_y C_M,$$

where

$$C_M = 0.833 \text{ from Table 3.7 or AITC 117 [18].}$$

Thus,

$$E' = (1{,}600 \text{ psi})(0.833) = 1{,}333{,}000 \text{ psi}$$

$$I = \tfrac{1}{12}(48 \text{ in.})(8.75 \text{ in.})^3 = 2680 \text{ in}^4$$

and therefore

$$\Delta_{\text{LL}} = \frac{(0.896)(12{,}000 \text{ lb})(240 \text{ in.})^3}{(48)(1{,}333{,}000 \text{ psi})2680 \text{ in}^4} = 0.867 \text{ in.}$$

The limiting deflection is $\ell/300 = 240$ in./300 = 0.80 in.

Since $\Delta_{\text{LL}} = 0.867$ in. *is not* $\leq \ell/300 = 240$ in./300 = 0.80 in., the live load deflection is excessive. In this case a thicker deck panel, or members with additional stiffness, should be selected.

If Douglas fir combination 3(L2D) is used, which has a modulus of elasticity of 1,900,000 psi, the deflection becomes

$$\Delta_{\text{LLcomb3}} = 0.867 \left[\frac{(1{,}600{,}000)(0.833) \text{ psi}}{(1{,}900{,}000)(0.833) \text{ psi}}\right] = 0.730 \text{ in.}$$

which is acceptable.

The design values for Douglas fir combination 3 are as follows: E = 1,900,000 psi, $F_{c\perp} = 650$ psi, $F_{by} = 2100$ psi, and $F_{vyic} = 165$ psi (no multiple-piece lams; cyclic loading per bridges). Since the design values for the combination 3 layup are equal to or exceed the design values used to check bending and shear, the 48 in. by $8\frac{3}{4}$ in. section in the combination 3 layup is still satisfactory.

Stiffeners: For longitudinal deck bridges, AASHTO [22] requires stiffeners at midspan and at intervals sufficient to prevent differential panel movement, but such intervals may not exceed 10 ft, and the stiffness, EI, of the members is not to be less than 80,000 kip-in^2 (80,000,000 lb-in^2). In this example, one stiffener should be used at midspan. A shallow-profile cross section will be investigated. Considering DF combination 2, $6\frac{3}{4}$ in. × 4.5 in., from AITC 117 [18], E_x = 1,600,000 psi and $I = \frac{1}{12}$ (6.75 in.)(4.5 in.)^3 = 51.3 in^4. The wet service factor will not be applied to the stiffeners; thus $E' = E_x$ = 1,600,000 psi and

$$E'I = (1{,}600{,}000 \text{ psi})(51.3 \text{ in}^4) = 82{,}000{,}000 \text{ lb-in}^2$$

which is adequate.

Overload: For overload, the bridge system will be considered as to be designed for one traffic lane. The load fraction for bending moment becomes the greater of [22]

$$\frac{W_p}{4.25 + L/28} \quad \text{or} \quad \frac{W_p}{5.50}$$

where

$$\frac{4.0}{4.25 + 20/28} = 0.806 \quad \text{and} \quad \frac{4.0}{5.50} = 0.727$$

Thus, use 0.806.

From AASHTO [22] Section 3.5.1, the live load is doubled and the basic unit stress may be increased by 50%; thus,

$$M = M_{\text{DL}} + M_{\text{LL}} = 14{,}800 \text{ lb-ft} + \frac{(0.806)(2)(12{,}000 \text{ lb})(20 \text{ ft})}{4}$$

$$= 111{,}500 \text{ lb-ft} = 1{,}338{,}000 \text{ lb-in.}$$

The bending stress is

$$f_b = \frac{M}{S} = \frac{1{,}338{,}000 \text{ lb-in.}}{612.5 \text{ in}^3} = 2185 \text{ psi}$$

For the combination 3 layup, F_{by} = 2100 psi; thus,

$$F'_b = [(1.5)(2100 \text{ psi})](1.15)(0.80)(1.036) = 3002 \text{ psi}$$

Since $f_b = 2185$ psi $\leq F'_b = 3002$ psi, the longitudinal stringers are acceptable in overload with respect to bending.

For shear,

$$V_{\text{TL}} = 2740 \text{ lb} + (1.0)[(2)(12{,}970 \text{ lb})] = 28{,}680 \text{ lb}$$

$$f_v = \frac{3}{2}\left(\frac{V}{A}\right) = \frac{3}{2}\left[\frac{28{,}680 \text{ lb}}{(48 \text{ in.})(8.75 \text{ in.})}\right] = 102 \text{ psi}$$

The design value for shear for combination 3 is the same as that for combination 2 except that it may be increased 50% in the overload condition; thus,

$$F'_v = (1.5)(166 \text{ psi}) = 249 \text{ psi}$$

Since $f_v = 102$ psi $\leq F'_v = 249$ psi, the longitudinal deck is also satisfactory in overload with regard to shear.

Answer: Eight 48 in. × $8\frac{3}{4}$ in. Douglas fir combination 3(L2D) panels will be used. One $6\frac{3}{4}$ in. × 4.5 in. DF combination 2(L2) stiffener will be used at midspan. Curb, rail, and post design for timber bridges is discussed in the following section.

Discussion: Depending on the availability of size and grade, other combinations and section sizes could be investigated. Whereas the deflection in the beam above was found to be satisfactory with regard to the recommendations of Table 3.11, AASHTO [22] recommends limiting live load deflections for timber bridges to $\ell/500$. In this regard, the suitability of the $\ell/300$ criterion used above should be verified with the permitting agency.

6.7.3 Rail, Post, and Curb Design

Rail, post, and curb systems may be designed using the principles in this book. Typical design considerations for rail, post, and curb systems include outward, upward, and downward impact loads on rails, plus outward, upward, downward, and longitudinal impact loads on posts, and outward impact loads on curbs. Rails, curbs, and posts are typically designed assuming wet conditions of use with impact load duration factors. Alternatively, a number of systems have been crash tested for which the construction requirements and details are available [25]. AASHTO provides both design and crash loading criteria [22].

An example of a crash-tested rail, curb, and post system for a longitudinal deck timber bridge is shown in Figure 6.11. A typical rail, curb, and post

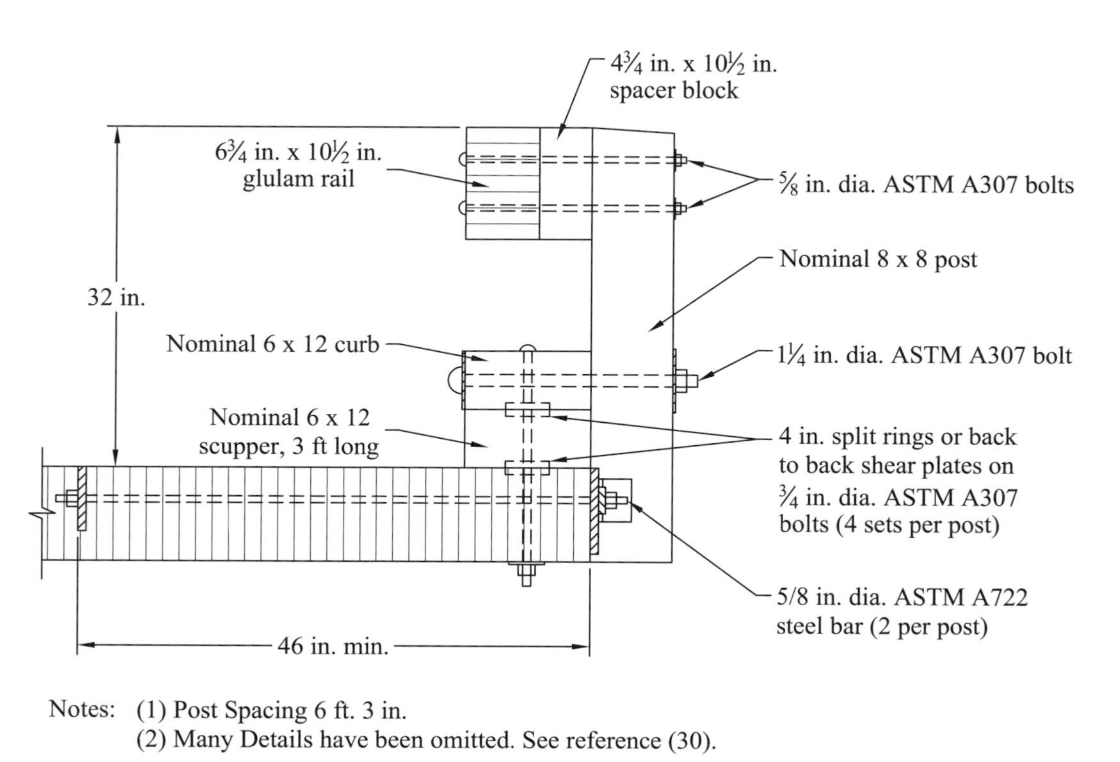

Notes: (1) Post Spacing 6 ft. 3 in.
(2) Many Details have been omitted. See reference (30).

Figure 6.11 Crash tested rail and curb system for a longitudinal deck timber highway bridge.

system for a transverse deck is shown in Figure 6.12. In the system shown in Figure 6.12, the post must be braced adequately in reaction to the outward load, typically by attachment to the outside girder, and the girder must be adequately braced. A design example of rail, post, and curb for a nondoweled transverse deck with longitudinal stringers follows. Additional post, rail, and curb design information and examples are included in *Glued Laminated Timber Bridge Systems* [24].

Example 6.6 Rail, Post, and Curb Design

Given: The traffic rail, post, and curb system illustrated in Figure 6.13 for a longitudinal stringer bridge with a transverse deck.

Determine: Design the traffic rail, post, and curb using southern pine glued laminated timber, combination 47.

Approach: AASHTO standard specifications related to timber bridge rail, post, and curb design are as follows:

Load duration factor: C_D = 1.65 (traffic railing only).

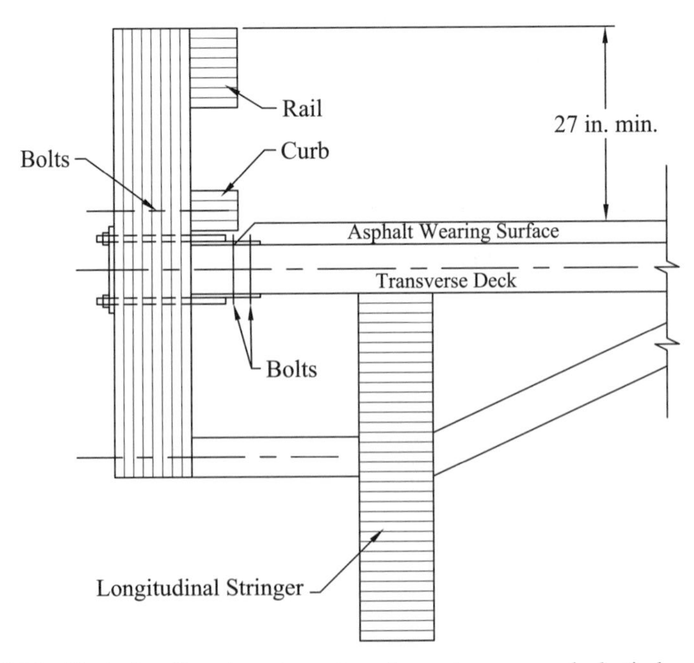

Figure 6.12 Typical rail and curb system for a transverse deck timber highway bridge.

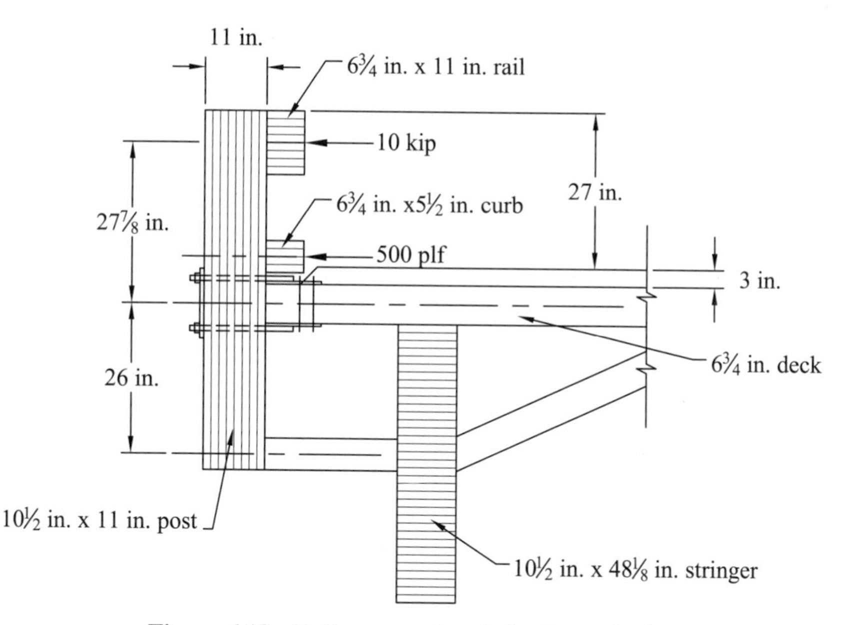

Figure 6.13 Rail, post, and curb for Example 6.6.

General: Minimum height of top of railing above roadway 2 ft 3 in.; for railing heights not exceeding 2 ft 9 in., the transverse outward load to be resisted is $P = 10$ kips.

Railings: Transverse load P applied horizontally outward applied at mid-depth; AASHTO recommended equation for moment: $M = PL/4$, where L is the center-to-center post spacing.

Posts: P applied outward from rail acting simultaneous with $0.5P$ applied longitudinally distributed over four posts. Each post should be designed to resist an independently applied inward or outward load of $P/4$.

Railing attachment: Should be designed to resist inward, upward, or downward load of $P/4$.

Curbs: 500 plf outward.

Solution: Proposed member dimensions:

Rail: $6\frac{3}{4}$ in. $\times$ 11 in.
Post: $10\frac{1}{2}$ in. $\times$ 11 in.
Curb: $6\frac{3}{4}$ in. $\times$ 5.5 in.
Posts at 8 ft o.c.

Design values from AITC 117, Table 8.2 [18], are as follows:

$E = 1{,}400{,}000$ psi

$F_{bx} = 1400$ psi (two lams to 15 in. deep; no tension lams)

$F_{by} = 1750$ psi (four or more lams)

$F_{vx} = 215$ psi

$F_{vy} = 190$ psi (four or more lams; no multiple-piece laminations)

$F_c = 1950$ psi (four or more lams)

$F_{c\perp} = 650$ psi

The wet service factor (AITC 117) is as follows:

$C_M = 0.833$ (modulus of elasticity)

$= 0.80$ (bending)

$= 0.875$ (shear parallel to the grain)

$= 0.73$ (compression parallel to the grain)

$= 0.53$ (compression perpendicular to the grain)

$= 0.70$ (dowel-type fasteners; Table 5.4 of this book)

Rail Design:

Bending: The rail oriented as illustrated in Figure 6.13 will resist the transverse (outward) load P in bending with regard to the y-direction.

$$M = \frac{PL}{6} = \frac{(10{,}000 \text{ lb})(8 \text{ ft})}{6} = 13{,}333 \text{ lb-ft} = 160{,}000 \text{ lb-in.}$$

$$S = \frac{bd^2}{6} = \frac{(11 \text{ in.})(6.75 \text{ in.})^2}{6} = 83.5 \text{ in}^3$$

$$f_b = \frac{M}{S} = \frac{160{,}000 \text{ lb-in.}}{83.5 \text{ in}^3} = 1915 \text{ psi}$$

$$C_{fu} = \left(\frac{12 \text{ in.}}{d}\right)^{1/9} = \left(\frac{12 \text{ in.}}{6.75 \text{ in.}}\right)^{1/9} = 1.066$$

$$F'_b = F_{by} C_D C_M C_{fu}$$

$$= (1750 \text{ psi})(1.65)(0.80)(1.066) = 2460 \text{ psi}$$

Since $f_b = 1915 \text{ psi} \leq F'_b = 2460$ psi, the rail is satisfactory with regard to bending due to the outward transverse traffic load.

Shear: For shear parallel to the grain, the transverse traffic load will be placed near the support. Thus, $V = 10{,}000$ lb.

$$f_v = \frac{3}{2}\frac{V}{A} = \frac{3}{2}\frac{10{,}000\text{ lb}}{(11\text{ in.})(6.75\text{ in.})} = 202\text{ psi}$$

$$F'_v = F_{vy}C_DC_M = (190\text{ psi})(1.65)(0.875) = 274\text{ psi}$$

Since $f_v = 202\text{ psi} \le F'_v = 274$ psi, the rail is satisfactory with regard to shear from the transverse traffic load.

Bearing: Using the rail bearing on $10\frac{1}{2}$-in.-wide posts,

$$f_{c\perp} = \frac{P}{A} = \frac{10{,}000\text{ lb}}{(10.5\text{ in.})(11\text{ in.})} = 87\text{ psi}$$

$$F'_{c\perp} = F_{c\perp}C_MC_b$$

where C_b is the bearing area factor, in this case 1.00. Thus,

$$F'_{c\perp} = (650\text{ psi})(0.53)(1.00) = 345\text{ psi}$$

Since $f_{c\perp} = 87\text{ psi} \le F'_{c\perp} = 345$ psi, the bearing condition between rail and post is satisfactory for the full transverse load.

Rail to Post Connection: Attachment of rail to post must be sufficient to resist $P/4 = 2500$ lb upward, downward, and inward. Four $\frac{3}{4}$-in. A307 through bolts will be considered, as illustrated in Figure 6.14. For the downward vertical load, the self-weight of the rail will be added to the vehicle load of 2500 lb; thus, the downward load is

$$2500\text{ lb} + \left(\frac{6.75}{12}\text{ ft}\right)\left(\frac{11}{12}\text{ ft}\right)(50\text{ pcf})(8\text{ ft}) = 2500\text{ lb} + 206\text{ lb} = 2706\text{ lb}$$

The allowable load on one $\frac{3}{4}$-in. bolt, assuming that all spacing and edge distance requirements are met for full design values, is

$$Z' = ZC_gC_DC_M$$

where Z is calculated from equations (5.6) to be 1088 lb, C_g will be assumed to be at or near unity ($C_g = 1.00$), and $C_M = 0.70$. Thus,

$$Z' = (1088\text{ lb})(1.00)(1.65)(0.70) = 1257\text{ lb.}$$

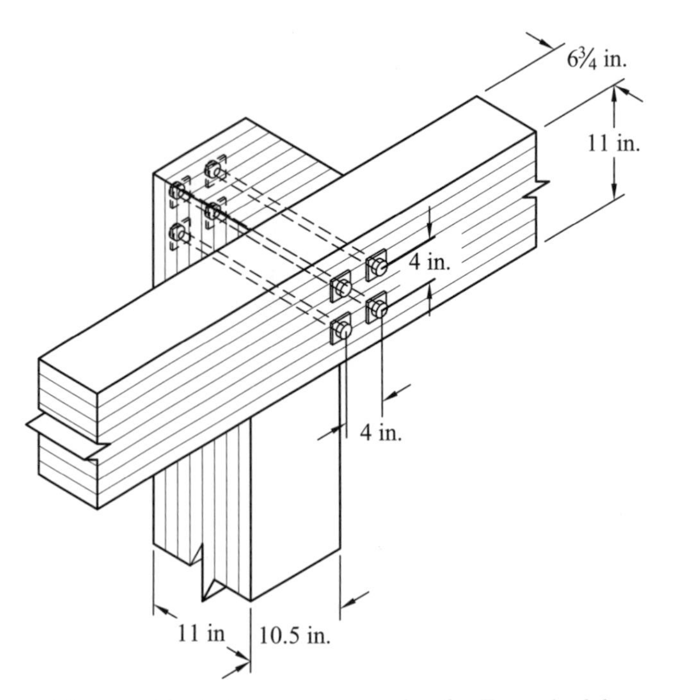

Figure 6.14 Rail-to-post connection for Example 6.6.

For four bolts, the allowable load is (4)(1257 lb) = 5027 lb. Since 2706 lb is less than 5027 lb, the attachment is satisfactory with regard to the downward vertical vehicle design load.

For the upward vertical load, the end distance in the vertical post is 3.5 in. The minimum end distance for reduced design value is $3.5D$ = (3.5)(0.75 in.) = 2.75 in. and the minimum end distance for full design load is $7.0D$ = 5.25 in. Thus, the geometry factor must be applied to the case of the upward design load; from equation (3.9),

$$C_{\Delta n} = \frac{\text{actual end distance}}{\text{minimum end distance for full design value}} = \frac{3.5 \text{ in.}}{5.25 \text{ in.}} = 0.667$$

and is applied to all bolts in the group. The design load is 2500 lb, upward. Since 2500 lb $\leq$ (4)(1088 lb)(1.65)(0.70)(0.667) = 3352 lb, the attachment is satisfactory with regard to the upward design load.

With regard to the inward design load of 2500 lb, each of the four $\frac{3}{4}$-in. bolts have a tensile capacity of (8800 lb) [25], which may be increased by 33% for transient loading. Thus, the tensile capacity of the bolts is more than sufficient; however, adequate bearing area must be provided by the washers.

Considering four 2 in. $\times$ 2 in. timber washers, with $\frac{7}{8}$-in. holes, the total bearing area is

$$A = (4)\left[\frac{(2 \text{ in.})(2 \text{ in.}) - \pi(0.875 \text{ in.})^2}{4}\right] = 13.6 \text{ in}^2$$

The compression perpendicular to the grain from the washers is

$$f_{c\perp} = \frac{P}{A} = \frac{2500 \text{ lb}}{13.6 \text{ in}^2} = 183 \text{ psi}$$

The bearing area factor, C_b may be applied to the bearing area provided by the washers. Using equation (3.7)

$$C_b = \frac{\ell_b + 0.375 \text{ in.}}{\ell_b}$$

where $\ell_b = 2$ in. Thus,

$$C_b = \frac{2.0 \text{ in.} + 0.375 \text{ in.}}{2.0 \text{ in.}} = 1.188$$

and

$$F'_{c\perp} = (650 \text{ psi})(0.53)(1.00)(1.188) = 410 \text{ psi}$$

Since $f_{c\perp} = 183 \text{ psi} \leq F'_{c\perp} = 410$ psi, the bearing provided by the washers is adequate.

Curb: The curb will be checked with regard to a maximum bending moment of $M = \omega\ell^2/8$ and a shear of $V = \omega\ell/2$. Using 500 plf,

$$M = \frac{(500 \text{ plf})(8 \text{ ft})^2}{8} = 4000 \text{ lb-ft} = 48{,}000 \text{ lb-in.}$$

$$S = \frac{(5.5 \text{ in.})(6.75 \text{ in.})^2}{6} = 41.8 \text{ in}^3$$

$$f_b = \frac{M}{S} = \frac{48{,}000 \text{ lb-in.}}{41.8 \text{ in}^3} = 1149 \text{ psi}$$

$$F'_b = 2460 \text{ psi (calculated previously)}$$

$$V = \frac{(500 \text{ plf})(8 \text{ ft})}{2} = 2000 \text{ lb}$$

$$f_v = \frac{3}{2}\left[\frac{2000 \text{ lb}}{(6.75 \text{ in.})(5.5 \text{ in.})}\right] = 81 \text{ psi}$$

$$F'_v = 274 \text{ psi (calculated previously)}$$

Since $f_b \leq F'_b$ and $f_v \leq F'_v$, the curb is also satisfactory. The curb is not subject to the same upward, downward, and inward loads as the rail. In this example, two $\frac{3}{4}$-in. through bolts in a single horizontal row should be sufficient to attach the curb to the post.

Post: The post must be designed to resist a load of 10,000 lb outward at the mid-depth of the rail acting simultaneously with a longitudinal load of 5000 lb distributed over four posts, or 1250 lb, thus producing biaxial loading in the post. Figure 6.15 *a* and *b* show the outward-acting load and reaction and resulting shear and moment diagrams for the post, due to the outward load at the middepth of the rail and the outward load on the curb. The load from the rail governs. Figure 6.15*c* shows the longitudinal load on a single post and the resulting reactions, shear, and moment.

The maximum shear and bending moment due to the outward forces at the rail are $V = V_x = 10{,}720$ lb and $M = M_x = 278{,}800$ lb-in., respectively. The maximum shear and bending moment due to the longitudinal force acting at the rail middepth are $V = V_y = 1250$ lb and $M = M_y = (1250 \text{ lb})(27.875 \text{ in.}) = 34{,}840$ lb-in., respectively (Figure 6.15*c*). The section properties for the post are

$$I_x = \tfrac{1}{12}(10.5 \text{ in.})(11 \text{ in.})^3 = 1271 \text{ in}^4$$

$$S_x = \frac{I}{c} = \frac{1271 \text{ in}^4}{11 \text{ in.}/2} = 231 \text{ in}^3$$

$$A = (10.5 \text{ in.})(11 \text{ in.}) = 115.5 \text{ in}^2$$

and

$$I_y = \tfrac{1}{12}(11 \text{ in.})(10.5 \text{ in.})^3 = 1061 \text{ in}^4$$

$$S_y = \frac{I}{c} = \frac{1061 \text{ in}^4}{10.5 \text{ in.}/2} = 202 \text{ in}^3$$

$$f_{bx} = \frac{M_x}{S_x} = \frac{278{,}800 \text{ lb-in.}}{231 \text{ in}^3} = 1210 \text{ psi}$$

$$f_{by} = \frac{M_y}{S_y} = \frac{34{,}840 \text{ lb-in.}}{202 \text{ in}^3} = 172 \text{ psi}$$

For the biaxial bending stress check, C_D, C_M, and C_V or C_L are applicable in the *x*-direction, and C_D, C_M, and C_{fu} in the *y* (longitudinal)-direction.

$$C_V = \left(\frac{5.125 \text{ in.}}{10.5 \text{ in.}}\right)^{\frac{1}{20}} \left(\frac{12 \text{ in.}}{11 \text{ in.}}\right)^{\frac{1}{20}} \left(\frac{21 \text{ ft}}{27/12 \text{ ft}}\right)^{\frac{1}{20}} \leq 1.00 = 1.00$$

For C_L,

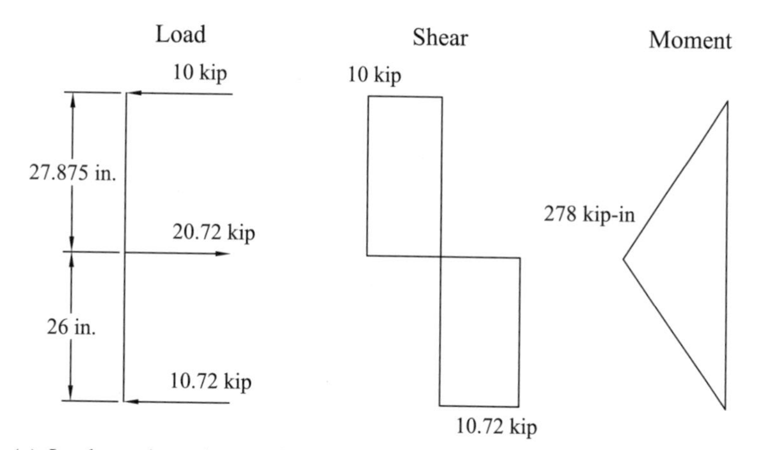

(*a*) Load, reactions, shear, and moment due to outward vehicle load on post at mid-depth of rail

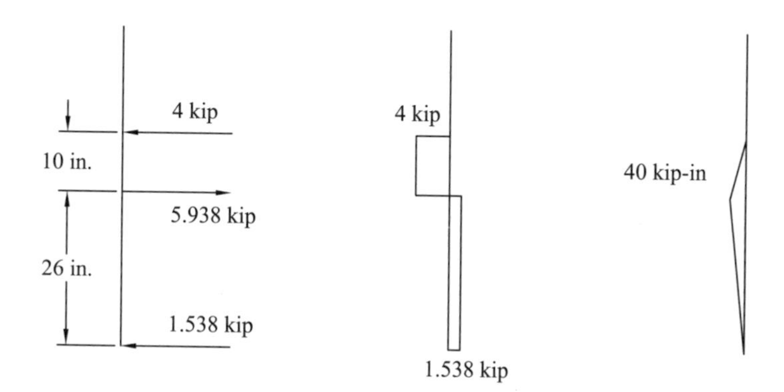

(*b*) Load, reactions, shear, and moment due to outward load on post at top of curb

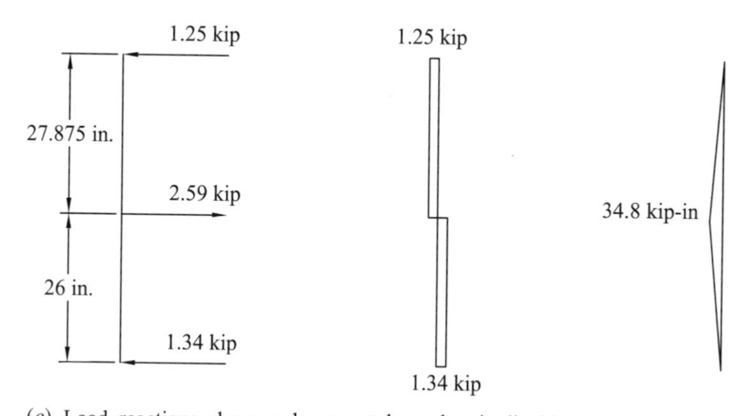

(*c*) Load, reactions, shear, and moment due to longitudinal load on post at mid-depth of rail

Figure 6.15 Load, reaction, shear, and moment diagrams for loads on post for Example 6.6.

$$R_B = \sqrt{\frac{\ell_e d}{b^2}}$$

where $\ell_e = 1.87\ell_u = (1.87)(27 \text{ in.}) = 50.5$ in. for $\ell_u/d = 27 \text{ in.}/11 \text{ in.} = 2.45 < 7$ from Table 4.3. Thus,

$$R_B = \sqrt{\frac{(50.5 \text{ in.})(11 \text{ in.})}{(10.5 \text{ in.})^2}} = 2.24$$

and

$$K_{bE} = 0.610 \text{ for glulam}$$

$$E' = EC_M = (1{,}400{,}000 \text{ psi})(0.833) = 1{,}166{,}000 \text{ psi}$$

which yield

$$F_{bE} = \frac{K_{bE}E'}{R_B^2} = \frac{(0.610)(1{,}166{,}000 \text{ psi})}{(2.24)^2} = 142{,}000 \text{ psi}$$

then

$$C_L = \frac{1 + (F_{bE}/F_b^*)}{1.9} - \sqrt{\left[\frac{1 + (F_{bE}/F_b^*)}{1.9}\right]^2 - \frac{F_{bE}/F_b^*}{0.95}}$$

$$= \frac{1 + (142{,}000 \text{ psi}/1848 \text{ psi})}{1.9} - \sqrt{\left(\frac{1 + (142{,}000 \text{ psi}/1848 \text{ psi})}{1.9}\right)^2}$$

$$- \frac{142{,}000 \text{ psi}/1848 \text{ psi}}{0.95} = 0.999$$

Thus,

$$F'_{bx} = (1400 \text{ psi})(1.65)(0.80)(0.999) = 1846 \text{ psi}$$

For F'_{by},

$$C_{fu} = \left(\frac{12 \text{ in.}}{11 \text{ in.}}\right)^{1/9} = 1.01$$

Thus,

$$F'_{by} = (1750 \text{ psi})(1.65)(0.80)(1.01) = 2333 \text{ psi}$$

and the biaxial stress check [equation (4.7)] is

$$\frac{f_{b1}}{F'_{b1}} + \frac{f_{b2}}{F'_{b2}[1 - (f_{b1}/F_{bE})^2]} = \frac{1210 \text{ psi}}{1846 \text{ psi}} + \frac{172 \text{ psi}}{(2333 \text{ psi})[1 - (1210 \text{ psi}/142{,}000 \text{ psi})^2]}$$

$$= 0.66 + 0.07 = 0.73 \leq 1.00 \qquad \text{good}$$

The post is checked for shear parallel to the grain with respect to the outward rail force as follows:

$$f_v = \frac{3}{2}\frac{V}{A} = \frac{3}{2}\left(\frac{10{,}720 \text{ lb}}{115.5 \text{ in}^2}\right) = 139 \text{ psi}$$

$$F'_v = F_{vx}C_DC_M = (215 \text{ psi})(1.65)(0.875) = 310 \text{ psi}$$

Since $f_v = 139$ psi $\leq F'_v = 310$ psi, the post is satisfactory with regard to shear from the outward railing force.

Post Bracket to Deck One means of fastening the post to the deck is illustrated in Figure 6.16. Four $\frac{7}{8}$-in. steel rods fasten the post to the deck through a back plate at the post, welded to the top and bottom plates on the deck. Four 1-in. through bolts are used to attach the top and bottom plates to the deck. The reaction at the deck due to the outward load is 20,720 lb, from Figure 6.15*a*. The longitudinal reaction is 2590 lb (Figure 6.15*c*). The four 1-in. through bolts are loaded at an angle to the grain with the deck, due to the combined outward and longitudinal loads. The 1-in. bolts are checked as follows.

Assuming a bracket plate thickness of at least $\frac{1}{4}$ in., a deck thickness of $6\frac{3}{4}$ in., a specific gravity of 0.55 (for southern pine), and a group action factor, C_g, of 1.00, the design values for the bolts may be determined from the equations of Chapter 5 or may be approximated using tabulated values for loading parallel and perpendicular to the grain from the *National Design Specification®* [7] and the Hankinson formula (*NDS®* , Appendix J), as follows.

From the *National Design Specification®* [7], $Z_{\parallel} = 5960$ lb and $Z_{\perp} = 3180$ lb. Thus,

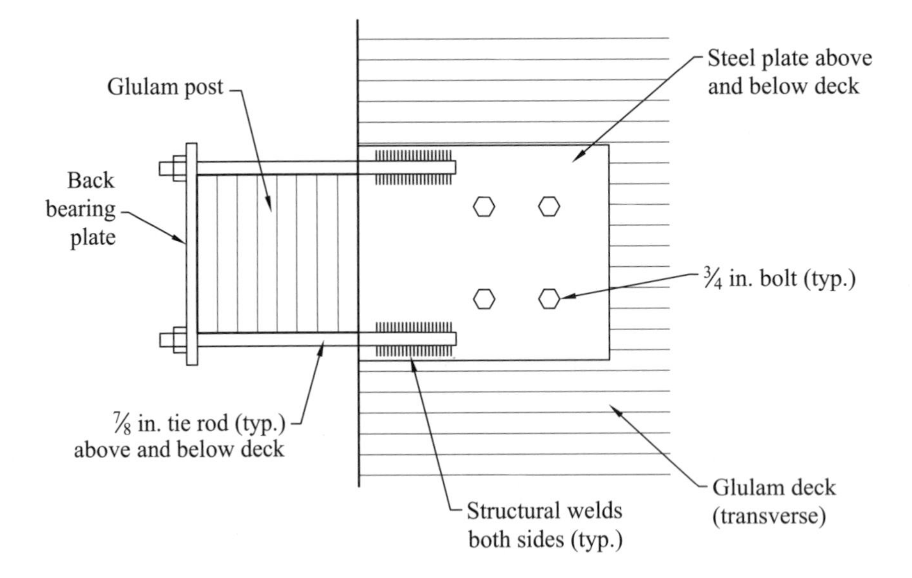

Figure 6.16 Post bracket for Example 6.6.

$$Z'_{\parallel} = Z_{\parallel} C_g C_D C_M = (5960 \text{ lb})(1.00)(1.65)(0.7) = 6880 \text{ lb}$$

$$Z'_{\perp} = Z_{\perp} C_g C_D C_M = (3180 \text{ lb})(1.00)(1.65)(0.7) = 3670 \text{ lb}$$

The angle of load to the grain is $\theta = \tan^{-1}(2590 \text{ lb}/20{,}720 \text{ lb}) = 7.1°$. Using the Hankinson formula, equation (6.2),

$$Z_{\theta} = \frac{Z_{\parallel}\,(Z_{\perp})}{Z_{\parallel} \sin^2\theta + Z_{\perp} \cos^2\theta}$$

$$Z_{7.1°} = \frac{(6880 \text{ lb})(3670 \text{ lb})}{6800 \text{ lb} \sin^2 7.1° + 3670 \text{ lb} \cos^2 7.1°} = 6790 \text{ lb}$$

The combined outward and longitudinal reaction is

$$R_{7.1°} = \sqrt{(20{,}720 \text{ lb})^2 + (2590 \text{ lb})^2} = 20{,}880 \text{ lb}$$

Since $R = 20{,}880 \text{ lb} \leq (4)(6790 \text{ lb}) = 27{,}200 \text{ lb}$, the four 1-in. through bolts should be satisfactory.

The metal fasteners, plate thickness, weld size and lengths must be checked according to the requirements of the American Institute of Steel Construction

[26]. Group and row tear-out of the four 1-in. bolts should also be investigated following the guidelines in Chapter 5. The bottom of the post must be braced against the adjacent stringer as illustrated in Figure 6.12, and the stringer further bridged (braced) to the next stringer or deck. Furthermore, the post must be braced longitudinally to prevent rotation under longitudinal load.

Answer: The following member dimensions were checked and found satisfactory for posts of southern pine combination 47 spaced 8 ft o.c. Special tension laminations are not required.

Rail: $6\frac{3}{4}$ in. × 11 in.

Post: $10\frac{1}{2}$ in. × 11 in.

Curb: $6\frac{3}{4}$ in. × 5.5 in.

Rail-to-post connection: four $\frac{3}{4}$-in. A307 bolts

Curb-to-post connection: four $\frac{3}{4}$-in. A307 bolts

Post bracket-to-Deck connection: four 1-in. A307 bolts

Discussion: The reactions, shear, and bending moment due to the load on the curb in this example are less than those from the vehicle load on the railing and thus were not checked independently. In the case of a transverse deck, the combined outward and longitudinal load is nearly parallel to the grain for the through bolts connecting the bracket to the deck. For a longitudinal deck, similar fasteners would have considerably less capacity as the primarily outward load would be oriented nearly perpendicular to the grain. A complete design of the rail, post, and curb system would involve further design checks on the shear plate connections, steel parts and fastenings, and the bracing described above.

CHAPTER 7

LOAD AND RESISTANCE FACTOR DESIGN

7.1 INTRODUCTION

Load and resistance factor design (LRFD) as used in timber engineering is discussed in this chapter. LRFD is similar to allowable stress design (ASD), covered in this book, in that proper design results in structural members and connections that are capable of resisting the loads or forces that can be expected in the life of the structure with some factor of safety and without loss of serviceability.

In allowable stress design, trial members are selected and stresses are calculated due to design loads. The stresses are compared to allowable stresses for the members, and trial grades and sizes are considered acceptable when all design stresses due to applied loads do not exceed the allowable stresses. The allowable stresses are determined from design values adjusted to end-use conditions. These design values incorporate internal factors of safety and are thus lower than the actual failure stresses of the materials. These factors of safety account implicity for overload (legitimate uncertainty in the predicted load), possible understrength (of the member), and the relative importance of the member in the structure (consequences and nature of possible failure).

In load and resistance factor design, resistances of the structural members and connections are determined based on material strength properties and end-use conditions. The resistance values are multiplied by resistance factors that explicitly consider possible understrength and consequence of failure. Similarly, anticipated loads are multiplied by load factors that explicitly consider the uncertainty in the loads. Acceptable members are those for which factored resistance values are equal to or greater than all appropriate factored loads. Because material strength values used with the LRFD approach are not reduced by internal factors of safety, stress and load levels appear higher than those in the ASD approach.

Since allowable stress design and load and resistance factor design use very different design values (upper limit versus safe), it is important that the approaches and design values not be confused. For this reason in this book we use material strength and member resistance values in terms of kips (thousands of pounds), kip-in. (thousands of pound-inches), and ksi (kips per square inch), compared with the lb, lb-in., and psi values associated with ASD. In LRFD, unfactored loads are used in deflection calculations. Thus, in both LRFD and ASD approaches, serviceability considerations are made with respect to the service load conditions. The same modulus of elasticity value, E, is used for both approaches, although in different units. For stability calculations, load and resistance factor design uses fifth percentile values for modulus of elasticity, E_{05}, whereas the ASD approach uses mean E values and adjusts for fifth percentile in the calculation of K_{cE} and K_{bE}.

7.2 DESIGN VALUES

Design values for LRFD are given in the product supplements of the *Load and Resistance Factor Design Manual for Engineered Wood Construction* (*LRFD Manual*) [1] or may be established by product testing in accordance with ASTM D5456 [2]. Design values may also be obtained by soft conversion of ASD design values from *National Design Specification®* [3] or AITC 117 [4] values using the factors of Table 7.1.

7.3 LOAD COMBINATIONS AND LOAD FACTORS

In load and resistance factor design, all applicable loads are combined and factored in accordance with equations (7.1) through (7.6) in Table 7.2. These combinations and factors account for uncertainty in design loads, the probability that only fractions of some loads are expected to occur simultaneously,

TABLE 7.1 Conversion Factors for Use of ASD Design Values in LRFD

Property or Design Value	Conversion Factor (ASD to LRFD)[a]
F_b, F_t, F_v, F_c, F_{rt}, F_s	$2.16/1000\phi$
$F_{c\perp}$	$1.875/1000\phi$
Z, W (connections)	$2.16/1000\phi$
E (stability only)[b]	$E_{05} = E\ (1 - 1.645\ \mathrm{COV}_E)/1000$

Source: ASTM D5457 [5].

[a] Factor includes conversion of units of psi and lb (ASD) to ksi and kips (LRFD). Values for resistance factor ϕ are provided in Section 7.4.

[b] Where tabulated mean modulus of elasticity E values have been adjusted for shear deflection, the value of E_{05} may be further increased by the ratio of the shear-free E to tabulated E.

and the potential for underweight systems being counted upon for resistance to uplift or overturning.

7.4 RESISTANCE FACTORS

Resistance factors for wood members and wood-based products and various design checks are defined and provided in Table 7.3. These factors take into consideration variability in member strength as well as the nature and consequence of failure.

7.5 TIME EFFECT FACTORS

The effect of duration of load on strength for wood members and connections is accounted for by the time effect factor, λ. In contrast to the load duration

TABLE 7.2 Load Combinations and Load Factors (LRFD)[a]

Load Combination	Eq.
$1.4(D + F)$	(7.1)
$1.2(D + F + T) + 1.6(L + H) + 0.5(L_r \text{ or } S \text{ or } R)$	(7.2)
$1.2D + 1.6(L_r \text{ or } S \text{ or } R) + (L \text{ or } 0.8W)$	(7.3)
$1.2D + 1.6W + L + 0.5(L_r \text{ or } S \text{ or } R)$	(7.4)
$1.2D + 1.0E + L + 0.2S$	(7.5)
$0.9D + 1.6W + 1.6H$	(7.6)
$0.9D + 1.0E + 1.6H$	(7.7)

Source: ASCE 7-02 [6].

[a] D = dead load,
E = earthquake load,
F = fluid pressure load,
H = lateral earth pressure,
L = live load caused by storage, occupancy, or impact,
L_r = roof live load,
R = load caused by initial rainwater and/or ice,
S = snow load,
T = self-straining load,
W = wind load.

TABLE 7.3 Resistance Factors for Wood Members and Wood-Based Products

ϕ_c	Compression	0.90
ϕ_b	Flexure	0.85
ϕ_s	Stability	0.85
ϕ_t	Tension	0.80
ϕ_v	Shear/torsion	0.75
ϕ_z	Connections	0.65

Source: AF&PA/ASCE 16-95 [7].

factor in allowable stress design, time effect factors are unity for wind and earthquake loads and are values less than unity for longer-duration loads. Time effect factors for the loads and load combinations of Table 7.2 are provided in Table 7.4.

7.6 ADJUSTMENT FACTORS

In addition to the time effect factors, load and resistance factor design requires use of adjustment factors that consider the end use of structural wood members. These factors are generally similar to the adjustment factors used in allowable stress design. Adjustment factors that are particular to individual types of products, such as wet service, temperature, size, and other factors, are generally published in the product supplements of the *LRFD Manual* [1]. Adjustment factors that are generic to product type but that depend on conditions such as geometry, lateral support, or end constraint, such the column stability, beam stability, and stress interaction factors, are computed using the methods and equations in the AF&PA/ASCE 16-95 standard itself [7].

TABLE 7.4 Load Combinations and Time Effect Factors

Load Combination	Time Effect Factor, λ
$1.4(D + F)$	0.6
$1.2(D + F + T) + 1.6(L + H) + 0.5(L_r \text{ or } S \text{ or } R)$	0.7 when L is from storage 0.8 when L is from occupancy 1.25 when L is from impact
$1.2D + 1.6(L_r \text{ or } S \text{ or } R) + (L \text{ or } 0.8W)$	0.8
$1.2D + 1.6W + L + 0.5(L_r \text{ or } S \text{ or } R)$	1.0
$1.2D + 1.0E + L + 0.2S$	1.0
$0.9D + 1.6W + 1.6H$	1.0
$0.9D + 1.0E + 1.6H$	1.0

Source: American Forest & Paper Association [7,8].

7.7 DESIGN CHECKS

Design checks for wood members using the load and resistance factor approach may be performed in several ways. When service conditions of particular structural members match prescribed reference conditions for the members, factored resistance values may be obtained directly from factored resistance tables for various individual wood products. Acceptable members are those for which all factored resistance values are at least as great as all loads computed from the applicable factored load combinations. For members such as beams and joists and columns, acceptable members may also be selected from load and span tables for the individual products. For beams and joists, the design checks require selection of members that satisfy both strength considerations based on factored loads and serviceability considerations based on unfactored loads.

Reference conditions for tabulated resistances are generally shown on the applicable table and typically include dry service, normal temperature range, live or snow duration of load ($\lambda = 0.80$), full lateral support (bending members), pin–pin end conditions (columns), and untreated material. When service conditions do not meet the prescribed reference conditions for the members, factored resistance values must be calculated based on reference strength values, member section properties, and appropriate adjustment factors. Example design checks are included in the following sections.

7.8 DESIGN EXAMPLE

A design example follows that illustrates load and resistance factor design. The example relates to a previous example in this book and illustrates the ways in which LRFD can be used in timber design, particularly with regard to glued laminated timber beams. More comprehensive examples and examples using additional wood products may be found in *Load and Resistance Factor Design Manual for Engineered Wood Construction* [1] and *Load and Resistance Factor Design: Example Problems for Wood Structures* [8].

Example 7.1 Glued Laminated Timber Beam

Given: The conditions and requirements of Example 4.1: a residential live load of 40 psf, a dead load of 15 psf, a tributary span of 12 ft increased 25% for continuous framing over the beam, and deflection limitations per Table 3.11. The beam is fully supported by framing on the compression side.

Determine: Suitable 24F-1.8E Douglas fir and southern pine glued laminated timber sizes using the load and resistance factor design approach (LRFD).

Approach: In this example, the solution of Example 4.1 will be verified in three ways: (1) using the load and span tables in the *Structural Glued Laminated Timber Supplement* to the *Load and Resistance Factor Design Manual for Engineered Wood Construction* [1]; (2) using the reference strength and section properties from that supplement; and (3) using tabulated reference resistance and reference stiffness values for glued laminated timber obtained in the product supplement of the *LRFD Manual.*

Solution (Load and Span Tables): Table 7.1 of the *Structural Glued Laminated Timber Supplement* to the *Load and Resistance Factor Design Manual* [1] provides design loads for simple-span Douglas Fir–Larch (24F-V4/WS) and Southern Pine (24F-V3/SP) glued laminated timber beams for both strength and serviceability considerations. The reference conditions for these tables include $\lambda = 0.8$ (occupancy load).

Considering strength, factored loads are used. From Table 7.4, the appropriate factored load combination $1.2D + 1.6L$ and time effect factor is $\lambda = 0.8$ (occupancy). From Example 4.1, the uniform loads on the beam were found to be DL = 314 plf (including the beam self-weight), LL = 800 plf, and total load = 1114 plf. Since the time effect factor for the example problem is identical to that for *Supplement* Table 7.1, the table may be used directly. The factored uniform load on the beam is (1.2) (314 plf) + (1.6) (800 plf) = 1657 plf.

From Table 7.1 in the *LRFD Manual* [1] *Supplement,* for a $5\frac{1}{8}$-in. width and a span of 12 ft, beam depths of 10.5 and 12 in. are found to be acceptable for Douglas Fir–Larch (having maximum factored total loads of 1808 and 2362 plf, respectively). The footnotes to *Supplement* Table 7.1 indicate that both sizes are governed by bending strength (not shear). For Southern Pine 5 in. width, acceptable depths of 11 and $12\frac{3}{8}$ in. are found, also governed by bending.

For deflection, first under live load, the unfactored live load of 800 plf is considered. For the Douglas Fir–Larch with a beam width of $5\frac{1}{8}$ in., the 10.5-in. depth is found not to be satisfactory, with a maximum unfactored load of 763 plf for a 12-ft span. The 12-in. depth is satisfactory, however, with a maximum unfactored load of 1139 plf. For the Southern Pine beam 5-in. width, both the 11- and $12\frac{3}{8}$-in. depths are satisfactory.

Table 3.11 also requires that the deflection due to applied load (in this case, live) plus dead load not exceed $\ell/240$. *Supplement* Table 7.1 is based on $\ell/360$. The footnotes to the table indicate that for $\ell/240$, the values are to be multiplied by 360/240 = 1.5. Alternatively, to obtain a table value, the total unfactored load for the example problem may be divided by 1.5 to obtain an equivalent maximum unfactored table value of 1114 plf/1.5 = 743 plf. For the $5\frac{1}{8}$-in.-wide Douglas fir beam, the 12-in. depth is acceptable. For the 5-in.-wide southern pine beam, both the 11- and $12\frac{3}{8}$-in. depths are acceptable.

The acceptable sizes meeting the strength consideration and both serviceability considerations are thus:

Douglas Fir–larch:

$5\frac{1}{8}$ in. × 12 in., F_b = 6.10 ksi; F_v = 0.545 ksi; E_x = 1800 ksi (24F-V4/WS)

Southern pine:

5 in. × 11 in., F_b = 6.10 ksi; F_v = 0.575 ksi; E_x = 1800 ksi (24F-V3/SP)

Alternative Solution (Reference Strength): The 5 in. × 11 in. southern pine beam will also be checked using section properties and reference strength values. Bending and shear will be checked by this method. The serviceability (deflection) check is identical with that of allowable stress design. From the AF&PA/ASCE 16-95 standard [7], for bending,

$$\lambda\phi_b M' = \lambda\phi_b F'_b S \geq M_u$$

where

λ = 0.8 (occupancy live load),
ϕ_b = 0.85 (flexure, Table 7.3),
M' = adjusted moment resistance, = $F'_b\ S$,
M_u = factored moment,
$F'_b = F_b C_M C_t C_L C_{fu} C_r C_{FV} \ldots,$
F_b = reference bending strength,

and $C_M C_t C_L C_{fu} C_r C_F C_V \ldots$ are the applicable adjustment factors.

The reference bending strength is obtained as F_b = 6.10 ksi from Table 3.1 in the *Structural Glued Laminated Timber Supplement* to the *LRFD Manual* [1]. In this example the applicable adjustment factors reduce to C_V (C_L = 1.00). The volume factor, C_V, is identical to the volume factor in allowable stress design and may be calculated or obtained from tabular values also in the *Supplement*. From Table 4.8 in the Supplement, the volume factor is found to be 1.00, based on a width, depth, and span of 5 in., 11 in., and 12 ft, respectively. Thus,

$$F'_b = F_b = 6.10 \text{ ksi}$$

The section properties for standard-size glued laminated timber beams may be found in the Appendix, AITC 117 [4], and the *Structural Glued Laminated Timber Supplement* to the *LRFD Manual* [1], or may be calculated directly as follows:

$$A = bd = (5 \text{ in.})(11 \text{ in.}) = 55 \text{ in.}^2$$

$$I = \tfrac{1}{12}\, bh^3 = \tfrac{1}{12}\,(5 \text{ in.})(11 \text{ in.})^3 = 555 \text{ in}^4$$

$$S = I/c$$

where $c = d/2$ = 11 in./2 = 5.5 in.; thus, S = 555 in^4/5.5 in. = 101 in^3. Therefore,

$$\lambda\phi_b M' = \lambda\phi_b F'_b S = (0.80)(0.85)(6.10 \text{ ksi})(101 \text{ in}^3) = 419 \text{ kip-in.}$$

The factored bending load may be calculated as follows: $M_u = \omega_u \ell^2/8$, where $\omega_u = 1657$ plf, from above, giving

$$M_u = \frac{(1657 \text{ plf})\ (12 \text{ ft})^2}{8} = 29{,}830 \text{ lb-ft} = 358{,}000 \text{ lb-in.} = 358 \text{ kip-in.}$$

Since $\lambda\phi_b M' = 419$ kip-in. $\geq M_u = 358$ kip-in., the 5 in. $\times$ 11 in. southern pine member is acceptable in bending.

The design check for shear using reference strength values is, from the AF&PA/ASCE 16-95 standard,

$$\lambda\phi_v V' \geq V_u$$

where

$\lambda = 0.80$ (from before),
$\phi_v = 0.75$ (from Table 7.3),
V' = adjusted shear resistance parallel to the grain,

and

V_u = the factored shear load.

For rectangular sections,

$$\lambda\phi_v V' = \lambda\phi_v \left(\tfrac{2}{3} F'_v A\right)$$

where

$F'_v = F_v C_M C_t \ldots,$
F_v = reference shear strength,
$C_M C_t \ldots$ are the applicable adjustment factors (in this case, all unity or not applicable),
A = the cross-sectional area, in this example 55 in^2.

From Table 3.1 in the *Structural Glued Laminated Timber Supplement* to the *LRFD Manual* [1], $F_v = 0.575$ ksi. Thus,

$$\begin{aligned}\lambda\phi_v V' &= \lambda\phi_v \left(\tfrac{2}{3} F'_v A\right) = (0.80)(0.75)\left[\tfrac{2}{3}\ (0.575 \text{ ksi})\ (55 \text{ in}^2)\right] \\ &= 12.65 \text{ kips.}\end{aligned}$$

The factored shear load, omitting loads within distance d of the support, is

$$V_u = \frac{\omega_u \ell}{2} - \omega_u d = \frac{(1657 \text{ plf}) \ (12 \text{ ft})}{2} - (1657 \text{ plf}) \ (11 \text{ in.})(1 \text{ ft}/12 \text{ in.})$$

$$= 8423 \text{ lb} = 8.42 \text{ kips}.$$

Since $\lambda\phi_v V' = 12.65$ kips $\geq V_u = 8.42$ kips, the 5 in. × 11 in. southern pine section is also adequate with respect to shear.

Checking deflection with respect to live load,

$$\Delta = \frac{5\omega\ell^4}{384EI}$$

where E is obtained from Table 3.1 in the *Structural Glued Laminated Timber Supplement* to the *LRFD Manual* [1] as 1800 ksi (identical to the allowable stress design value). Hence, the live load deflection check becomes

$$\Delta_{\text{LL}} = \frac{5\omega_{\text{LL}}\ell^4}{384EI} = \frac{(5)(0.80 \text{ kip/ft})(1 \text{ ft}/12 \text{ in.})[12 \text{ ft } (12 \text{ in./ft})]^4}{(384)(1800 \text{ ksi})(555 \text{ in}^4)} = 0.37 \text{ in.}$$

Since 0.37 in. $\leq \ell/360$ = (12)(12) in./360 = 0.40 in.; the 5 in. × 11 in. Southern Pine beam is acceptable considering live load deflection.

The total load deflection check may be performed as follows.

$$\Delta_{\text{TL}} = \frac{5\omega_{\text{TL}}\ell^4}{384EI} = \Delta_{\text{LL}} \frac{\omega_{\text{TL}}}{\omega_{\text{LL}}} = 0.37 \text{ in.} \left(\frac{1.114 \text{ kips/ft}}{0.80 \text{ kip/ft}}\right) = 0.52 \text{ in.}$$

Since 0.52 in. $\leq \ell/240$ = [(12)(12) in.]/240 = 0.60 in.; the 5 in. × 11 in. southern pine beam is also acceptable considering total load deflection.

Alternative Solution: (*Reference Resistance and Reference Stiffness*): Design checks and beam selection for the bending moment in wood members may be performed using

$$\lambda\phi_b M' \geq M_u$$

where $\lambda\phi_b M'$ is the factored resistance and may be obtained from tables in the product supplements to the *LRFD Manual* [1] or from the product manufacturer.

The factored resistance values are typically a function of member grade and section properties. When the service conditions match the reference conditions of the tables, the factored resistance may be read directly. In the case of glued laminated timber, since member resistance is also a function of member length, factored resistance values are not tabulated directly. Instead, reference resistance values, $\lambda\phi_b M$, are published and must be adjusted for length (volume) and all other applicable adjustment factors. As such, for glued laminated timbers,

$$\lambda\phi_b M' = (\lambda\phi_b M)C_M C_t C_L C_V C_{fu} \ldots \quad \text{(size factor } C_F \text{ and form factor } C_f \text{ not applicable)}$$

In the present example, the 5 in. × 11 in. 24F southern pine beam, from Table 5.2 in the *Structural Glued Laminated Timber Supplement* of the *LRFD Manual* [1], has a factored reference resistance, $\lambda\phi_b M = 418$ kip-in. This value must be multiplied by the volume factor, which has already been shown to be 1.00 in this example. All other adjustment factors in this example are also 1.00 (or not applicable). Thus,

$$\lambda\phi_b M' = (\lambda\phi_b M)C_M C_t C_L C_V C_{fu} \ldots = (418 \text{ kip-in.})\,(1.00)(1.00) \ldots$$
$$= 418 \text{ kip-in.}$$

This value is identical to the value obtained using the reference bending strength and section properties above.

Similarly, for shear design checks or member selection,

$$\lambda\phi_v V' \geq V_u$$

where

$$\lambda\phi_v V'_u = (\lambda\phi_v V)C_M C_t \ldots$$

and where the factored reference shear resistance values ($\lambda\phi_v V$) are published and must be adjusted by all applicable adjustment factors. In the present example, for the 5 in. × 11 in. southern pine beam, for $F_v = 0.575$ ksi, the factored reference shear resistance is 12.65 kips, identical to the value calculated above.

For deflection calculations, member stiffness (EI) may also be obtained from the tables in the product supplements to the *LRFD Manual* [1]. In this example, for the 5 in. × 11 in. southern pine beam with $E = 1800$ ksi, the corresponding EI is $(998.3)(10)^3$ kip-in^2, which is identical to the product of E and I in the calculations above; (1800 ksi)(555 in^4) = 998,000 kip-in^2.

Answer: For the conditions of Example 4.1, $5\frac{1}{8}$ in. × 12 in. Douglas fir (24F-V4/WS) and 5 in. × 11 in. Southern Pine (24F-V3) glued laminated timbers were found acceptable using the load and resistance factor design approach. Members were checked using load and span tables, reference strength values, and factored reference resistance values, all producing identical results.

Discussion: The load and span tables for *Structural Glued Laminated Timber Supplement* for the *Load and Resistance Factor Design Manual for Engineered Wood Construction* [1] are applicable for simply supported uniformly loaded members with specific reference conditions. These tables already take the volume factor into consideration. For members subject to other support and loading conditions, design checks using reference strength values and section properties or factored reference resistance values must be used. In the latter cases the volume factor must be calculated or obtained separately in the design check process.

7.9 CONCLUSION

In many cases the load and resistance factor design (LRFD) approach will produce identical or nearly identical results to those using the allowable stress design approach (ASD), particularly since structural members are manufactured in discrete sizes and commonly accepted grades. In cases where dead loads account for large proportions of total loads, the load and resistance factor design approach may result in smaller members (greater economy) since the LRFD approach more accurately distinguishes load uncertainty between live and dead loads (e.g., load factors of 1.2 for dead load and 1.6 for live load, as in the preceding example). The differences between the application of the time effect factor (LRFD) and the load duration factor (ASD) may also lead to differences in member selection. However, in the design of any single system, only one approach, either ASD or LRFD, should be used. Although the LRFD approach more accurately accounts for uncertainty in loads (and resistance) and might thus be considered to provide "better" results. The trade-off is that the method is more complex.

CHAPTER 8

REFERENCE INFORMATION

8.1 TYPICAL CONSTRUCTION DETAILS

AMERICAN INSTITUTE OF TIMBER CONSTRUCTION

7012 South Revere Parkway - Suite 140 - Englewood, Colorado 80112 - Telephone 303/792-9559

AITC 104-2003
TYPICAL CONSTRUCTION DETAILS

Adopted as Recommendations April 27, 2003

Copyright 2003 by American Institute of Timber Construction

CONTENTS

1. INTRODUCTION

1.1 These typical construction details are intended as guides for architects and engineers. They have been developed and used by the engineered timber construction industry and, being based on judgement and experience, will help to assure a high quality of construction.

1.2 **Warnings:** Because the details are to be used only as guides, dimensions have not been included and the drawings should not be scaled. Quantities and sizes of bolts, connectors and other fastening hardware are illustrative only. The actual quantities and sizes required will depend on the loads to be carried and the member sizes. End and edge distances, as well as spacing between fasteners, should be in accordance with the *National Design Specification for Wood Construction* by the American Forest & Paper Association. Sufficient clearance must be provided between sides of steel connection hardware and wood members to permit installation. This clearance should not exceed the member width plus 1/4 in.

1.3 **Designing for Strength.** Connection details must effectively transfer loads, utilize durable materials, and be as free from maintenance as possible. The strength of wood is different in the parallel and perpendicular grain directions. Wood also has much less strength in tension perpendicular to grain than in compression perpendicular to grain. These facts influence design details.

Vertical loads should be transferred so as to take advantage of the high compression perpendicular to grain strength of wood. For example, a beam should bear on the top of a column or wall or be seated in a shoe or hanger. Such a detail is preferred to the support of a beam by bolts at its end, particularly where there are large numbers of bolts.

Beams should be anchored at the ends in order to carry induced horizontal and vertical loads. Vertical loads may be either gravity loads or net uplift loads. The connections typically shown in this standard are primarily for vertical gravity loads. Provisions should be made to resist uplift or lateral loads as required. The bolts or fasteners at the beam ends must be located near the bottom bearing of the beam to minimize the effect of shrinkage of the wood member between the bottom of the beam and the fasteners.

In many cases, individual details do not include all structural elements, such as lateral bracing ties to connect all of the components of the building together.

Loads suspended from glued laminated timber beams or girders should preferably be suspended from the top of the member or above the neutral axis.

1.4 Consideration of Shrinkage and Swelling. In addition to designing connections to transfer loads, effort should be made to avoid splitting the member due to expansion and contraction of the wood. Consideration must be given to wood swelling and shrinking due to moisture content changes in service, similar to the consideration given to details in metal construction that must accommodate the expanding and contracting metal due to changes in temperature.

Because wood swells and shrinks (primarily in the perpendicular to grain directions) due to moisture content changes, connections should not restrain this movement. Figure 1.1 illustrates typical shrinkage in a sawn member when drying from green to 8% moisture content and a glued laminated timber drying from 12% to 8% moisture content. Even in covered structures, large laminated timbers may shrink after installation due to moisture loss from low relative humidity conditions. Long rows of bolts perpendicular to grain fastened to a single cover plate should be avoided. Although relatively dry at time or manufacture, glued laminated timber can still shrink to reach equilibrium moisture content in service.

When possible, designers should avoid joint details that could loosen in service due to wood shrinking or that could cause problems when wood expands due to increased moisture content. Machine bolts should be used rather than lag bolts whenever possible. When lag bolts are used, correct lead hole sizes are important. Connections should be detailed to avoid loading lag bolts in withdrawal whenever possible.

Figure 1.1 Shrinkage due to moisture loss.

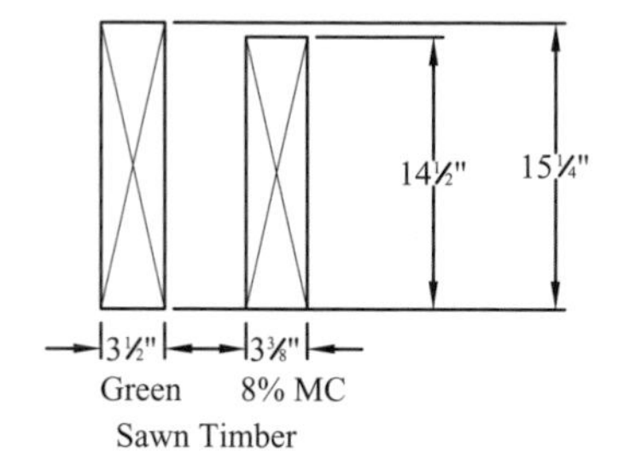

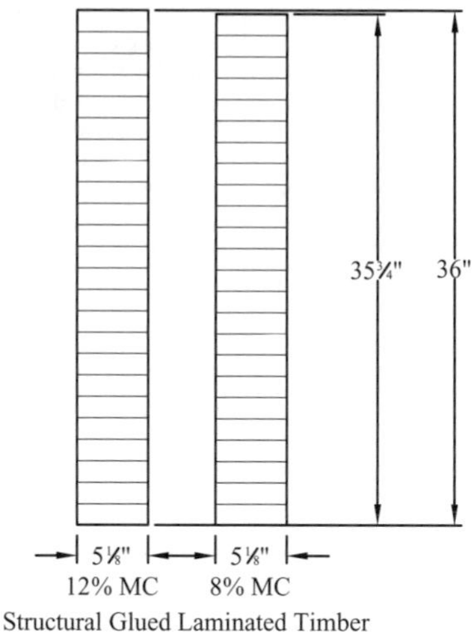

1.5 Designing to Avoid Tension Perpendicular to Grain Stresses. Whenever possible, joints should be designed to avoid causing tension perpendicular to grain stresses in wood members. Examples of connections that induce tension perpendicular to grain stresses are simple beams, which have been notched at the ends on the tension side (see Detail A5).

Long lines of fasteners spaced close together along the grain should be avoided, particularly if the bolts are in tightly drilled holes. These types of connections may induce tension perpendicular to grain stresses due to prying actions from secondary moments.

1.6 Consideration of Decay. When proper construction details are used and other good design and construction practices are followed, wood is a permanent construction material. Moisture barriers, flashings and other protective features should be used to prevent moisture or free water being trapped (see Section 11, "Details to Protect Against Decay"). Preservative treatments are recommended when wood is fully exposed to the weather without roof cover. Materials should be protected during construction. Arch and column bases should not be embedded below finished concrete floor levels. Arch and column bases should be elevated a minimum of one inch above the concrete floor level if there is potential for wetting of the floor.

1.7 Other Considerations. It is important to provide adequate site drainage to prevent moisture problems. Fasteners and connection hardware should be protected from corrosion by using corrosive-resistant metals or resistant coatings or platings. Additional guidelines for specifying and designing with structural glued laminated timber are outlined in other AITC standards and publications.

1.8 End Rotation of Beams. Consideration should be given to end rotation of beams resulting from vertical load deflection. Location of fasteners that tend to create end fixity should be avoided. Splitting at fasteners can result unless such a connection is designed to develop a fixed end moment sufficient to resist end rotation due to deflection.

2. BEAM TO MASONRY ANCHORAGES

2.1 Figures 2.1 through 2.10 illustrate various beam to masonry anchorages. The seats illustrated may be anchored in the concrete or masonry with one or more anchors. An important point illustrated in several of the figures is that the timber member should be separated from the masonry or concrete by a minimum of 1/2 in. clearance on all sides. In the case of net uplift loads, notched beam shear should be checked.

2.2 For beams sloped where a 1/8 in. or greater gap might occur, the beam bearing should be detailed to obtain full contact between bearing surfaces (see Figures 2.6 and 2.7). Seat cuts at the top end of the slope should be checked for notched beam effect.

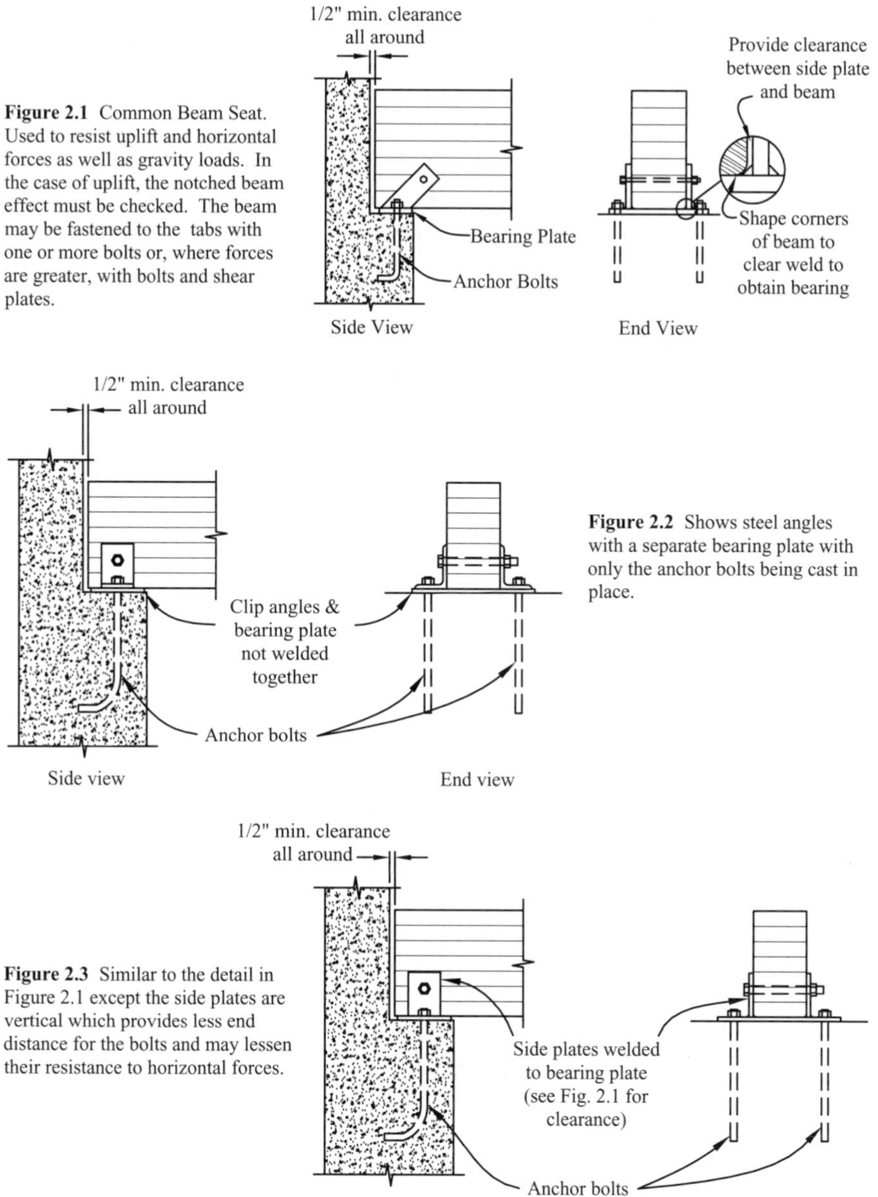

Figure 2.1 Common Beam Seat. Used to resist uplift and horizontal forces as well as gravity loads. In the case of uplift, the notched beam effect must be checked. The beam may be fastened to the tabs with one or more bolts or, where forces are greater, with bolts and shear plates.

Figure 2.2 Shows steel angles with a separate bearing plate with only the anchor bolts being cast in place.

Figure 2.3 Similar to the detail in Figure 2.1 except the side plates are vertical which provides less end distance for the bolts and may lessen their resistance to horizontal forces.

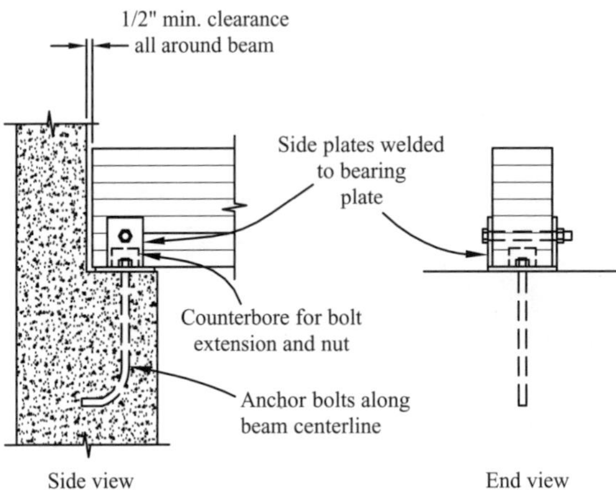

Figure 2.4 This detail may be used when the pilaster is not wide enough for outside anchor bolts. The anchor(s) may be welded to the underside of the bearing plate or the bolts and nuts may be located in holes counterbored into the bottom of the beam. This detail is for use where the width of the connection must be minimized or appearance is important.

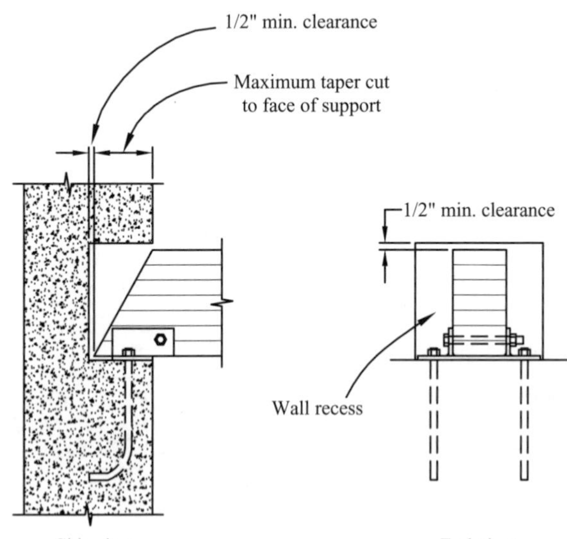

Figure 2.5 This detail illustrates a typical taper end cut sometimes referred to as a fire cut. The end of the beam is tapered so that the top of the beam does not hit the top of the wall recess if the beam deflects excessively during a fire.

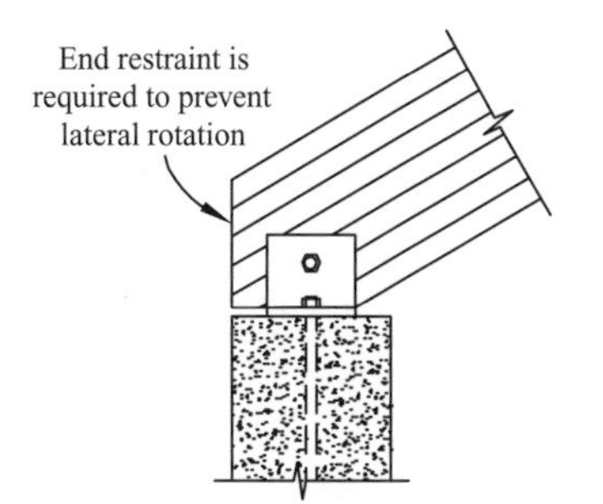

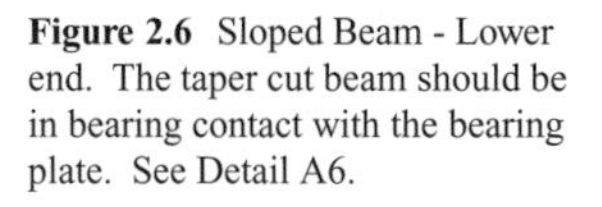
Figure 2.6 Sloped Beam - Lower end. The taper cut beam should be in bearing contact with the bearing plate. See Detail A6.

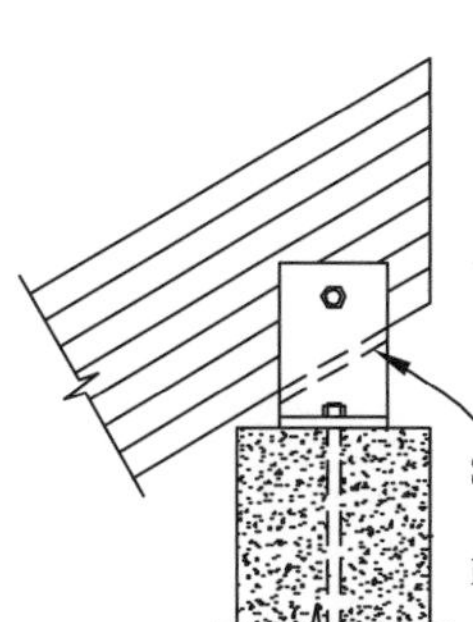

Figure 2.7 Sloped Beam - Upper End. The support at the top end of a sloped member should be designed with a sloping seat rather than a notched end. The bolt must be designed to resist the parallel-to-grain component of the vertical beam reaction. See Detail A7.

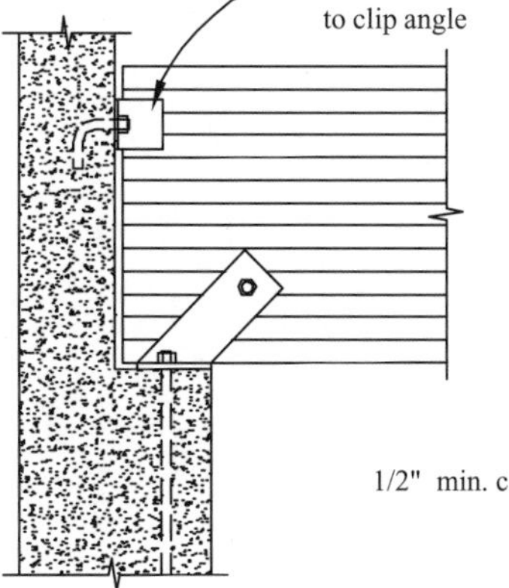

Figure 2.8 Lateral support of the ends of beams can be provided with clip angles anchored to the wall but without a connection to the beam. This will not restrain vertical movement due to end rotation or beam shrinkage. See Detail A8.

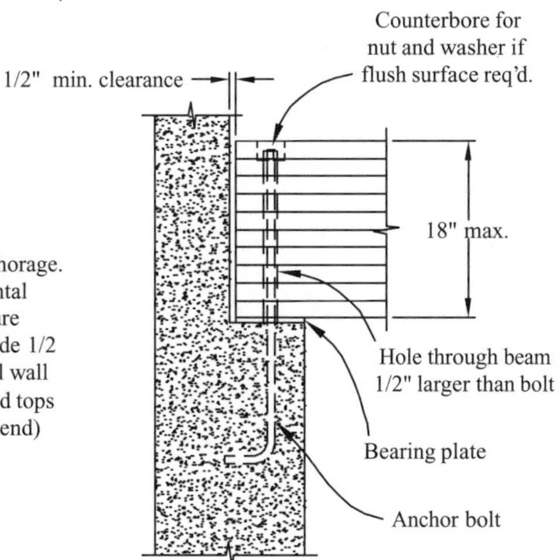

Figure 2.9 Simple Beam Anchorage. Resists small uplift and horizontal forces. Bearing plate or moisture barrier is recommended. Provide 1/2 in. minimum clearance from all wall contact surfaces, ends, sides and tops (if masonry exists above beam end)

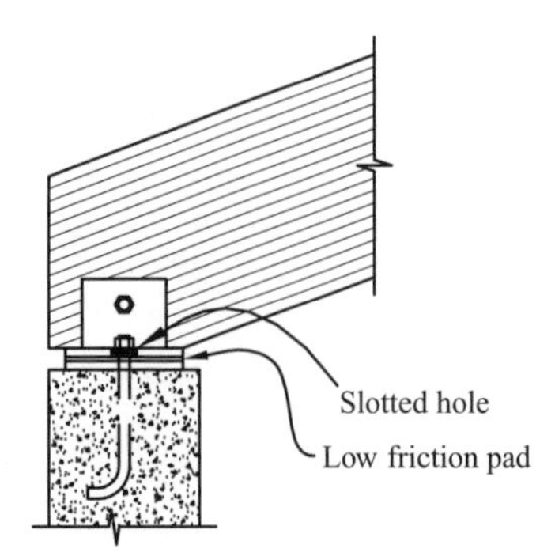

Figure 2.10 Curved or Pitched Beam Anchorage - Typical Slip Joint. Slotted or oversize holes at one or both ends of beam permit horizontal movement under lateral deflection or deformation. Length of slotted or oversized hole is based on calculated maximum horizontal deflection. Position bolt in slot to allow for anticipated movement. Bolts in slotted holes should be hand tightened only, so that movement is permitted. See Detail A6.

Wind or seismic loading may not permit use of this connection.

3. CANTILEVER BEAM CONNECTIONS

Figure 3.1 Cantilever Hinge Connection. See Detail A9.

This is a common type cantilever beam connector. The details that follow are examples using this connector.

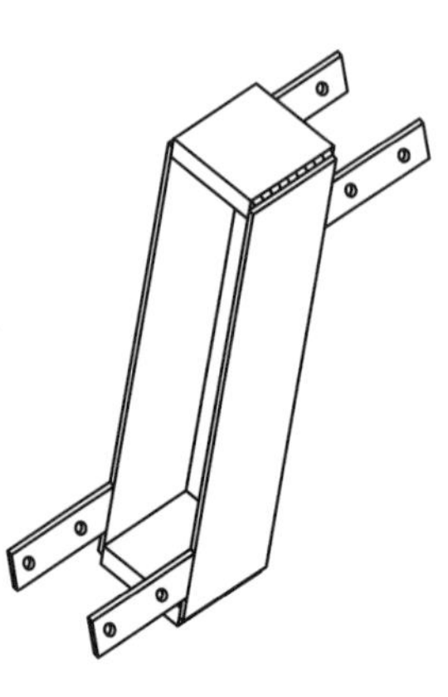

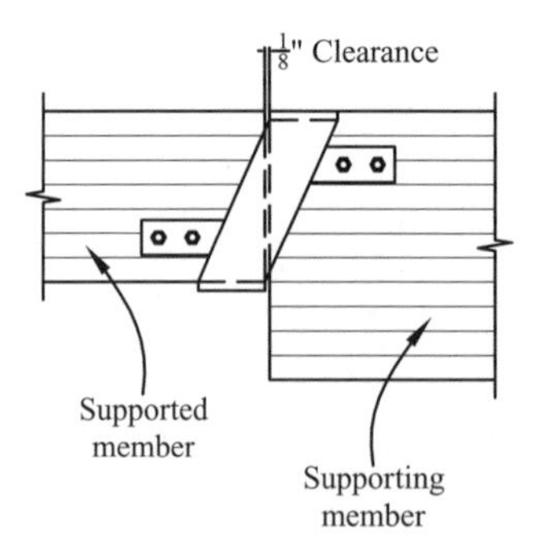

Figure 3.2 The vertical reaction of the supported member is carried by the side plates and transferred in bearing perpendicular grain to the supporting member. The rotation due to the eccentric loading is resisted by the bolts through the tabs at the top and bottom. The connector may be installed with the top (and bottom) bearing plates dapped into the members to obtain a flush surface or may be installed without daps. Notching on the tension side should be minimized.

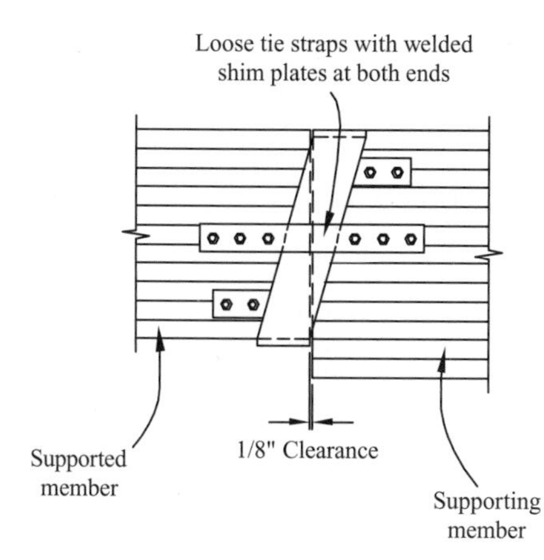

Figure 3.3 Where horizontal forces must be resisted by a hinge connection, loose tension ties may be installed on both sides of the beam. The tie shown is not fastened to the cantilever hanger.

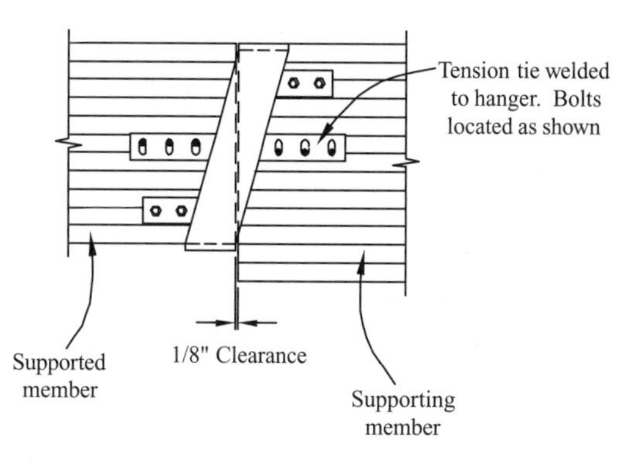

Figure 3.4 If tension ties are fastened to the cantilever hanger, vertically slotted holes are required in the tie and careful location of the bolts in the end of the slot farthest from the bearing seat is required to prevent splitting due to shrinkage and seating deformations. Bolts in slotted holes should be hand tightened only, so that movement is permitted.

4. **BEAM AND PURLIN HANGERS FOR ROOF SYSTEMS.** In Figure 4.1 and similar details, locate fasteners as close as practical to the bearing surface to minimize splitting due to shrinkage (See Section 1.4). For floor systems, additional restrictions may be necessary to minimize the effects of differential shrinkage of connected members. See Detail A1.

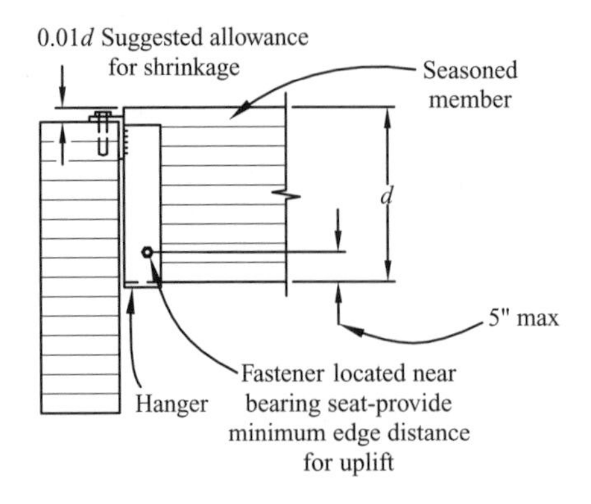

Figure 4.1 Seasoned Members. When supported members are of seasoned material, the top of the supported member may be set approximately flush with the top of the supporting member.

Figure 4.2 Unseasoned Members. When supported members are of unseasoned material, the hangers should be so dimensioned that the top edge of the supported member is raised above the top of the supporting member or the top of the hanger strap to allow for shrinkage as the members season in place. For supported members with moisture content at or above fiber saturation point when installed, the distance raised should be about 6% of the member's depth above its bearing point.

Note: For main members loaded on one side as in Figures 4.1, 4.2, and 4.3 provide a tie between the beam and purlin to restrain potential rotation of the beam due to eccentricity of the hanger load.

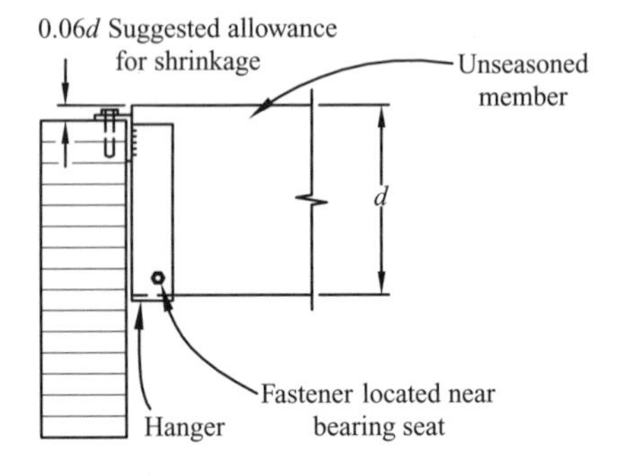

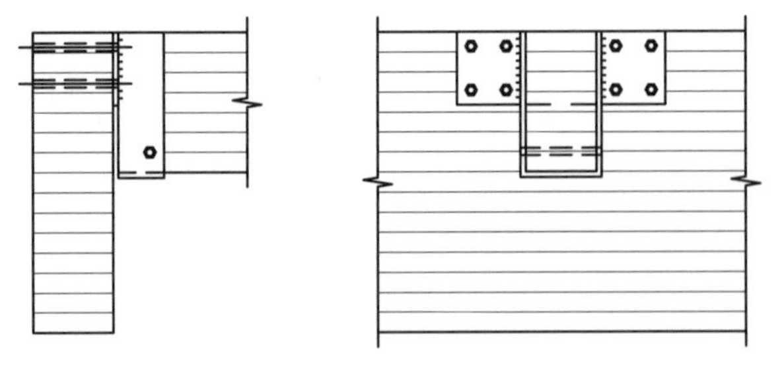

Figure 4.3 Welded Face Hanger. See Detail A2.

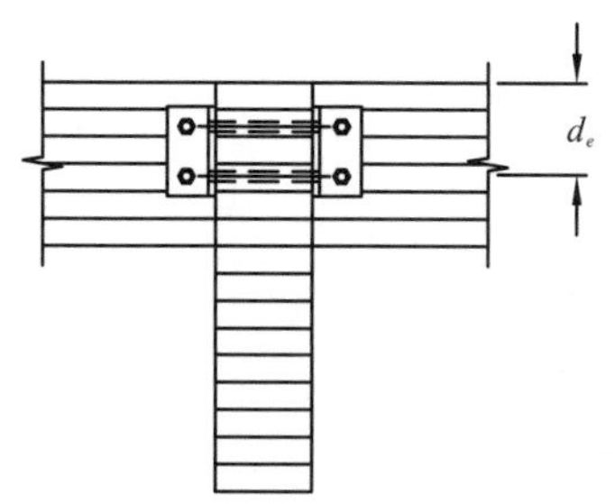

Figure 4.4 A clip angle connection without a bearing seat may be used for small beams and light loads. The connection should be designed for gravity loads as a notched beam using d_e as shown in the notched beam formula. The distance between bolts should be checked for possible effects of shrinkage. A bearing connection as shown in Figure 4.1 is preferred to the support of the beam by bolts. See Detail A3.

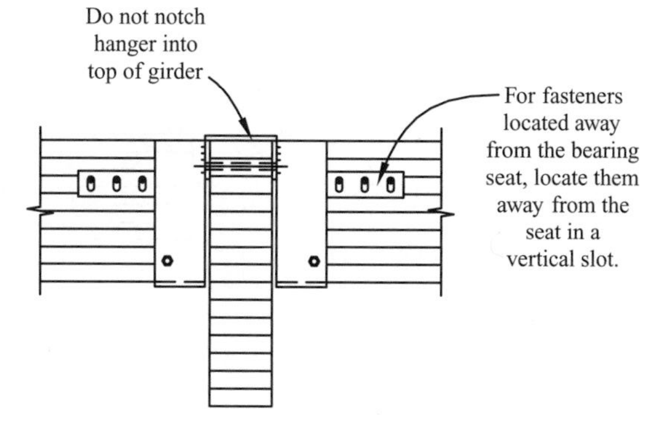

Figure 4.5 Welded and Bent Strap Hanger. A separate tension tie may be used across the top in lieu of the tabs to resist lateral forces.

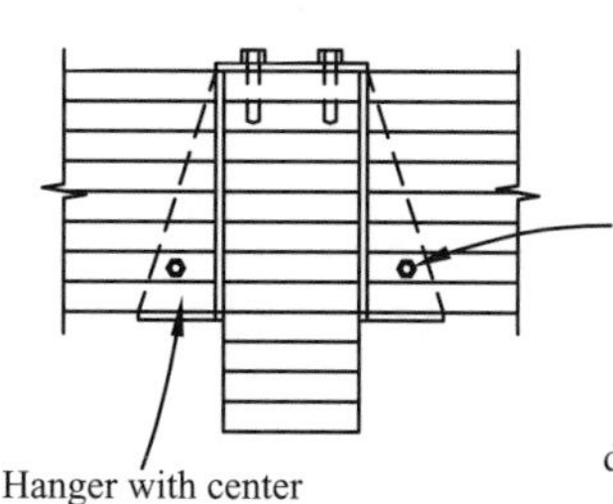

Figure 4.6 Partially Concealed Type. For moderate loads, base may be let in flush with bottoms of purlins. See Detail A4.

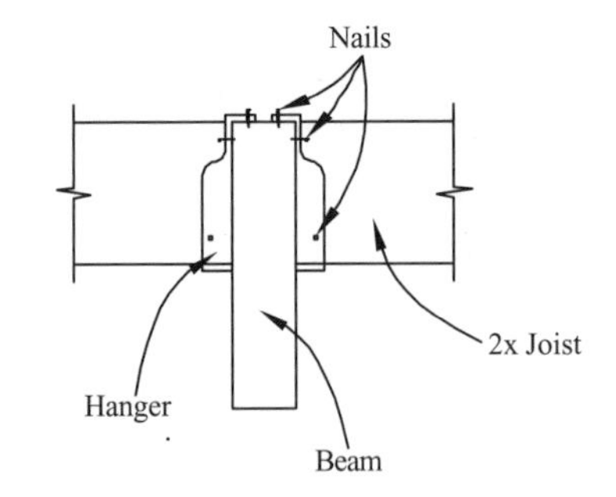

Figure 4.7 Stamped Joist Hanger for Light Loads. Stamped from light-gage metal.

5. BEAM TO COLUMN CONNECTIONS

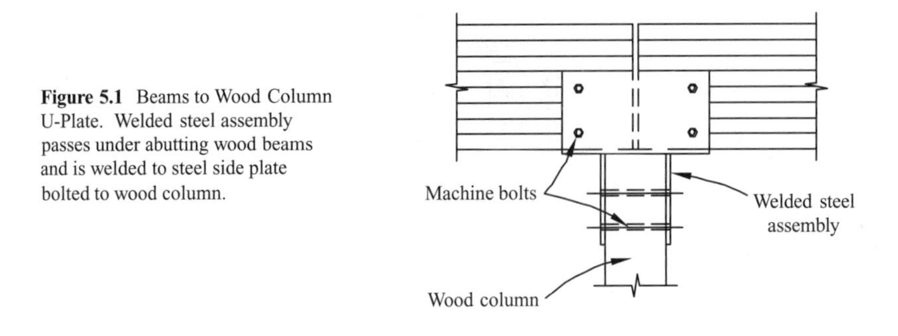

Figure 5.1 Beams to Wood Column U-Plate. Welded steel assembly passes under abutting wood beams and is welded to steel side plate bolted to wood column.

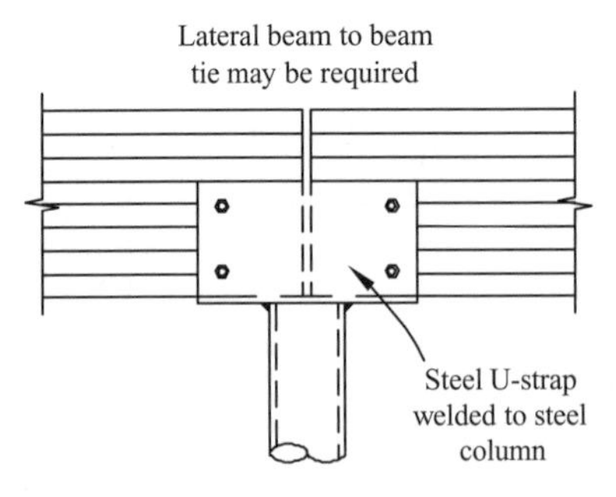

Figure 5.2 Beams to Steel Column. Similar to Figure 5.1.

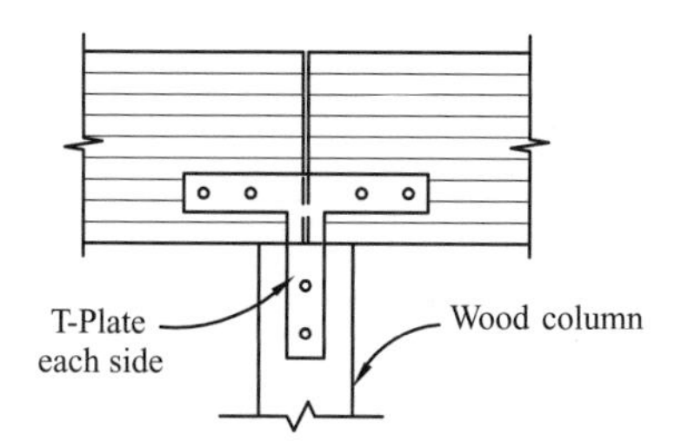

Figure 5.3 Beam to Wood Column T-Plates. Steel T-plate is bolted to abutting wood beams and to wood column. Loose bearing plate may be used where column cross sectional area is insufficient to provide bearing for beams in compression perpendicular to grain. Beams should be checked for notch shear with net uplift design.

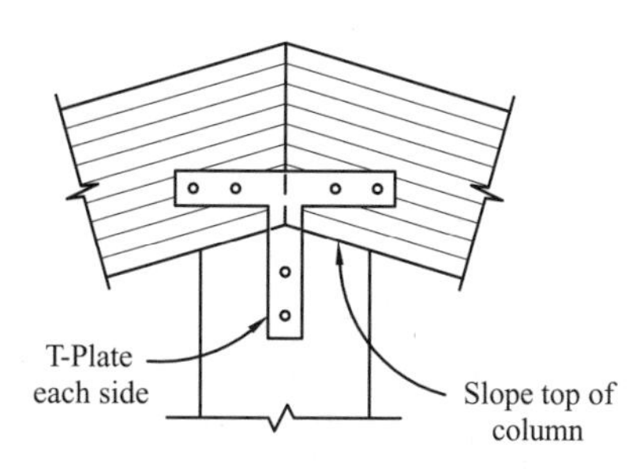

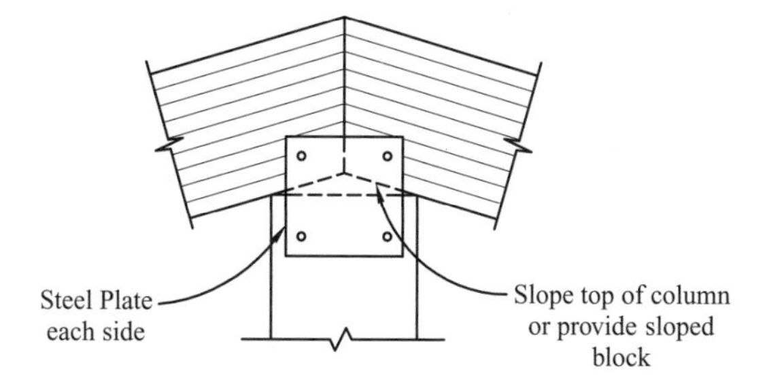

Figure 5.4 Shed Roof Type End Detail. Beams may be notched at bottom to rest on column, but notched beam shear must be checked. Beams may be pitched away from both sides or from only one side of column. See also Beam-to-Masonry Anchorages, Figure 2.7.

Figure 5.5 Beam to Wood Column. Metal bearing plate may be used where column cross-sectional area is insufficient to provide bearing for beam in compression perpendicular to grain. If connection is to be used for net uplift, beam must be checked for notched shear design.

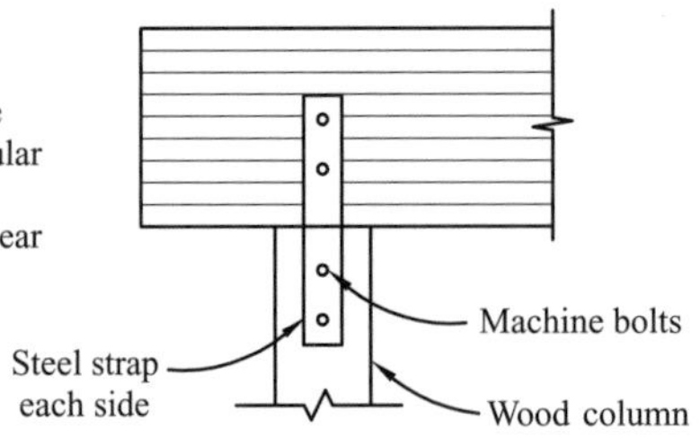

Machine bolts
Steel strap each side
Wood column

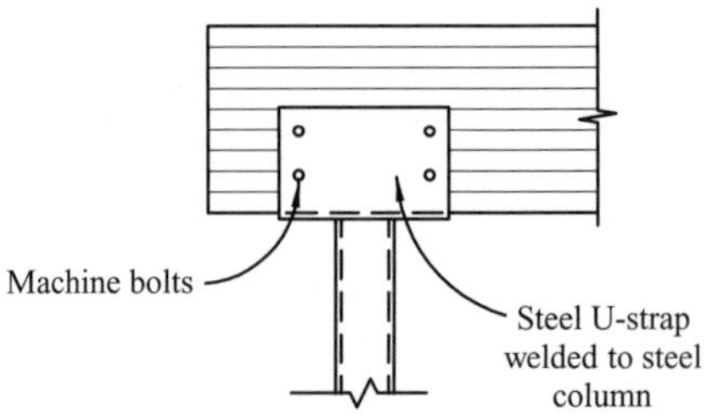

Figure 5.6 Beam to Steel Column. Steel U-strap passes under timber beam and is welded to top of steel column.

Machine bolts
Steel U-strap welded to steel column

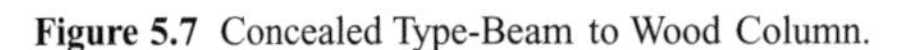

Figure 5.7 Concealed Type-Beam to Wood Column.

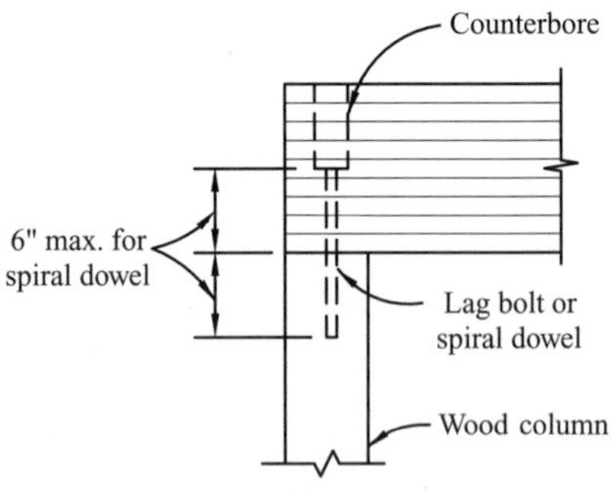

6. **COLUMN ANCHORAGES.** Column bearing elevation should be raised a minimum of one inch above finished floors that may be subjected to high moisture conditions. In locations where column base anchorages are subject to damage by moving vehicles, protection of the columns from such damage should be considered. Columns exposed to the weather or repeated wetting should be treated in accordance with *Standard for Preservative Treatment of Structural Glued Laminated Timber, AITC 109.*

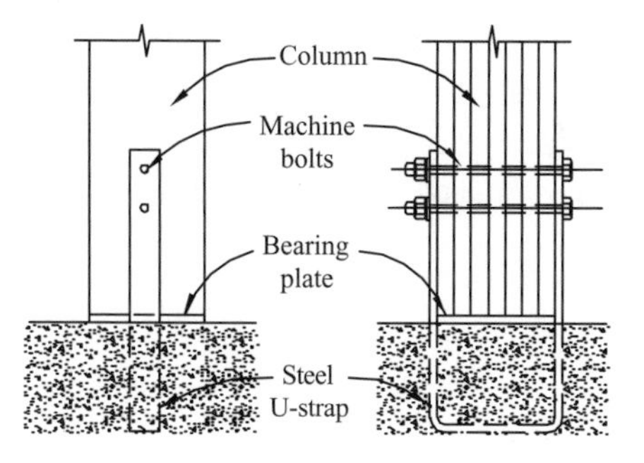

Figure 6.1 U-Strap Anchorage. Resists both horizontal forces and uplift. Bearing plate or moisture barrier is required. May be used with shear plates.

Do not place column below finished concrete floor level.

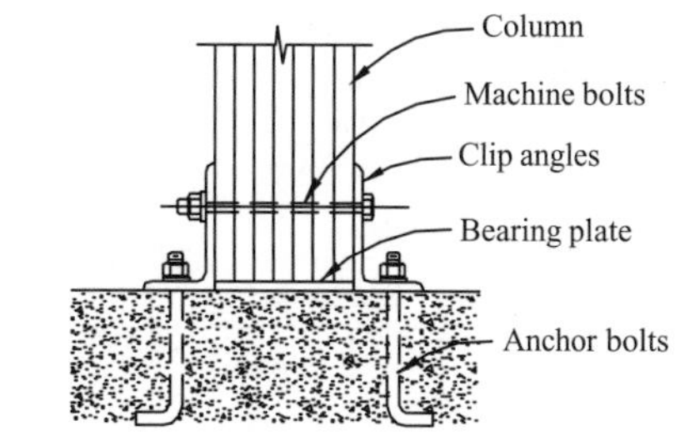

Figure 6.2 Clip Angle Anchorage to Concrete Base. Resists both horizontal forces and uplift. Bearing plate or moisture barrier is required.

Do not place column below finished concrete floor level.

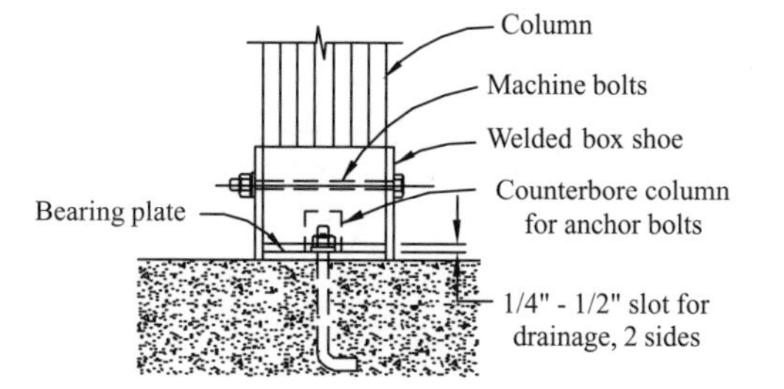

Figure 6.3 Box Shoe. For use when bottom of box is flush with top of concrete floor.

Do not place column below finished concrete floor level.

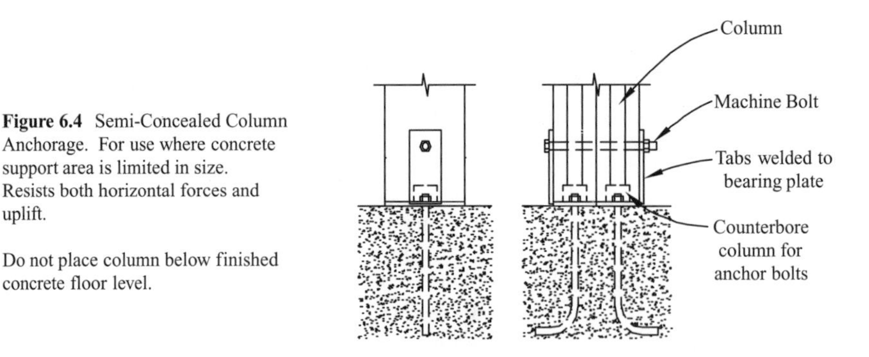

Figure 6.4 Semi-Concealed Column Anchorage. For use where concrete support area is limited in size. Resists both horizontal forces and uplift.

Do not place column below finished concrete floor level.

7. ARCH ANCHORAGES

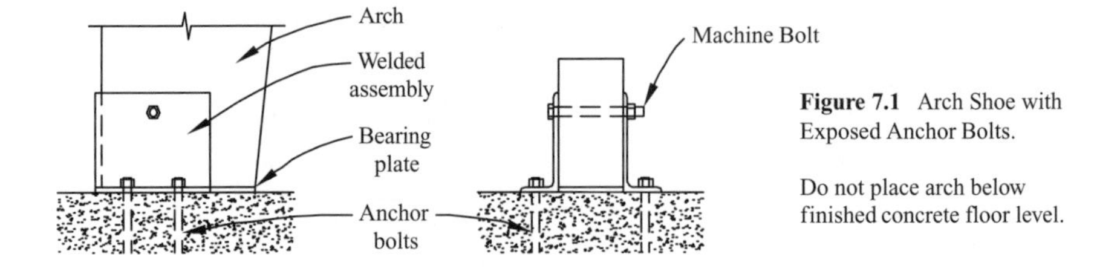

Figure 7.1 Arch Shoe with Exposed Anchor Bolts.

Do not place arch below finished concrete floor level.

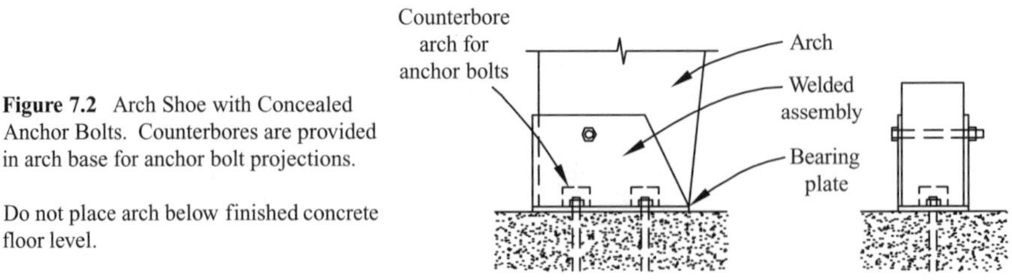

Figure 7.2 Arch Shoe with Concealed Anchor Bolts. Counterbores are provided in arch base for anchor bolt projections.

Do not place arch below finished concrete floor level.

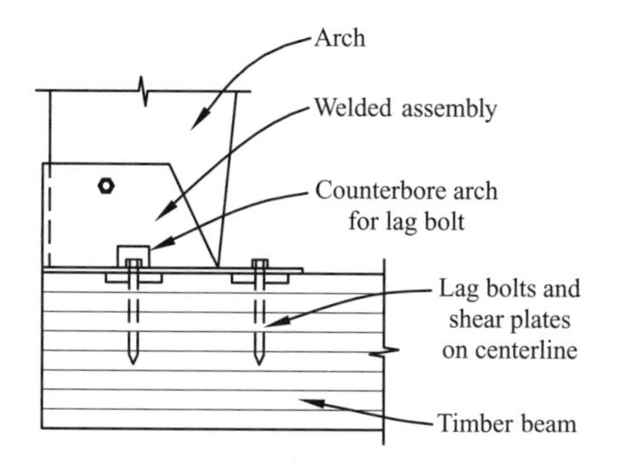

Figure 7.3 Arch Anchorage to Timber Beam. Vertical load is taken directly by bearing into timber beam. Vertical uplift and thrust are taken by the lag bolts and shear plates into the beam tie.

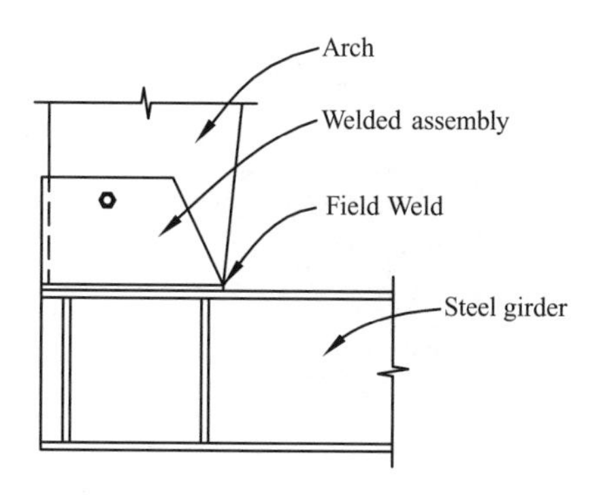

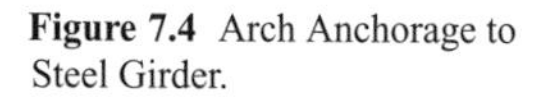

Figure 7.4 Arch Anchorage to Steel Girder.

Do not place arch below finished concrete floor level.

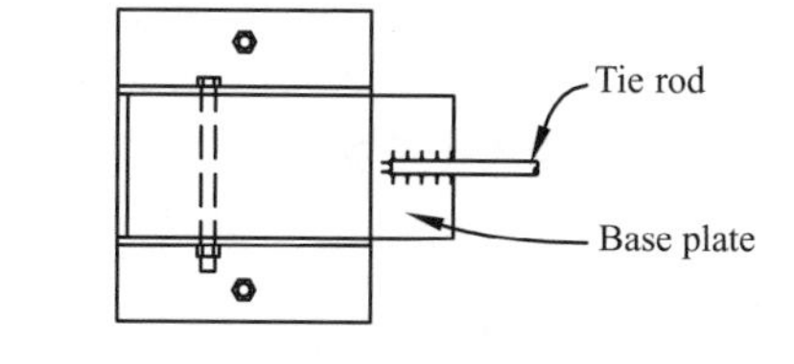

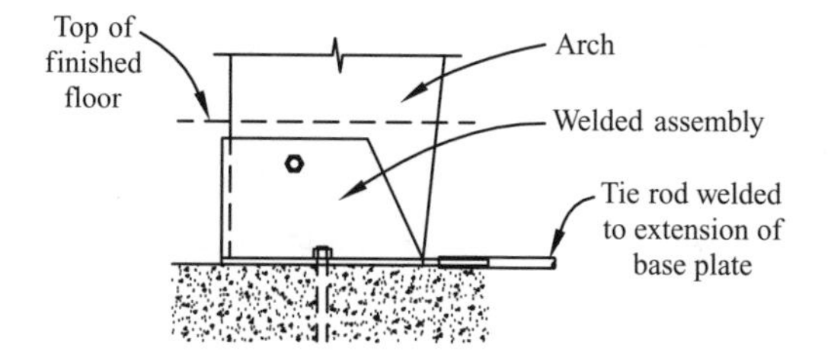

Figure 7.5 Tie Rod to Arch Shoe. Horizontal thrust is taken directly by the tie rod welded to the arch shoe. This detail is intended for use with a raised joist floor where the tie rod can be concealed.

Do not place arch below finished concrete floor level.

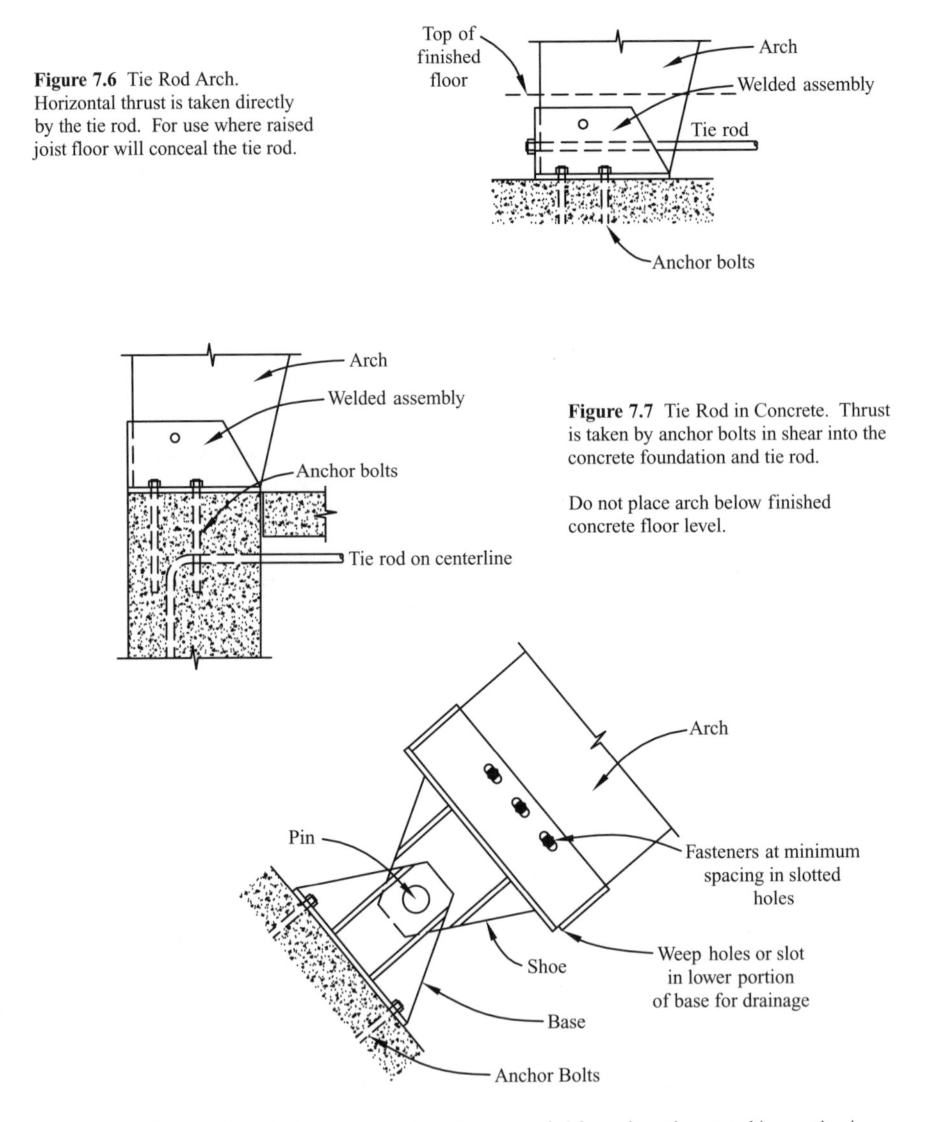

Figure 7.6 Tie Rod Arch. Horizontal thrust is taken directly by the tie rod. For use where raised joist floor will conceal the tie rod.

Figure 7.7 Tie Rod in Concrete. Thrust is taken by anchor bolts in shear into the concrete foundation and tie rod.

Do not place arch below finished concrete floor level.

Figure 7.8 True Hinge Anchorage for Arches. Recommended for arches where true hinge action is desired (see Figure 11.6 for protection considerations).

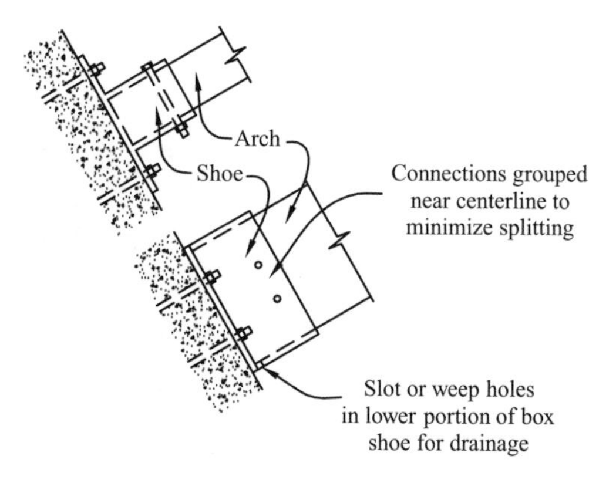

Figure 7.9 Arch Anchorage Where True Hinge Is Not Required. Recommended for arches where a true hinge is not required. Base shoe is anchored directly to buttress (see Figure 11.6 for protection considerations). Do not embed arch in concrete floor.

8. ARCH CONNECTIONS

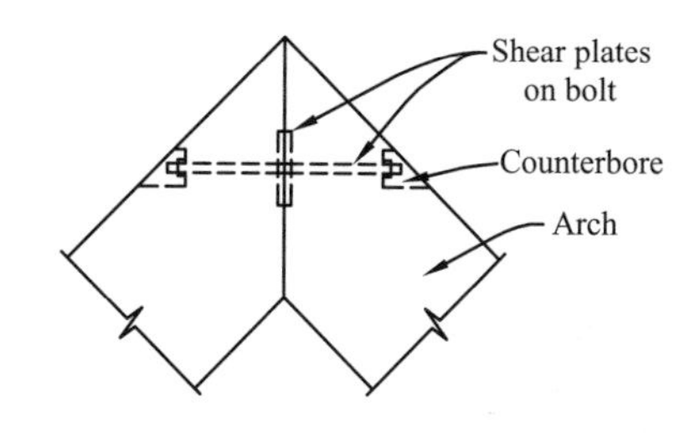

Figure 8.1 Arch Peak. For Arches with slopes of 3:12 and greater. This connection will transfer both vertical forces (shear) and horizontal forces tension and compression).

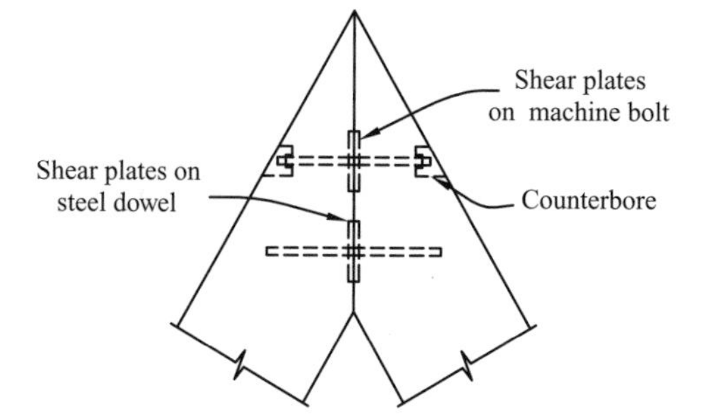

Figure 8.2 Arch Peak. When the vertical shear is too great for one pair of shear plates, or when deep sections would require extra shear plates for alignment, additional pairs of shear plates centered on dowels or machine bolts may be used.

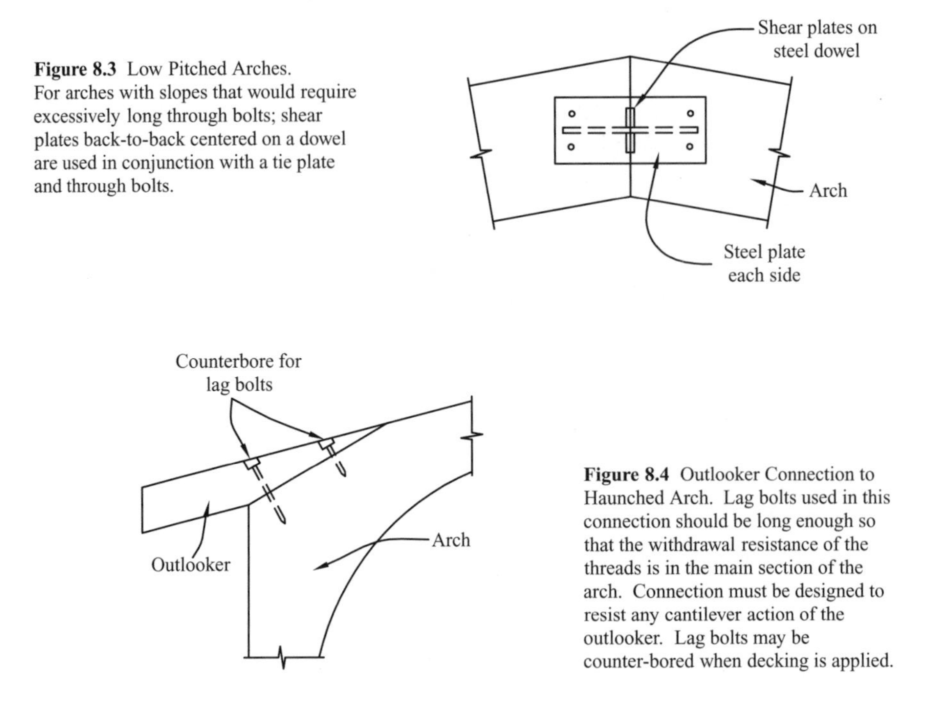

Figure 8.3 Low Pitched Arches.
For arches with slopes that would require excessively long through bolts; shear plates back-to-back centered on a dowel are used in conjunction with a tie plate and through bolts.

Figure 8.4 Outlooker Connection to Haunched Arch. Lag bolts used in this connection should be long enough so that the withdrawal resistance of the threads is in the main section of the arch. Connection must be designed to resist any cantilever action of the outlooker. Lag bolts may be counter-bored when decking is applied.

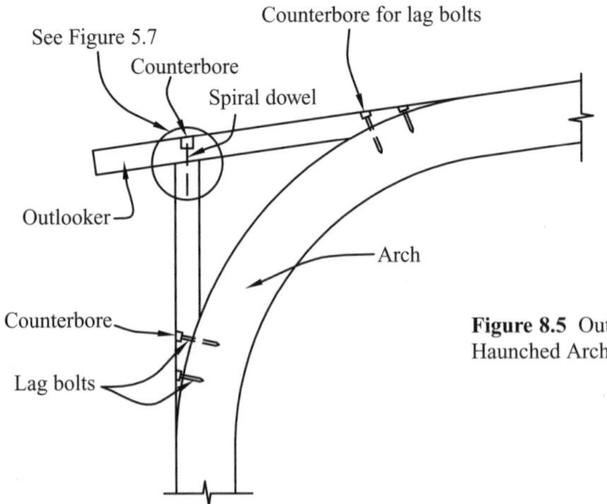

Figure 8.5 Outlooker Connection to Open Haunched Arch.

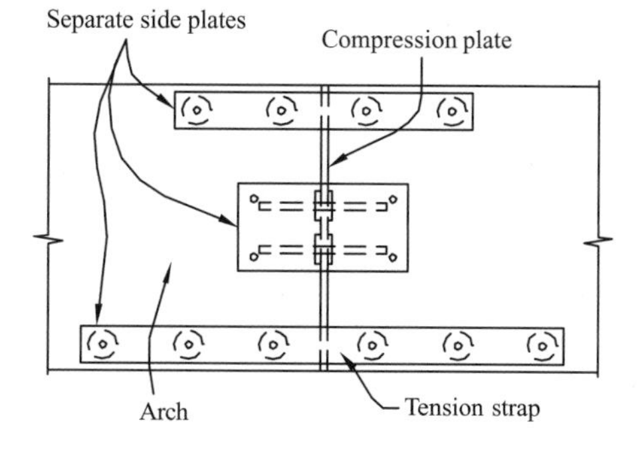

Figure 8.6 Arch Moment Splice. Drawing shows a typical moment splice. Compression stress is taken in bearing on the wood through a steel compression plate. Tension is taken across the splice by means of steel straps and shear plates. Side plates and straps are used to hold sides and tops of members in position. Shear is taken by shear plates in end grain.

9. **TRUSS CONNECTIONS.** When unseasoned lumber is used in truss construction, periodic inspection is recommended along with retightening of hardware and connections as necessary.

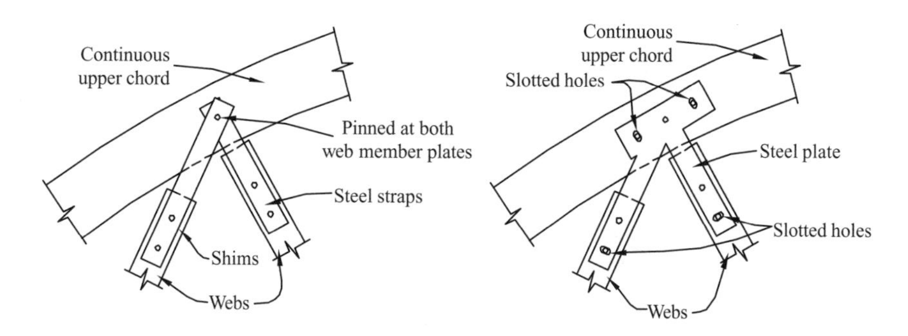

Figure 9.1 Monochord-Steel Straps. For trusses with continuous upper chord. Provide clearance between web ends and chord. Provide shims at web to prevent bending of straps.

Figure 9.2 Slotted Gusset Plates. Alternate detail to Figure 9.1. Slotted holes allow rotation of joint, reducing tension perpendicular to grain stresses.

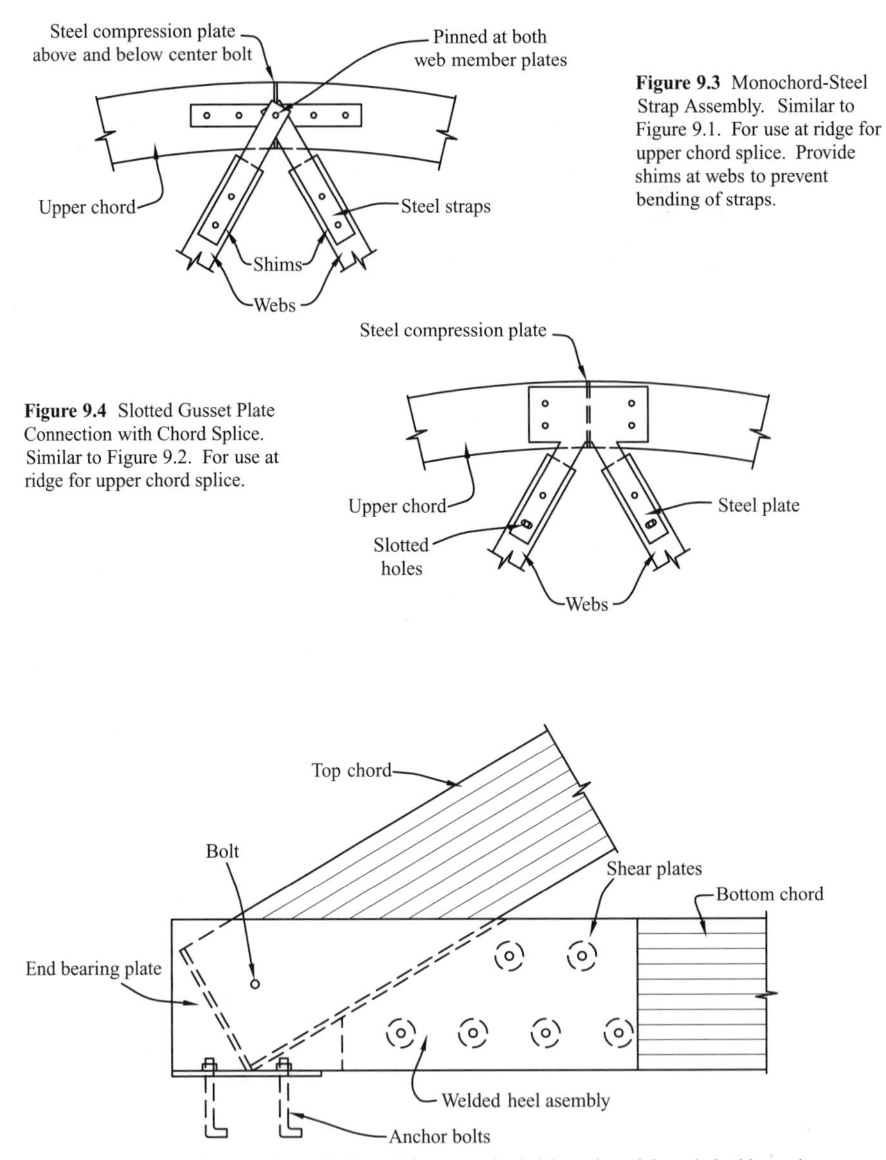

Figure 9.3 Monochord-Steel Strap Assembly. Similar to Figure 9.1. For use at ridge for upper chord splice. Provide shims at webs to prevent bending of straps.

Figure 9.4 Slotted Gusset Plate Connection with Chord Splice. Similar to Figure 9.2. For use at ridge for upper chord splice.

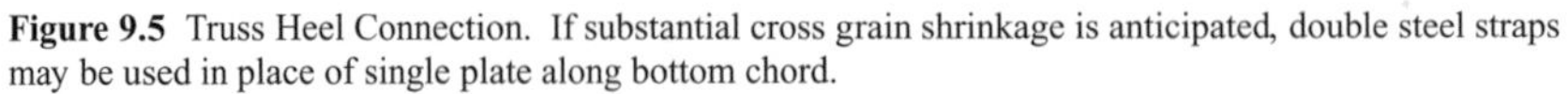

Figure 9.5 Truss Heel Connection. If substantial cross grain shrinkage is anticipated, double steel straps may be used in place of single plate along bottom chord.

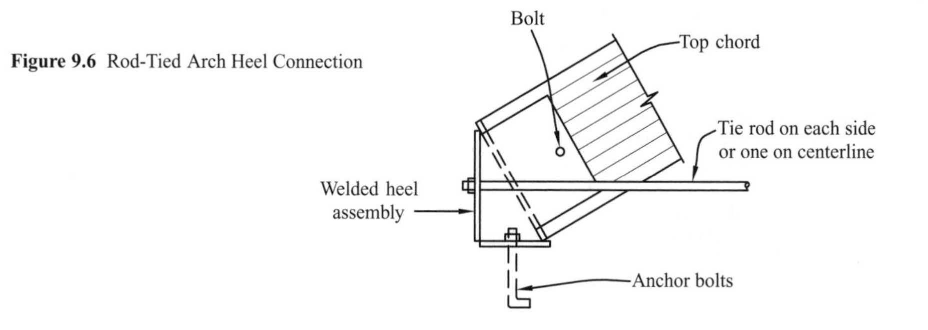

Figure 9.6 Rod-Tied Arch Heel Connection

10. SUSPENDED LOADS

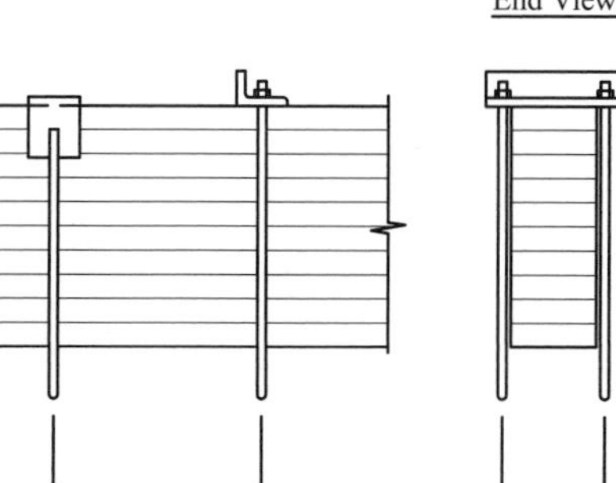

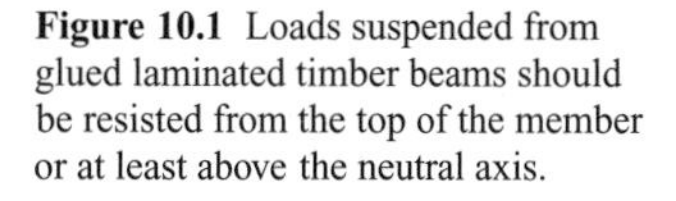

Figure 10.1 Loads suspended from glued laminated timber beams should be resisted from the top of the member or at least above the neutral axis.

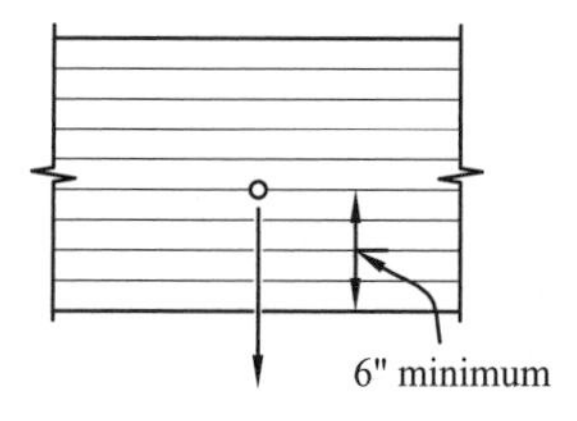

Figure 10.2 Light loads such as small conduit may be suspended with small fasteners near the bottom of glued laminated timber beams as shown.

11. DETAILS TO PROTECT AGAINST DECAY

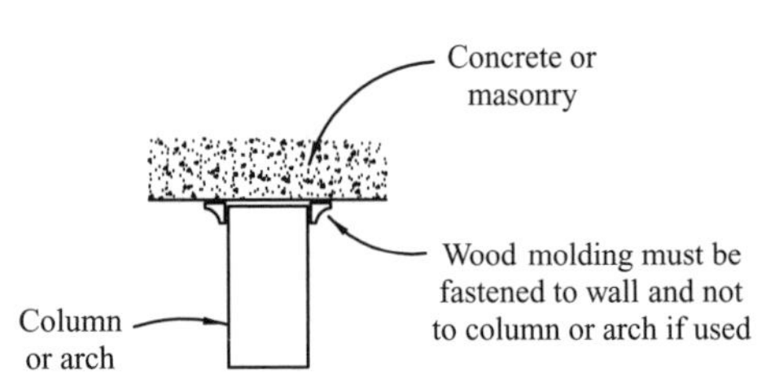

Figure 11.1 Wood Member near Continuous Masonry Wall. Minimum of 1/2 in. air space between member and wall or adequate moisture barrier must be provided. For arches, additional space may be required to permit outward deflection of the arch leg.

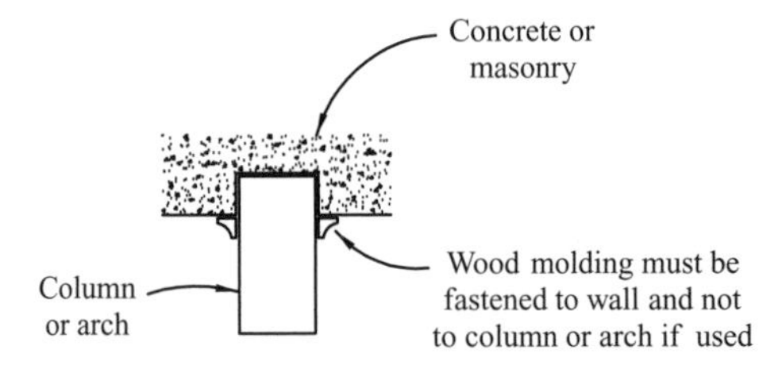

Figure 11.2 Wood Member Set in Masonry Wall Pocket. Minimum of 1/2 in. air space between member and wall pocket or adequate moisture barrier must be provided. For arches, additional space may be required to permit outward deflection of the arch leg.

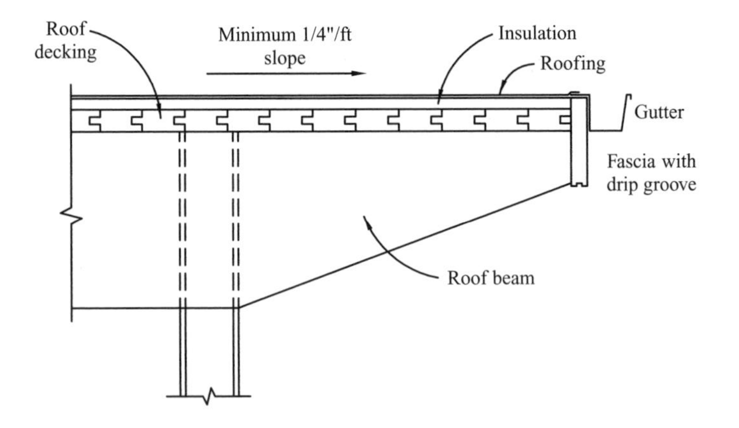

Figure 11.3 Protection Considerations for Buildings with Covered Overhang. Beam is protected from direct exposure to weather by fascia. Roof should be sloped for drainage and designed to prevent ponding of water. Fascia should be preservatively treated or made from decay-resistant species. Taper cut should be sealed.

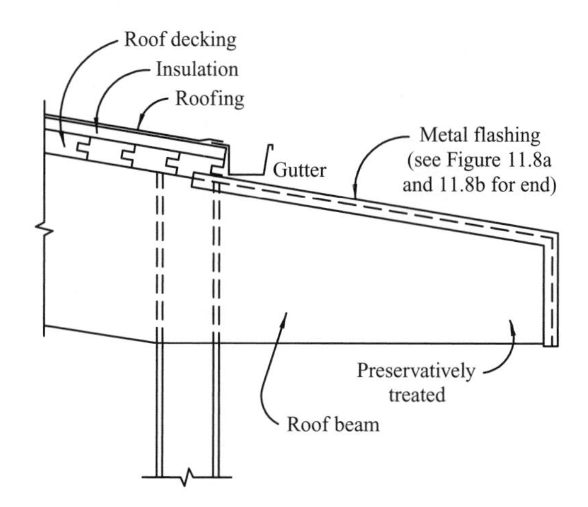

Figure 11.4 Protection Considerations for Buildings With Uncovered Overhang. Portion of beam extending outside of building should be protected by metal cap and preservative treatment. Periodic refinishing of the surfaces exposed to the weather will be required to maintain appearance.

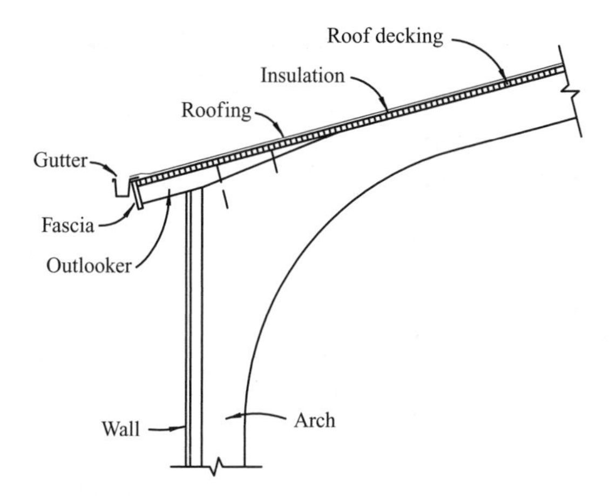

Figure 11.5 Protection Considerations for Arch Outlooker Overhang. Outlooker is protected from direct exposure to weather. Arch is protected by wall from direct exposure to the weather.

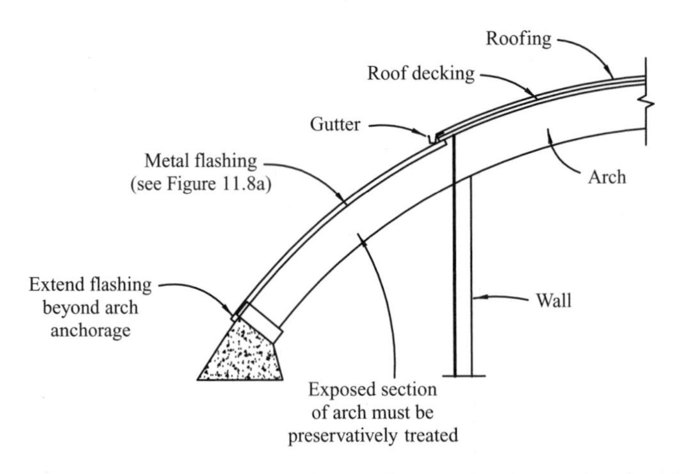

Figure 11.6 Protection Considerations for Partially Exposed Arches. Portion of arch leg extending outside of building should be protected by metal flashing and preservative. At least 12 in. clearance must be provided between arch base and grade. Preservative treatment in accordance with *Standard for Preservative Treatment of Structural Glued Laminated Timber*, AITC 109, must be used for exposed portion of arch.

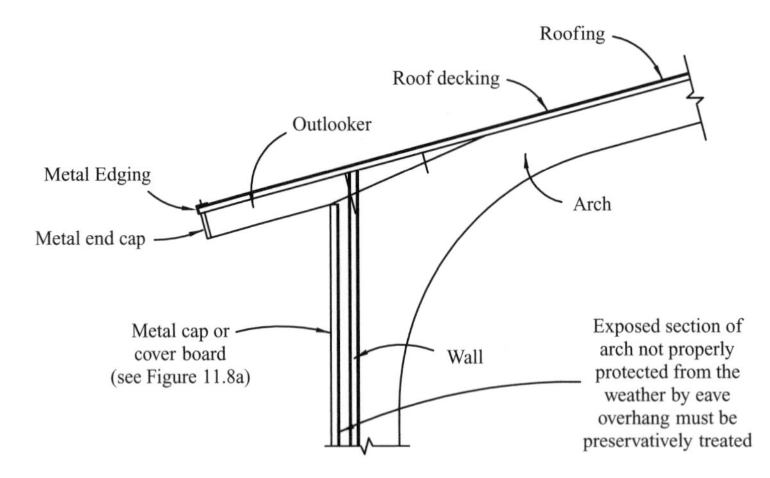

Figure 11.7 Arch Leg Protection. Metal end cap or treated cover board should be used on edge of exterior portion of arch in conjunction with preservative treatment of the arch leg. Metal cap is as illustrated in Figure 11.8. Cover board should be vertical grain material set in building sealant and attached with weatherproof nails or screws. All wood with exterior exposure must be adequately protected and maintained.

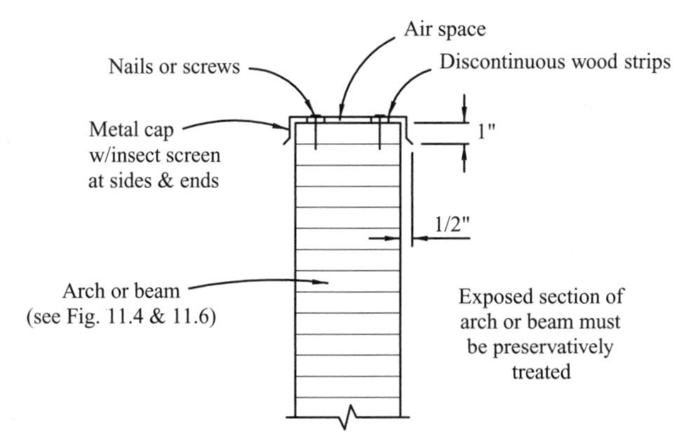

a) Top cap for horizontal or sloped members

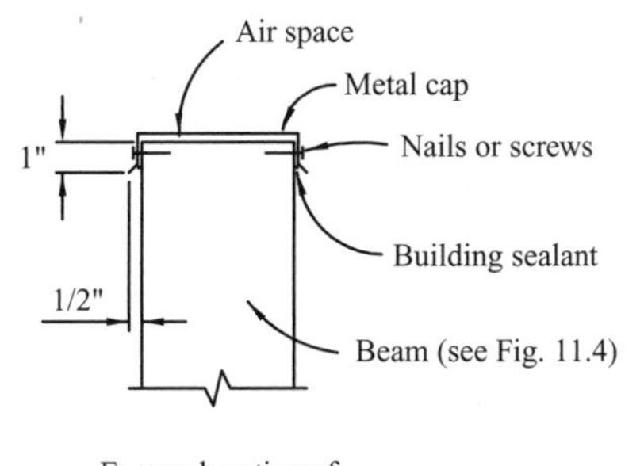

b) End cap for exposed beams or vertical members

Figure 11.8 Protection Metal Cap or Flashing Details. Caps or flashings are made of 20-gage minimum thickness weatherproof metal. Nails or screws are weatherproofed and heads are sealed with building sealant or neoprene washers. A minimum of 1/2 in. air space must be provided between cap and the face of the wood section. For vertical use conditions, a continuous bead of building sealant is required.

APPENDIX TO AITC 104
CONNECTION DETAILS TO BE AVOIDED

The following are examples of poor detailing practice and suggestions for improvement of the poor details.

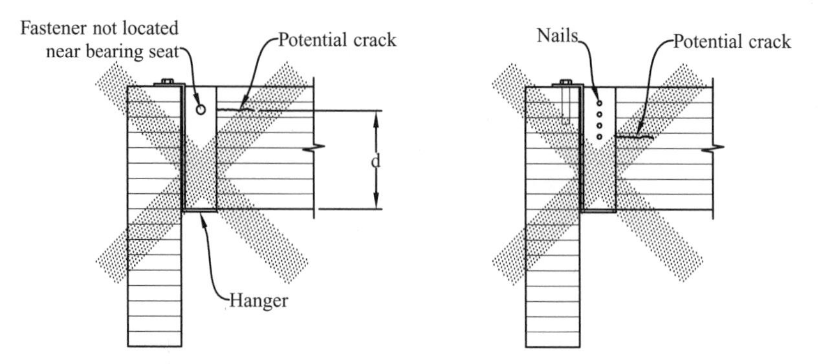

Detail A1 - Glued laminated timbers, although relatively dry at the time of manufacture, may shrink as they reach equilibrium moisture content in service. When fasteners are not located near the bearing seat but in the upper portion of the beam, shrinkage in the beam over the depth, d, can cause the beam reaction to be carried by the fasteners rather than in the bearing on the hanger. This induces notch shear and tension perpendicular to grain stresses that can cause splitting along the beam as shown.

SUGGESTED REVISION - Detail the connection as shown in figure 4.1 with the fasteners located near the bearing seat, or slot the hole in the steel hanger and place the fastener in the top of the slot.

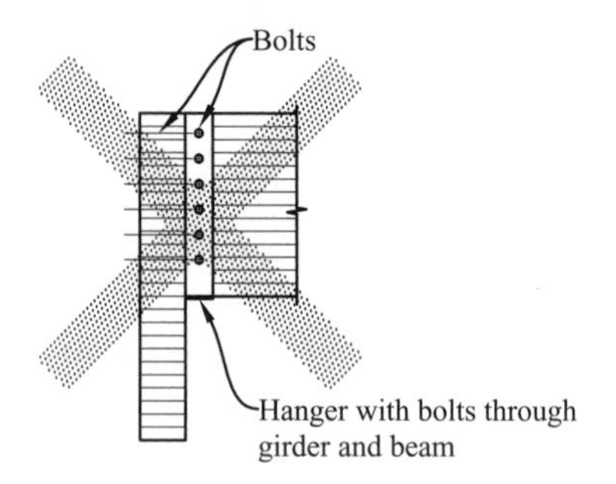

Detail A2 - This detail is similar to Detail A1 in that shrinkage in the beam can result in the bearing being carried by the fasteners rather than the bearing on the hanger seat which, in turn, results in notch shear and tension perpendicular to grain stress. Also, see Detail A3 concerning long rows of fasteners perpendicular to grain as it relates to the hanger-to-girder connection. Section 1.8 discusses the effect of end rotation on the fasteners in this type of connection.

SUGGESTED REVISION - Detail the connection as illustrated in Figure 4.3 with the fasteners in the beam located near the bearing seat and the fasteners in the girder grouped near the top of the girder.

Detail A3 - End connections which include long rows of fasteners perpendicular to grain through steel side members should be avoided. Shrinkage of the wood will be restrained by the steel resulting in tension perpendicular to grain stresses. Because the beam is not supported by a bearing plate, notch shear and tension perpendicular to grain stresses at the end of the beam will also result. The individual or combined effect may cause splitting of the member. See also section 1.4

SUGGESTED REVISION - Change to a bearing connection as shown in Figure 4.1 with the beam being supported in bearing and the fasteners located only near the bearing seat.

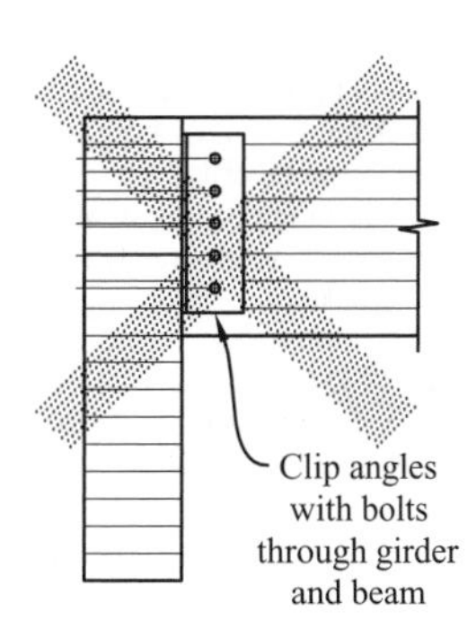

Fish plate in kerf in
center of beam with
bolts through beam

Detail A4 - This detail is similar to Detail A3 except that the beam is supported by bolts through the plate located in a saw kerf in the center beam. See also Section 1.4

SUGGESTED REVISION - Detail the connection as shown in Figure 4.6 where a bearing seat has been added and the bolts away from the bearing seat have been omitted.

Detail A5 - An abrupt notch in the end of a wood member creates two problems. One is that the effective shear strength of the member is reduced because of the end notch. The other is that the exposure of end grain in the notch will permit a more rapid migration of moisture in the upper portion of the member. The individual or combined effect may cause the indicated split.

SUGGESTED REVISION - Detail the connection as shown in Figure 2.1 without the end notch. Notches are not recommended on the tension side of glued laminated timber, but if used, they should not exceed the lesser of 10 % of the depth or 3 inches and shear stresses should be checked by the notched beam formula.

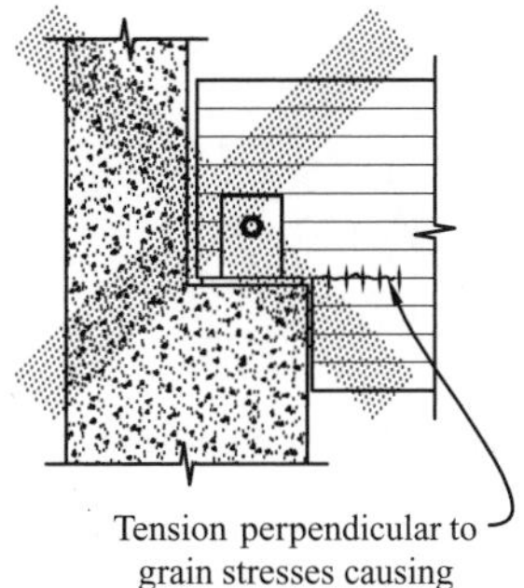

Detail A6 - This condition is similar to that shown in Detail A5. The shear strength of the end of the member is reduced, tension perpendicular to grain stresses are induced and the exposed end grain may result in splitting because of rapid drying.

SUGGESTED REVISION - Revise the taper cut to provide bearing as shown in Figure 2.6.

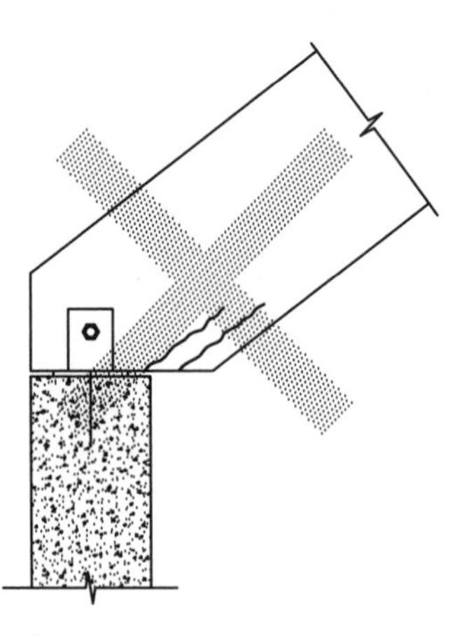

Detail A7 - This detail at the upper end of a sloped beam is similar to the notched beam detail shown in Detail A5.

SUGGESTED REVISION - Provide a sloping seat as shown in Figure 2.7.

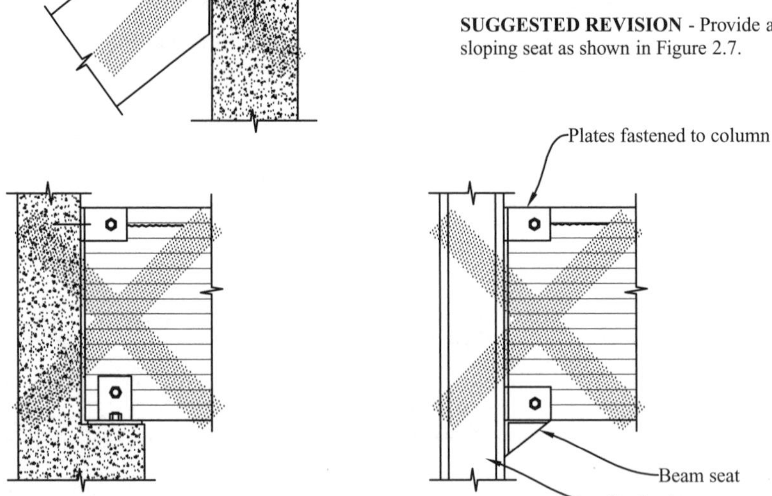

Detail A8 - In this situation, the beam is bearing on the beam seat and the top is laterally supported by clip angles or similar hardware. In a deep beam, the shrinkage due to drying reduces the depth of the beam and will create a split at the upper connection. This connection will also resist deflection of the beam, creating a horizontal reaction force that may cause damage to the wall or the beam.

SUGGESTED REVISION - Provide restraint against lateral rotation without restraining the member against shrinkage or deflection as shown in Figure 2.8.

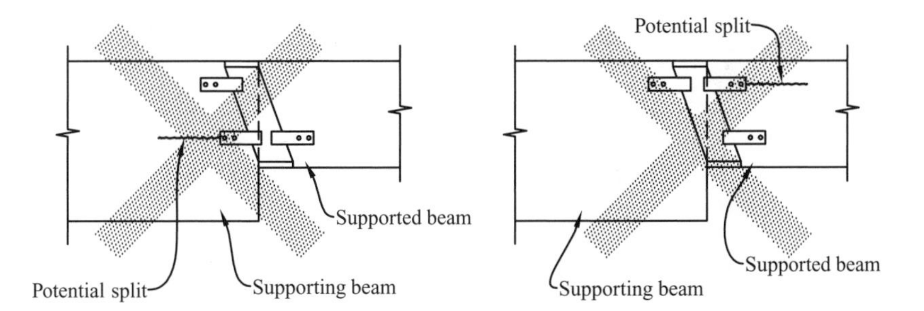

Detail A9 - When a tension connection across a cantilever beam hanger is designed using integral tabs either at the top or bottom of the hanger, splitting may occur due to shrinkage between the bearing point of the hanger and the bolts as shown.

SUGGESTED REVISION - Detail as shown in Figure 3.1 for suggested cantilever hanger with a loose tension tie. If tabs are provided as shown, detail vertically slotted holes in the tabs and locate the bolts in the end of the slot farthest from the bearing seat. Bolts in slotted holes should be hand tightened only so that movement is permitted.

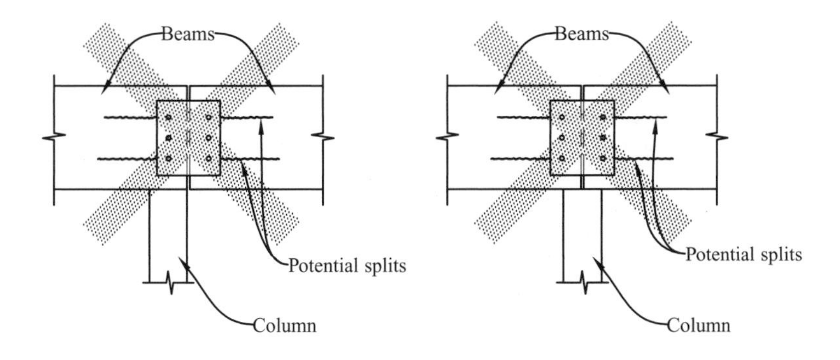

Detail A10 - This situation is similar to Detail A3 where deep splice plates are applied to both faces of the beam. This may be a splice over the column or a situation where one beam is supporting the next one. As the wood shrinks, the steel side plates resist the shrinkage effect causing splits in the beams. This condition is particularly hazardous if one beam is supporting the next one as shown on the left or as a cantilever connection, because the splits at the bolt holes will reduce the effective strength of the beam.

SUGGESTED REVISION - See details in section 5 for recommended column connections and Figure 8.6 for recommended moment splice if continuity over column is desired. Moment splices are not recommended for beams.

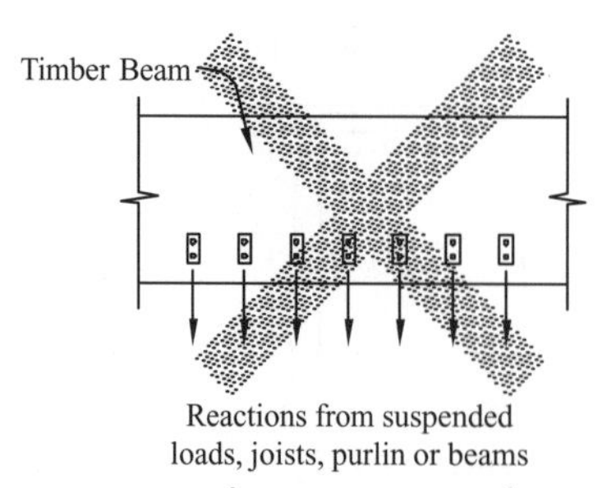

Detail A11 - Loads suspended from beams as shown induce tension perpendicular to grain stresses.

Suggested Revision - See Section 10 for recommendations for supporting loads from beams.

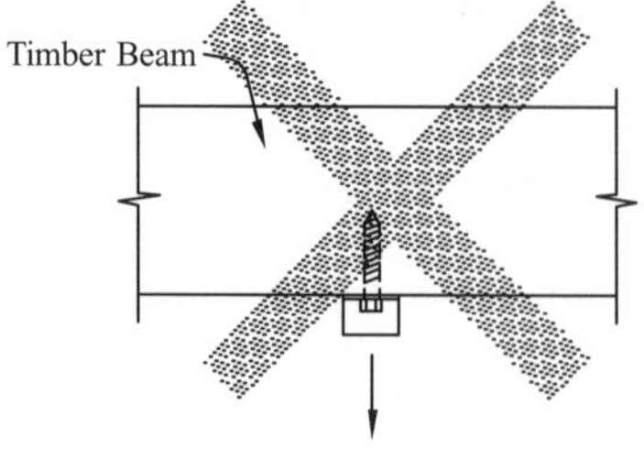

NOTE - This type of connection is to be used to carry light loads only and must be designed by a qualified design professional.

Detail A12 - This shows a general condition where, particularly in continuous framing, the top tension fibers have been cut to provide for a recessed hardware connection or for the passage of conduit or other elements over the top of the beam. This is particularly serious in glulam construction since the tension laminations are critical to the proper performance of the structure.

SUGGESTED REVISION - Detail the connection as shown in Figure 4.5 without the notch in the tension side of the cantilever member. If recessed hardware is required, provide mechanically fastened blocking.

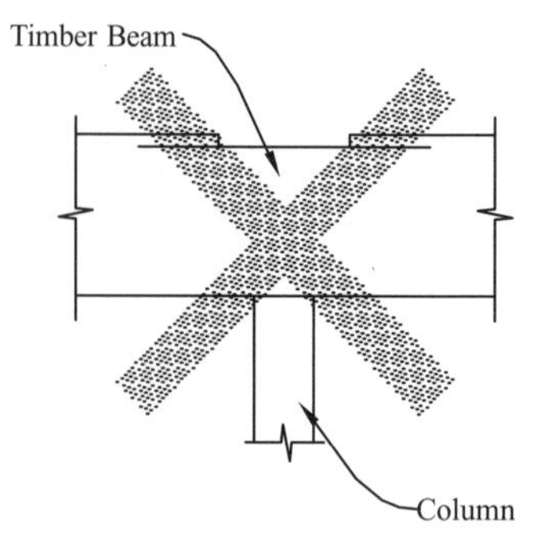

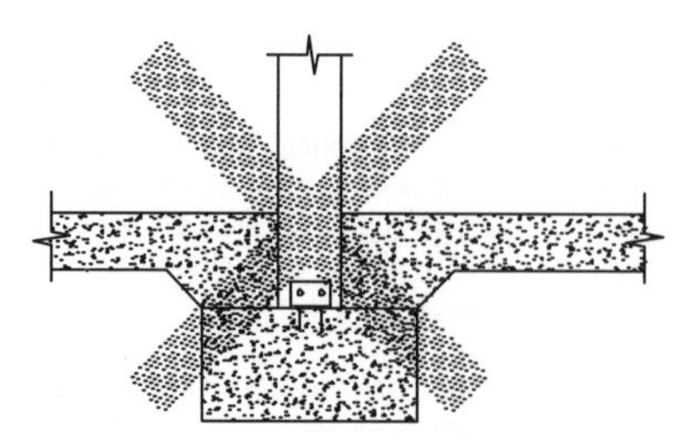

Detail A13 - Some designers try to conceal the base of a column or an arch by placing concrete around the connection. Moisture will migrate into the lower part of the wood and cause decay.

SUGGESTED REVISION - Detail column bases as shown in Section 6 or arch anchorages as shown in Section 7.

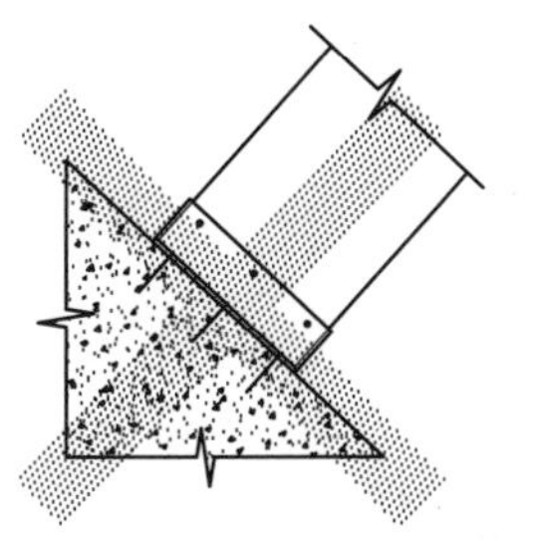

Detail A14 - Similar to Detail A13 in that the base of an arch or column is placed in a closed steel box where moisture will accumulate and cause decay.

Suggested Revision - Detail arch base as shown in Figure 7.9 with connections grouped near the center of the arch and drainage provided to prevent collection of water in the shoe.

Detail A15 - When the centerlines of members do not intersect at a common point in a truss, considerable shear and moment stresses may result in the bottom chord. When these are combined with the presumably high tension stress in the member, failure may occur.

Suggested Revision - Detail web to chord connections to be concentric as shown in Figure 9.1.

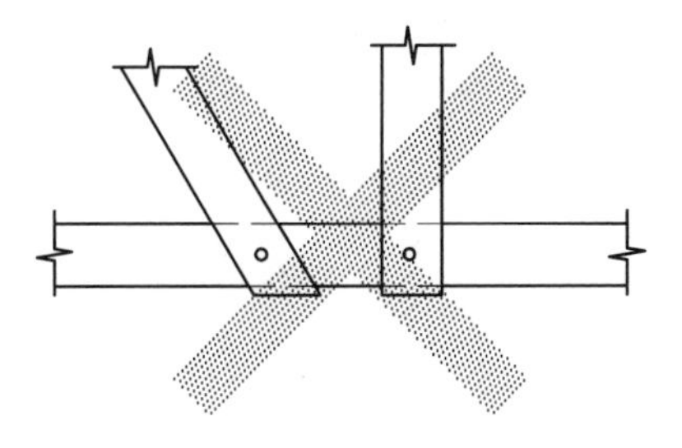

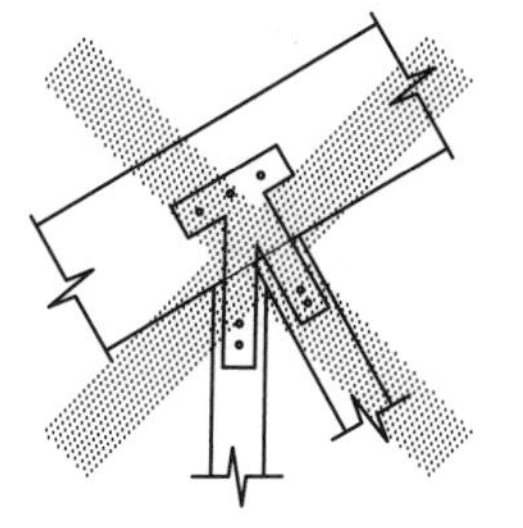

Detail A16 - Truss chord to web connection made from a single plate or plates welded rigidly together. Not recommended when truss deflections could produce rotation of members, which could cause splitting.

Suggested Revision - Detail as shown in Figure 9.1 with pinned connection at the web and chord with clearance provided between webs and chords. Alternatively, slotted bolt holes can be utilized to allow for joint rotation as shown in Figure 9.2.

Detail A17 - Truss heel connection with eccentric force lines that cause prying action which may result in splitting of the members.

Suggested Revision - Detail connection similar to Figure 9.5.

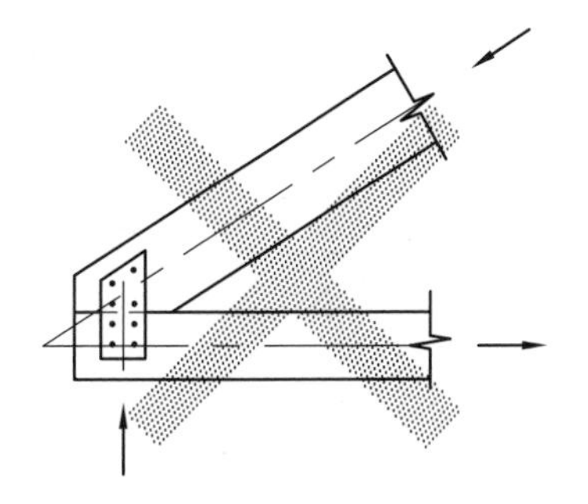

8.2 STANDARD APPEARANCE GRADES FOR STRUCTURAL GLUED LAMINATED TIMBER

AMERICAN INSTITUTE OF TIMBER CONSTRUCTION

7012 South Revere Parkway Suite 140 • Englewood, Colorado 80112
Telephone (303) 792-9559 http://www.aitc-glulam.org

AITC 110-2001

STANDARD APPEARANCE GRADES FOR STRUCTURAL GLUED LAMINATED TIMBER

Adopted as Recommendations October, 2001
Copyright 2001 by American Institute of Timber Construction

CONTENTS

1. INTRODUCTION

1.1 Various grades of appearance are desired for different uses. The appearance grades apply to the surfaces of glued laminated members and include such items as growth characteristics, void filling and surfacing operations but not laminating procedures, stains, varnishes, or other finishes, nor wrappings, cratings, or other protective coverings. The appearance grades do not modify the design stresses, fabrication controls, grades of lumber used and other provisions of the standards for structural glued laminated timber.

1.2 The appearance grades are for the guidance of the designer so that a product consistent with the use of the structure may be specified and provide a suitable appearance at appropriate cost. The designer should specify the desired appearance grade to give a clear understanding between buyer and seller. Requirements given in appearance descriptions are intended to achieve a general and distinctive uniformity of appearance, and reasonable tolerance is permitted. Appearance grading should reflect good judgment. Often the natural growth characteristics of the wood enhance the beauty of the member and help avoid an artificial appearance.

1.3 Four appearance grades F raming, Industrial, Architectural and Premiumare applicable to all species or mixtures of species used in laminating. Some combinations in AITC 117, *Standard Specifications for Structural Glued Laminated Timber of Softwood Species*, permit the mixing of species within a given member; subsequently, the potential for differences in color or grain of adjacent laminations must be recognized. For those appearance applications where differences in color or grain might be important, the designer may specify a single species or species group This may restrict availability.

1.4 When members containing wood filler are stained, the filler may not accept the stain in the same manner as the adjacent wood. In addition, filler exposed to the elements may weather differently than the adjacent wood. The buyer should check with the supplier concerning past experience in these matters before specifying.

1.5 Preservative treatment may affect the appearance of the finished laminated timber. See AITC 109, *Standard for Preservative Treatment of Structural Glued Laminated Timber,* for detailed information.

1.6 When cover boards are not used, a sloping end cut exposed to view shall have the same appearance requirements for loose knots, knot holes, etc., that are required for the sides of members.

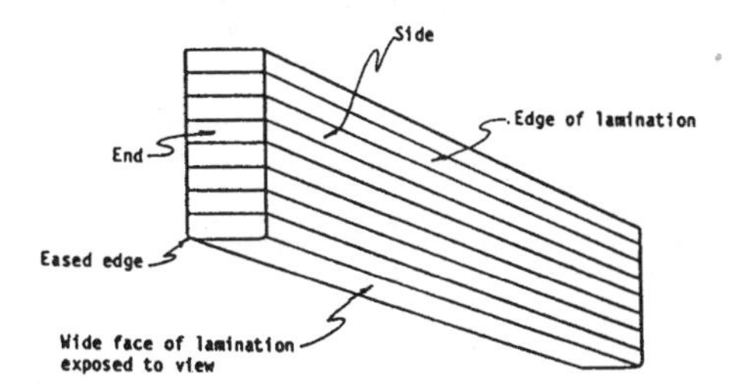

FIGURE 1. Illustration of Terms Related to Appearance Grades

2. TEXTURED SURFACES

2.1 When specified by the designer, a textured surface may be used. Textured surfaces are produced by a variety of methods and the buyer should check with the supplier before specifying.

2.2 Textured surfaces may change the net finished sizes and tolerances given in AITC 113, *Standard for Dimensions of Glued Laminated Structural Members*. Depending upon the degree of texturing, it may be necessary for the designer to compensate for the resulting loss of cross section.

2.3 Texturing will change the appearance grade specified. Additional voids and splintering may result. Textured surfaces will readily absorb stain, but adjacent areas of wood filler may not accept stain in the same manner. Eased edges will not normally be furnished.

3. FRAMING APPEARANCE GRADE

3.1 APPLICATION Framing grade is ordinarily suitable for construction in conventional framing applications such as windows and door headers where appearance is of no concern and the size of the glued laminated timber matches the size of the framing.

3.2 DESCRIPTION

3.2.1 Laminations are permitted to possess the natural growth characteristics of the lumber grade.

3.2.2 Voids appearing on the edge of laminations need not be filled.

3.2.3 Loose knots and open knot holes in the wide face of laminations exposed to view need not be filled. Gaps in edge joints appearing on the wide face of laminations exposed to view need not be filled.

3.2.4 Members are required to be surfaced hit or miss on two sides only to match conventional framing sizes. The following appearance requirements apply only to these two sides. Misses and low laminations are permitted. The maximum area of low laminations shall not exceed 25% of the surface area of a side. Wane (limited to a maximum of 1/4 in. measured across the width) is permitted on a cumulative basis. The accumulative depth of hit or miss and wane shall not exceed 10% of the width of the member at any glueline. The frequency of occurrence shall not exceed one in 10 pieces of lumber used.

3.2.5 In accordance with the provisions in 3.2.4, wane (limited to 1/4 in. measured across the width) is permitted in all combinations and is not limited in length. Occasional wane approximately one foot in length and not exceeding the permissible depth of a low lamination shall be permitted in all combinations without regard to the cumulative effects indicated in 3.2.4. Wane permitted in specific laminating combinations up to 1/6 the lumber width on each side is not limited in length.

3.2.6 Hit or miss surfaces may change the net finished sizes and tolerances given in AITC 113, *Standard for Dimensions of Glued Laminated Structural Members*, and ANSI/AITC A190.1. Depending upon the degree of hit or miss, it may be necessary for the designer to compensate for the resulting loss of cross section.

4. INDUSTRIAL APPEARANCE GRADE

4.1 APPLICATION Industrial appearance grade is ordinarily suitable for construction in industrial applications, warehouses, garages, and for other uses where appearance is not of primary concern.

4.2 DESCRIPTION

4.2.1 Laminations are permitted to possess the natural growth characteristics of the lumber grade.

4.2.2 Voids appearing on the edge of laminations need not be filled.

4.2.3 Loose knots and open knot holes in the wide face of laminations exposed to view shall be filled. This restriction does not apply for glued laminated timber truss members. Gaps in edge joints appearing on the wide face of laminations exposed to view need not be filled.

4.2.4 Members are required to be surfaced on two sides only. The following appearance requirements apply only to these two sides. Occasional misses, low laminations or wane (limited to a maximum of 1/4 in. measured across the width) are permitted on a cumulative basis. The cumulative depth of the misses, low laminations and wane shall not exceed 10% of the width of the member at any glue line. The frequency of occurrence shall not exceed one in 10 pieces of lumber used. The maximum area of low laminations shall not exceed 5% of the surface area of a side; and no more than two low laminations shall be adjacent to one another.

4.2.5 In accordance with provisions in 4.2.4, wane (limited to 1/4 in. measured across the width) is permitted in all combinations and is not limited in length. Wane permitted in specific laminating combinations up to 1/6 the lumber width on each side is not limited in length. Occasional wane approximately one foot in length and not exceeding the permissible depth of a low lamination shall be permitted in all combinations without regard to the cumulative effects indicated in 4.2.4. Wane permitted in specific laminating combinations up to 1/6 the lumber width on each side is not limited in length.

5. ARCHITECTURAL APPEARANCE GRADE

5.1 APPLICATION Architectural appearance grade is ordinarily suitable for construction where appearance is an important requirement.

5.2 DESCRIPTION

5.2.1 Laminations are permitted to possess the natural growth characteristics of the lumber grade.

5.2.2 In exposed surfaces, voids measuring over 3/4 in. shall be filled by the fabricator with a wood-tone colored filler which reasonably blends with the final product or with clear wood inserts selected for similarity to the grain and color of the adjacent wood.

5.2.2.1 For appearance grading purposes, measurement of voids shall be in the direction of the length of the lamination and shall not exceed 3/4 in. except that a void may be longer than 3/4 in. if its area does not exceed 1/2 sq. in. Void measurement limitations apply only to the surfaces of the member exposed in the final structure. All characteristics shall be considered with respect to their effects on general appearance.

5.2.3 The wide face of laminations exposed to view shall be free of loose knots. Open knot holes shall be filled. Voids greater than 1/16 in. wide in edge joints appearing on the wide face of laminations exposed to view shall be filled.

5.2.4 Exposed faces shall be surfaced smooth. Misses and wane are not permitted.

5.2.5 The corners of the member exposed to view in the final structure shall be eased with a minimum relief of 1/8 in. Other dimensions for eased edges may be agreed upon between buyer and seller.

6. PREMIUM APPEARANCE GRADE

6.1 APPLICATION Premium appearance grade is the highest standard appearance grade.

6.2 DESCRIPTION

6.2.1 Laminations are permitted to possess the natural growth characteristics of the lumber grade.

6.2.2 In exposed surfaces, voids shall be filled by the fabricator with a wood-tone colored filler that reasonably blends with the final product or with clear wood inserts selected similarity to the grain and color of the adjacent wood.

6.2.3 The wide face of laminations exposed to view shall be selected for appearance and shall be free of loose knots. Voids shall be filled. Knot size shall be limited to 20% of the net face width of the lamination. Not over two maximum size knots or their equivalent shall occur in a 6 ft length. Voids greater than 1/16 in. wide in edge joints appearing on the wide face of laminations exposed to view shall be filled.

6.2.4 Exposed faces shall be surfaced smooth. Misses and wane are not permitted.

6.2.5 The corners of the member exposed to view in the final structure shall be eased with a minimum relief of 1/8 in. Other dimensions for eased edges may be agreed upon between buyer and seller.

7. SPECIAL APPEARANCE REQUIREMENTS

7.1 For special applications, the buyer may specify other requirements. Such requirements may limit availability.

8. SUMMARY OF APPEARANCE GRADE SPECIFICATIONS

Description Item	Framing Appearance Grade	Industrial Appearance Grade	Architectural Appearance Grade	Premium Appearance Grade
Natural growth characteristics of lumber grade	Allowed	Allowed	Allowed	Allowed
Paragraph reference	3.2.1	4.2.1	5.2.1	6.2.1
Filling of voids on edge of laminations	Not required	Not required	Required for voids over 3/4 in.	Required for all voids
Paragraph reference	3.2.2	4.2.2	5.2.2, 5.2.2.1	6.2.2
Wide face of laminations exposed to view	Loose knots, open knot holes permitted	Void filling required except for trusses	Free of loose knots, void filling required	Selected for appearance, free of loose knots, void filling required, knot sizes limited
Paragraph reference	3.2.3	4.2.3	5.2.3	6.2.3
Edge joints appearing on wide faces of laminations exposed to view	Filling not required	Filling not required	Filling required for voids over 1/16 in. wide	Filling required for voids over 1/16 in. wide
Paragraph reference	3.2.3	4.2.3	5.2.3	6.2.3
Surfacing of sides	Required H it or Miss, Low laminations permitted. Limited amount of wane permitted	Required. Limited amounts of misses, low laminations and wane permitted.	Required. Misses not permitted	Required. Misses not permitted
Paragraph reference	3.2.4, 3.2.5	4.2.4, 4.2.5	5.2.4	6.2.4
Surfacing of wide face of laminations exposed to view	Not required	Not required	Required. Misses not permitted	Required. Misses not permitted
Paragraph reference	3.2.4	4.2.4	5.2.4	6.2.4
Eased edges	Not required	Not required	Required	Required
Paragraph reference			5.2.5	6.2.5

8.3 GUIDE FOR SPECIFYING STRUCTURAL GLUED LAMINATED TIMBER

AITC - TECHNICAL NOTE 24

GUIDE FOR SPECIFYING STRUCTURAL GLUED LAMINATED TIMBER

May 2003

American Institute of Timber Construction
7012 S. Revere Parkway Suite 140 Englewood, CO 80112
Phone 303-792-9559 •Fax 303-792-0669
•Email: info@aitc-glulam.org •Website: www.aitc-glulam.org

This document is intended to provide basic guidelines for specifying structural glued laminated timber. The first section is a general commentary intended to provide brief background information to the designer and references for more detailed information. A sample specification follows the general commentary.

Structural glued laminated timber may be designed and specified by the engineer or architect for a project, or it may be designed by the manufacturer. All glued laminated timber shall be designed according to the applicable provisions in *AITC 117-Design*, AITC 119, the *Timber Construction Manual*, and *the National Design Specification® for Wood Construction* (*NDS®*). For specialty products, such as arches and pitched-and-tapered curved beams, the manufacturer should be contacted as early as possible in the design process, whether the members are designed by the manufacturer or by the project engineer.

Critical Specification Information

For each structural glued laminated timber member, the following information should be communicated to the manufacturer.

1. **Quality Mark-** All structural glued laminated timber should be manufactured in conformance with standard *ANSI/AITC A190.1* in facilities inspected by the American Institute of Timber Construction (AITC). AITCs quality mark ensures that the manufacturer meets stringent quality control guidelines to ensure consistent quality and reliable performance. For members that will be exposed on all sides, where a quality stamp may have an undesirable appearance, members should be accompanied by a certificate of conformance. More information regarding the AITC Quality Inspection Program and Quality Mark is available in AITC Technical Note 10.

2. **Stress Class (or Combination Symbol) and Species-** Structural glued laminated timber can be manufactured from a number of species and grades of lumber in various combinations. To ensure adequate structural performance for the intended use, it is necessary to specify the appropriate stress class (or combination symbol) and species required. Specification by stress class simplifies design and allows manufacturers to choose from several combinations with similar properties based on their lumber resources, thus increasing availability and possibly reducing price. *AITC 117-Design* and *NDS®* include design values for softwood structural glued laminated timber stress classes and combinations. *AITC 119* and *NDS®* include design values for hardwood structural glued laminated timber combinations.

3. **End Use Requirements-** For most applications, structural glued laminated timber does not require preservative treatment. However, structural glued laminated timber for use in outdoor exposures (not protected by roof covering), in ground contact, in environments with high humidity, or in contact with water should be treated with a suitable wood preservative in accordance with *AITC 109* or manufactured from the heartwood of a naturally decay-resistant species. Some species do not treat well with certain preservatives, and availability may be limited. Available treatments also vary between manufacturers. It is recommended that the

manufacturer be contacted to discuss options for decay resistance if required. Where possible, it is recommended that glued laminated timbers be designed and detailed to keep them dry, rather than relying on preservative treatment to prevent decay. AITC *Technical Note 12* provides guidelines for designing structural glued laminated timber for permanence.

4. **Geometry and Dimensions**- Structural glued laminated timber can be manufactured in many sizes and shapes. The size and shape are determined from architectural and engineering requirements. Most glued laminated timber is manufactured with rectangular cross sections, but other special cross section shapes may also be manufactured. Specification of non-rectangular cross sections should be accompanied by drawings with all applicable information. Standard cross sectional dimensions are given in *AITC 113*. In addition to cross sectional shape, the geometry of the member along the length should be specified. For beams, radius of curvature or camber requirements should be specified as required. Specification for pitched and curved beams, arches, tapered beams, trusses, and other custom members should be accompanied by drawings from the designer. All required dimensions should be specified.

5. **Special Fire Code Requirements**- Large exposed timbers are recognized in model building codes as having excellent fire resistance. Minimum timber sizes and other requirements for heavy timber construction are specified in the codes and allow for increased building areas. Exposed structural glued laminated timber is also permitted to be used where building codes require one or two hour fire resistive construction. To meet code requirements for one and two hour fire resistance ratings, the members are required to be oversized in accordance with AITC *Technical Note 7* or the 2001 *NDS®* to increase their fire endurance time. The layup of the glued laminated timber is also required to be modified by the manufacturer to ensure adequate fire performance. The quality stamp will indicate that the manufacturer has made required layup modifications. Because the increased dimensions and the special layup requirements will increase the cost of the members, fire resistance ratings are not typically specified unless required by the building code. Steel members and fasteners used in fire rated construction are also required to be protected and must be detailed by the designer.

6. **Appearance Grade**- AITC has standardized four appearance grades: Framing, Industrial, Architectural, and Premium. Complete descriptions of these grades are provided in *AITC 110*. Special appearance requirements that are not covered by the standard appearance grades should be specified. Some examples include special surfacing requirements such as rough sawn texturing or special lamination thickness for visual reasons. If curved members with different radii are used in close proximity, it may be desirable to specify that all members be manufactured with identical lamination thickness, so nearby members will match visually. Mixed species requirements or prohibitions should also be communicated. Any special requirements may limit availability.

7. **Finish Requirements**- Certain manufacturers provide coatings, paints, or sealants to finish and/or protect the structural glued laminated timber from incidental moisture exposure. Individual manufacturers should be contacted to determine the coating options available.

8. **Fabrication Details**- Fabrication details should accompany the order for structural glued laminated timber. It should be made clear who will perform the fabrication and who is responsible for the structural analysis of the connection details.

9. **Wrapping Requirements**- *AITC 111* discusses recommendations for protecting structural glued laminated timber during transit, erection, and storage. Appropriate specifications should be chosen from that standard.

10. **Other Requirements**- Any other requirements should be clearly communicated to the manufacturer in the specifications.

Sample Specification

Scope
All structural glued laminated timber shall be furnished as shown detailed on the plans and specified herein.

Quality Mark
All structural glued laminated timber shall be manufactured in accordance with *ANSI/AITC A190.1-2002.* Each timber shall be marked with the AITC Quality Mark or be accompanied with an AITC Certificate of Conformance. Only timbers specified to be surfaced on all four sides for appearance requirements shall not be grade marked, but shall be accompanied by an AITC Certificate of Conformance.

Stress Class (or Combination Symbol) and Species
All softwood structural glued laminated timber beams shall meet the requirements of Stress Class _________ from *AITC 117-2001 Design* or *2001 National Design Specification®*. Unless otherwise specified beams shall have [balanced, unbalanced] layup.

All hardwood structural glued laminated timber beams shall meet the requirements for Combination Symbol _________ from *AITC 119-96 or 2001 National Design Specification®*.

All structural glued laminated timber columns shall meet the requirements for Combination Symbol _________ from *AITC 117-2001 Design*, *AITC 119-96*, or *2001 National Design Specification®*. Special tension laminations shall not be required unless otherwise noted.

All softwood structural glued laminated arches shall meet the requirements for Combination Symbol_________ from *AITC 117-2001* or *2001 National Design Specification®*.

All structural glued laminated timbers shall be manufactured from [any species meeting stress requirements, Douglas Fir-Larch, Southern Pine, Alaska Cedar]

> Note: *Other species, including hardwoods, are available on a limited basis. Contact AITC or the structural glued laminated timber manufacturer for information.*

End Use Requirements
Structural glued laminated timbers will be subject to [dry use, wet use]. Structural glued laminated timbers shall be [untreated, pressure treated with ______ preservative to a retention of _____ lb/ft^3 in accordance with AITC 109-98].

Special Fire Code Requirements
Structural glued laminated timber combinations [shall, shall not] be modified to meet the layup requirements for [one hour, two hour] fire resistance rating and marked accordingly as required by with ANSI/AITC A190.1-2002.

> Note: *If fire resistance rating is required, timbers must be sized by the designer according to AITC Technical Note 7 or Chapter 16 of 2001 National Design Specification® in addition to having layup modifications.*

Appearance Grade
Appearance of members shall meet the requirements of the [Premium, Architectural, Industrial, Framing] Appearance Grade as described in AITC 110-2001:

Finish Requirements
Members shall be [uncoated/unsealed, coated/sealed with _________].

Fabrication Details
Fabrication detail shop drawings [are, are not] attached. Fabrication shall be performed by [manufacturer, other than manufacturer]. Connection hardware shall be supplied by _________. Steel fasteners shall be conform to ASTM A307 unless otherwise noted. Other Steel shall conform to ASTM A36 unless otherwise noted.

Wrapping Requirements
Members shall be [not wrapped, load wrapped, bundle wrapped, individually wrapped].

References

AITC Quality Control Program. Technical Note 10. American Institute of Timber Construction. Englewood, Colorado.

Calculation of Fire Resistance of Glued Laminated Timbers. Technical Note 7. American Institute of Timber Construction. Englewood, Colorado.

Designing Structural Glued Laminated Timber for Permanence. Technical Note 12. American Institute of Timber Construction. Englewood, Colorado.

National Design Specification® for Wood Construction. 2001. American Forest & Paper Association. Washington, D. C.

Standard Appearance Grades for Structural Glued Laminated Timber. AITC 110-2001. American Institute of Timber Construction. Englewood, Colorado

Standard for Dimensions of Structural Glued Laminated Timber. AITC 113-2001. American Institute of Timber Construction. Englewood, Colorado.

Standard for Preservative Treatment of Structural Glued Laminated Timber. AITC 109-98. American Institute of Timber Construction. Englewood, Colorado.

Standard Specifications for Structural Glued Laminated Timber of Softwood Species. AITC 117-2001-*Design*. American Institute of Timber Construction. Englewood, Colorado.

Structural Glued Laminated Timber. ANSI/AITC A190.1-2002. American Institute of Timber Construction. Englewood, Colorado.

Timber Construction Manual. 4th edition. 1994. American Institute of Timber Construction. Englewood, Colorado.

APPENDIX

A.1 SECTION PROPERTIES

A.1.1 Standard Dressed (S4S) Sawn Lumber

TABLE A.1.1 Section Properties of Standard Dressed (S4S) Sawn Lumber

Nominal Size b × d	Standard Dressed Size (S4S) b × d inches × inches	Area of Section A in²	X-X AXIS Section Modulus S_{xx} in³	X-X AXIS Moment of Inertia I_{xx} in⁴	Y-Y AXIS Section Modulus S_{yy} in³	Y-Y AXIS Moment of Inertia I_{yy} in⁴	Approximate weight in pounds per linear foot (lb/ft) of piece when density of wood equals: 25 lb/ft³	30 lb/ft³	35 lb/ft³	40 lb/ft³	45 lb/ft³	50 lb/ft³
1 × 3	3/4 × 2-1/2	1.875	0.781	0.977	0.234	0.088	0.326	0.391	0.456	0.521	0.586	0.651
1 × 4	3/4 × 3-1/2	2.625	1.531	2.680	0.328	0.123	0.456	0.547	0.638	0.729	0.820	0.911
1 × 6	3/4 × 5-1/2	4.125	3.781	10.40	0.516	0.193	0.716	0.859	1.003	1.146	1.289	1.432
1 × 8	3/4 × 7-1/4	5.438	6.570	23.82	0.680	0.255	0.944	1.133	1.322	1.510	1.699	1.888
1 × 10	3/4 × 9-1/4	6.938	10.70	49.47	0.867	0.325	1.204	1.445	1.686	1.927	2.168	2.409
1 × 12	3/4 × 11-1/4	8.438	15.82	88.99	1.055	0.396	1.465	1.758	2.051	2.344	2.637	2.930
2 × 3	1-1/2 × 2-1/2	3.750	1.563	1.953	0.938	0.703	0.651	0.781	0.911	1.042	1.172	1.302
2 × 4	1-1/2 × 3-1/2	5.250	3.063	5.359	1.313	0.984	0.911	1.094	1.276	1.458	1.641	1.823
2 × 5	1-1/2 × 4-1/2	6.750	5.063	11.39	1.688	1.266	1.172	1.406	1.641	1.875	2.109	2.344
2 × 6	1-1/2 × 5-1/2	8.250	7.563	20.80	2.063	1.547	1.432	1.719	2.005	2.292	2.578	2.865
2 × 8	1-1/2 × 7-1/4	10.88	13.14	47.63	2.719	2.039	1.888	2.266	2.643	3.021	3.398	3.776
2 × 10	1-1/2 × 9-1/4	13.88	21.39	98.93	3.469	2.602	2.409	2.891	3.372	3.854	4.336	4.818
2 × 12	1-1/2 × 11-1/4	16.88	31.64	178.0	4.219	3.164	2.930	3.516	4.102	4.688	5.273	5.859
2 × 14	1-1/2 × 13-1/4	19.88	43.89	290.8	4.969	3.727	3.451	4.141	4.831	5.521	6.211	6.901
3 × 4	2-1/2 × 3-1/2	8.750	5.104	8.932	3.646	4.557	1.519	1.823	2.127	2.431	2.734	3.038
3 × 5	2-1/2 × 4-1/2	11.25	8.438	18.98	4.688	5.859	1.953	2.344	2.734	3.125	3.516	3.906
3 × 6	2-1/2 × 5-1/2	13.75	12.60	34.66	5.729	7.161	2.387	2.865	3.342	3.819	4.297	4.774
3 × 8	2-1/2 × 7-1/4	18.13	21.90	79.39	7.552	9.440	3.147	3.776	4.405	5.035	5.664	6.293
3 × 10	2-1/2 × 9-1/4	23.13	35.65	164.9	9.635	12.04	4.015	4.818	5.621	6.424	7.227	8.030
3 × 12	2-1/2 × 11-1/4	28.13	52.73	296.6	11.72	14.65	4.883	5.859	6.836	7.813	8.789	9.766
3 × 14	2-1/2 × 13-1/4	33.13	73.15	484.6	13.80	17.25	5.751	6.901	8.051	9.201	10.35	11.50
3 × 16	2-1/2 × 15-1/4	38.13	96.90	738.9	15.89	19.86	6.619	7.943	9.266	10.59	11.91	13.24
4 × 4	3-1/2 × 3-1/2	12.25	7.146	12.51	7.146	12.51	2.127	2.552	2.977	3.403	3.828	4.253
4 × 5	3-1/2 × 4-1/2	15.75	11.81	26.58	9.188	16.08	2.734	3.281	3.828	4.375	4.922	5.469
4 × 6	3-1/2 × 5-1/2	19.25	17.65	48.53	11.23	19.65	3.342	4.010	4.679	5.347	6.016	6.684
4 × 8	3-1/2 × 7-1/4	25.38	30.66	111.1	14.80	25.90	4.405	5.286	6.168	7.049	7.930	8.811
4 × 10	3-1/2 × 9-1/4	32.38	49.91	230.8	18.89	33.05	5.621	6.745	7.869	8.993	10.12	11.24
4 × 12	3-1/2 × 11-1/4	39.38	73.83	415.3	22.97	40.20	6.836	8.203	9.570	10.94	12.30	13.67
4 × 14	3-1/2 × 13-1/4	46.38	102.4	678.5	27.05	47.34	8.051	9.661	11.27	12.88	14.49	16.10
4 × 16	3-1/2 × 15-1/4	53.38	135.7	1034	31.14	54.49	9.266	11.12	12.97	14.83	16.68	18.53
5 × 5	4-1/2 × 4-1/2	20.25	15.19	34.17	15.19	34.17	3.516	4.219	4.922	5.625	6.328	7.031
6 × 6	5-1/2 × 5-1/2	30.25	27.73	76.26	27.73	76.26	5.252	6.302	7.352	8.403	9.453	10.50
6 × 8	5-1/2 × 7-1/2	41.25	51.56	193.4	37.81	104.0	7.161	8.594	10.03	11.46	12.89	14.32
6 × 10	5-1/2 × 9-1/2	52.25	82.73	393.0	47.90	131.7	9.071	10.89	12.70	14.51	16.33	18.14
6 × 12	5-1/2 × 11-1/2	63.25	121.2	697.1	57.98	159.4	10.98	13.18	15.37	17.57	19.77	21.96
6 × 14	5-1/2 × 13-1/2	74.25	167.1	1128	68.06	187.2	12.89	15.47	18.05	20.63	23.20	25.78
6 × 16	5-1/2 × 15-1/2	85.25	220.2	1707	78.15	214.9	14.80	17.76	20.72	23.68	26.64	29.60
6 × 18	5-1/2 × 17-1/2	96.25	280.7	2456	88.23	242.6	16.71	20.05	23.39	26.74	30.08	33.42
6 × 20	5-1/2 × 19-1/2	107.3	348.6	3398	98.31	270.4	18.62	22.34	26.07	29.79	33.52	37.24
6 × 22	5-1/2 × 21-1/2	118.3	423.7	4555	108.4	298.1	20.53	24.64	28.74	32.85	36.95	41.06
6 × 24	5-1/2 × 23-1/2	129.3	506.2	5948	118.5	325.8	22.44	26.93	31.41	35.90	40.39	44.88
8 × 8	7-1/2 × 7-1/2	56.25	70.31	263.7	70.31	263.7	9.766	11.72	13.67	15.63	17.58	19.53
8 × 10	7-1/2 × 9-1/2	71.25	112.8	535.9	89.06	334.0	12.37	14.84	17.32	19.79	22.27	24.74
8 × 12	7-1/2 × 11-1/2	86.25	165.3	950.5	107.8	404.3	14.97	17.97	20.96	23.96	26.95	29.95
8 × 14	7-1/2 × 13-1/2	101.3	227.8	1538	126.6	474.6	17.58	21.09	24.61	28.13	31.64	35.16
8 × 16	7-1/2 × 15-1/2	116.3	300.3	2327	145.3	544.9	20.18	24.22	28.26	32.29	36.33	40.36
8 × 18	7-1/2 × 17-1/2	131.3	382.8	3350	164.1	615.2	22.79	27.34	31.90	36.46	41.02	45.57
8 × 20	7-1/2 × 19-1/2	146.3	475.3	4634	182.8	685.5	25.39	30.47	35.55	40.63	45.70	50.78
8 × 22	7-1/2 × 21-1/2	161.3	577.8	6211	201.6	755.9	27.99	33.59	39.19	44.79	50.39	55.99
8 × 24	7-1/2 × 23-1/2	176.3	690.3	8111	220.3	826.2	30.60	36.72	42.84	48.96	55.08	61.20
10 × 10	9-1/2 × 9-1/2	90.25	142.9	678.8	142.9	678.8	15.67	18.80	21.94	25.07	28.20	31.34
10 × 12	9-1/2 × 11-1/2	109.3	209.4	1204	173.0	821.7	18.97	22.76	26.55	30.35	34.14	37.93
10 × 14	9-1/2 × 13-1/2	128.3	288.6	1948	203.1	964.5	22.27	26.72	31.17	35.63	40.08	44.53
10 × 16	9-1/2 × 15-1/2	147.3	380.4	2948	233.1	1107	25.56	30.68	35.79	40.90	46.02	51.13
10 × 18	9-1/2 × 17-1/2	166.3	484.9	4243	263.2	1250	28.86	34.64	40.41	46.18	51.95	57.73
10 × 20	9-1/2 × 19-1/2	185.3	602.1	5870	293.3	1393	32.16	38.59	45.03	51.46	57.89	64.32
10 × 22	9-1/2 × 21-1/2	204.3	731.9	7868	323.4	1536	35.46	42.55	49.64	56.74	63.83	70.92
10 × 24	9-1/2 × 23-1/2	223.3	874.4	10270	353.5	1679	38.76	46.51	54.26	62.01	69.77	77.52

TABLE A.1.1 (*Continued*)

Nominal Size b×d	Standard Dressed Size (S4S) b×d inches × inches	Area of Section A in.²	X-X Axis Section Modulus S_x in.³	X-X Axis Moment of Inertia I_x in.⁴	Y-Y Axis Section Modulus S_y in.³	Y-Y Axis Moment of Inertia I_y in.⁴	Approximate weight in pounds per linear foot (lb./ft.) of piece when density of wood equals: 25 lb./ft.³	30 lb./ft.³	35 lb./ft.³	40 lb./ft.³	45 lb./ft.³	50 lb./ft.³
12 × 12	11-1/2 × 11-1/2	132.3	253.5	1458	253.5	1458	22.96	27.55	32.14	36.74	41.33	45.92
12 × 14	11-1/2 × 13-1/2	155.3	349.3	2358	297.6	1711	26.95	32.34	37.73	43.13	48.52	53.91
12 × 16	11-1/2 × 15-1/2	178.3	460.5	3569	341.6	1964	30.95	37.14	43.32	49.51	55.70	61.89
12 × 18	11-1/2 × 17-1/2	201.3	587.0	5136	385.7	2218	34.94	41.93	48.91	55.90	62.89	69.88
12 × 20	11-1/2 × 19-1/2	224.3	728.8	7106	429.8	2471	38.93	46.72	54.51	62.29	70.08	77.86
12 × 22	11-1/2 × 21-1/2	247.3	886.0	9524	473.9	2725	42.93	51.51	60.10	68.68	77.27	85.85
12 × 24	11-1/2 × 23-1/2	270.3	1058	12440	518.0	2978	46.92	56.30	65.69	75.07	84.45	93.84
14 × 14	13-1/2 × 13-1/2	182.3	410.1	2768	410.1	2768	31.64	37.97	44.30	50.63	56.95	63.28
14 × 16	13-1/2 × 15-1/2	209.3	540.6	4189	470.8	3178	36.33	43.59	50.86	58.13	65.39	72.66
14 × 18	13-1/2 × 17-1/2	236.3	689.1	6029	531.6	3588	41.02	49.22	57.42	65.63	73.83	82.03
14 × 20	13-1/2 × 19-1/2	263.3	855.6	8342	592.3	3998	45.70	54.84	63.98	73.13	82.27	91.41
14 × 22	13-1/2 × 21-1/2	290.3	1040	11180	653.1	4408	50.39	60.47	70.55	80.63	90.70	100.8
14 × 24	13-1/2 × 23-1/2	317.3	1243	14600	713.8	4818	55.08	66.09	77.11	88.13	99.14	110.2
16 × 16	15-1/2 × 15-1/2	240.3	620.6	4810	620.6	4810	41.71	50.05	58.39	66.74	75.08	83.42
16 × 18	15-1/2 × 17-1/2	271.3	791.1	6923	700.7	5431	47.09	56.51	65.93	75.35	84.77	94.18
16 × 20	15-1/2 × 19-1/2	302.3	982.3	9578	780.8	6051	52.47	62.97	73.46	83.96	94.45	104.9
16 × 22	15-1/2 × 21-1/2	333.3	1194	12840	860.9	6672	57.86	69.43	81.00	92.57	104.1	115.7
16 × 24	15-1/2 × 23-1/2	364.3	1427	16760	941.0	7293	63.24	75.89	88.53	101.2	113.8	126.5
18 × 18	17-1/2 × 17-1/2	306.3	893.2	7816	893.2	7816	53.17	63.80	74.44	85.07	95.70	106.3
18 × 20	17-1/2 × 19-1/2	341.3	1109	10810	995.3	8709	59.24	71.09	82.94	94.79	106.6	118.5
18 × 22	17-1/2 × 21-1/2	376.3	1348	14490	1097	9602	65.32	78.39	91.45	104.5	117.6	130.6
18 × 24	17-1/2 × 23-1/2	411.3	1611	18930	1199	10500	71.40	85.68	99.96	114.2	128.5	142.8
20 × 20	19-1/2 × 19-1/2	380.3	1236	12050	1236	12050	66.02	79.22	92.42	105.6	118.8	132.0
20 × 22	19-1/2 × 21-1/2	419.3	1502	16150	1363	13280	72.79	87.34	101.9	116.5	131.0	145.6
20 × 24	19-1/2 × 23-1/2	458.3	1795	21090	1489	14520	79.56	95.47	111.4	127.3	143.2	159.1
22 × 22	21-1/2 × 21-1/2	462.3	1656	17810	1656	17810	80.25	96.30	112.4	128.4	144.5	160.5
22 × 24	21-1/2 × 23-1/2	505.3	1979	23250	1810	19460	87.72	105.3	122.8	140.3	157.9	175.4
24 × 24	23-1/2 × 23-1/2	552.3	2163	25420	2163	25420	95.88	115.1	134.2	153.4	172.6	191.8

Source: Reprinted with permission from *National Design Specification® Supplement.* Copyright © 2001, American Forest & Paper Association, Inc.

A.1.2 Structural Glued Laminated Timber

A.1.2.1 Western Species (Based on $1\frac{1}{2}$-in.-Thick Laminations)

Beam Size			X-X Axis		Y-Y Axis		DF-L Weight per foot (based on 35 lb/ft³) lb/ft	HF Weight per foot (based on 27 lb/ft³) lb/ft
b Width in.	d Depth in.	A Area in.²	S_x Section Modulus in.³	I_x Moment of Inertia in.⁴	S_y Section Modulus in.³	I_y Moment of Inertia in.⁴		
$2\frac{1}{2}$	× 3	7.500	3.750	5.625	3.125	3.906	1.823	1.406
$2\frac{1}{2}$	× $4\frac{1}{2}$	11.25	8.438	18.98	4.688	5.859	2.734	2.109
$2\frac{1}{2}$	× 6	15.00	15.00	45.00	6.250	7.813	3.646	2.813
$2\frac{1}{2}$	× $7\frac{1}{2}$	18.75	23.44	87.89	7.813	9.766	4.557	3.516
$2\frac{1}{2}$	× 9	22.50	33.75	151.9	9.375	11.72	5.469	4.219
$2\frac{1}{2}$	× $10\frac{1}{2}$	26.25	45.94	241.2	10.94	13.67	6.380	4.922
$2\frac{1}{2}$	× 12	30.00	60.00	360.0	12.50	15.63	7.292	5.625
$2\frac{1}{2}$	× $13\frac{1}{2}$	33.75	75.94	512.6	14.06	17.58	8.203	6.328
$2\frac{1}{2}$	× 15	37.50	93.75	703.1	15.63	19.53	9.115	7.031
$2\frac{1}{2}$	× $16\frac{1}{2}$	41.25	113.4	935.9	17.19	21.48	10.03	7.734
$2\frac{1}{2}$	× 18	45.00	135.0	1215	18.75	23.44	10.94	8.438
$2\frac{1}{2}$	× $19\frac{1}{2}$	48.75	158.4	1545	20.31	25.39	11.85	9.141
$3\frac{1}{8}$	× $4\frac{1}{2}$	14.06	10.55	23.73	7.324	11.44	3.418	2.637
$3\frac{1}{8}$	× 6	18.75	18.75	56.25	9.766	15.26	4.557	3.516
$3\frac{1}{8}$	× $7\frac{1}{2}$	23.44	29.30	109.9	12.21	19.07	5.697	4.395
$3\frac{1}{8}$	× 9	28.13	42.19	189.8	14.65	22.89	6.836	5.273
$3\frac{1}{8}$	× $10\frac{1}{2}$	32.81	57.42	301.5	17.09	26.70	7.975	6.152
$3\frac{1}{8}$	× 12	37.50	75.00	450.0	19.53	30.52	9.115	7.031
$3\frac{1}{8}$	× $13\frac{1}{2}$	42.19	94.92	640.7	21.97	34.33	10.25	7.910
$3\frac{1}{8}$	× 15	46.88	117.2	878.9	24.41	38.15	11.39	8.789
$3\frac{1}{8}$	× $16\frac{1}{2}$	51.56	141.8	1170	26.86	41.96	12.53	9.668
$3\frac{1}{8}$	× 18	56.25	168.8	1519	29.30	45.78	13.67	10.55
$3\frac{1}{8}$	× $19\frac{1}{2}$	60.94	198.0	1931	31.74	49.59	14.81	11.43
$3\frac{1}{8}$	× 21	65.63	229.7	2412	34.18	53.41	15.95	12.30
$3\frac{1}{8}$	× $22\frac{1}{2}$	70.31	263.7	2966	36.62	57.22	17.09	13.18
$3\frac{1}{8}$	× 24	75.00	300.0	3600	39.06	61.04	18.23	14.06
$3\frac{1}{2}$	× $4\frac{1}{2}$	15.75	11.81	26.58	9.188	16.08	3.828	2.953
$3\frac{1}{2}$	× 6	21.00	21.00	63.00	12.25	21.44	5.104	3.938
$3\frac{1}{2}$	× $7\frac{1}{2}$	26.25	32.81	123.0	15.31	26.80	6.380	4.922
$3\frac{1}{2}$	× 9	31.50	47.25	212.6	18.38	32.16	7.656	5.906
$3\frac{1}{2}$	× $10\frac{1}{2}$	36.75	64.31	337.6	21.44	37.52	8.932	6.891
$3\frac{1}{2}$	× 12	42.00	84.00	504.0	24.50	42.88	10.21	7.875
$3\frac{1}{2}$	× $13\frac{1}{2}$	47.25	106.3	717.6	27.56	48.23	11.48	8.859
$3\frac{1}{2}$	× 15	52.50	131.3	984.4	30.63	53.59	12.76	9.844
$3\frac{1}{2}$	× $16\frac{1}{2}$	57.75	158.8	1310	33.69	58.95	14.04	10.83
$3\frac{1}{2}$	× 18	63.00	189.0	1710	36.75	64.31	15.31	11.81
$3\frac{1}{2}$	× $19\frac{1}{2}$	68.25	221.8	2163	39.81	69.67	16.59	12.80
$3\frac{1}{2}$	× 21	73.50	257.3	2701	42.88	75.03	17.86	13.78
$3\frac{1}{2}$	× $22\frac{1}{2}$	78.75	295.3	3322	45.94	80.39	19.14	14.77
$3\frac{1}{2}$	× 24	84.00	336.0	4032	49.00	85.75	20.42	15.75
$3\frac{1}{2}$	× $25\frac{1}{2}$	89.25	379.3	4836	52.06	91.11	21.69	16.73
$3\frac{1}{2}$	× 27	94.50	425.3	5741	55.13	96.47	22.97	17.72

A.1.2.1 Western Species (Continued)

Beam Size			X-X Axis		Y-Y Axis		DF-L Weight per foot (based on 35 lb/ft³)	HF Weight per foot (based on 27 lb/ft³)
b Width in.	d Depth in.	A Area in.²	S_x Section Modulus in.³	I_x Moment of Inertia in.⁴	S_y Section Modulus in.³	I_y Moment of Inertia in.⁴	lb/ft	lb/ft
$5\frac{1}{8}$	× 6	30.75	30.75	92.25	26.27	67.31	7.474	5.766
$5\frac{1}{8}$	× $7\frac{1}{2}$	38.44	48.05	180.2	32.83	84.13	9.342	7.207
$5\frac{1}{8}$	× 9	46.13	69.19	311.3	39.40	101.0	11.21	8.648
$5\frac{1}{8}$	× $10\frac{1}{2}$	53.81	94.17	494.4	45.96	117.8	13.08	10.09
$5\frac{1}{8}$	× 12	61.50	123.0	738.0	52.53	134.6	14.95	11.53
$5\frac{1}{8}$	× $13\frac{1}{2}$	69.19	155.7	1051	59.10	151.4	16.82	12.97
$5\frac{1}{8}$	× 15	76.88	192.2	1441	65.66	168.3	18.68	14.41
$5\frac{1}{8}$	× $16\frac{1}{2}$	84.56	232.5	1919	72.23	185.1	20.55	15.86
$5\frac{1}{8}$	× 18	92.25	276.8	2491	78.80	201.9	22.42	17.30
$5\frac{1}{8}$	× $19\frac{1}{2}$	99.94	324.8	3167	85.36	218.7	24.29	18.74
$5\frac{1}{8}$	× 21	107.6	376.7	3955	91.93	235.6	26.16	20.18
$5\frac{1}{8}$	× $22\frac{1}{2}$	115.3	432.4	4865	98.50	252.4	28.03	21.62
$5\frac{1}{8}$	× 24	123.0	492.0	5904	105.1	269.2	29.90	23.06
$5\frac{1}{8}$	× $25\frac{1}{2}$	130.7	555.4	7082	111.6	286.0	31.76	24.50
$5\frac{1}{8}$	× 27	138.4	622.7	8406	118.2	302.9	33.63	25.95
$5\frac{1}{8}$	× $28\frac{1}{2}$	146.1	693.8	9887	124.8	319.7	35.50	27.39
$5\frac{1}{8}$	× 30	153.8	768.8	11530	131.3	336.5	37.37	28.83
$5\frac{1}{8}$	× $31\frac{1}{2}$	161.4	847.5	11350	137.9	353.4	39.24	30.27
$5\frac{1}{8}$	× 33	169.1	930.2	15350	144.5	370.2	41.11	31.71
$5\frac{1}{8}$	× $34\frac{1}{2}$	176.8	1017	17540	151.0	387.0	42.98	33.15
$5\frac{1}{8}$	× 36	184.5	1107	19930	157.6	403.8	44.84	34.59
$5\frac{1}{8}$	× $37\frac{1}{2}$	192.2	1201	22520	164.2	420.7	46.71	36.04
$5\frac{1}{8}$	× 39	199.9	1299	25330	170.7	437.5	48.58	37.48
$5\frac{1}{8}$	× $40\frac{1}{2}$	207.6	1401	28370	177.3	454.3	50.45	38.92

A.1.2.1 Western Species (Continued)

Beam Size				X-X Axis		Y-Y Axis		DF-L Weight per foot (based on 35 lb/ft³) lb/ft	HF Weight per foot (based on 27 lb/ft³) lb/ft
b Width in.		d Depth in.	A Area in.²	S_x Section Modulus in.³	I_x Moment of Inertia in.⁴	S_y Section Modulus in.³	I_y Moment of Inertia in.⁴		
$5\frac{1}{2}$	×	6	33.00	33.00	99.00	30.25	83.19	8.021	6.188
$5\frac{1}{2}$	×	$7\frac{1}{2}$	41.25	51.56	193.4	37.81	104.0	10.03	7.734
$5\frac{1}{2}$	×	9	49.50	74.25	334.1	45.38	124.8	12.03	9.281
$5\frac{1}{2}$	×	$10\frac{1}{2}$	57.75	101.1	530.6	52.94	145.6	14.04	10.83
$5\frac{1}{2}$	×	12	66.00	132.0	792.0	60.50	166.4	16.04	12.38
$5\frac{1}{2}$	×	$13\frac{1}{2}$	74.25	167.1	1128	68.06	187.2	18.05	13.92
$5\frac{1}{2}$	×	15	82.50	206.3	1547	75.63	208.0	22.05	15.47
$5\frac{1}{2}$	×	$16\frac{1}{2}$	90.75	249.6	2059	83.19	228.8	22.06	17.02
$5\frac{1}{2}$	×	18	99.00	297.0	2673	90.75	249.6	24.06	18.56
$5\frac{1}{2}$	×	$19\frac{1}{2}$	107.3	348.6	3398	98.31	270.4	26.07	20.11
$5\frac{1}{2}$	×	21	115.5	404.3	4245	105.9	291.2	28.07	21.66
$5\frac{1}{2}$	×	$22\frac{1}{2}$	123.8	464.1	5221	113.4	312.0	30.08	23.20
$5\frac{1}{2}$	×	24	132.0	528.0	6336	121.0	332.8	32.08	24.75
$5\frac{1}{2}$	×	$25\frac{1}{2}$	140.3	596.1	7600	128.6	353.5	34.09	26.30
$5\frac{1}{2}$	×	27	148.5	668.3	9021	136.1	374.3	36.09	27.84
$5\frac{1}{2}$	×	$28\frac{1}{2}$	156.8	744.6	10610	143.7	395.1	38.10	29.39
$5\frac{1}{2}$	×	30	165.0	825.0	12380	151.3	415.9	40.10	30.94
$5\frac{1}{2}$	×	$31\frac{1}{2}$	173.3	909.6	14330	158.8	436.7	42.11	32.48
$5\frac{1}{2}$	×	33	181.5	998.3	16470	166.4	457.5	44.11	34.03
$5\frac{1}{2}$	×	$34\frac{1}{2}$	189.8	1091	18820	173.9	478.3	46.12	35.58
$5\frac{1}{2}$	×	36	198.0	1188	21380	181.5	499.1	48.13	37.13
$5\frac{1}{2}$	×	$37\frac{1}{2}$	206.3	1289	24170	189.1	519.9	50.13	38.67
$5\frac{1}{2}$	×	39	214.5	1394	27190	196.6	540.7	52.14	40.22
$5\frac{1}{2}$	×	$40\frac{1}{2}$	222.8	1504	30450	204.2	561.5	54.14	41.77
$5\frac{1}{2}$	×	42	231.0	1617	33960	211.8	582.3	56.15	43.31
$5\frac{1}{2}$	×	$43\frac{1}{2}$	239.3	1735	37730	219.3	603.1	58.15	44.86

A.1.2.1 Western Species (Continued)

Beam Size			X-X Axis		Y-Y Axis		DF-L	HF
b Width in.	d Depth in.	A Area in.^{2}	S_x Section Modulus in.^{3}	I_x Moment of Inertia in.^{4}	S_y Section Modulus in.^{3}	I_y Moment of Inertia in.^{4}	Weight per foot (based on 35 lb/ft^3) lb/ft	Weight per foot (based on 27 lb/ft^3) lb/ft
$6\frac{3}{4}$	× $7\frac{1}{2}$	50.63	63.28	237.3	56.95	192.2	12.30	9.492
$6\frac{3}{4}$	× 9	60.75	91.13	410.1	68.34	230.7	14.77	11.39
$6\frac{3}{4}$	× $10\frac{1}{2}$	70.88	124.0	651.2	79.73	269.1	17.23	13.29
$6\frac{3}{4}$	× 12	81.00	162.0	972.0	91.13	307.5	19.69	15.39
$6\frac{3}{4}$	× $13\frac{1}{2}$	91.13	205.0	1384	102.5	346.0	22.15	17.09
$6\frac{3}{4}$	× 15	101.3	253.1	1898	113.9	384.4	24.61	18.98
$6\frac{3}{4}$	× $16\frac{1}{2}$	111.4	306.3	2527	125.3	422.9	27.07	20.88
$6\frac{3}{4}$	× 18	121.5	364.5	3281	136.7	461.3	29.53	22.78
$6\frac{3}{4}$	× $19\frac{1}{2}$	131.6	427.8	4171	148.1	499.8	31.99	24.68
$6\frac{3}{4}$	× 21	141.8	496.1	5209	159.5	538.2	34.45	26.58
$6\frac{3}{4}$	× $22\frac{1}{2}$	151.9	569.5	6407	170.9	576.7	36.91	28.48
$6\frac{3}{4}$	× 24	162.0	648.0	7776	182.3	615.1	39.38	30.38
$6\frac{3}{4}$	× $25\frac{1}{2}$	172.1	731.5	9327	193.6	653.5	41.84	32.27
$6\frac{3}{4}$	× 27	182.3	820.1	11070	205.0	692.0	44.30	34.17
$6\frac{3}{4}$	× $28\frac{1}{2}$	192.4	913.8	13020	216.4	730.4	46.76	36.07
$6\frac{3}{4}$	× 30	202.5	1013	15190	227.8	768.9	49.22	37.97
$6\frac{3}{4}$	× $31\frac{1}{2}$	212.6	1116	17580	239.2	807.3	51.68	39.87
$6\frac{3}{4}$	× 33	222.8	1225	20210	250.6	845.8	54.14	41.77
$6\frac{3}{4}$	× $34\frac{1}{2}$	232.9	1339	23100	262.0	884.2	56.60	43.66
$6\frac{3}{4}$	× 36	243.0	1458	26240	273.4	922.6	59.06	45.56
$6\frac{3}{4}$	× $37\frac{1}{2}$	253.1	1582	29660	284.8	961.1	61.52	47.46
$6\frac{3}{4}$	× 39	263.3	1711	33370	296.2	999.5	63.98	49.36
$6\frac{3}{4}$	× $40\frac{1}{2}$	273.4	1845	37370	307.5	1038	66.45	51.26
$6\frac{3}{4}$	× 42	283.5	1985	41670	318.9	1076	68.91	53.16
$6\frac{3}{4}$	× $43\frac{1}{2}$	293.6	2129	46300	330.3	1115	71.37	55.05
$6\frac{3}{4}$	× 45	303.8	2278	51260	341.7	1153	73.83	56.95
$6\frac{3}{4}$	× $46\frac{1}{2}$	313.9	2433	56560	353.1	1192	76.29	58.85
$6\frac{3}{4}$	× 48	324.0	2592	62210	364.5	1230	78.75	60.75
$6\frac{3}{4}$	× $49\frac{1}{2}$	334.1	2757	68220	375.9	1269	81.21	62.65
$6\frac{3}{4}$	× 51	344.3	2926	74620	387.3	1307	83.67	64.55
$6\frac{3}{4}$	× $52\frac{1}{2}$	354.4	3101	81400	398.7	1346	86.13	66.45
$6\frac{3}{4}$	× 54	364.5	3281	88570	410.1	1384	88.59	68.34

A.1.2.1 Western Species (Continued)

Beam Size			X-X Axis		Y-Y Axis		DF-L Weight per foot (based on 35 lb/ft³) lb/ft	HF Weight per foot (based on 27 lb/ft³) lb/ft
b Width in.	d Depth in.	A Area in.²	S_x Section Modulus in.³	I_x Moment of Inertia in.⁴	S_y Section Modulus in.³	I_y Moment of Inertia in.⁴		
$8\frac{3}{4}$	× 9	78.75	118.1	531.6	114.8	502.4	19.14	14.77
$8\frac{3}{4}$	× $10\frac{1}{2}$	91.88	160.8	844.1	134.0	586.2	22.33	17.23
$8\frac{3}{4}$	× 12	105.0	210.0	1260	153.1	669.9	25.52	19.69
$8\frac{3}{4}$	× $13\frac{1}{2}$	118.1	265.8	1794	172.3	753.7	28.71	22.15
$8\frac{3}{4}$	× 15	131.3	328.1	2461	191.4	837.4	31.90	24.61
$8\frac{3}{4}$	× $16\frac{1}{2}$	144.4	397	3276	210.5	921.1	35.09	27.07
$8\frac{3}{4}$	× 18	157.5	472.5	4253	229.7	1005	38.28	29.53
$8\frac{3}{4}$	× $19\frac{1}{2}$	170.6	554.5	5407	248.8	1089	41.47	31.99
$8\frac{3}{4}$	× 21	183.8	643.1	6753	268.0	1172	44.66	34.45
$8\frac{3}{4}$	× $22\frac{1}{2}$	196.9	738.3	8306	287.1	1256	47.85	36.91
$8\frac{3}{4}$	× 24	210.0	840.0	10080	306.3	1340	51.04	39.38
$8\frac{3}{4}$	× $25\frac{1}{2}$	223.1	948.3	12090	325.4	1424	54.23	41.84
$8\frac{3}{4}$	× 27	236.3	1063	14350	344.5	1507	57.42	44.30
$8\frac{3}{4}$	× $28\frac{1}{2}$	249.4	1185	16880	363.7	1591	60.61	46.76
$8\frac{3}{4}$	× 30	262.5	1313	19690	382.8	1675	63.80	49.22
$8\frac{3}{4}$	× $31\frac{1}{2}$	275.6	1447	22790	402.0	1759	66.99	51.68
$8\frac{3}{4}$	× 33	288.8	1588	26200	421.1	1842	70.18	54.14
$8\frac{3}{4}$	× $34\frac{1}{2}$	301.9	1736	29940	440.2	1926	73.37	56.60
$8\frac{3}{4}$	× 36	315.0	1890	34020	459.4	2010	76.56	59.06
$8\frac{3}{4}$	× $37\frac{1}{2}$	328.1	2051	38450	478.5	2094	79.75	61.52
$8\frac{3}{4}$	× 39	341.3	2218	43250	497.7	2177	82.94	63.98
$8\frac{3}{4}$	× $40\frac{1}{2}$	354.4	2392	48440	516.8	2261	86.13	66.45
$8\frac{3}{4}$	× 42	367.5	2573	54020	535.9	2345	89.32	68.91
$8\frac{3}{4}$	× $43\frac{1}{2}$	380.6	2760	60020	555.1	2428	92.51	71.37
$8\frac{3}{4}$	× 45	393.8	2953	66450	574.2	2512	95.70	73.83
$8\frac{3}{4}$	× $46\frac{1}{2}$	406.9	3153	73310	593.4	2596	98.89	76.29
$8\frac{3}{4}$	× 48	420.0	3360	80640	612.5	2680	102.1	78.75
$8\frac{3}{4}$	× $49\frac{1}{2}$	433.1	3573	88440	631.6	2763	105.3	81.21
$8\frac{3}{4}$	× 51	446.3	3793	96720	650.8	2847	108.5	83.67
$8\frac{3}{4}$	× $52\frac{1}{2}$	459.4	4020	105500	669.9	2931	111.7	86.13
$8\frac{3}{4}$	× 54	472.5	4253	114800	689.1	3051	114.8	88.59
$8\frac{3}{4}$	× $55\frac{1}{2}$	485.6	4492	124700	708.2	3098	118.0	91.05
$8\frac{3}{4}$	× 57	498.8	4738	135000	727.3	3182	121.2	93.52
$8\frac{3}{4}$	× $58\frac{1}{2}$	511.9	4991	146000	746.5	3266	124.4	95.98
$8\frac{3}{4}$	× 60	525.0	5250	157500	765.6	3350	127.6	98.44

A.1.2.1 Western Species (Continued)

Beam Size			X-X Axis		Y-Y Axis		DF-L Weight per foot (based on 35 lb/ft³) lb/ft	HF Weight per foot (based on 27 lb/ft³) lb/ft
b Width in.	**d** Depth in.	**A** Area in.²	S_x Section Modulus in.³	I_x Moment of Inertia in.⁴	S_y Section Modulus in.³	I_y Moment of Inertia in.⁴		
$10\frac{3}{4}$	× 12	129.0	258	1548	231.1	1242	31.35	24.19
$10\frac{3}{4}$	× $13\frac{1}{2}$	145.1	326.5	2204	260.0	1398	35.27	27.21
$10\frac{3}{4}$	× 15	161.3	403.1	3023	288.9	1553	39.19	30.23
$10\frac{3}{4}$	× $16\frac{1}{2}$	177.4	487.8	4024	317.8	1708	43.11	33.26
$10\frac{3}{4}$	× 18	193.5	580.5	5225	346.7	1863	47.03	36.28
$10\frac{3}{4}$	× $19\frac{1}{2}$	209.6	681.3	6642	375.6	2019	50.95	39.30
$10\frac{3}{4}$	× 21	225.8	790.1	8296	404.5	2174	54.87	42.33
$10\frac{3}{4}$	× $22\frac{1}{2}$	241.9	907	10200	433.4	2329	58.79	45.35
$10\frac{3}{4}$	× 24	258.0	1032	12380	462.3	2485	62.71	48.38
$10\frac{3}{4}$	× $25\frac{1}{2}$	274.1	1165	14850	491.1	2640	66.63	51.40
$10\frac{3}{4}$	× 27	290.3	1306	17630	520.0	2795	70.55	54.42
$10\frac{3}{4}$	× $28\frac{1}{2}$	306.4	1455	20740	548.9	2950	74.47	57.45
$10\frac{3}{4}$	× 30	322.5	1613	24190	577.8	3106	78.39	60.47
$10\frac{3}{4}$	× $31\frac{1}{2}$	338.6	1778	28000	606.7	3261	82.30	63.49
$10\frac{3}{4}$	× 33	354.8	1951	32190	635.6	3416	86.22	66.52
$10\frac{3}{4}$	× $34\frac{1}{2}$	370.9	2133	36790	664.5	3572	90.14	69.54
$10\frac{3}{4}$	× 36	387.0	2322	41800	693.4	3727	94.06	72.56
$10\frac{3}{4}$	× $37\frac{1}{2}$	403.1	2520	47240	722.3	3882	97.98	75.59
$10\frac{3}{4}$	× 39	419.3	2725	53140	751.2	4037	101.9	78.61
$10\frac{3}{4}$	× $40\frac{1}{2}$	435.4	2939	59510	780.0	4193	105.8	81.63
$10\frac{3}{4}$	× 42	451.5	3161	66370	808.9	4348	109.7	84.66
$10\frac{3}{4}$	× $43\frac{1}{2}$	467.6	3390	73740	837.8	4503	113.7	87.68
$10\frac{3}{4}$	× 45	483.8	3628	81630	866.7	4659	117.6	90.70
$10\frac{3}{4}$	× $46\frac{1}{2}$	499.9	3874	90070	895.6	4814	121.5	93.73
$10\frac{3}{4}$	× 48	516.0	4128	99070	924.5	4969	125.4	96.75
$10\frac{3}{4}$	× $49\frac{1}{2}$	532.1	4390	108700	953.4	5124	129.3	99.77
$10\frac{3}{4}$	× 51	548.3	4660	118800	982.3	5280	133.3	102.8
$10\frac{3}{4}$	× $52\frac{1}{2}$	564.4	4938	129600	1011	5435	137.2	105.8
$10\frac{3}{4}$	× 54	580.5	5225	141100	1040	5590	141.1	108.8
$10\frac{3}{4}$	× $55\frac{1}{2}$	596.6	5519	153100	1069	5746	145.0	111.9
$10\frac{3}{4}$	× 57	612.8	5821	165900	1098	5901	148.9	114.9
$10\frac{3}{4}$	× $58\frac{1}{2}$	628.9	6132	179300	1127	6056	152.9	117.9
$10\frac{3}{4}$	× 60	645.0	6450	193500	1156	6211	156.8	120.9

A.1.2.1 Western Species (Continued)

Beam Size			X-X Axis		Y-Y Axis		DF-L Weight per foot (based on 35 lb/ft³) lb/ft	HF Weight per foot (based on 27 lb/ft³) lb/ft
b Width in.	d Depth in.	A Area in.²	S_x Section Modulus in.³	I_x Moment of Inertia in.⁴	S_y Section Modulus in.³	I_y Moment of Inertia in.⁴		
12¼	× 13½	165.4	372.1	2512	337.6	2068	40.20	31.01
12¼	× 15½	183.8	459.4	3445	375.2	2298	44.66	34.45
12¼	× 16½	202.1	555.8	4586	412.7	2528	49.13	37.90
12¼	× 18	220.5	661.5	5954	450.2	2757	53.59	41.34
12¼	× 19½	238.9	776.3	7569	487.7	2987	58.06	44.79
12¼	× 21	257.3	900.4	9454	525.2	3217	62.53	48.23
12¼	× 22½	275.6	1034	11630	562.7	3447	66.99	51.68
12¼	× 24	294.0	1176	14110	600.3	3667	71.46	55.13
12¼	× 25½	312.4	1328	16930	637.8	3906	75.92	58.57
12¼	× 27	330.8	1488	20090	675.3	4136	80.39	62.02
12¼	× 28½	349.1	1658	23630	712.8	4366	84.86	65.46
12¼	× 30	367.5	1838	27560	750.3	4596	89.32	68.91
12¼	× 31½	385.9	2026	31910	787.8	4825	93.79	72.35
12¼	× 33	404.3	2223	36690	825.3	5055	98.26	75.80
12¼	× 34½	422.6	2430	41920	862.9	5285	102.7	79.24
12¼	× 36	441.0	2646	47630	900.4	5515	107.2	82.69
12¼	× 37½	459.4	2871	53830	937.9	5745	111.7	86.13
12¼	× 39	477.8	3105	60550	975.4	5974	116.1	89.58
12¼	× 40½	496.1	3349	67810	1013	6204	120.6	93.02
12¼	× 42	514.5	3602	75630	1050	6434	125.1	96.47
12¼	× 43½	532.9	3863	84030	1088	6664	129.5	99.91
12¼	× 45	551.3	4134	93020	1125	6893	134.0	103.4
12¼	× 46½	569.6	4415	102600	1163	7123	138.5	106.8
12¼	× 48	588.0	4704	112900	1201	7353	142.9	110.3
12¼	× 49½	606.4	5003	123800	1238	7583	147.4	113.7
12¼	× 51	624.8	5310	135400	1276	7813	151.8	117.1
12¼	× 52½	643.1	5627	147700	1313	8042	156.3	120.6
12¼	× 54	661.5	5954	160700	1351	8272	160.8	124.0
12¼	× 55½	679.9	6289	174500	1388	8502	165.2	127.5
12¼	× 57	698.3	6633	189100	1426	8732	169.7	130.9
12¼	× 58½	716.6	6987	204400	1463	8962	174.2	134.4
12¼	× 60	735.0	7350	220500	1501	9191	178.6	137.8

A.1.2.1 Western Species (Continued)

Beam Size			X-X Axis		Y-Y Axis		DF-L Weight per foot (based on 35 lb/ft³) lb/ft	HF Weight per foot (based on 27 lb/ft³) lb/ft
b Width in.	d Depth in.	A Area in.²	S_x Section Modulus in.³	I_x Moment of Inertia in.⁴	S_y Section Modulus in.³	I_y Moment of Inertia in.⁴		
14¼	× 15	213.8	534.4	4008	507.7	3617	51.95	40.08
14¼	× 16½	235.1	646.6	5334	558.4	3979	57.15	44.09
14¼	× 18	256.5	769.5	6926	609.2	4340	62.34	48.09
14¼	× 19½	277.9	903.1	8805	660.0	4702	67.54	52.10
14¼	× 21	299.3	1047	11000	710.7	5064	72.73	56.11
14¼	× 22½	320.6	1202	13530	761.5	5426	77.93	60.12
14¼	× 24	342.0	1368	16420	812.3	5787	83.13	64.13
14¼	× 25½	363.4	1544	19690	863.0	6149	88.32	68.13
14¼	× 27	384.8	1731	23370	913.8	6511	93.52	72.14
14¼	× 28½	406.1	1929	27490	964.5	6872	98.71	76.15
14¼	× 30	427.5	2138	32060	1015	7234	103.9	80.16
14¼	× 31½	448.9	2357	37120	1066	7596	109.1	84.16
14¼	× 33	470.3	2586	42680	1117	7958	114.3	88.17
14¼	× 34½	491.6	2827	48760	1168	8319	119.5	92.18
14¼	× 36	513.0	3078	55400	1218	8681	124.7	96.19
14¼	× 37½	534.4	3340	62620	1269	9043	129.9	100.2
14¼	× 39	555.8	3612	70440	1320	9404	135.1	104.2
14¼	× 40½	577.1	3896	78890	1371	9766	140.3	108.2
14¼	× 42	598.5	4190	87980	1421	10130	145.5	112.2
14¼	× 43½	619.9	4494	97750	1472	10490	150.7	116.2
14¼	× 45	641.3	4809	108200	1523	10850	155.9	120.2
14¼	× 46½	662.6	5135	119400	1574	11210	161.1	124.2
14¼	× 48	684.0	5472	131300	1625	11570	166.3	128.3
14¼	× 49½	705.4	5819	144000	1675	11940	171.4	132.3
14¼	× 51	726.8	6177	157500	1726	12300	176.6	136.3
14¼	× 52½	748.1	6546	171800	1777	12660	181.8	140.3
14¼	× 54	769.5	6926	187000	1828	13020	187.0	144.3
14¼	× 55½	790.9	7316	203000	1878	13380	192.2	148.3
14¼	× 57	812.3	7716	219900	1929	13740	197.4	152.3
14¼	× 58½	833.6	8128	237700	1980	14110	202.6	156.3
14¼	× 60	855.0	8550	256500	2031	14470	207.8	160.3

A.1.2.2 Southern Pine (Based on $1\frac{3}{8}$-in.-Thick Laminations)

Beam Size				X-X Axis		Y-Y Axis		SP
b Width in.		**d** Depth in.	**A** Area in.^{2}	$\mathbf{S_x}$ Section Modulus in.^{3}	$\mathbf{I_x}$ Moment of Inertia in.^{4}	$\mathbf{S_y}$ Section Modulus in.^{3}	$\mathbf{I_y}$ Moment of Inertia in.^{4}	Weight per foot (based on 36 lb/ft^3) lb/ft
$2\frac{1}{2}$	×	$2\frac{3}{4}$	6.875	3.151	4.333	2.865	3.581	1.7
$2\frac{1}{2}$	×	$4\frac{1}{8}$	10.31	7.090	14.62	4.297	5.371	2.5
$2\frac{1}{2}$	×	$5\frac{1}{2}$	13.75	12.60	34.66	5.729	7.161	3.3
$2\frac{1}{2}$	×	$6\frac{7}{8}$	17.19	19.69	67.70	7.161	8.952	4.2
$2\frac{1}{2}$	×	$8\frac{1}{4}$	20.63	28.36	117.0	8.594	10.74	5.0
$2\frac{1}{2}$	×	$9\frac{5}{8}$	24.06	38.60	185.8	10.03	12.53	5.8
$2\frac{1}{2}$	×	11	27.50	50.42	277.3	11.46	14.32	6.7
$2\frac{1}{2}$	×	$12\frac{3}{8}$	30.94	63.81	394.8	12.89	16.11	7.5
$2\frac{1}{2}$	×	$13\frac{3}{4}$	34.38	78.78	541.6	14.32	17.90	8.4
$2\frac{1}{2}$	×	$15\frac{1}{8}$	37.81	95.32	720.9	15.76	19.69	9.2
$2\frac{1}{2}$	×	$16\frac{1}{2}$	41.25	113.4	935.9	17.19	21.48	10.0
$2\frac{1}{2}$	×	$17\frac{7}{8}$	44.69	133.1	1190	18.62	23.27	10.9
$2\frac{1}{2}$	×	$19\frac{1}{4}$	48.13	154.4	1486	20.05	25.07	11.7
3	×	$4\frac{1}{8}$	12.38	8.508	17.55	6.188	9.281	3.0
3	×	$5\frac{1}{2}$	16.50	15.13	41.59	8.250	12.38	4.0
3	×	$6\frac{7}{8}$	20.63	23.63	81.24	10.31	15.47	5.0
3	×	$8\frac{1}{4}$	24.75	34.03	140.4	12.38	18.56	6.0
3	×	$9\frac{5}{8}$	28.88	46.32	222.9	14.44	21.66	7.0
3	×	11	33.00	60.50	332.8	16.50	24.75	8.0
3	×	$12\frac{3}{8}$	37.13	76.57	473.8	18.56	27.84	9.0
3	×	$13\frac{3}{4}$	41.25	94.53	649.9	20.63	30.94	10.0
3	×	$15\frac{1}{8}$	45.38	114.4	865.0	22.69	34.03	11.0
3	×	$16\frac{1}{2}$	49.50	136.1	1123	24.75	37.13	12.0
3	×	$17\frac{7}{8}$	53.63	159.8	1428	26.81	40.22	13.0
3	×	$19\frac{1}{4}$	57.75	185.3	1783	28.88	43.31	14.0
3	×	$20\frac{5}{8}$	61.88	212.7	2193	30.94	46.41	15.0
3	×	22	66.00	242.0	2662	33.00	49.50	16.0
3	×	$23\frac{3}{8}$	70.13	273.2	3193	35.06	52.59	17.0

A.1.2.2 Southern Pine (Continued)

Beam Size				X-X Axis		Y-Y Axis		
b Width in.		d Depth in.	A Area in.^{2}	S_x Section Modulus in.^{3}	I_x Moment of Inertia in.^{4}	S_y Section Modulus in.^{3}	I_y Moment of Inertia in.^{4}	SP Weight per foot (based on 36 lb/ft^3) lb/ft
$3\frac{1}{8}$	×	$4\frac{1}{8}$	12.89	8.862	18.28	6.714	10.49	3.1
$3\frac{1}{8}$	×	$5\frac{1}{2}$	17.19	15.76	43.33	8.952	13.99	4.2
$3\frac{1}{8}$	×	$6\frac{7}{8}$	21.48	24.62	84.62	11.19	17.48	5.2
$3\frac{1}{8}$	×	$8\frac{1}{4}$	25.78	35.45	146.2	13.43	20.98	6.3
$3\frac{1}{8}$	×	$9\frac{5}{8}$	30.08	48.25	232.2	15.67	24.48	7.3
$3\frac{1}{8}$	×	11	34.38	63.02	346.6	17.90	27.97	8.4
$3\frac{1}{8}$	×	$12\frac{3}{8}$	38.67	79.76	493.5	20.14	31.47	9.4
$3\frac{1}{8}$	×	$13\frac{3}{4}$	42.97	98.47	677	22.38	34.97	10.4
$3\frac{1}{8}$	×	$15\frac{1}{8}$	47.27	119.1	901.1	24.62	38.46	11.5
$3\frac{1}{8}$	×	$16\frac{1}{2}$	51.56	141.8	1170	26.86	41.96	12.5
$3\frac{1}{8}$	×	$17\frac{7}{8}$	55.86	166.4	1487	29.09	45.46	13.6
$3\frac{1}{8}$	×	$19\frac{1}{4}$	60.16	193.0	1858	31.33	48.96	14.6
$3\frac{1}{8}$	×	$20\frac{5}{8}$	64.45	221.6	2285	33.57	52.45	15.7
$3\frac{1}{8}$	×	22	68.75	252.1	2773	35.81	55.95	16.7
$3\frac{1}{8}$	×	$23\frac{3}{8}$	73.05	284.6	3326	38.05	59.45	17.8
$3\frac{1}{8}$	×	$24\frac{3}{4}$	77.34	319.0	3948	40.28	62.94	18.8
$3\frac{1}{2}$	×	$4\frac{1}{8}$	14.44	9.926	20.47	8.422	14.74	3.5
$3\frac{1}{2}$	×	$5\frac{1}{8}$	19.25	17.65	48.53	11.23	19.65	4.7
$3\frac{1}{2}$	×	$6\frac{7}{8}$	24.06	27.57	94.78	14.04	24.56	5.8
$3\frac{1}{2}$	×	$8\frac{1}{4}$	28.88	39.70	163.8	16.84	29.48	7.0
$3\frac{1}{2}$	×	$9\frac{5}{8}$	33.69	54.04	260.1	19.65	34.39	8.2
$3\frac{1}{2}$	×	11	38.50	70.58	388.2	22.46	39.30	9.4
$3\frac{1}{2}$	×	$12\frac{3}{8}$	43.31	89.33	552.7	25.27	44.21	10.5
$3\frac{1}{2}$	×	$13\frac{3}{4}$	48.13	110.3	758.2	28.07	49.13	11.7
$3\frac{1}{2}$	×	$15\frac{1}{8}$	52.94	133.4	1009	30.88	54.04	12.9
$3\frac{1}{2}$	×	$16\frac{1}{2}$	57.75	158.8	1310	33.69	58.95	14.0
$3\frac{1}{2}$	×	$17\frac{7}{8}$	62.56	186.4	1666	36.49	63.87	15.2
$3\frac{1}{2}$	×	$19\frac{1}{4}$	67.38	216.2	2081	39.30	68.78	16.4
$3\frac{1}{2}$	×	$20\frac{5}{8}$	72.19	248.1	2559	42.11	73.69	17.5
$3\frac{1}{2}$	×	22	77.00	282.3	3106	44.92	78.60	18.7
$3\frac{1}{2}$	×	$23\frac{3}{8}$	81.81	318.7	3725	47.72	83.52	19.9
$3\frac{1}{2}$	×	$24\frac{3}{4}$	86.63	357.3	4422	50.53	88.43	21.1
$3\frac{1}{2}$	×	$26\frac{1}{8}$	91.44	398.1	5201	53.34	93.34	22.2
$3\frac{1}{2}$	×	$27\frac{1}{2}$	96.25	441.1	6066	56.15	98.26	23.4

A.1.2.2 Southern Pine (Continued)

Beam Size				X-X Axis		Y-Y Axis		SP
b Width in.		d Depth in.	A Area in.^{2}	S_x Section Modulus in.^{3}	I_x Moment of Inertia in.^{4}	S_y Section Modulus in.^{3}	I_y Moment of Inertia in.^{4}	Weight per foot (based on 36 lb/ft^3) lb/ft
5	×	$5\frac{1}{2}$	27.50	25.21	69.32	22.92	57.29	6.7
5	×	$6\frac{7}{8}$	34.38	39.39	135.4	28.65	71.61	8.4
5	×	$8\frac{1}{4}$	41.25	56.72	234.0	34.38	85.94	10.0
5	×	$9\frac{5}{8}$	48.13	77.20	371.5	40.10	100.3	11.7
5	×	11	55.00	100.8	554.6	45.83	114.6	13.4
5	×	$12\frac{3}{8}$	61.88	127.6	789.6	51.56	128.9	15.0
5	×	$13\frac{3}{4}$	68.75	157.6	1083	57.29	143.2	16.7
5	×	$15\frac{1}{8}$	75.63	190.6	1442	63.02	157.6	18.4
5	×	$16\frac{1}{2}$	82.50	226.9	1872	68.75	171.9	20.1
5	×	$17\frac{7}{8}$	89.38	266.3	2380	74.48	186.2	21.7
5	×	$19\frac{1}{4}$	96.25	308.8	2972	80.21	200.5	23.4
5	×	$20\frac{5}{8}$	103.1	354.5	3656	85.94	214.8	25.1
5	×	22	110.0	403.3	4437	91.67	229.2	26.7
5	×	$23\frac{3}{8}$	116.9	455.3	5322	97.40	243.5	28.4
5	×	$24\frac{3}{4}$	123.8	510.5	6317	103.1	257.8	30.1
5	×	$26\frac{1}{8}$	130.6	568.8	7429	108.9	272.1	31.7
5	×	$27\frac{1}{2}$	137.5	630.2	8665	114.6	286.5	33.4
5	×	$28\frac{7}{8}$	144.4	694.8	10030	120.3	300.8	35.1
5	×	$30\frac{1}{4}$	151.3	762.6	11530	126.0	315.1	36.8
5	×	$31\frac{5}{8}$	158.1	833.5	13180	131.8	329.4	38.4
5	×	33	165.0	907.5	14970	137.5	343.8	40.1
5	×	$34\frac{3}{8}$	171.9	984.7	16920	143.2	358.1	41.8
5	×	$35\frac{3}{4}$	178.8	1065	19040	149.0	372.4	43.4
5	×	$37\frac{1}{8}$	185.6	1149	21320	154.7	386.7	45.1
5	×	$38\frac{1}{2}$	192.5	1235	23780	160.4	401.0	46.8
5	×	$39\frac{7}{8}$	199.4	1325	26420	166.1	415.4	48.5

A.1.2.2 Southern Pine (Continued)

Beam Size				X-X Axis		Y-Y Axis		SP
b Width in.		d Depth in.	A Area in.^{2}	S_x Section Modulus in.^{3}	I_x Moment of Inertia in.^{4}	S_y Section Modulus in.^{3}	I_y Moment of Inertia in.^{4}	Weight per foot (based on 36 lb/ft^3) lb/ft
$5\frac{1}{8}$	×	$5\frac{1}{2}$	28.19	25.84	71.06	24.08	61.70	6.9
$5\frac{1}{8}$	×	$6\frac{7}{8}$	35.23	40.37	138.8	30.10	77.12	8.6
$5\frac{1}{8}$	×	$8\frac{1}{4}$	42.28	58.14	239.8	36.12	92.55	10.3
$5\frac{1}{8}$	×	$9\frac{5}{8}$	49.33	79.13	380.8	42.13	108.0	12.0
$5\frac{1}{8}$	×	11	56.38	103.4	568.4	48.15	123.4	13.7
$5\frac{1}{8}$	×	$12\frac{3}{8}$	63.42	130.8	809.4	54.17	138.8	15.4
$5\frac{1}{8}$	×	$13\frac{3}{4}$	70.47	161.5	1110	60.19	154.2	17.1
$5\frac{1}{8}$	×	$15\frac{1}{8}$	77.52	195.4	1478	66.21	169.7	18.8
$5\frac{1}{8}$	×	$16\frac{1}{2}$	84.56	232.5	1919	72.23	185.1	20.6
$5\frac{1}{8}$	×	$17\frac{7}{8}$	91.61	272.9	2439	78.25	200.5	22.3
$5\frac{1}{8}$	×	$19\frac{1}{4}$	98.66	316.5	3047	84.27	215.9	24.0
$5\frac{1}{8}$	×	$20\frac{5}{8}$	105.7	363.4	3747	90.29	231.4	25.7
$5\frac{1}{8}$	×	22	112.8	413.4	4548	96.31	246.8	27.4
$5\frac{1}{8}$	×	$23\frac{3}{8}$	119.8	466.7	5455	102.3	262.2	29.1
$5\frac{1}{8}$	×	$24\frac{3}{4}$	126.8	523.2	6475	108.3	277.6	30.8
$5\frac{1}{8}$	×	$26\frac{1}{8}$	133.9	583.0	7615	114.4	293.1	32.5
$5\frac{1}{8}$	×	$27\frac{1}{2}$	140.9	646.0	8882	120.4	308.5	34.3
$5\frac{1}{8}$	×	$28\frac{7}{8}$	148.0	712.2	10280	126.4	323.9	36.0
$5\frac{1}{8}$	×	$30\frac{1}{4}$	155.0	781.6	11820	132.4	339.3	37.7
$5\frac{1}{8}$	×	$31\frac{5}{8}$	162.1	854.3	13510	138.4	354.8	39.4
$5\frac{1}{8}$	×	33	169.1	930.2	15350	144.5	370.2	41.1
$5\frac{1}{8}$	×	$34\frac{3}{8}$	176.2	1009	17350	150.5	385.6	42.8
$5\frac{1}{8}$	×	$35\frac{3}{4}$	183.2	1092	19510	156.5	401.0	44.5
$5\frac{1}{8}$	×	$37\frac{1}{8}$	190.3	1177	21850	162.5	416.5	46.2
$5\frac{1}{8}$	×	$38\frac{1}{2}$	197.3	1266	24370	168.5	431.9	48.0
$5\frac{1}{8}$	×	$39\frac{7}{8}$	204.4	1358	27080	174.6	447.3	49.7

A.1.2.2 Southern Pine (Continued)

Beam Size				X-X Axis		Y-Y Axis		SP
b Width in.		d Depth in.	A Area in.^{2}	S_x Section Modulus in.^{3}	I_x Moment of Inertia in.^{4}	S_y Section Modulus in.^{3}	I_y Moment of Inertia in.^{4}	Weight per foot (based on 36 lb/ft^3) lb/ft
$5\frac{1}{2}$	×	$5\frac{1}{2}$	30.25	27.73	76.26	27.73	76.26	7.4
$5\frac{1}{2}$	×	$6\frac{7}{8}$	37.81	43.33	148.9	34.66	95.32	9.2
$5\frac{1}{2}$	×	$8\frac{1}{4}$	45.38	62.39	257.4	41.59	114.4	11.0
$5\frac{1}{2}$	×	$9\frac{5}{8}$	52.94	84.92	408.7	48.53	133.4	12.9
$5\frac{1}{2}$	×	11	60.50	110.9	610.0	55.46	152.5	14.7
$5\frac{1}{2}$	×	$12\frac{3}{8}$	68.06	140.4	868.6	62.39	171.6	16.5
$5\frac{1}{2}$	×	$13\frac{3}{4}$	75.63	173.3	1191	69.32	190.6	18.4
$5\frac{1}{2}$	×	$15\frac{1}{8}$	83.19	209.7	1586	76.26	209.7	20.2
$5\frac{1}{2}$	×	$16\frac{1}{2}$	90.75	249.6	2059	83.19	228.8	22.1
$5\frac{1}{2}$	×	$17\frac{7}{8}$	98.31	292.9	2618	90.12	247.8	23.9
$5\frac{1}{2}$	×	$19\frac{1}{4}$	105.9	339.7	3269	97.05	266.9	25.7
$5\frac{1}{2}$	×	$20\frac{5}{8}$	113.4	389.9	4021	104.0	286.0	27.6
$5\frac{1}{2}$	×	22	121.0	443.7	4880	110.9	305.0	29.4
$5\frac{1}{2}$	×	$23\frac{3}{8}$	128.6	500.9	5854	117.8	324.1	31.2
$5\frac{1}{2}$	×	$24\frac{3}{4}$	136.1	561.5	6949	124.8	343.1	33.1
$5\frac{1}{2}$	×	$26\frac{1}{8}$	143.7	625.6	8172	131.7	362.2	34.9
$5\frac{1}{2}$	×	$27\frac{1}{2}$	151.3	693.2	9532	138.6	381.3	36.8
$5\frac{1}{2}$	×	$28\frac{7}{8}$	158.8	764.3	11030	145.6	400.3	38.6
$5\frac{1}{2}$	×	$30\frac{1}{4}$	166.4	838.8	12690	152.5	419.4	40.4
$5\frac{1}{2}$	×	$31\frac{5}{8}$	173.9	916.8	14500	159.4	438.5	42.3
$5\frac{1}{2}$	×	33	181.5	998.3	16470	166.4	457.5	44.1
$5\frac{1}{2}$	×	$34\frac{3}{8}$	189.1	1083	18620	173.3	476.6	46.0
$5\frac{1}{2}$	×	$35\frac{3}{4}$	196.6	1172	20940	180.2	495.7	47.8
$5\frac{1}{2}$	×	$37\frac{1}{8}$	204.2	1263	23450	187.2	514.7	49.6
$5\frac{1}{2}$	×	$38\frac{1}{2}$	211.8	1359	26160	194.1	533.8	51.5
$5\frac{1}{2}$	×	$39\frac{7}{8}$	219.3	1458	29060	201.0	552.9	53.3
$5\frac{1}{2}$	×	$41\frac{1}{4}$	226.9	1560	32170	208.0	571.9	55.1
$5\frac{1}{2}$	×	$42\frac{5}{8}$	234.4	1665	35500	214.9	591.0	57.0
$5\frac{1}{2}$	×	44	242.0	1775	39040	221.8	610.0	58.8

A.1.2.2 Southern Pine (*Continued*)

Beam Size				X-X Axis		Y-Y Axis		SP
b Width in.		**d** Depth in.	**A** Area in.^{2}	$\mathbf{S_x}$ Section Modulus in.^{3}	$\mathbf{I_x}$ Moment of Inertia in.^{4}	$\mathbf{S_y}$ Section Modulus in.^{3}	$\mathbf{I_y}$ Moment of Inertia in.^{4}	Weight per foot (based on 36 lb/ft^3) lb/ft
$6\frac{3}{4}$	×	$6\frac{7}{8}$	46.41	53.17	182.8	52.21	176.2	11.3
$6\frac{3}{4}$	×	$8\frac{1}{4}$	55.69	76.57	315.9	62.65	211.4	13.5
$6\frac{3}{4}$	×	$9\frac{5}{8}$	64.97	104.2	501.6	73.09	246.7	15.8
$6\frac{3}{4}$	×	11	74.25	136.1	748.7	83.53	281.9	18.0
$6\frac{3}{4}$	×	$12\frac{3}{8}$	83.53	172.3	1066	93.97	317.2	20.3
$6\frac{3}{4}$	×	$13\frac{3}{4}$	92.81	212.7	1462	104.4	352.4	22.6
$6\frac{3}{4}$	×	$15\frac{1}{8}$	102.1	257.4	1946	114.9	387.6	24.8
$6\frac{3}{4}$	×	$16\frac{1}{2}$	111.4	306.3	2527	125.3	422.9	27.1
$6\frac{3}{4}$	×	$17\frac{7}{8}$	120.7	359.5	3213	135.7	458.1	29.3
$6\frac{3}{4}$	×	$19\frac{1}{4}$	129.9	416.9	4012	146.2	493.4	31.6
$6\frac{3}{4}$	×	$20\frac{5}{8}$	139.2	478.6	4935	156.6	528.6	33.8
$6\frac{3}{4}$	×	22	148.5	544.5	5990	167.1	563.8	36.1
$6\frac{3}{4}$	×	$23\frac{3}{8}$	157.8	614.7	7184	177.5	599.1	38.3
$6\frac{3}{4}$	×	$24\frac{3}{4}$	167.1	689.1	8528	187.9	634.3	40.6
$6\frac{3}{4}$	×	$26\frac{1}{8}$	176.3	767.8	10030	198.4	669.6	42.9
$6\frac{3}{4}$	×	$27\frac{1}{2}$	185.6	850.8	11700	208.8	704.8	45.1
$6\frac{3}{4}$	×	$28\frac{7}{8}$	194.9	938.0	13540	219.3	740.0	47.4
$6\frac{3}{4}$	×	$30\frac{1}{4}$	204.2	1029	15570	229.7	775.3	49.6
$6\frac{3}{4}$	×	$31\frac{5}{8}$	213.5	1125	17790	240.2	810.5	51.9
$6\frac{3}{4}$	×	33	222.8	1225	20210	250.6	845.8	54.1
$6\frac{3}{4}$	×	$34\frac{3}{8}$	232.0	1329	22850	261.0	881.0	56.4
$6\frac{3}{4}$	×	$35\frac{3}{4}$	241.3	1438	25700	271.5	916.2	58.7
$6\frac{3}{4}$	×	$37\frac{1}{8}$	250.6	1551	28780	281.9	951.5	60.9
$6\frac{3}{4}$	×	$38\frac{1}{2}$	259.9	1668	32100	292.4	986.7	63.2
$6\frac{3}{4}$	×	$39\frac{7}{8}$	269.2	1789	35660	302.8	1022	65.4
$6\frac{3}{4}$	×	$41\frac{1}{4}$	278.4	1914	39480	313.2	1057	67.7
$6\frac{3}{4}$	×	$42\frac{5}{8}$	287.7	2044	43560	323.7	1092	69.9
$6\frac{3}{4}$	×	44	297.0	2178	47920	334.1	1128	72.2
$6\frac{3}{4}$	×	$45\frac{3}{8}$	306.3	2316	52550	344.6	1163	74.4
$6\frac{3}{4}$	×	$46\frac{3}{4}$	315.6	2459	57470	355	1198	76.7
$6\frac{3}{4}$	×	$48\frac{1}{8}$	324.8	2606	62700	365.4	1233	79.0
$6\frac{3}{4}$	×	$49\frac{1}{2}$	334.1	2757	68220	375.9	1269	81.2
$6\frac{3}{4}$	×	$50\frac{7}{8}$	343.4	2912	74070	386.3	1304	83.5
$6\frac{3}{4}$	×	$52\frac{1}{4}$	352.7	3071	80240	396.8	1339	85.7
$6\frac{3}{4}$	×	$53\frac{5}{8}$	362.0	3235	86740	407.2	1374	88.0

A.1.2.2 Southern Pine (Continued)

Beam Size				X-X Axis		Y-Y Axis		SP
b Width in.		**d** Depth in.	**A** Area in.^{2}	$\mathbf{S_x}$ Section Modulus in.^{3}	$\mathbf{I_x}$ Moment of Inertia in.^{4}	$\mathbf{S_y}$ Section Modulus in.^{3}	$\mathbf{I_y}$ Moment of Inertia in.^{4}	Weight per foot (based on 36 lb/ft^3) lb/ft
$8\frac{1}{2}$	×	$9\frac{5}{8}$	81.81	131.2	631.6	115.9	492.6	19.9
$8\frac{1}{3}$	×	11	93.50	171.4	942.8	132.5	562.9	22.7
$8\frac{1}{3}$	×	$12\frac{3}{8}$	105.2	216.9	1342	149.0	633.3	25.6
$8\frac{1}{3}$	×	$11\frac{3}{4}$	116.9	267.8	1841	165.6	703.7	28.4
$8\frac{1}{3}$	×	$15\frac{1}{8}$	128.6	324.1	2451	182.1	774.1	31.2
$8\frac{1}{3}$	×	$16\frac{1}{2}$	140.3	385.7	3182	198.7	844.4	34.1
$8\frac{1}{3}$	×	$17\frac{7}{8}$	151.9	452.6	4046	215.2	914.8	36.9
$8\frac{1}{3}$	×	$19\frac{1}{4}$	163.6	525.0	5053	231.8	985.2	39.8
$8\frac{1}{3}$	×	$20\frac{5}{8}$	175.3	602.6	6215	248.4	1056	42.6
$8\frac{1}{3}$	×	22	187.0	685.7	7542	264.9	1126	45.5
$8\frac{1}{3}$	×	$23\frac{3}{8}$	198.7	774.1	9047	281.5	1196	48.3
$8\frac{1}{2}$	×	$22\frac{3}{4}$	210.4	867.8	10740	298.0	1267	51.1
$8\frac{1}{2}$	×	$26\frac{1}{8}$	222.1	966.9	12630	314.6	1337	54.0
$8\frac{1}{2}$	×	$27\frac{1}{2}$	233.8	1071	14730	331.1	1407	56.8
$8\frac{1}{2}$	×	$28\frac{7}{8}$	245.4	1181	17050	347.7	1478	59.7
$8\frac{1}{2}$	×	$30\frac{1}{4}$	257.1	1296	19610	364.3	1548	62.5
$8\frac{1}{2}$	×	$31\frac{5}{8}$	268.8	1417	22400	380.8	1618	65.3
$8\frac{1}{2}$	×	33	280.5	1543	25460	397.4	1689	68.2
$8\frac{1}{2}$	×	$34\frac{3}{8}$	292.2	1674	28770	413.9	1759	71.0
$8\frac{1}{2}$	×	$35\frac{3}{4}$	303.9	1811	32360	430.5	1830	73.9
$8\frac{1}{2}$	×	$37\frac{1}{8}$	315.6	1953	36240	447	1900	76.7
$8\frac{1}{2}$	×	$38\frac{1}{2}$	327.3	2100	40420	463.6	1970	79.5
$8\frac{1}{2}$	×	$39\frac{7}{8}$	338.9	2253	44910	480.2	2041	82.4
$8\frac{1}{2}$	×	$41\frac{1}{4}$	350.6	2411	49720	496.7	2111	85.2
$8\frac{1}{2}$	×	$42\frac{5}{8}$	362.3	2574	54860	513.3	2181	88.1
$8\frac{1}{2}$	×	44	374.0	2743	60340	529.8	2252	90.9
$8\frac{1}{2}$	×	$45\frac{3}{8}$	385.7	2917	66170	546.4	2322	93.7
$8\frac{1}{2}$	×	$46\frac{3}{4}$	397.4	3096	72370	562.9	2393	96.6
$8\frac{1}{2}$	×	$48\frac{1}{8}$	409.1	3281	78950	579.5	2463	99.4
$8\frac{1}{2}$	×	$49\frac{1}{2}$	420.8	3471	85910	596.1	2533	102.3
$8\frac{1}{2}$	×	$50\frac{7}{8}$	432.4	3667	93270	612.6	2604	105.1
$8\frac{1}{2}$	×	$52\frac{1}{4}$	444.1	3868	101000	629.2	2674	107.9
$8\frac{1}{2}$	×	$53\frac{5}{8}$	455.8	4074	109200	645.7	2744	110.8
$8\frac{1}{2}$	×	55	467.5	4285	117800	662.3	2815	113.6
$8\frac{1}{2}$	×	$56\frac{3}{8}$	479.2	4502	126900	678.8	2885	116.5
$8\frac{1}{2}$	×	$57\frac{3}{4}$	490.9	4725	136400	695.4	2955	119.3
$8\frac{1}{2}$	×	$59\frac{1}{8}$	502.6	4952	146400	712	3026	122.2

A.1.2.2 Southern Pine (Continued)

Beam Size				X-X Axis		Y-Y Axis		SP
b Width in.		**d** Depth in.	**A** Area in.^{2}	**S_x** Section Modulus in.^{3}	**I_x** Moment of Inertia in.^{4}	**S_y** Section Modulus in.^{3}	**I_y** Moment of Inertia in.^{4}	Weight per foot (based on 36 lb/ft^3) lb/ft
$10\frac{1}{2}$	×	11	115.5	211.8	1165	202.1	1061	28.1
$10\frac{1}{2}$	×	$12\frac{3}{8}$	129.9	268.0	1658	227.4	1194	31.6
$10\frac{1}{2}$	×	$13\frac{3}{4}$	144.4	330.9	2275	252.7	1326	35.1
$10\frac{1}{2}$	×	$15\frac{1}{8}$	158.8	400.3	3028	277.9	1459	38.6
$10\frac{1}{2}$	×	$16\frac{1}{2}$	173.3	476.4	3931	303.2	1592	42.1
$10\frac{1}{2}$	×	$17\frac{7}{8}$	187.7	559.2	4997	328.5	1724	45.6
$10\frac{1}{2}$	×	$19\frac{1}{4}$	202.1	648.5	6242	353.7	1857	49.1
$10\frac{1}{2}$	×	$20\frac{5}{8}$	216.6	744.4	7677	379.0	1990	52.6
$10\frac{1}{2}$	×	22	231.0	847.0	9317	404.3	2122	56.1
$10\frac{1}{2}$	×	$23\frac{3}{8}$	245.4	956.2	11180	429.5	2255	59.7
$10\frac{1}{2}$	×	$24\frac{3}{4}$	259.9	1072	13270	454.8	2388	63.2
$10\frac{1}{2}$	×	$26\frac{1}{8}$	274.3	1194	15600	480.0	2520	66.7
$10\frac{1}{2}$	×	$27\frac{1}{2}$	288.8	1323	18200	505.3	2653	70.2
$10\frac{1}{2}$	×	$28\frac{7}{8}$	303.2	1459	21070	530.6	2786	73.7
$10\frac{1}{2}$	×	$30\frac{1}{4}$	317.6	1601	24220	555.8	2918	77.2
$10\frac{1}{2}$	×	$31\frac{5}{8}$	332.1	1750	27680	581.1	3051	80.7
$10\frac{1}{2}$	×	33	346.5	1906	31440	606.4	3183	84.2
$10\frac{1}{2}$	×	$34\frac{3}{8}$	360.9	2068	35540	631.6	3316	87.7
$10\frac{1}{2}$	×	$35\frac{3}{4}$	375.4	2237	39980	656.9	3449	91.2
$10\frac{1}{2}$	×	$37\frac{1}{8}$	389.8	2412	44770	682.2	3581	94.7
$10\frac{1}{2}$	×	$38\frac{1}{2}$	404.3	2594	49930	707.4	3714	98.3
$10\frac{1}{2}$	×	$39\frac{7}{8}$	418.7	2783	55480	732.7	3847	101.8
$10\frac{1}{2}$	×	$41\frac{1}{4}$	433.1	2978	61420	758	3979	105.3
$10\frac{1}{2}$	×	$42\frac{5}{8}$	447.6	3180	67760	783.2	4112	108.8
$10\frac{1}{2}$	×	44	462.0	3388	74540	808.5	4245	112.3
$10\frac{1}{2}$	×	$45\frac{3}{8}$	476.4	3603	81740	833.8	4377	115.8
$10\frac{1}{2}$	×	$46\frac{3}{4}$	490.9	3825	89400	859	4510	119.3
$10\frac{1}{2}$	×	$48\frac{1}{8}$	505.3	4053	97530	884.3	4643	122.8
$10\frac{1}{2}$	×	$49\frac{1}{2}$	519.8	4288	106100	909.6	4775	126.3
$10\frac{1}{2}$	×	$50\frac{7}{8}$	534.2	4529	115200	934.8	4908	129.8
$10\frac{1}{2}$	×	$52\frac{1}{4}$	548.6	4778	124800	960.1	5040	133.3
$10\frac{1}{2}$	×	$53\frac{5}{8}$	563.1	5032	134900	985.4	5173	136.9
$10\frac{1}{2}$	×	55	577.5	5294	145600	1011	5306	140.4
$10\frac{1}{2}$	×	$56\frac{3}{8}$	591.9	5562	156800	1036	5438	143.9
$10\frac{1}{2}$	×	$57\frac{3}{4}$	606.4	5836	168500	1061	5571	147.4
$10\frac{1}{2}$	×	$59\frac{1}{8}$	620.8	6118	180900	1086	5704	150.9

A.1.2.2 Southern Pine (Continued)

Beam Size				X-X Axis		Y-Y Axis		SP
b Width in.		d Depth in.	A Area in.^{2}	S_x Section Modulus in.^{3}	I_x Moment of Inertia in.^{4}	S_y Section Modulus in.^{3}	I_y Moment of Inertia in.^{4}	Weight per foot (based on 36 lb/ft^3) lb/ft
12	×	$12\frac{3}{8}$	148.5	306.3	1895	297.0	1782	36.1
12	×	$13\frac{3}{4}$	165.0	378.1	2600	330.0	1980	40.1
12	×	$15\frac{1}{8}$	181.5	457.5	3460	363.0	2178	44.1
12	×	$16\frac{1}{2}$	198.0	544.5	4492	396.0	2376	48.1
12	×	$17\frac{7}{8}$	214.5	639.0	5711	429.0	2574	52.1
12	×	$19\frac{1}{4}$	231.0	741.1	7133	462.0	2772	56.1
12	×	$20\frac{5}{8}$	247.5	850.8	8774	495.0	2970	60.2
12	×	22	264.0	968.0	10650	528.0	3168	64.2
12	×	$23\frac{3}{8}$	280.5	1093	12770	561.0	3366	68.2
12	×	$24\frac{3}{4}$	297.0	1225	15160	594.0	3564	72.2
12	×	$26\frac{1}{8}$	313.5	1365	17830	627.0	3762	76.2
12	×	$27\frac{1}{2}$	330.0	1513	20800	660.0	3960	80.2
12	×	$28\frac{7}{8}$	346.5	1668	24070	693.0	4158	84.2
12	×	$30\frac{1}{4}$	363.0	1830	27680	726.0	4356	88.2
12	×	$31\frac{5}{8}$	379.5	2000	31630	759.0	4554	92.2
12	×	33	396.0	2178	35940	792.0	4752	96.3
12	×	$34\frac{3}{8}$	412.5	2363	40620	825.0	4950	100.3
12	×	$35\frac{3}{4}$	429.0	2556	45690	858.0	5148	104.3
12	×	$37\frac{1}{8}$	445.5	2757	51170	891.0	5346	108.3
12	×	$38\frac{1}{2}$	462.0	2965	57070	924.0	5544	112.3
12	×	$39\frac{7}{8}$	478.5	3180	63400	957.0	5742	116.3
12	×	$41\frac{1}{4}$	495.0	3403	70190	990.0	5940	120.3
12	×	$42\frac{5}{8}$	511.5	3634	77440	1023	6138	124.3
12	×	44	528.0	3872	85180	1056	6336	128.3
12	×	$45\frac{3}{8}$	544.5	4118	93420	1089	6534	132.3
12	×	$46\frac{3}{4}$	561.0	4371	102200	1122	6732	136.4
12	×	$48\frac{1}{8}$	577.5	4632	111500	1155	6930	140.4
12	×	$49\frac{1}{2}$	594.0	4901	121300	1188	7128	144.4
12	×	$50\frac{7}{8}$	610.5	5177	131700	1221	7326	148.4
12	×	$52\frac{1}{4}$	627.0	5460	142600	1254	7524	152.4
12	×	$53\frac{5}{8}$	643.5	5751	154200	1287	7722	156.4
12	×	55	660.0	6050	166400	1320	7920	160.4
12	×	$56\frac{3}{8}$	676.5	6356	179200	1353	8118	164.4
12	×	$57\frac{3}{4}$	693.0	6670	192600	1386	8316	168.4
12	×	$59\frac{1}{8}$	709.5	6992	206700	1419	8514	172.4

A.1.2.2 Southern Pine (Continued)

Beam Size				X-X Axis		Y-Y Axis		SP
b Width in.		d Depth in.	A Area in.^{2}	S_x Section Modulus in.^{3}	I_x Moment of Inertia in.^{4}	S_y Section Modulus in.^{3}	I_y Moment of Inertia in.^{4}	Weight per foot (based on 36 lb/ft^3) lb/ft
14	×	$15\frac{1}{8}$	211.8	533.8	4037	494.1	3459	51.5
14	×	$16\frac{1}{2}$	231.0	635.3	5241	539.0	3773	56.1
14	×	$17\frac{7}{8}$	250.3	745.5	6663	583.9	4087	60.8
14	×	$19\frac{1}{4}$	269.5	864.6	8322	628.8	4402	65.5
14	×	$20\frac{5}{8}$	288.8	992.6	10240	673.8	4716	70.2
14	×	22	308.0	1129	12420	718.7	5031	74.9
14	×	$23\frac{3}{8}$	327.3	1275	14900	763.6	5345	79.5
14	×	$24\frac{3}{4}$	346.5	1429	17690	808.5	5660	84.2
14	×	$26\frac{1}{8}$	365.8	1593	20800	853.4	5974	88.9
14	×	$27\frac{1}{2}$	385.0	1765	24260	898.3	6288	93.6
14	×	$28\frac{7}{8}$	404.3	1945	28090	943.3	6603	98.3
14	×	$30\frac{1}{4}$	423.5	2135	32290	988.2	6917	102.9
14	×	$31\frac{5}{8}$	442.8	2334	36900	1033	7232	107.6
14	×	33	462.0	2541	41930	1078	7546	112.3
14	×	$34\frac{3}{8}$	481.3	2757	47390	1123	7860	117.0
14	×	$35\frac{3}{4}$	500.5	2982	53310	1168	8175	121.6
14	×	$37\frac{1}{8}$	519.8	3216	59700	1213	8489	126.3
14	×	$38\frac{1}{2}$	539.0	3459	66580	1258	8804	131.0
14	×	$39\frac{7}{8}$	558.3	3710	73970	1303	9118	135.7
14	×	$41\frac{1}{4}$	577.5	3970	81890	1348	9433	140.4
14	×	$42\frac{5}{8}$	596.8	4239	90350	1392	9747	145.0
14	×	44	616.0	4517	99380	1437	10060	149.7
14	×	$45\frac{3}{8}$	635.3	4804	109000	1482	10380	154.4
14	×	$46\frac{3}{4}$	654.5	5100	119200	1527	10690	159.1
14	×	$48\frac{1}{8}$	673.8	5404	130000	1572	11000	163.8
14	×	$49\frac{1}{2}$	693.0	5717	141500	1617	11320	168.4
14	×	$50\frac{7}{8}$	712.3	6039	153600	1662	11630	173.1
14	×	$52\frac{1}{4}$	731.5	6370	166400	1707	11950	177.8
14	×	$53\frac{5}{8}$	750.8	6710	179900	1752	12260	182.5
14	×	55	770.0	7058	194100	1797	12580	187.2
14	×	$56\frac{3}{8}$	789.3	7416	209000	1842	12890	191.8
14	×	$57\frac{3}{4}$	808.5	7782	224700	1887	13210	196.5
14	×	$59\frac{1}{8}$	827.8	8157	241100	1931	13520	201.2

A.2 BEAM DIAGRAMS AND FORMULAS

A.2.1 Single-Beam Diagrams and Formulas

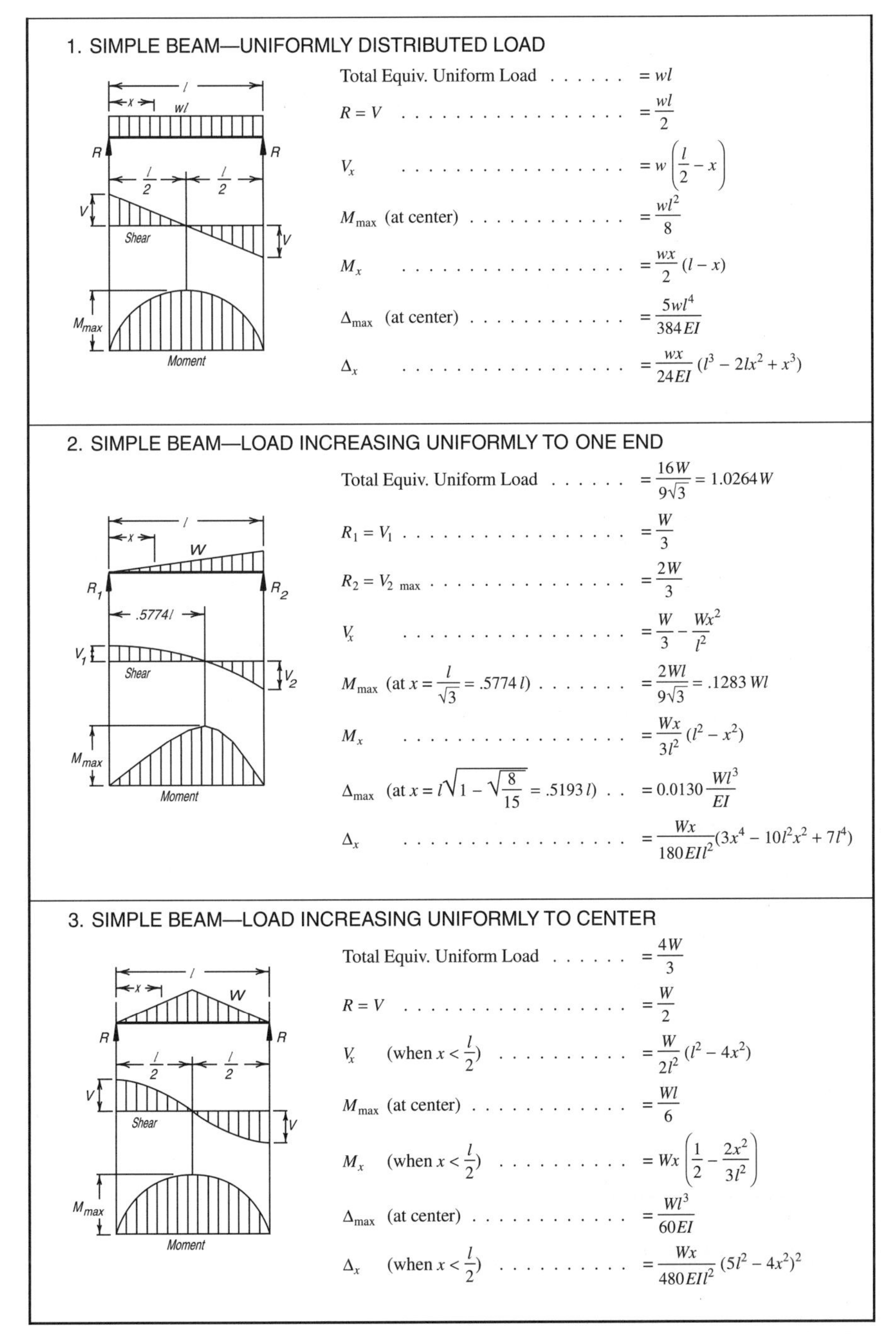

1. SIMPLE BEAM—UNIFORMLY DISTRIBUTED LOAD

Total Equiv. Uniform Load ...... $= wl$

$R = V$ .... $= \frac{wl}{2}$

V_x .... $= w\left(\frac{l}{2} - x\right)$

$M_{\max}$ (at center) .... $= \frac{wl^2}{8}$

M_x .... $= \frac{wx}{2}(l - x)$

$\Delta_{\max}$ (at center) .... $= \frac{5wl^4}{384EI}$

Δ_x .... $= \frac{wx}{24EI}(l^3 - 2lx^2 + x^3)$

2. SIMPLE BEAM—LOAD INCREASING UNIFORMLY TO ONE END

Total Equiv. Uniform Load ...... $= \frac{16W}{9\sqrt{3}} = 1.0264W$

$R_1 = V_1$ .... $= \frac{W}{3}$

$R_2 = V_{2\ \max}$ .... $= \frac{2W}{3}$

V_x .... $= \frac{W}{3} - \frac{Wx^2}{l^2}$

$M_{\max}$ (at $x = \frac{l}{\sqrt{3}} = .5774\,l$) .... $= \frac{2Wl}{9\sqrt{3}} = .1283\,Wl$

M_x .... $= \frac{Wx}{3l^2}(l^2 - x^2)$

$\Delta_{\max}$ (at $x = l\sqrt{1 - \sqrt{\frac{8}{15}}} = .5193\,l$) .. $= 0.0130\frac{Wl^3}{EI}$

Δ_x .... $= \frac{Wx}{180EIl^2}(3x^4 - 10l^2x^2 + 7l^4)$

3. SIMPLE BEAM—LOAD INCREASING UNIFORMLY TO CENTER

Total Equiv. Uniform Load ...... $= \frac{4W}{3}$

$R = V$ .... $= \frac{W}{2}$

V_x (when $x < \frac{l}{2}$) .... $= \frac{W}{2l^2}(l^2 - 4x^2)$

$M_{\max}$ (at center) .... $= \frac{Wl}{6}$

M_x (when $x < \frac{l}{2}$) .... $= Wx\left(\frac{1}{2} - \frac{2x^2}{3l^2}\right)$

$\Delta_{\max}$ (at center) .... $= \frac{Wl^3}{60EI}$

Δ_x (when $x < \frac{l}{2}$) .... $= \frac{Wx}{480EIl^2}(5l^2 - 4x^2)^2$

Source: AISC [2]; 29a from AITC. Copyright © American Institute of Steel Construction, Inc. Reprinted with permission. All rights reserved.

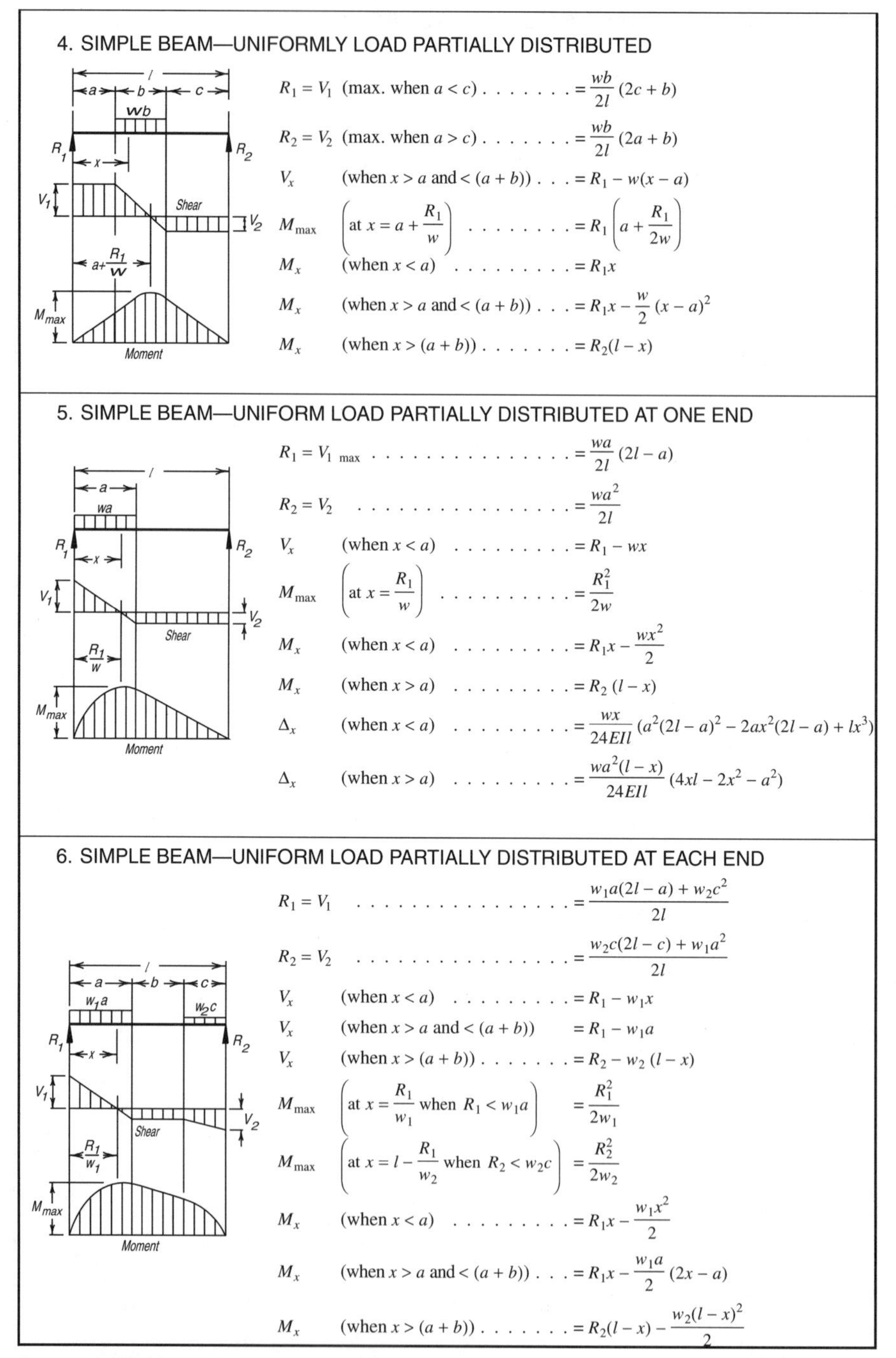

4. SIMPLE BEAM—UNIFORMLY LOAD PARTIALLY DISTRIBUTED

$R_1 = V_1$ (max. when $a < c$) . . . . $= \dfrac{wb}{2l}(2c + b)$

$R_2 = V_2$ (max. when $a > c$) . . . . $= \dfrac{wb}{2l}(2a + b)$

V_x (when $x > a$ and $< (a + b)$) . . . $= R_1 - w(x - a)$

$M_{\max}$ $\left(\text{at } x = a + \dfrac{R_1}{w}\right)$ . . . . $= R_1\left(a + \dfrac{R_1}{2w}\right)$

M_x (when $x < a$) . . . . $= R_1 x$

M_x (when $x > a$ and $< (a + b)$) . . . $= R_1 x - \dfrac{w}{2}(x - a)^2$

M_x (when $x > (a + b)$) . . . . $= R_2(l - x)$

5. SIMPLE BEAM—UNIFORM LOAD PARTIALLY DISTRIBUTED AT ONE END

$R_1 = V_1$ max . . . . $= \dfrac{wa}{2l}(2l - a)$

$R_2 = V_2$ . . . . $= \dfrac{wa^2}{2l}$

V_x (when $x < a$) . . . . $= R_1 - wx$

$M_{\max}$ $\left(\text{at } x = \dfrac{R_1}{w}\right)$ . . . . $= \dfrac{R_1^2}{2w}$

M_x (when $x < a$) . . . . $= R_1 x - \dfrac{wx^2}{2}$

M_x (when $x > a$) . . . . $= R_2(l - x)$

Δ_x (when $x < a$) . . . . $= \dfrac{wx}{24EIl}(a^2(2l - a)^2 - 2ax^2(2l - a) + lx^3)$

Δ_x (when $x > a$) . . . . $= \dfrac{wa^2(l - x)}{24EIl}(4xl - 2x^2 - a^2)$

6. SIMPLE BEAM—UNIFORM LOAD PARTIALLY DISTRIBUTED AT EACH END

$R_1 = V_1$ . . . . $= \dfrac{w_1 a(2l - a) + w_2 c^2}{2l}$

$R_2 = V_2$ . . . . $= \dfrac{w_2 c(2l - c) + w_1 a^2}{2l}$

V_x (when $x < a$) . . . . $= R_1 - w_1 x$

V_x (when $x > a$ and $< (a + b)$) $= R_1 - w_1 a$

V_x (when $x > (a + b)$) . . . . $= R_2 - w_2(l - x)$

$M_{\max}$ $\left(\text{at } x = \dfrac{R_1}{w_1} \text{ when } R_1 < w_1 a\right)$ $= \dfrac{R_1^2}{2w_1}$

$M_{\max}$ $\left(\text{at } x = l - \dfrac{R_1}{w_2} \text{ when } R_2 < w_2 c\right)$ $= \dfrac{R_2^2}{2w_2}$

M_x (when $x < a$) . . . . $= R_1 x - \dfrac{w_1 x^2}{2}$

M_x (when $x > a$ and $< (a + b)$) . . . $= R_1 x - \dfrac{w_1 a}{2}(2x - a)$

M_x (when $x > (a + b)$) . . . . $= R_2(l - x) - \dfrac{w_2(l - x)^2}{2}$

Source: AISC [2]; 29a from AITC. Copyright © American Institute of Steel Construction, Inc. Reprinted with permission. All rights reserved.

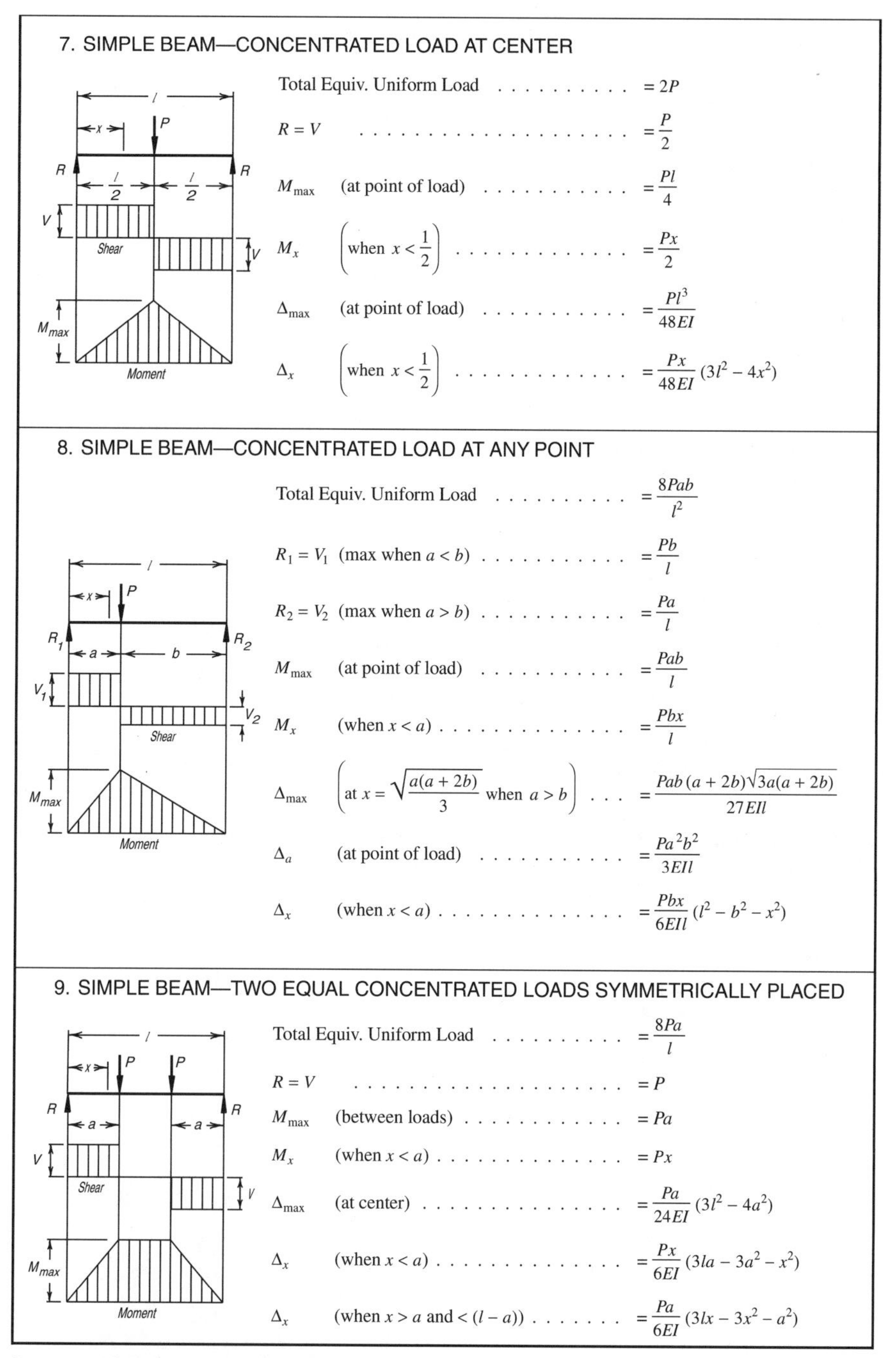

7. SIMPLE BEAM—CONCENTRATED LOAD AT CENTER

Total Equiv. Uniform Load . . . . $= 2P$

$R = V$ . . . . $= \frac{P}{2}$

M_{max} (at point of load) . . . . $= \frac{Pl}{4}$

M_x $\left(\text{when } x < \frac{1}{2}\right)$ . . . . $= \frac{Px}{2}$

Δ_{max} (at point of load) . . . . $= \frac{Pl^3}{48EI}$

Δ_x $\left(\text{when } x < \frac{1}{2}\right)$ . . . . $= \frac{Px}{48EI}(3l^2 - 4x^2)$

8. SIMPLE BEAM—CONCENTRATED LOAD AT ANY POINT

Total Equiv. Uniform Load . . . . $= \frac{8Pab}{l^2}$

$R_1 = V_1$ (max when $a < b$) . . . . $= \frac{Pb}{l}$

$R_2 = V_2$ (max when $a > b$) . . . . $= \frac{Pa}{l}$

M_{max} (at point of load) . . . . $= \frac{Pab}{l}$

M_x (when $x < a$) . . . . $= \frac{Pbx}{l}$

Δ_{max} $\left(\text{at } x = \sqrt{\frac{a(a+2b)}{3}} \text{ when } a > b\right)$ . . . $= \frac{Pab(a+2b)\sqrt{3a(a+2b)}}{27EIl}$

Δ_a (at point of load) . . . . $= \frac{Pa^2b^2}{3EIl}$

Δ_x (when $x < a$) . . . . $= \frac{Pbx}{6EIl}(l^2 - b^2 - x^2)$

9. SIMPLE BEAM—TWO EQUAL CONCENTRATED LOADS SYMMETRICALLY PLACED

Total Equiv. Uniform Load . . . . $= \frac{8Pa}{l}$

$R = V$ . . . . $= P$

M_{max} (between loads) . . . . $= Pa$

M_x (when $x < a$) . . . . $= Px$

Δ_{max} (at center) . . . . $= \frac{Pa}{24EI}(3l^2 - 4a^2)$

Δ_x (when $x < a$) . . . . $= \frac{Px}{6EI}(3la - 3a^2 - x^2)$

Δ_x (when $x > a$ and $< (l - a)$) . . . . $= \frac{Pa}{6EI}(3lx - 3x^2 - a^2)$

Source: AISC [2]; 29a from AITC. Copyright © American Institute of Steel Construction, Inc. Reprinted with permission. All rights reserved.

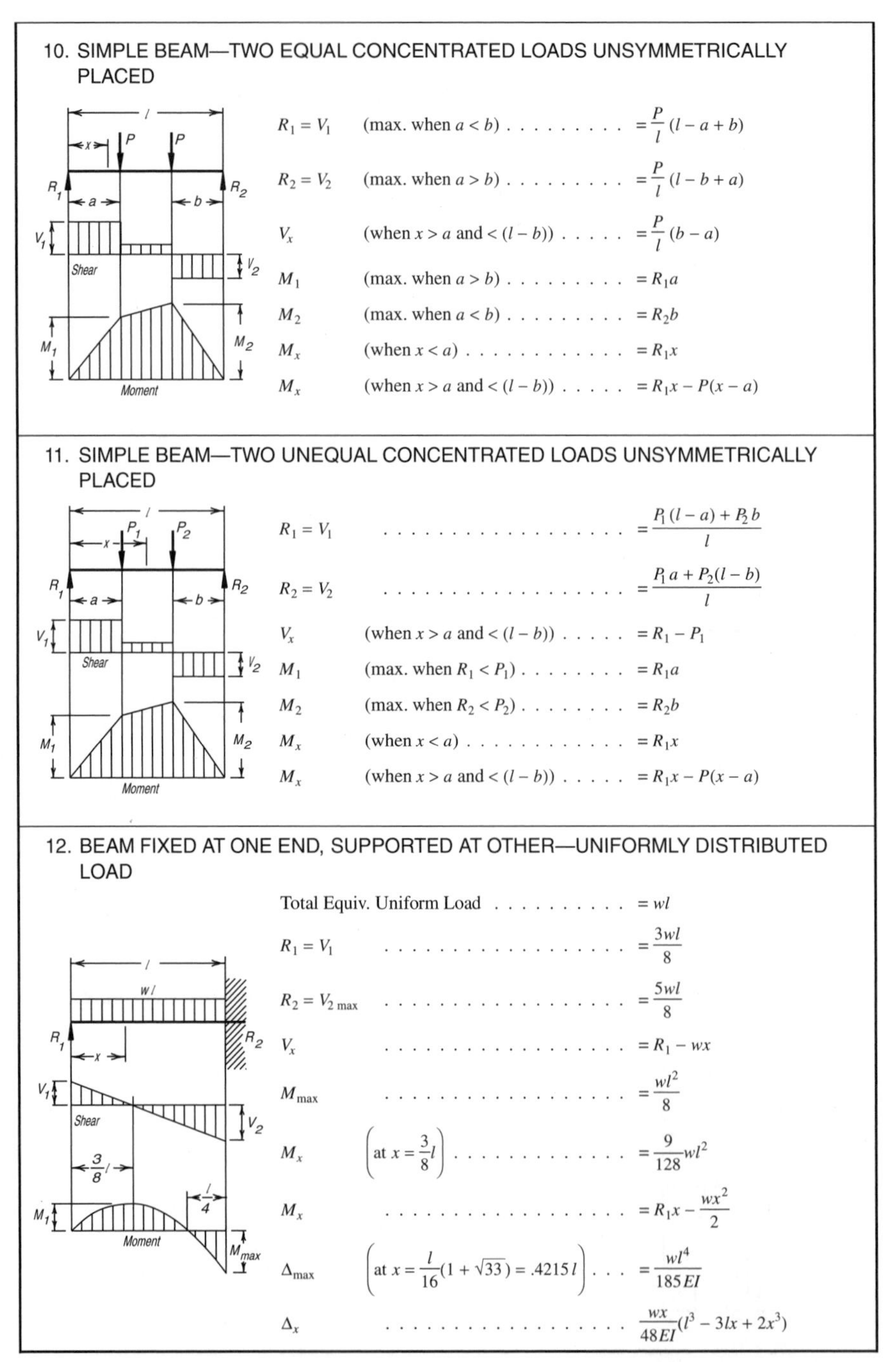

10. SIMPLE BEAM—TWO EQUAL CONCENTRATED LOADS UNSYMMETRICALLY PLACED

$R_1 = V_1$	(max. when $a < b$)	$= \frac{P}{l}(l - a + b)$
$R_2 = V_2$	(max. when $a > b$)	$= \frac{P}{l}(l - b + a)$
V_x	(when $x > a$ and $< (l - b)$)	$= \frac{P}{l}(b - a)$
M_1	(max. when $a > b$)	$= R_1 a$
M_2	(max. when $a < b$)	$= R_2 b$
M_x	(when $x < a$)	$= R_1 x$
M_x	(when $x > a$ and $< (l - b)$)	$= R_1 x - P(x - a)$

11. SIMPLE BEAM—TWO UNEQUAL CONCENTRATED LOADS UNSYMMETRICALLY PLACED

$R_1 = V_1$	. . . .	$= \frac{P_1(l - a) + P_2 b}{l}$
$R_2 = V_2$	. . . .	$= \frac{P_1 a + P_2(l - b)}{l}$
V_x	(when $x > a$ and $< (l - b)$)	$= R_1 - P_1$
M_1	(max. when $R_1 < P_1$)	$= R_1 a$
M_2	(max. when $R_2 < P_2$)	$= R_2 b$
M_x	(when $x < a$)	$= R_1 x$
M_x	(when $x > a$ and $< (l - b)$)	$= R_1 x - P(x - a)$

12. BEAM FIXED AT ONE END, SUPPORTED AT OTHER—UNIFORMLY DISTRIBUTED LOAD

Total Equiv. Uniform Load	. . . .	$= wl$
$R_1 = V_1$	. . . .	$= \frac{3wl}{8}$
$R_2 = V_{2\,max}$	. . . .	$= \frac{5wl}{8}$
V_x	. . . .	$= R_1 - wx$
M_{max}	. . . .	$= \frac{wl^2}{8}$
M_x	$\left(\text{at } x = \frac{3}{8}l\right)$	$= \frac{9}{128}wl^2$
M_x	. . . .	$= R_1 x - \frac{wx^2}{2}$
Δ_{max}	$\left(\text{at } x = \frac{l}{16}(1 + \sqrt{33}) = .4215\,l\right)$ . . .	$= \frac{wl^4}{185EI}$
Δ_x	. . . .	$\frac{wx}{48EI}(l^3 - 3lx^2 + 2x^3)$

Source: AISC [2]; 29a from AITC. Copyright © American Institute of Steel Construction, Inc. Reprinted with permission. All rights reserved.

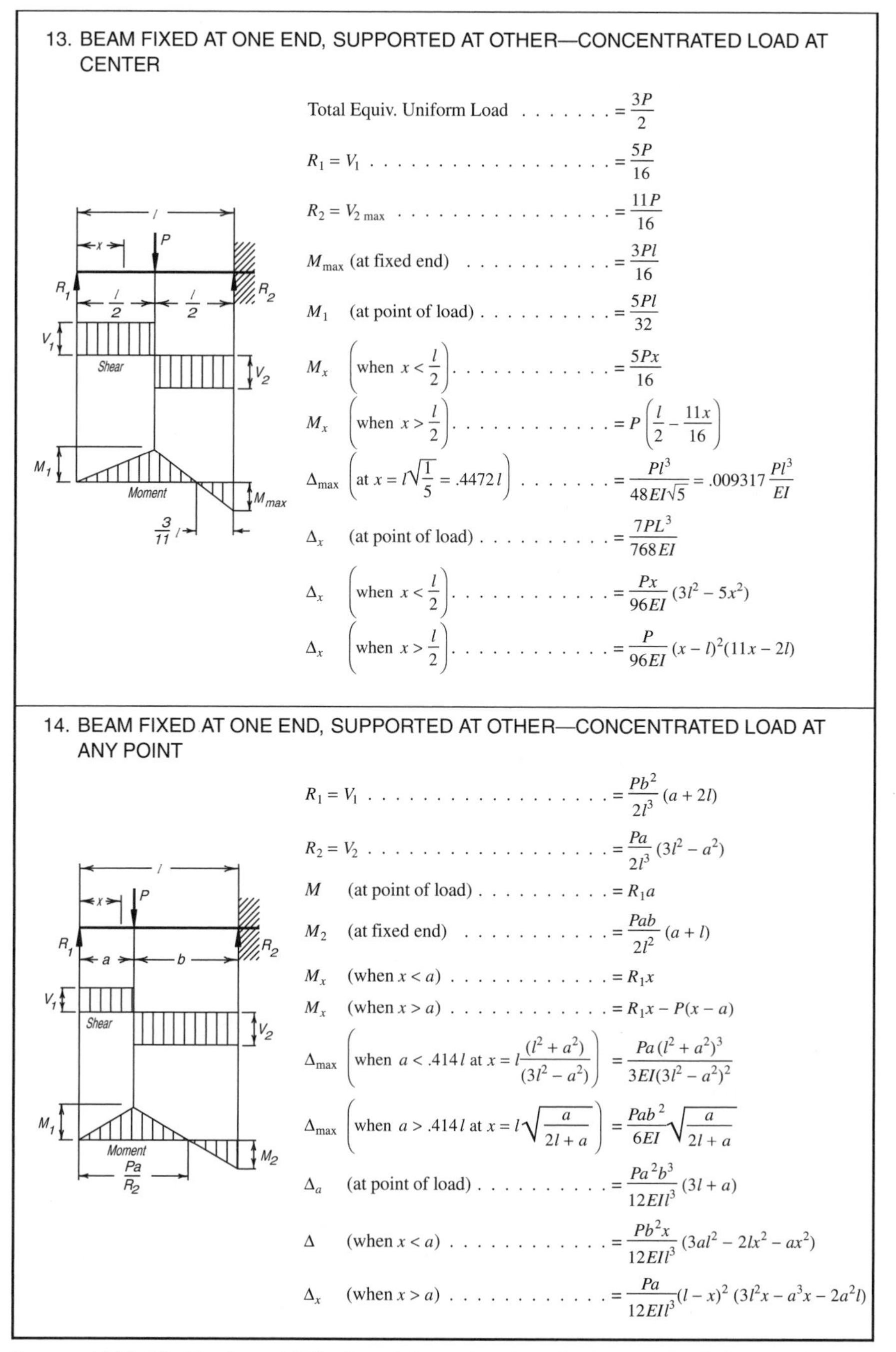

13. BEAM FIXED AT ONE END, SUPPORTED AT OTHER—CONCENTRATED LOAD AT CENTER

Total Equiv. Uniform Load . . . . $= \frac{3P}{2}$

$R_1 = V_1$ . . . . $= \frac{5P}{16}$

$R_2 = V_{2\,max}$ . . . . $= \frac{11P}{16}$

M_{max} (at fixed end) . . . . $= \frac{3Pl}{16}$

M_1 (at point of load) . . . . $= \frac{5Pl}{32}$

$M_x \left(\text{when } x < \frac{l}{2}\right)$ . . . . $= \frac{5Px}{16}$

$M_x \left(\text{when } x > \frac{l}{2}\right)$ . . . . $= P\left(\frac{l}{2} - \frac{11x}{16}\right)$

$\Delta_{max} \left(\text{at } x = l\sqrt{\frac{1}{5}} = .4472\,l\right)$ . . . . $= \frac{Pl^3}{48EI\sqrt{5}} = .009317\frac{Pl^3}{EI}$

Δ_x (at point of load) . . . . $= \frac{7PL^3}{768EI}$

$\Delta_x \left(\text{when } x < \frac{l}{2}\right)$ . . . . $= \frac{Px}{96EI}(3l^2 - 5x^2)$

$\Delta_x \left(\text{when } x > \frac{l}{2}\right)$ . . . . $= \frac{P}{96EI}(x - l)^2(11x - 2l)$

14. BEAM FIXED AT ONE END, SUPPORTED AT OTHER—CONCENTRATED LOAD AT ANY POINT

$R_1 = V_1$ . . . . $= \frac{Pb^2}{2l^3}(a + 2l)$

$R_2 = V_2$ . . . . $= \frac{Pa}{2l^3}(3l^2 - a^2)$

M (at point of load) . . . . $= R_1 a$

M_2 (at fixed end) . . . . $= \frac{Pab}{2l^2}(a + l)$

M_x (when $x < a$) . . . . $= R_1 x$

M_x (when $x > a$) . . . . $= R_1 x - P(x - a)$

$\Delta_{max} \left(\text{when } a < .414\,l \text{ at } x = l\frac{(l^2 + a^2)}{(3l^2 - a^2)}\right) = \frac{Pa(l^2 + a^2)^3}{3EI(3l^2 - a^2)^2}$

$\Delta_{max} \left(\text{when } a > .414\,l \text{ at } x = l\sqrt{\frac{a}{2l + a}}\right) = \frac{Pab^2}{6EI}\sqrt{\frac{a}{2l + a}}$

Δ_a (at point of load) . . . . $= \frac{Pa^2b^3}{12EIl^3}(3l + a)$

Δ (when $x < a$) . . . . $= \frac{Pb^2x}{12EIl^3}(3al^2 - 2lx^2 - ax^2)$

Δ_x (when $x > a$) . . . . $= \frac{Pa}{12EIl^3}(l - x)^2(3l^2x - a^3x - 2a^2l)$

Source: AISC [2]; 29a from AITC. Copyright © American Institute of Steel Construction, Inc. Reprinted with permission. All rights reserved.

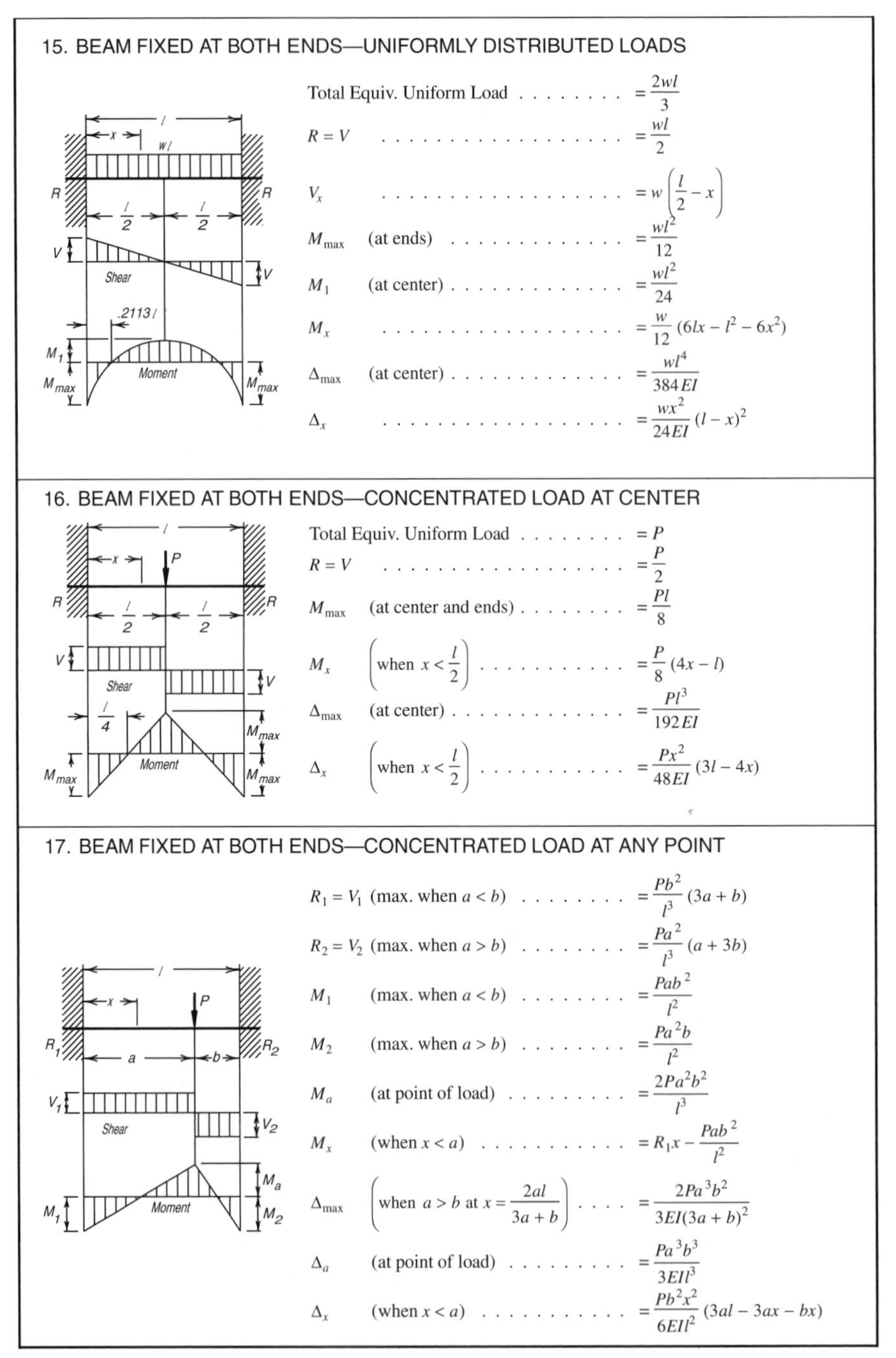

15. BEAM FIXED AT BOTH ENDS—UNIFORMLY DISTRIBUTED LOADS

Total Equiv. Uniform Load . . . . $= \frac{2wl}{3}$

$R = V$ . . . . $= \frac{wl}{2}$

V_x . . . . $= w\left(\frac{l}{2} - x\right)$

M_{max} (at ends) . . . . $= \frac{wl^2}{12}$

M_1 (at center) . . . . $= \frac{wl^2}{24}$

M_x . . . . $= \frac{w}{12}(6lx - l^2 - 6x^2)$

Δ_{max} (at center) . . . . $= \frac{wl^4}{384EI}$

Δ_x . . . . $= \frac{wx^2}{24EI}(l - x)^2$

16. BEAM FIXED AT BOTH ENDS—CONCENTRATED LOAD AT CENTER

Total Equiv. Uniform Load . . . . $= P$

$R = V$ . . . . $= \frac{P}{2}$

M_{max} (at center and ends) . . . . $= \frac{Pl}{8}$

M_x $\left(\text{when } x < \frac{l}{2}\right)$ . . . . $= \frac{P}{8}(4x - l)$

Δ_{max} (at center) . . . . $= \frac{Pl^3}{192EI}$

Δ_x $\left(\text{when } x < \frac{l}{2}\right)$ . . . . $= \frac{Px^2}{48EI}(3l - 4x)$

17. BEAM FIXED AT BOTH ENDS—CONCENTRATED LOAD AT ANY POINT

$R_1 = V_1$ (max. when $a < b$) . . . . $= \frac{Pb^2}{l^3}(3a + b)$

$R_2 = V_2$ (max. when $a > b$) . . . . $= \frac{Pa^2}{l^3}(a + 3b)$

M_1 (max. when $a < b$) . . . . $= \frac{Pab^2}{l^2}$

M_2 (max. when $a > b$) . . . . $= \frac{Pa^2b}{l^2}$

M_a (at point of load) . . . . $= \frac{2Pa^2b^2}{l^3}$

M_x (when $x < a$) . . . . $= R_1x - \frac{Pab^2}{l^2}$

Δ_{max} $\left(\text{when } a > b \text{ at } x = \frac{2al}{3a + b}\right)$ . . . . $= \frac{2Pa^3b^2}{3EI(3a + b)^2}$

Δ_a (at point of load) . . . . $= \frac{Pa^3b^3}{3EIl^3}$

Δ_x (when $x < a$) . . . . $= \frac{Pb^2x^2}{6EIl^2}(3al - 3ax - bx)$

Source: AISC [2]; 29a from AITC. Copyright © American Institute of Steel Construction, Inc. Reprinted with permission. All rights reserved.

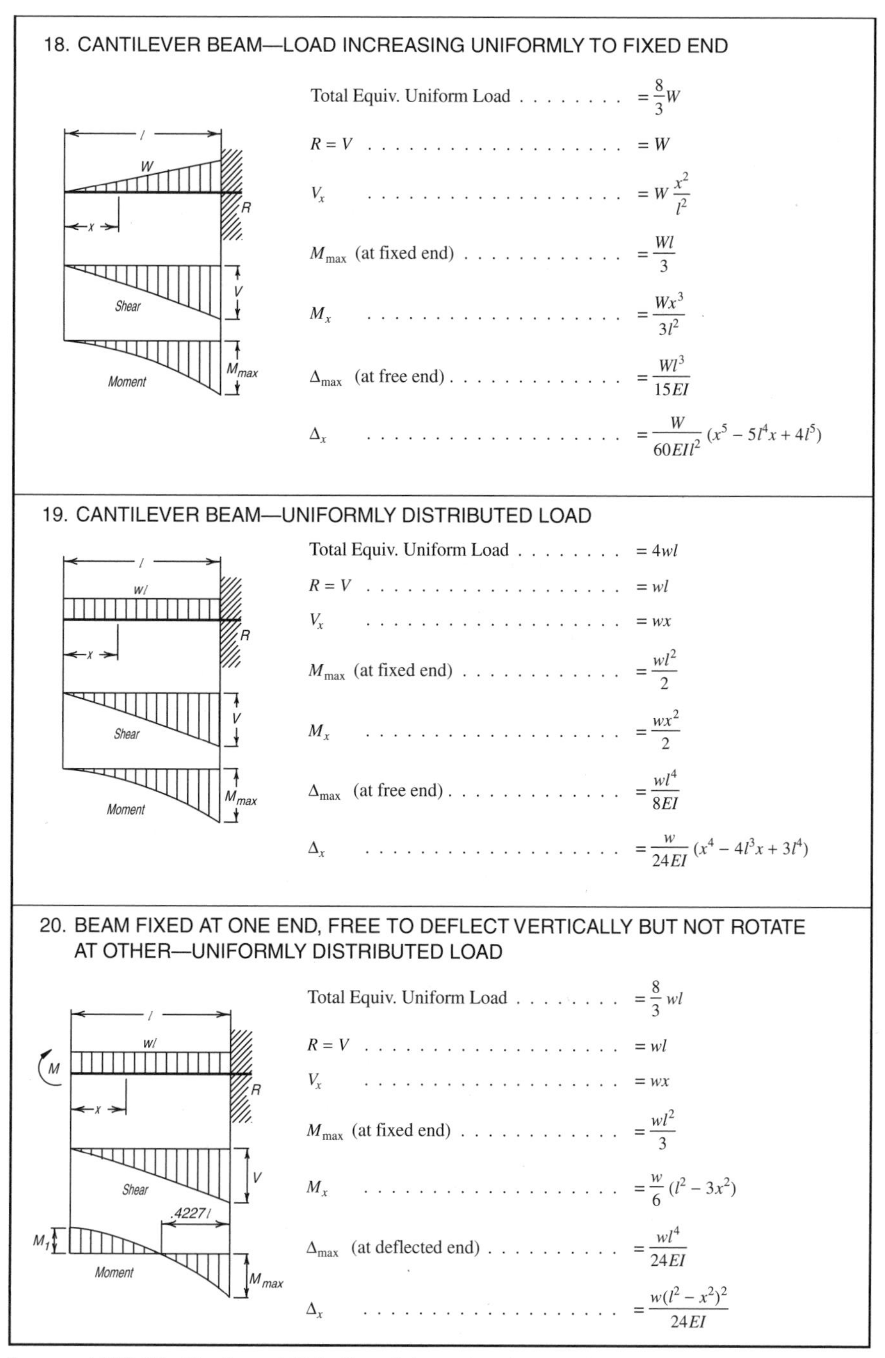

18. CANTILEVER BEAM—LOAD INCREASING UNIFORMLY TO FIXED END

Total Equiv. Uniform Load . . . . $= \frac{8}{3}W$

$R = V$ . . . . $= W$

V_x . . . . $= W\frac{x^2}{l^2}$

$M_{\max}$ (at fixed end) . . . . $= \frac{Wl}{3}$

M_x . . . . $= \frac{Wx^3}{3l^2}$

$\Delta_{\max}$ (at free end) . . . . $= \frac{Wl^3}{15EI}$

Δ_x . . . . $= \frac{W}{60EIl^2}(x^5 - 5l^4x + 4l^5)$

19. CANTILEVER BEAM—UNIFORMLY DISTRIBUTED LOAD

Total Equiv. Uniform Load . . . . $= 4wl$

$R = V$ . . . . $= wl$

V_x . . . . $= wx$

$M_{\max}$ (at fixed end) . . . . $= \frac{wl^2}{2}$

M_x . . . . $= \frac{wx^2}{2}$

$\Delta_{\max}$ (at free end) . . . . $= \frac{wl^4}{8EI}$

Δ_x . . . . $= \frac{w}{24EI}(x^4 - 4l^3x + 3l^4)$

20. BEAM FIXED AT ONE END, FREE TO DEFLECT VERTICALLY BUT NOT ROTATE AT OTHER—UNIFORMLY DISTRIBUTED LOAD

Total Equiv. Uniform Load . . . . $= \frac{8}{3}wl$

$R = V$ . . . . $= wl$

V_x . . . . $= wx$

$M_{\max}$ (at fixed end) . . . . $= \frac{wl^2}{3}$

M_x . . . . $= \frac{w}{6}(l^2 - 3x^2)$

$\Delta_{\max}$ (at deflected end) . . . . $= \frac{wl^4}{24EI}$

Δ_x . . . . $= \frac{w(l^2 - x^2)^2}{24EI}$

Source: AISC [2]; 29a from AITC. Copyright © American Institute of Steel Construction, Inc. Reprinted with permission. All rights reserved.

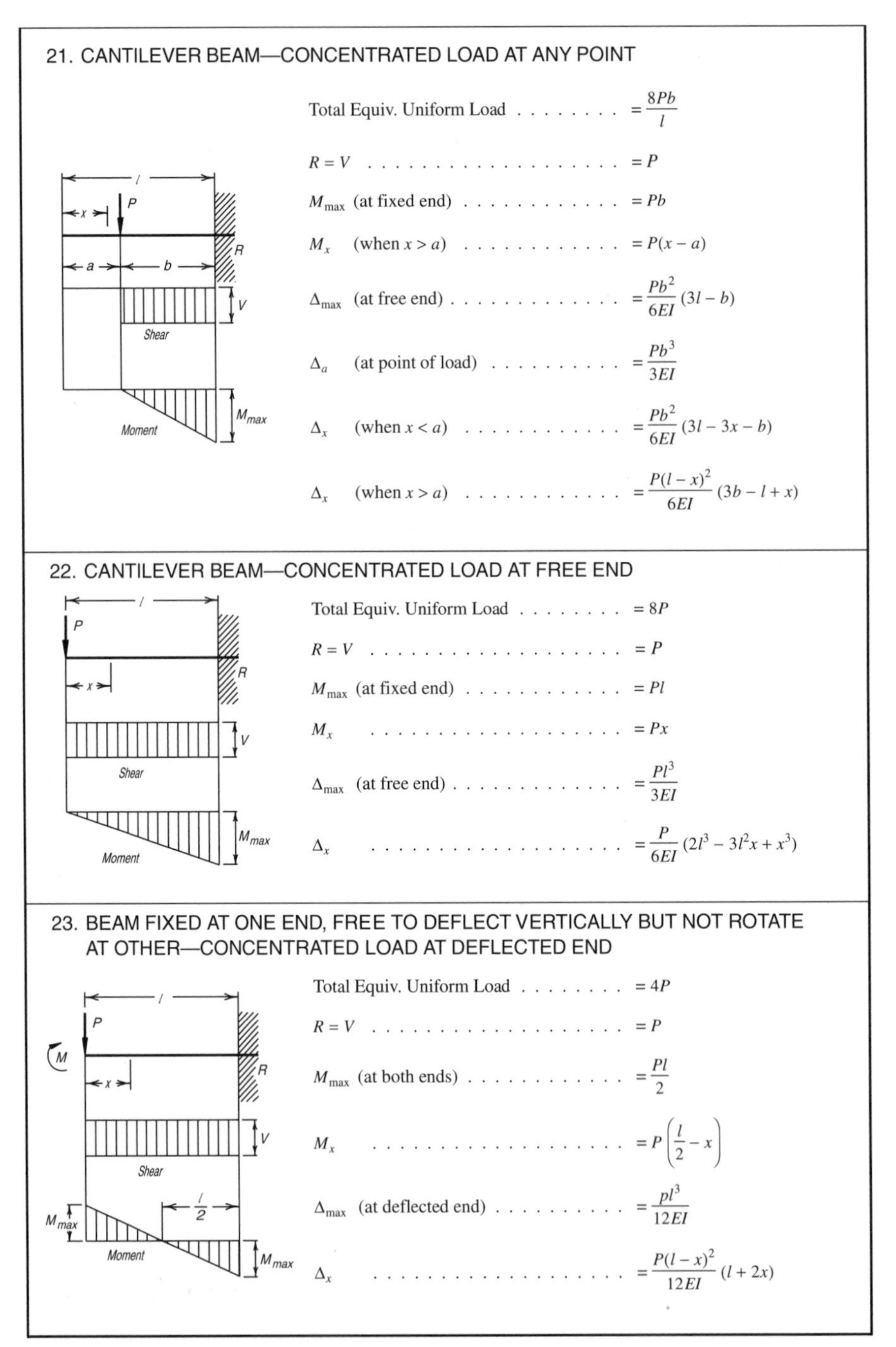

21. CANTILEVER BEAM—CONCENTRATED LOAD AT ANY POINT

Total Equiv. Uniform Load . . . . $= \frac{8Pb}{l}$

$R = V$ . . . . $= P$

M_{max} (at fixed end) . . . . $= Pb$

M_x (when $x > a$) . . . . $= P(x - a)$

Δ_{max} (at free end) . . . . $= \frac{Pb^2}{6EI}(3l - b)$

Δ_a (at point of load) . . . . $= \frac{Pb^3}{3EI}$

Δ_x (when $x < a$) . . . . $= \frac{Pb^2}{6EI}(3l - 3x - b)$

Δ_x (when $x > a$) . . . . $= \frac{P(l-x)^2}{6EI}(3b - l + x)$

22. CANTILEVER BEAM—CONCENTRATED LOAD AT FREE END

Total Equiv. Uniform Load . . . . $= 8P$

$R = V$ . . . . $= P$

M_{max} (at fixed end) . . . . $= Pl$

M_x . . . . $= Px$

Δ_{max} (at free end) . . . . $= \frac{Pl^3}{3EI}$

Δ_x . . . . $= \frac{P}{6EI}(2l^3 - 3l^2x + x^3)$

23. BEAM FIXED AT ONE END, FREE TO DEFLECT VERTICALLY BUT NOT ROTATE AT OTHER—CONCENTRATED LOAD AT DEFLECTED END

Total Equiv. Uniform Load . . . . $= 4P$

$R = V$ . . . . $= P$

M_{max} (at both ends) . . . . $= \frac{Pl}{2}$

M_x . . . . $= P\left(\frac{l}{2} - x\right)$

Δ_{max} (at deflected end) . . . . $= \frac{pl^3}{12EI}$

Δ_x . . . . $= \frac{P(l-x)^2}{12EI}(l + 2x)$

Source: AISC [2]; 29a from AITC. Copyright © American Institute of Steel Construction, Inc. Reprinted with permission. All rights reserved.

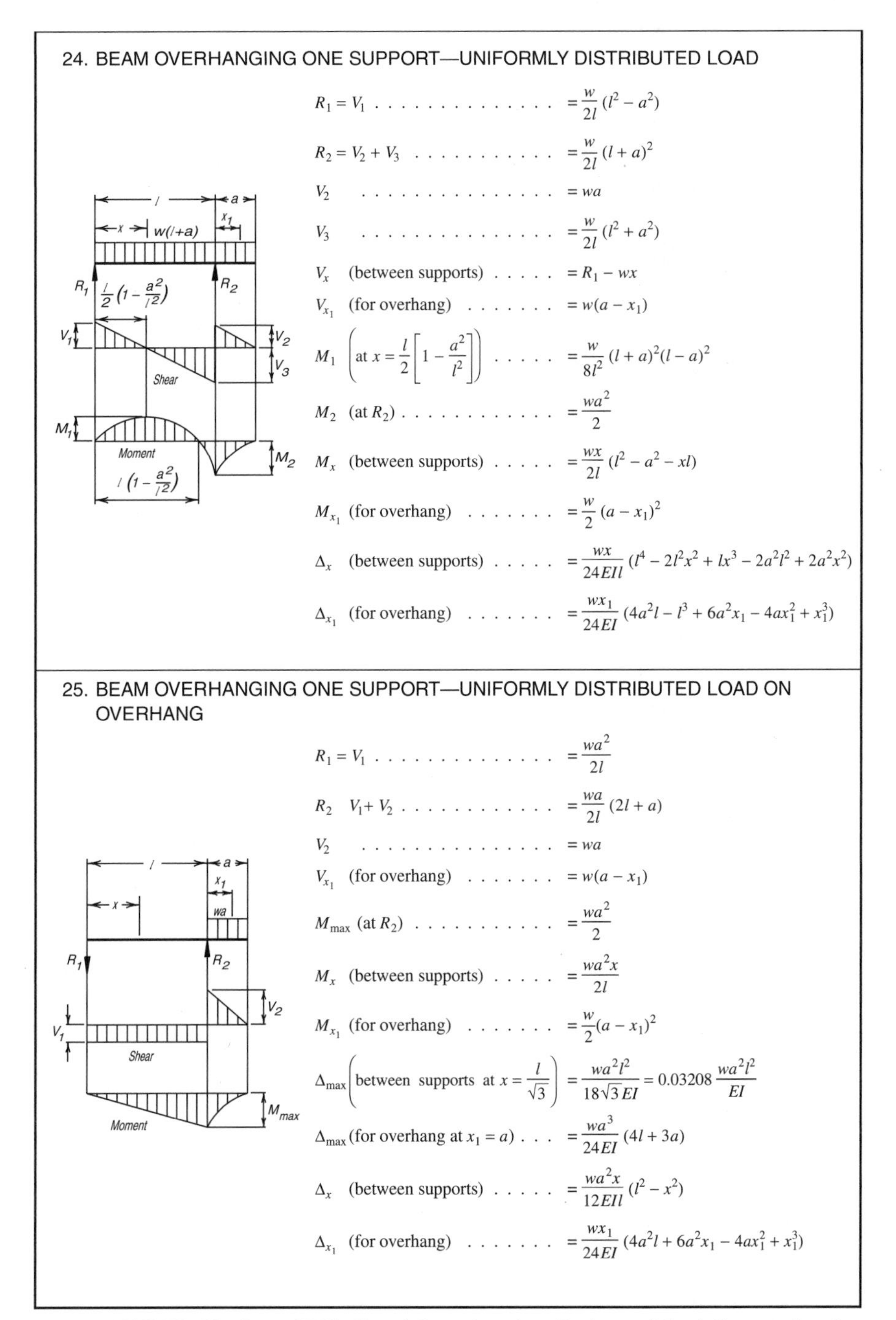

24. BEAM OVERHANGING ONE SUPPORT—UNIFORMLY DISTRIBUTED LOAD

$R_1 = V_1$ . . . . $= \frac{w}{2l}(l^2 - a^2)$

$R_2 = V_2 + V_3$ . . . . $= \frac{w}{2l}(l + a)^2$

V_2 . . . . $= wa$

V_3 . . . . $= \frac{w}{2l}(l^2 + a^2)$

V_x (between supports) . . . . . $= R_1 - wx$

V_{x_1} (for overhang) . . . . $= w(a - x_1)$

$M_1 \left(\text{at } x = \frac{l}{2}\left[1 - \frac{a^2}{l^2}\right]\right)$ . . . . . $= \frac{w}{8l^2}(l + a)^2(l - a)^2$

M_2 (at R_2) . . . . $= \frac{wa^2}{2}$

M_x (between supports) . . . . . $= \frac{wx}{2l}(l^2 - a^2 - xl)$

M_{x_1} (for overhang) . . . . $= \frac{w}{2}(a - x_1)^2$

Δ_x (between supports) . . . . . $= \frac{wx}{24EIl}(l^4 - 2l^2x^2 + lx^3 - 2a^2l^2 + 2a^2x^2)$

Δ_{x_1} (for overhang) . . . . $= \frac{wx_1}{24EI}(4a^2l - l^3 + 6a^2x_1 - 4ax_1^2 + x_1^3)$

25. BEAM OVERHANGING ONE SUPPORT—UNIFORMLY DISTRIBUTED LOAD ON OVERHANG

$R_1 = V_1$ . . . . $= \frac{wa^2}{2l}$

R_2 $V_1 + V_2$ . . . . $= \frac{wa}{2l}(2l + a)$

V_2 . . . . $= wa$

V_{x_1} (for overhang) . . . . $= w(a - x_1)$

$M_{\max}$ (at R_2) . . . . $= \frac{wa^2}{2}$

M_x (between supports) . . . . . $= \frac{wa^2x}{2l}$

M_{x_1} (for overhang) . . . . $= \frac{w}{2}(a - x_1)^2$

$\Delta_{\max}\left(\text{between supports at } x = \frac{l}{\sqrt{3}}\right) = \frac{wa^2l^2}{18\sqrt{3}EI} = 0.03208\frac{wa^2l^2}{EI}$

$\Delta_{\max}$ (for overhang at $x_1 = a$) . . . $= \frac{wa^3}{24EI}(4l + 3a)$

Δ_x (between supports) . . . . . $= \frac{wa^2x}{12EIl}(l^2 - x^2)$

Δ_{x_1} (for overhang) . . . . $= \frac{wx_1}{24EI}(4a^2l + 6a^2x_1 - 4ax_1^2 + x_1^3)$

Source: AISC [2]; 29a from AITC. Copyright © American Institute of Steel Construction, Inc. Reprinted with permission. All rights reserved.

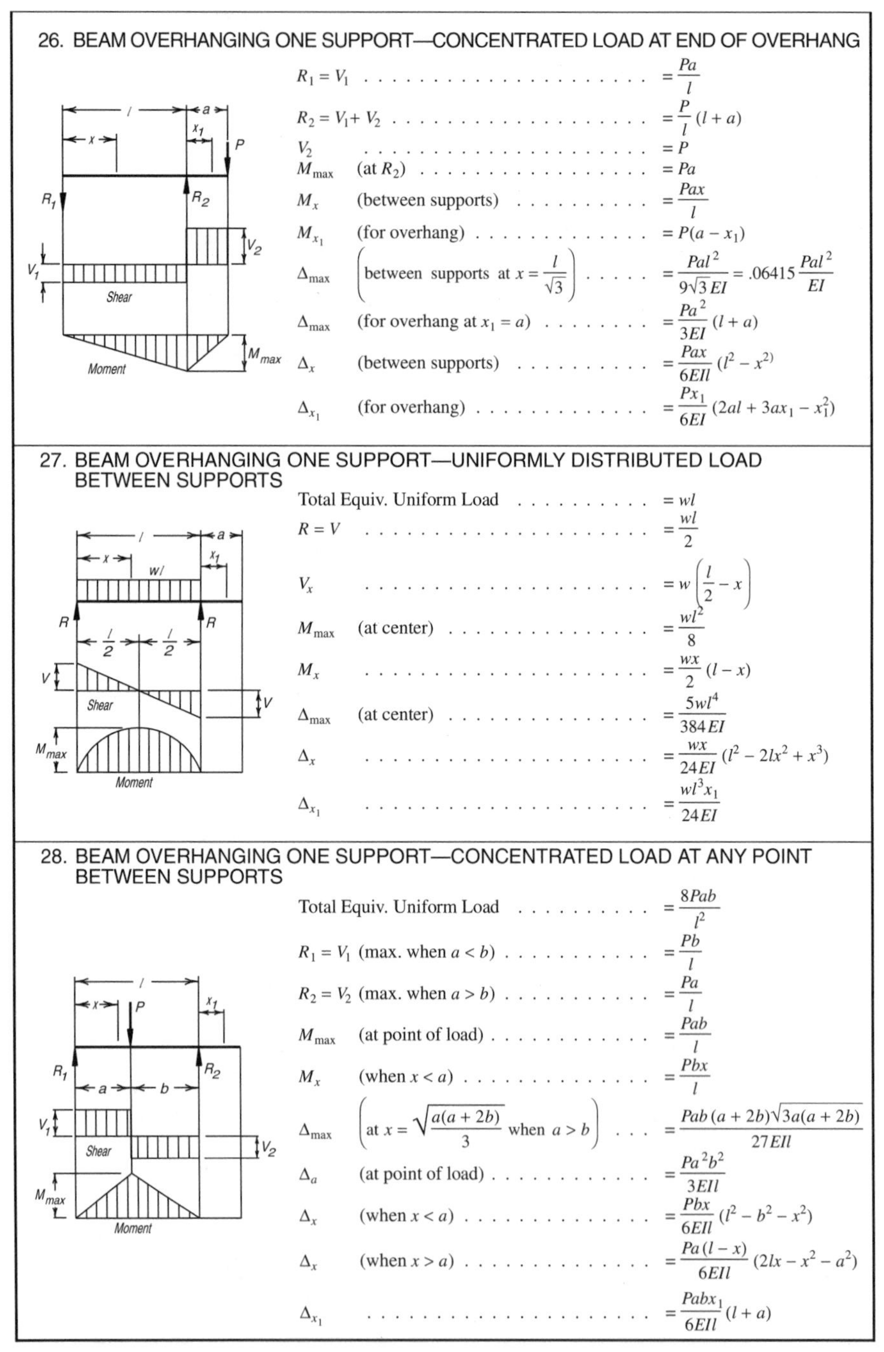

26. BEAM OVERHANGING ONE SUPPORT—CONCENTRATED LOAD AT END OF OVERHANG

$R_1 = V_1 \quad \dots\dots \quad = \dfrac{Pa}{l}$

$R_2 = V_1 + V_2 \quad \dots\dots \quad = \dfrac{P}{l}(l + a)$

$V_2 \quad \dots\dots \quad = P$

$M_{\max}$ (at R_2) $\dots\dots = Pa$

M_x (between supports) $\dots\dots = \dfrac{Pax}{l}$

M_{x_1} (for overhang) $\dots\dots = P(a - x_1)$

$\Delta_{\max}$ $\left(\text{between supports at } x = \dfrac{l}{\sqrt{3}}\right)$ $\dots\dots = \dfrac{Pal^2}{9\sqrt{3}\,EI} = .06415\dfrac{Pal^2}{EI}$

$\Delta_{\max}$ (for overhang at $x_1 = a$) $\dots\dots = \dfrac{Pa^2}{3EI}(l + a)$

Δ_x (between supports) $\dots\dots = \dfrac{Pax}{6EIl}(l^2 - x^2)$

Δ_{x_1} (for overhang) $\dots\dots = \dfrac{Px_1}{6EI}(2al + 3ax_1 - x_1^2)$

27. BEAM OVERHANGING ONE SUPPORT—UNIFORMLY DISTRIBUTED LOAD BETWEEN SUPPORTS

Total Equiv. Uniform Load $\dots\dots = wl$

$R = V \quad \dots\dots \quad = \dfrac{wl}{2}$

$V_x \quad \dots\dots \quad = w\left(\dfrac{l}{2} - x\right)$

$M_{\max}$ (at center) $\dots\dots = \dfrac{wl^2}{8}$

$M_x \quad \dots\dots \quad = \dfrac{wx}{2}(l - x)$

$\Delta_{\max}$ (at center) $\dots\dots = \dfrac{5wl^4}{384EI}$

$\Delta_x \quad \dots\dots \quad = \dfrac{wx}{24EI}(l^3 - 2lx^2 + x^3)$

$\Delta_{x_1} \quad \dots\dots \quad = \dfrac{wl^3x_1}{24EI}$

28. BEAM OVERHANGING ONE SUPPORT—CONCENTRATED LOAD AT ANY POINT BETWEEN SUPPORTS

Total Equiv. Uniform Load $\dots\dots = \dfrac{8Pab}{l^2}$

$R_1 = V_1$ (max. when $a < b$) $\dots\dots = \dfrac{Pb}{l}$

$R_2 = V_2$ (max. when $a > b$) $\dots\dots = \dfrac{Pa}{l}$

$M_{\max}$ (at point of load) $\dots\dots = \dfrac{Pab}{l}$

M_x (when $x < a$) $\dots\dots = \dfrac{Pbx}{l}$

$\Delta_{\max}$ $\left(\text{at } x = \sqrt{\dfrac{a(a + 2b)}{3}} \text{ when } a > b\right)$ $\dots = \dfrac{Pab(a + 2b)\sqrt{3a(a + 2b)}}{27EIl}$

Δ_a (at point of load) $\dots\dots = \dfrac{Pa^2b^2}{3EIl}$

Δ_x (when $x < a$) $\dots\dots = \dfrac{Pbx}{6EIl}(l^2 - b^2 - x^2)$

Δ_x (when $x > a$) $\dots\dots = \dfrac{Pa(l - x)}{6EIl}(2lx - x^2 - a^2)$

$\Delta_{x_1} \quad \dots\dots \quad = \dfrac{Pabx_1}{6EIl}(l + a)$

Source: AISC [2]; 29a from AITC. Copyright © American Institute of Steel Construction, Inc. Reprinted with permission. All rights reserved.

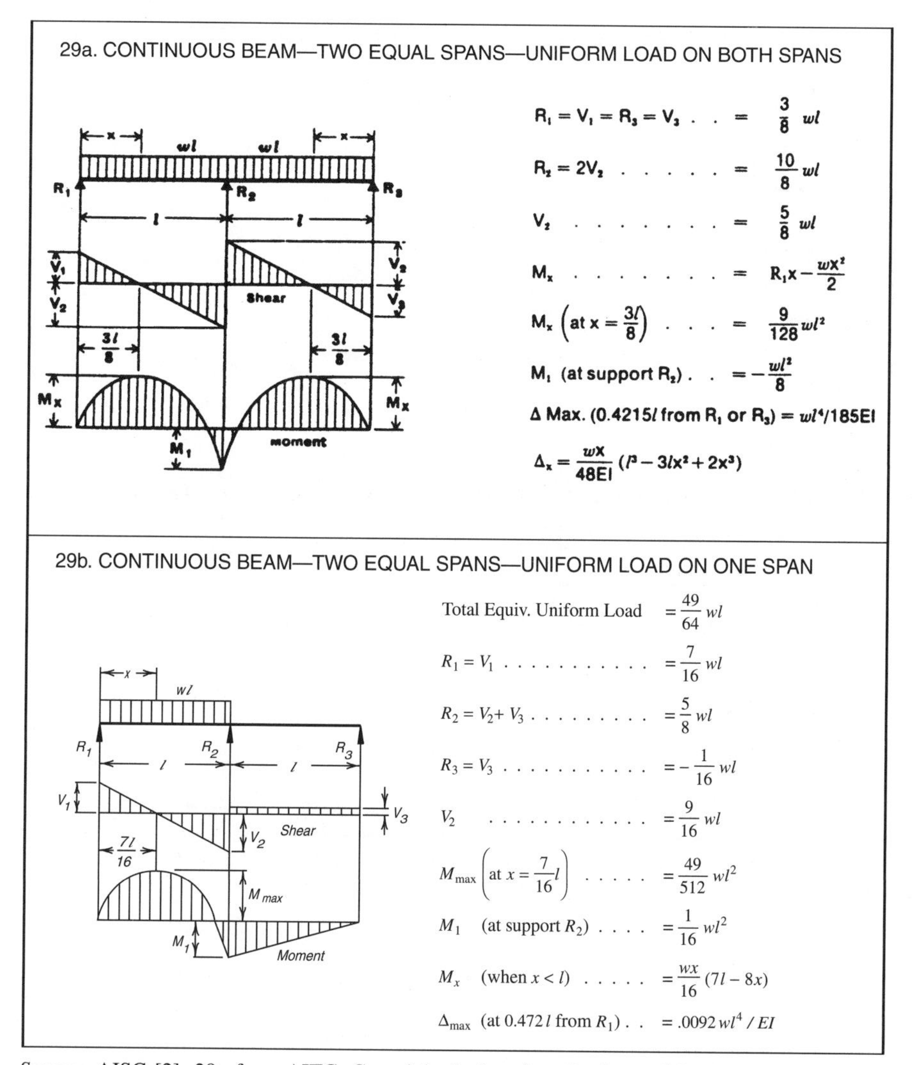

Source: AISC [2]; 29a from AITC. Copyright © American Institute of Steel Construction, Inc. Reprinted with permission. All rights reserved.

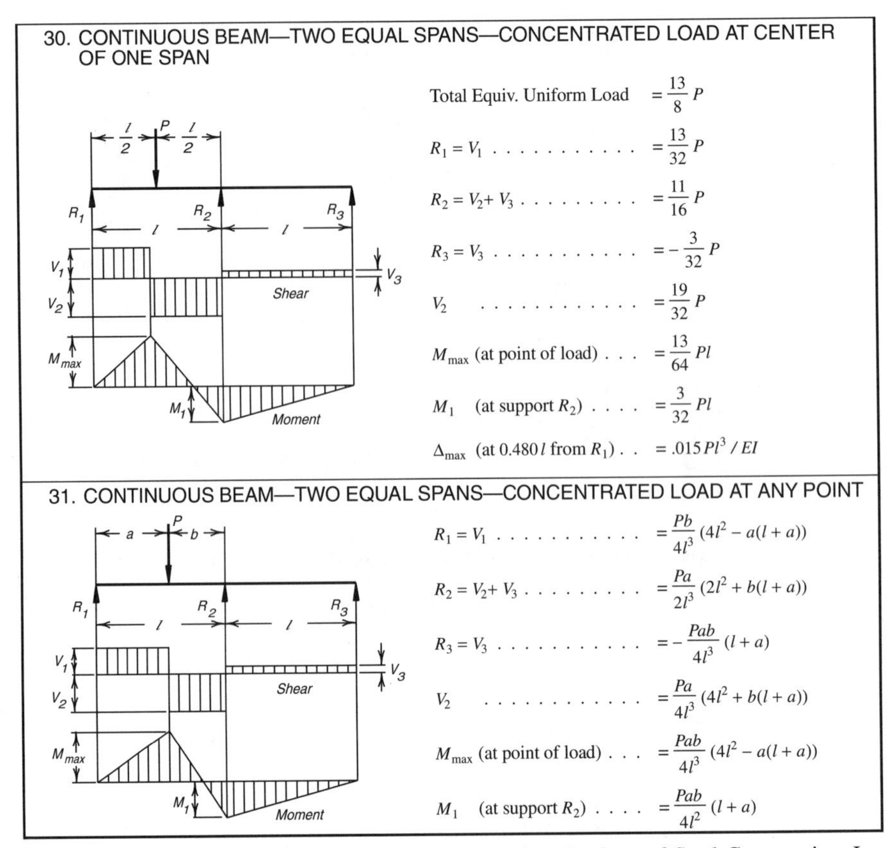

30. CONTINUOUS BEAM—TWO EQUAL SPANS—CONCENTRATED LOAD AT CENTER OF ONE SPAN

Total Equiv. Uniform Load $= \frac{13}{8} P$

$R_1 = V_1$ . . . . $= \frac{13}{32} P$

$R_2 = V_2 + V_3$ . . . . $= \frac{11}{16} P$

$R_3 = V_3$ . . . . $= -\frac{3}{32} P$

V_2 . . . . $= \frac{19}{32} P$

M_{max} (at point of load) . . . $= \frac{13}{64} Pl$

M_1 (at support R_2) . . . . $= \frac{3}{32} Pl$

Δ_{max} (at $0.480\,l$ from R_1) . . $= .015\, Pl^3 / EI$

31. CONTINUOUS BEAM—TWO EQUAL SPANS—CONCENTRATED LOAD AT ANY POINT

$R_1 = V_1$ . . . . $= \frac{Pb}{4l^3}(4l^2 - a(l + a))$

$R_2 = V_2 + V_3$ . . . . $= \frac{Pa}{2l^3}(2l^2 + b(l + a))$

$R_3 = V_3$ . . . . $= -\frac{Pab}{4l^3}(l + a)$

V_2 . . . . $= \frac{Pa}{4l^3}(4l^2 + b(l + a))$

M_{max} (at point of load) . . . $= \frac{Pab}{4l^3}(4l^2 - a(l + a))$

M_1 (at support R_2) . . . . $= \frac{Pab}{4l^2}(l + a)$

Source: AISC [2]; 29a from AITC. Copyright © American Institute of Steel Construction, Inc. Reprinted with permission. All rights reserved.

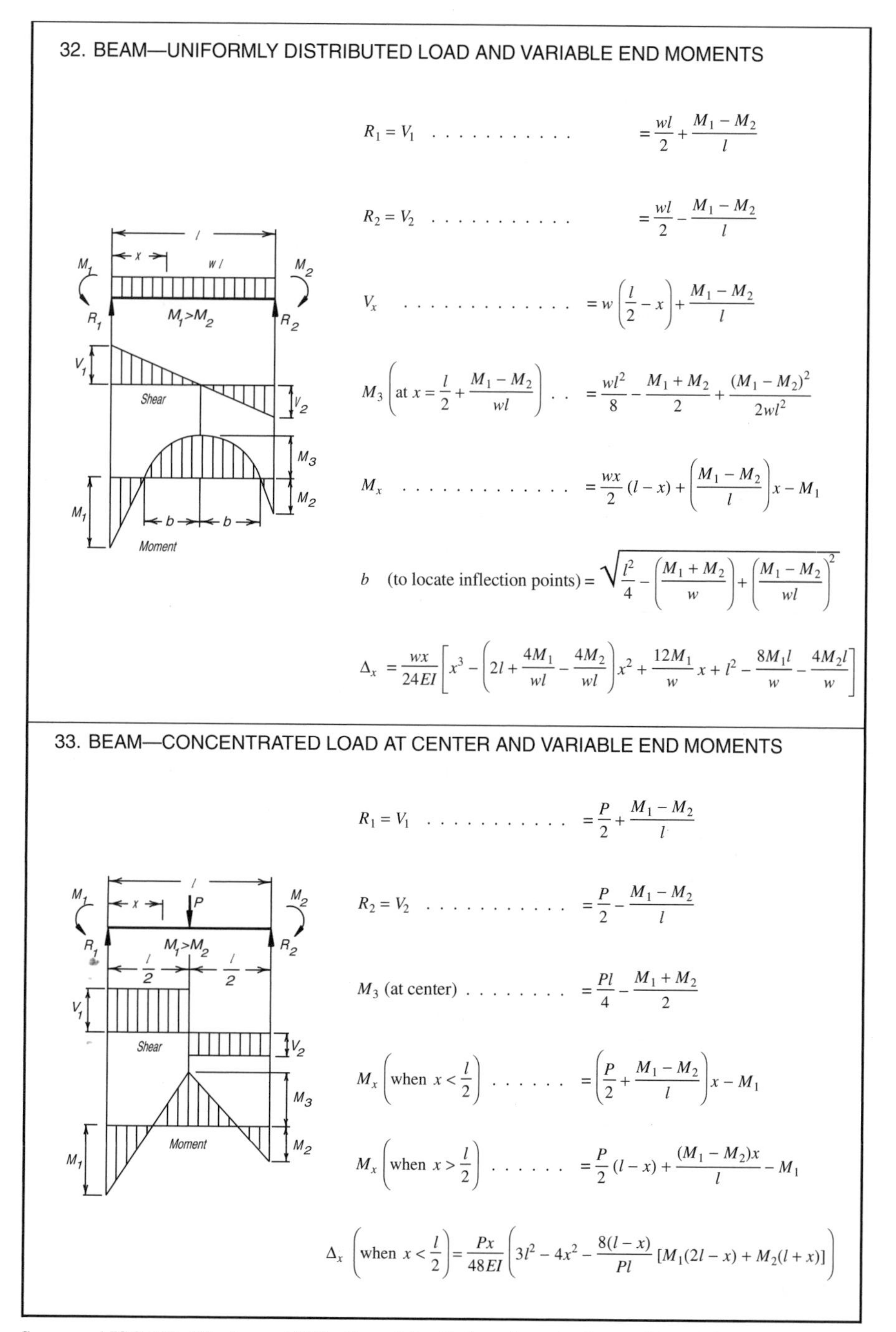

32. BEAM—UNIFORMLY DISTRIBUTED LOAD AND VARIABLE END MOMENTS

$$R_1 = V_1 \ldots\ldots = \frac{wl}{2} + \frac{M_1 - M_2}{l}$$

$$R_2 = V_2 \ldots\ldots = \frac{wl}{2} - \frac{M_1 - M_2}{l}$$

$$V_x \ldots\ldots = w\left(\frac{l}{2} - x\right) + \frac{M_1 - M_2}{l}$$

$$M_3 \left(\text{at } x = \frac{l}{2} + \frac{M_1 - M_2}{wl}\right) \ldots = \frac{wl^2}{8} - \frac{M_1 + M_2}{2} + \frac{(M_1 - M_2)^2}{2wl^2}$$

$$M_x \ldots\ldots = \frac{wx}{2}(l - x) + \left(\frac{M_1 - M_2}{l}\right)x - M_1$$

$$b \quad \text{(to locate inflection points)} = \sqrt{\frac{l^2}{4} - \left(\frac{M_1 + M_2}{w}\right) + \left(\frac{M_1 - M_2}{wl}\right)^2}$$

$$\Delta_x = \frac{wx}{24EI}\left[x^3 - \left(2l + \frac{4M_1}{wl} - \frac{4M_2}{wl}\right)x^2 + \frac{12M_1}{w}x + l^3 - \frac{8M_1 l}{w} - \frac{4M_2 l}{w}\right]$$

33. BEAM—CONCENTRATED LOAD AT CENTER AND VARIABLE END MOMENTS

$$R_1 = V_1 \ldots\ldots = \frac{P}{2} + \frac{M_1 - M_2}{l}$$

$$R_2 = V_2 \ldots\ldots = \frac{P}{2} - \frac{M_1 - M_2}{l}$$

$$M_3 \text{ (at center)} \ldots\ldots = \frac{Pl}{4} - \frac{M_1 + M_2}{2}$$

$$M_x \left(\text{when } x < \frac{l}{2}\right) \ldots\ldots = \left(\frac{P}{2} + \frac{M_1 - M_2}{l}\right)x - M_1$$

$$M_x \left(\text{when } x > \frac{l}{2}\right) \ldots\ldots = \frac{P}{2}(l - x) + \frac{(M_1 - M_2)x}{l} - M_1$$

$$\Delta_x \left(\text{when } x < \frac{l}{2}\right) = \frac{Px}{48EI}\left(3l^2 - 4x^2 - \frac{8(l - x)}{Pl}[M_1(2l - x) + M_2(l + x)]\right)$$

Source: AISC [2]; 29a from AITC. Copyright © American Institute of Steel Construction, Inc. Reprinted with permission. All rights reserved.

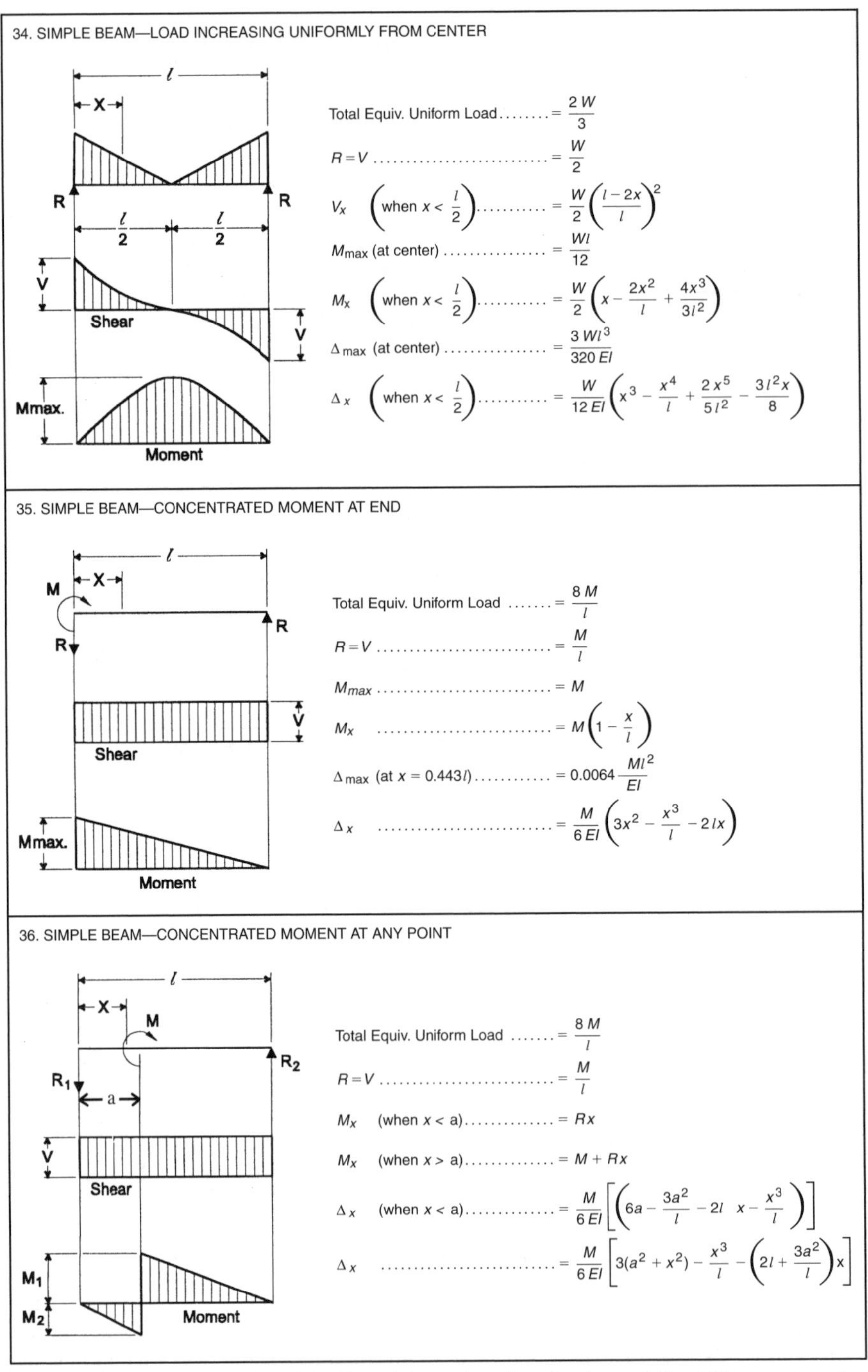

Source: AISC [2]; 29a from AITC. Copyright © American Institute of Steel Construction, Inc. Reprinted with permission. All rights reserved.

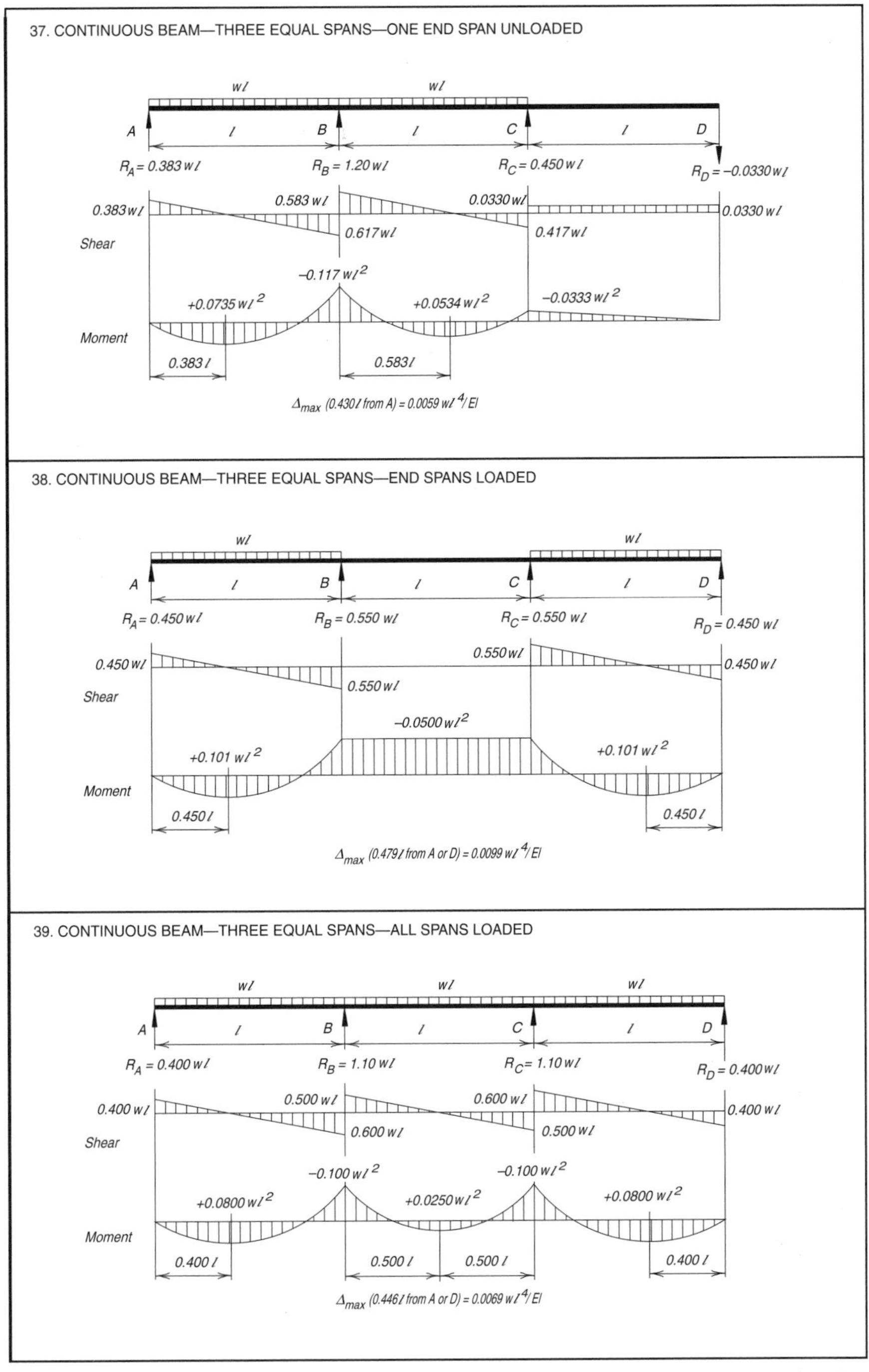

Source: AISC [2]; 29a from AITC. Copyright © American Institute of Steel Construction, Inc. Reprinted with permission. All rights reserved.

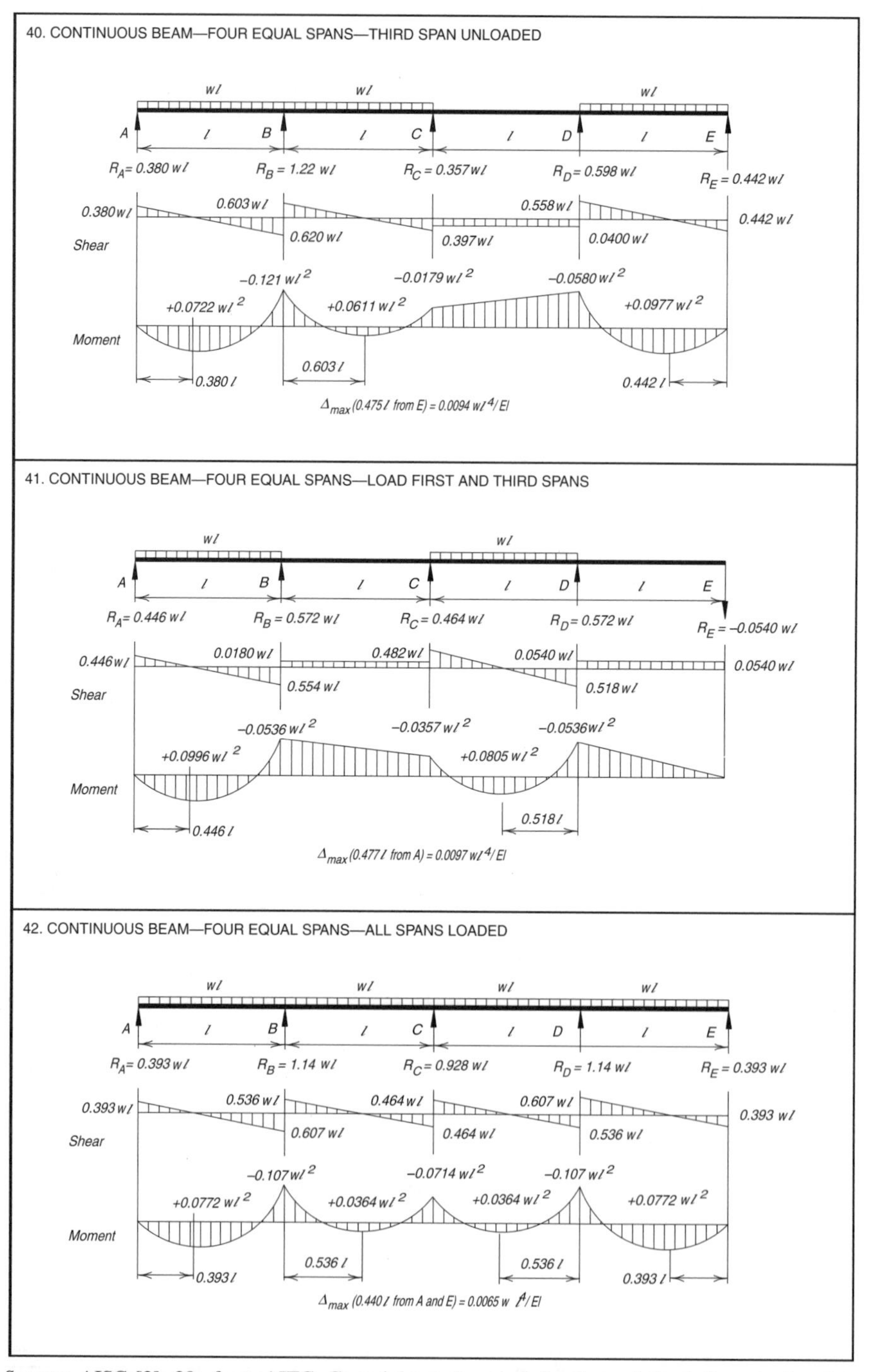

Source: AISC [2]; 29a from AITC. Copyright © American Institute of Steel Construction, Inc. Reprinted with permission. All rights reserved.

A.2.2 Cantilever Beam System Diagrams and Formulas

A.2.2.1 Two Equal Spans

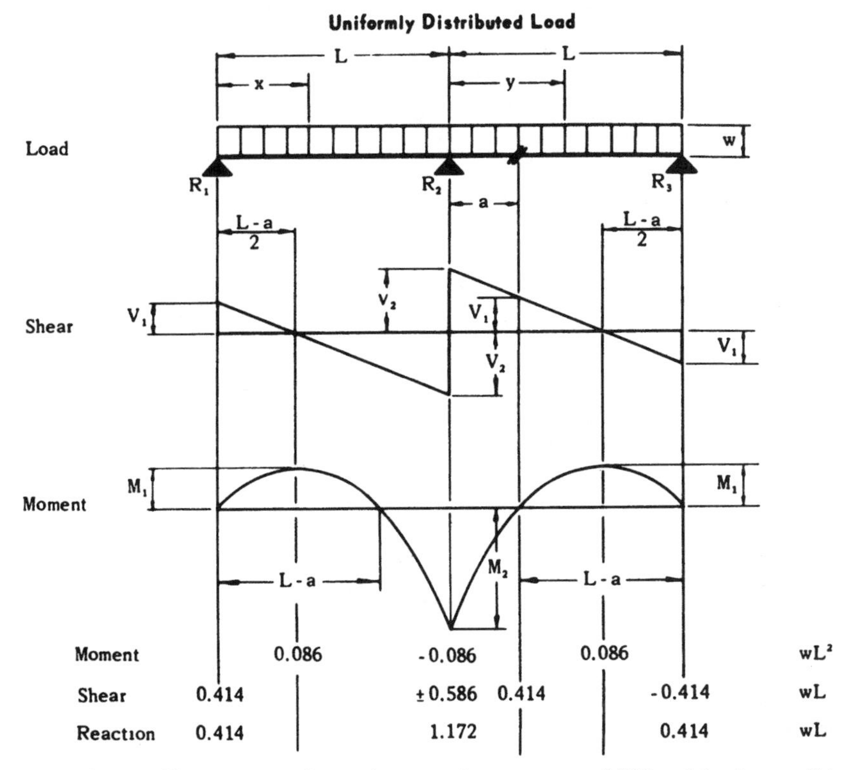

For maximum positive moment equal to maximum negative moment, $a = 0.172L$, and the above coefficients may be applied to wL^2 and wL to find the respective critical values of moment, shear, and reaction. Maximum deflection in either span will be

$$\Delta = 13.31 \frac{wL^4}{EI} \text{ in.}$$

General formulas are:

$$R_1 = R_3 = \frac{w}{2}(L - a) \qquad V_y = \frac{w}{2}(L + a - 2y)$$

$$R_2 = w(L + a) \qquad M_1 = \frac{w}{8}(L - a)^2$$

$$V_1 = \pm\frac{w}{2}(L - a) \qquad M_2 = -\frac{wLa}{2}$$

$$V_2 = \pm\frac{w}{2}(L + a) \qquad M_x = \frac{wx}{2}(L - a - x)$$

$$V_x = \frac{w}{2}(L - a - 2x) \qquad M_y = \frac{w}{2}(y - a)(L - y)$$

A.2.2.2 Two Unequal Spans

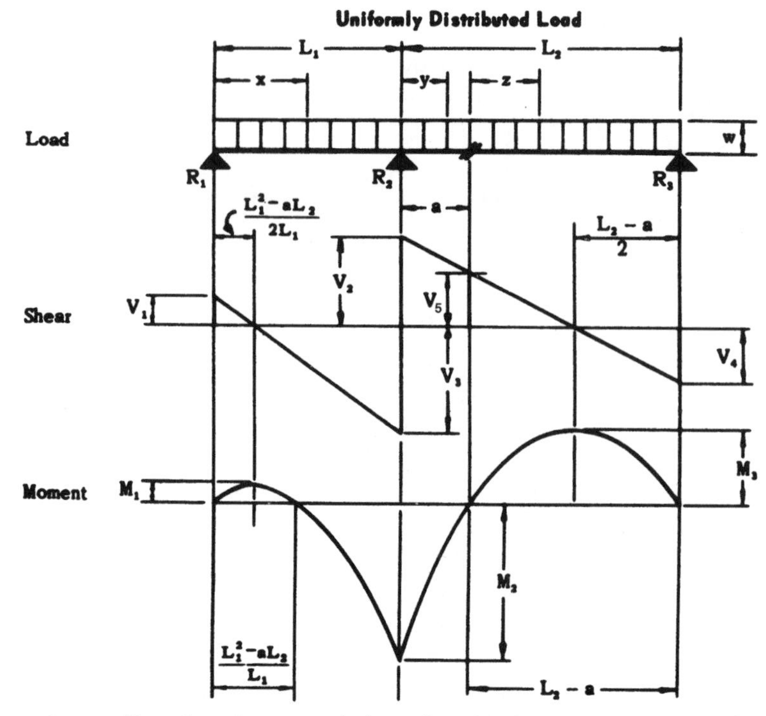

For maximum positive and negative moments in the cantilevered portion to be equal, $a = 0.172L_1^2/L_2$. Under these conditions, $M_1 = M_2 = 0.086wL_1^2$, $R_1 = V_1 = 0.414wL_1$, and $V_3 = -0.586wL_1$. Other coefficients can be determined for the above or other values of a from the general formulas following:

$$R_1 = \frac{w}{2L_1}(L_1^2 - aL_2)$$

$$R_2 = \frac{w}{2L_1}(L_1 + a)(L_1 + L_2)$$

$$R_3 = \frac{w}{2}(L_2 - a)$$

$$V_1 = \frac{w}{2L_1}(L_1^2 - aL_2)$$

$$V_2 = \frac{w}{2}(L_2 + a)$$

$$V_3 = -\frac{w}{2L_1}(L_1^2 - aL_2)$$

$$V_4 = -\frac{w}{2}(L_2 - a)$$

$$V_5 = \frac{w}{2}(L_2 - a) = -V_4$$

$$V_x = \frac{w}{2L_1}(L_1^2 - aL_2) - wx$$

$$V_y = \frac{w}{2}(L_2 + a) - wy$$

$$V_z = \frac{w}{2}(L_2 - a) - wz$$

$$M_1 = \frac{w}{8L_1^2}(L_1^2 - aL_2)^2$$

$$M_2 = -\frac{wL_2a}{2}$$

$$M_3 = \frac{w}{8}(L_2 - a)^2$$

$$M_x = \frac{wx}{2L_1}(L_1^2 - xL_1 - aL_2)$$

$$M_y = \frac{w}{2}(L_2 - y)(y - a)$$

$$M_z = \frac{wz}{2}(L_2 - a - z)$$

A.2.2.3 Three Equal Spans, Single Cantilever at Each End

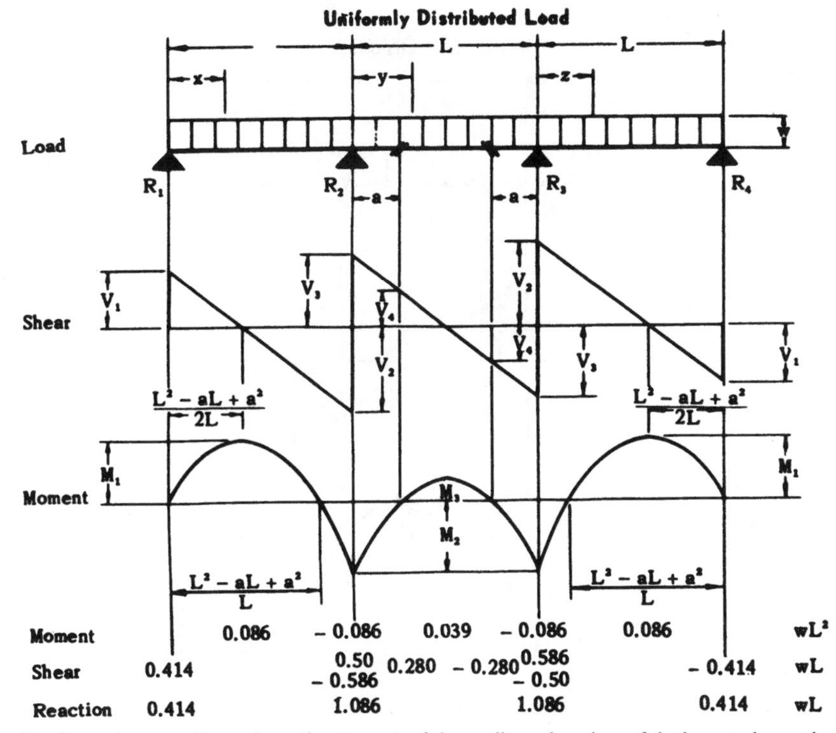

For the maximum positive and negative moments of the cantilevered portions of the beam to be equal, $a = 0.220L$, and the above coefficients may be applied to wL^2 and wL to find the respective critical values of moment, shear, and reaction. Maximum deflection in end spans will be

$$\Delta = 13.31 \frac{wL_4}{EI} \text{ in.}$$

General formulas are:

$$R_1 = R_4 = \frac{w}{2L}(L^2 - aL + a^2) \qquad M_1 = +\frac{w}{8L^2}(L^2 - aL + a^2)^2$$

$$R_2 = R_1 = \frac{w}{2L}(2L^2 + aL - a^2) \qquad M_2 = -\frac{w}{2}(aL - a^2)$$

$$V_1 = \pm\frac{w}{2L}(L^2 - aL + a^2) \qquad M_3 = \frac{w}{8}(L - 2a)^2$$

$$V_2 = \pm\frac{w}{2L}(L^2 + aL - a^2) \qquad V_x = \frac{w}{2L}(L^2 - aL + a^2) - wx$$

$$V_3 = \pm\frac{wL}{2} \qquad V_y = \frac{w}{2}(L - 2y)$$

$$V_4 = \pm\frac{w}{2}(L - 2a) \qquad V_z = \frac{w}{2L}(L^2 + aL - a^2) - wz$$

$$M_x = \frac{wx}{2L}(L^2 - aL + a^2) - \frac{wx^2}{2} \qquad M_y = \frac{w}{2}(y - a)(L - y - a)$$

$$M_z = \frac{w}{2}(L - z)\left(\frac{a^2}{L} + z - a\right)$$

A.2.2.4 Three Spans, End Spans Equal, Single Cantilever at Each End

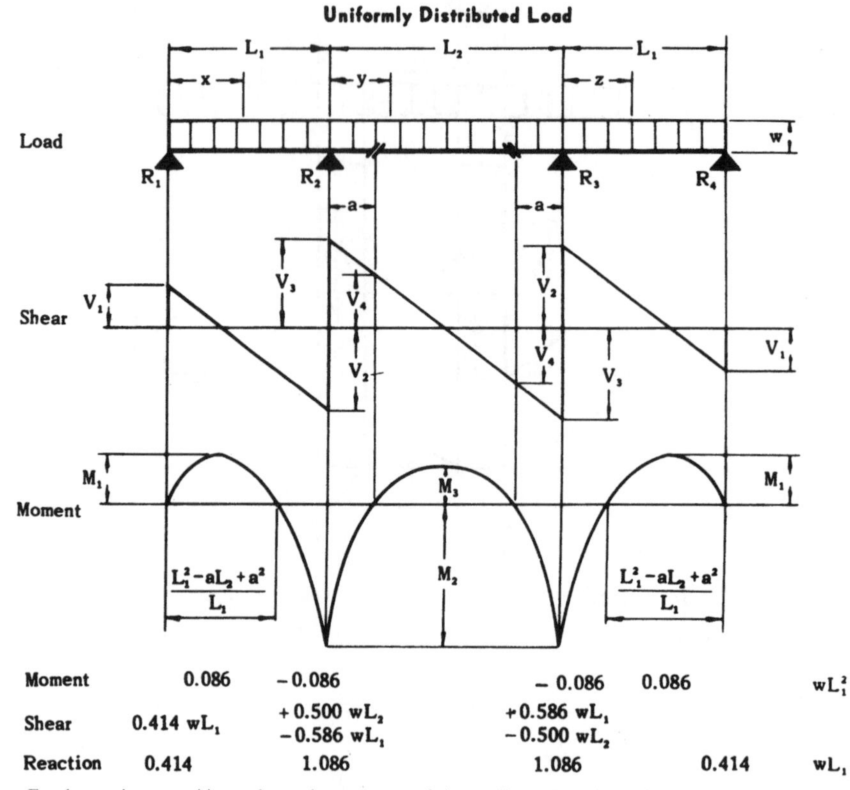

For the maximum positive and negative moments of the cantilevered portions of the beam to be equal, $a = \frac{1}{2}(L_2 - \sqrt{L_2^2 - 0.688L_1^2})$ and the above coefficients may be applied to find the respective critical values of moment, shear, and reaction. Coefficients are omitted when calculation using the general formula is simpler. General formulas are:

$$R_1 = R_4 = \frac{w}{2L_1}(L_1^2 - aL_2 + a^2)$$

$$V_4 = \pm\frac{w}{2}(L_2 - 2a)$$

$$M_2 = -\frac{w}{2}(aL_2 - a^2)$$

$$R_2 = R_3 = \frac{w}{2L_1}(L_1 + a)(L_1 + L_2 - a)$$

$$M_1 = \frac{w}{8L_1^2}(L_1^2 - aL_2 + a^2)^2$$

$$M_3 = \frac{w}{8}(L_2 - 2a)^2$$

$$V_1 = \pm\frac{w}{2L_1}(L_1^2 - aL_2 - a^2)$$

$$V_x = \frac{w}{2L_1}(L_1^2 - aL_2 + a^2) - wx$$

$$M_x = \frac{wx}{2L_1}(L_1^2 - aL_2 + a^2) - \frac{wx^2}{2}$$

$$V_2 = \pm\frac{w}{2L_1}(L_1^2 + aL_2 - a^2)$$

$$V_y = \frac{w}{2}(L_2 - 2y)$$

$$M_y = \frac{w}{2}(y - a)(L_2 - y - a)$$

$$V_3 = \pm\frac{wL_2}{2}$$

$$V_z = \frac{w}{2L_1}(L_1^2 + aL_2 + a^2) - wz$$

$$M_z = \frac{w}{2L_1}(L_1 z - aL_2 + a^2)(L_1 - z)$$

A.2.2.5 Three Equal Spans, Double Cantilever

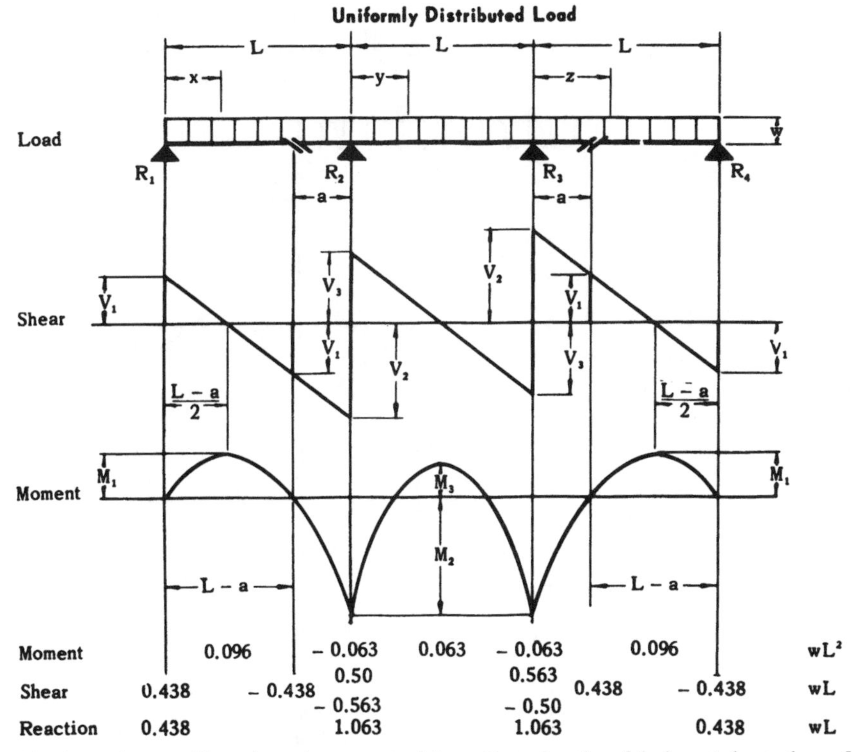

For the maximum positive and negative moments of the cantilevered portion of the beam to be equal, $a = L/8$, and the above coefficients may be applied to wL^2 and wL to find the respective critical values of moment, shear, and reaction. Maximum deflection in center span will be

$$\Delta = 8.99 \frac{wL^4}{EI}$$

General formulas are

$$R_1 = R_4 = \frac{w}{2}(L - a) \qquad V_x = \frac{w}{2}(L - a - 2x) \qquad M_3 = \frac{w}{8}(L^2 - 4aL)$$

$$R_2 = R_3 = \frac{w}{2}(2L + a) \qquad V_y = \frac{w}{2}(L - 2y) \qquad M_x = \frac{wx}{2}(L - a - x)$$

$$V_1 = \pm\frac{w}{2}(L - a) \qquad V_z = \frac{w}{2}(L + a - 2z) \qquad M_y = \frac{w}{2}(L_y - y^2 - La)$$

$$V_2 = \pm\frac{w}{2}(L + a) \qquad M_1 = \frac{w}{8}(L - a)^2 \qquad M_z = \frac{w}{2}(L - z)(z - a)$$

$$V_3 = \pm\frac{wL}{2} \qquad M_2 = -\frac{wLa}{2}$$

Points of zero moment in center span occur at

$$y = \frac{L_2 \pm \sqrt{L_2^2 - 4aL_1}}{2}$$

A.2.2.6 Three Spans, End Spans Equal, Double Cantilever

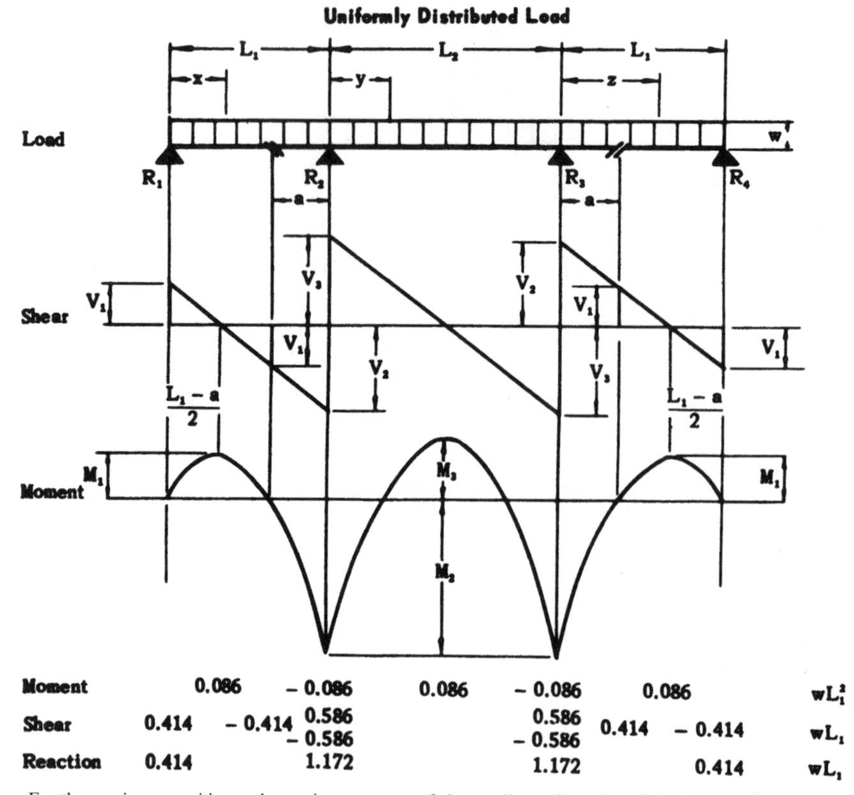

For the maximum positive and negative moments of the cantilevered portion of the beam to be equal, $a = 0.125L_2^2/L_1$. For the special case where all maximum positive and negative moments are equal, that is, $M_1 = M_2 = M_3$, $a = 0.172L_1$ and $L_2 = 1.172L_1$, and the above coefficients may be applied to wL_1^2 and wL_1 to find the respective critical values of moment, shear, and reaction. General formulas are

$$R_1 = R_4 = \frac{w}{2}(L_1 - a)$$

$$R_2 = R_3 = \frac{w}{2}(L_1 + L_2 + a)$$

$$V_1 = \pm\frac{w}{2}(L_1 - a)$$

$$V_2 = \pm\frac{w}{2}(L_1 + a)$$

$$V_3 = \pm\frac{wL_2}{2}$$

$$V_x = \frac{w}{2}(L_1 - a - 2x)$$

$$V_y = \frac{w}{2}(L_2 - 2y)$$

$$V_x = \frac{w}{2}(L_1 + a - 2z)$$

$$M_1 = \frac{w}{8}(L_1 - a)^2$$

$$M_2 = -\frac{wL_1a}{2}$$

$$M_3 = \frac{w}{8}(L_2^2 - 4aL_1)$$

$$M_x = \frac{wx}{2}(L_1 - a - x)$$

$$M_y = \frac{w}{2}(L_2y - y^2 - aL_1)$$

$$M_z = \frac{w}{2}(z - L_1)(a - z)$$

Points of zero moment in center span occur at

$$y = \frac{L_2 \pm \sqrt{L_2^2 - 4aL_1}}{2}$$

A.3 TYPICAL FASTENER DIMENSIONS AND YIELD STRENGTHS

A.3.1 Hex Bolts[1]

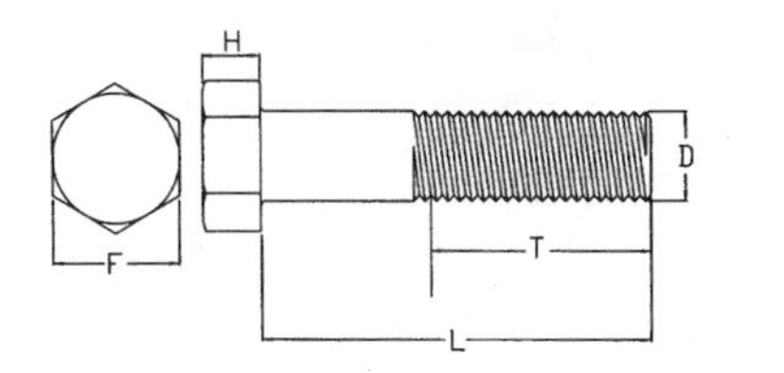

D = diameter
D_r = root diameter
T = thread length
L = bolt length

		Diameter, D							
		1/4"	**5/16"**	**3/8"**	**1/2"**	**5/8"**	**3/4"**	**7/8"**	**1"**
D_r		0.189"	0.245"	0.298"	0.406"	0.514"	0.627"	0.739"	0.847"
F		7/16"	1/2"	9/16"	3/4"	15/16"	1-1/8"	1-5/16"	1-1/2"
H		11/64"	7/32"	1/4"	11/32"	27/64"	1/2"	37/64"	43/64"
T	L ≤ 6 in.	3/4"	7/8"	1"	1-1/4"	1-1/2"	1-3/4"	2"	2-1/4"
	L > 6 in.	1"	1 1/8"	1-1/4"	1-1/2"	1-3/4"	2"	2-1/4"	2-1/2"

1. Tolerances specified in ANSI B18.2.1. Full body diameter bolt is shown. Root diameter based on UNC (coarse) thread series (see ANSI B1.1).

Source: Reprinted with permission from *National Design Specification® for Wood Construction.* Copyright © 2001. American Forest & Paper Association, Inc.

A.3.2 Hex Lag Screws[1]

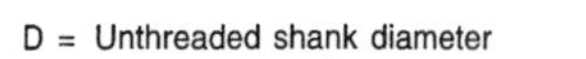

D = Unthreaded shank diameter
D_r = Root diameter of threaded portion
F = Width of head across flats
H = Height of head

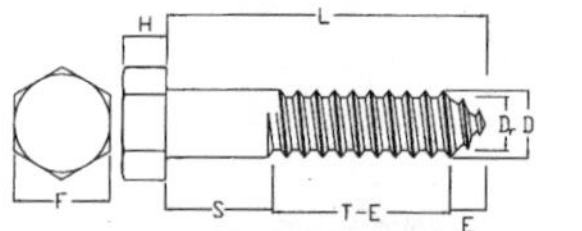

S = Unthreaded shank length
T = Minimum thread length[2]
E = Length of tapered tip
N = Number of threads/inch

Length L		Diameter, D										
		1/4"	5/16"	3/8"	7/16"	1/2"	5/8"	3/4"	7/8"	1"	1-1/8"	1-1/4"
	D_r	0.173"	0.227"	0.265"	0.328"	0.371"	0.471"	0.579"	0.683"	0.780"	0.887"	1.012"
	E	5/32"	3/16"	7/32"	9/32"	5/16"	13/32"	1/2"	19/32"	11/16"	25/32"	7/8"
	H	11/64"	7/32"	1/4"	19/64"	11/32"	27/64"	1/2"	37/64"	43/64"	3/4"	27/32"
	F	7/16"	1/2"	9/16"	5/8"	3/4"	15/16"	1-1/8"	1-5/16"	1-1/2"	1-11/16"	1-7/8"
	N	10	9	7	7	6	5	4-1/2	4	3-1/2	3-1/4	3-1/4
	S	1/4"	1/4"	1/4"	1/4"	1/4"						
1"	T	3/4"	3/4"	3/4"	3/4"	3/4"						
	T-E	19/32"	9/16"	17/32"	15/32"	7/16"						
	S	1/4"	1/4"	1/4"	1/4"	1/4"						
1-1/2"	T	1-1/4"	1-1/4"	1-1/4"	1-1/4"	1-1/4"						
	T-E	1-3/32"	1-1/16"	1-1/32"	31/32"	15/16"						
	S	1/2"	1/2"	1/2"	1/2"	1/2"	1/2"					
2"	T	1-1/2"	1-1/2"	1-1/2"	1-1/2"	1-1/2"	1-1/2"					
	T-E	1-11/32"	1-5/16"	1-9/32"	1-7/32"	1-3/16"	1-3/32"					
	S	3/4"	3/4"	3/4"	3/4"	3/4"	3/4"					
2-1/2"	T	1-3/4"	1-3/4"	1-3/4"	1-3/4"	1-3/4"	1-3/4"					
	T-E	1-19/32"	1-9/16"	1-17/32"	1-15/32"	1-7/16"	1-11/32"					
	S	1"	1"	1"	1"	1"	1"	1"	1"	1"		
3"	T	2"	2"	2"	2"	2"	2"	2"	2"	2"		
	T-E	1-27/32"	1-13/16"	1-25/32"	1-23/32"	1-11/16"	1-19/32"	1-1/2"	1-13/32"	1-5/16"		
	S	1-1/2"	1-1/2"	1-1/2"	1-1/2"	1-1/2"	1-1/2"	1-1/2"	1-1/2"	1-1/2"	1-1/2"	1-1/2"
4"	T	2-1/2"	2-1/2"	2-1/2"	2-1/2"	2-1/2"	2-1/2"	2-1/2"	2-1/2"	2-1/2"	2-1/2"	2-1/2"
	T-E	2-11/32"	2-5/16"	2-9/32"	2-7/32"	2-3/16"	2-3/32"	2"	1-29/32"	1-13/16"	1-23/32"	1-5/8"
	S	2"	2"	2"	2"	2"	2"	2"	2"	2"	2"	2"
5"	T	3"	3"	3"	3"	3"	3"	3"	3"	3"	3"	3"
	T-E	2-27/32"	2-13/16"	2-25/32"	2-23/32"	2-11/16"	2-19/32"	2-1/2"	2-13/32"	2-5/16"	2-7/32"	2-1/8"
	S	2-1/2"	2-1/2"	2-1/2"	2-1/2"	2-1/2"	2-1/2"	2-1/2"	2-1/2"	2-1/2"	2-1/2"	2-1/2"
6"	T	3-1/2"	3-1/2"	3-1/2"	3-1/2"	3-1/2"	3-1/2"	3-1/2"	3-1/2"	3-1/2"	3-1/2"	3-1/2"
	T-E	3-11/32"	3-5/16"	3-9/32"	3-7/32"	3-3/16"	3-3/32"	3"	2-29/32"	2-13/16"	2-23/32"	2-5/8"
	S	3"	3"	3"	3"	3"	3"	3"	3"	3"	3"	3"
7"	T	4"	4"	4"	4"	4"	4"	4"	4"	4"	4"	4"
	T-E	3-27/32"	3-13/16"	3-25/32"	3-23/32"	3-11/16"	3-19/32"	3-1/2"	3-13/32"	3-5/16"	3-7/32"	3-1/8"
	S	3-1/2"	3-1/2"	3-1/2"	3-1/2"	3-1/2"	3-1/2"	3-1/2"	3-1/2"	3-1/2"	3-1/2"	3-1/2"
8"	T	4-1/2"	4-1/2"	4-1/2"	4-1/2"	4-1/2"	4-1/2"	4-1/2"	4-1/2"	4-1/2"	4-1/2"	4-1/2"
	T-E	4-11/32"	4-5/16"	4-9/32"	4-7/32"	4-3/16"	4-3/32"	4"	3-29/32"	3-13/16"	3-23/32"	3-5/8"
	S	4"	4"	4"	4"	4"	4"	4"	4"	4"	4"	4"
9"	T	5"	5"	5"	5"	5"	5"	5"	5"	5"	5"	5"
	T-E	4-27/32"	4-13/16"	4-25/32"	4-23/32"	4-11/16"	4-19/32"	4-1/2"	4-13/32"	4-5/16"	4-7/32"	4-1/8"
	S	4-1/2"	4-1/2"	4-1/2"	4-1/2"	4-1/2"	4-1/2"	4-1/2"	4-1/2"	4-1/2"	4-1/2"	4-1/2"
10"	T	5-1/2"	5-1/2"	5-1/2"	5-1/2"	5-1/2"	5-1/2"	5-1/2"	5-1/2"	5-1/2"	5-1/2"	5-1/2"
	T-E	5-11/32"	5-5/16"	5-9/32"	5-7/32"	5-3/16"	5-3/32"	5"	4-29/32"	4-13/16"	4-23/32"	4-5/8"
	S	5"	5"	5"	5"	5"	5"	5"	5"	5"	5"	5"
11"	T	6"	6"	6"	6"	6"	6"	6"	6"	6"	6"	6"
	T-E	5-27/32"	5-13/16"	5-25/32"	5-23/32"	5-11/16"	5-19/32"	5-1/2"	5-13/32"	5-5/16"	5-7/32"	5-1/8"
	S	6"	6"	6"	6"	6"	6"	6"	6"	6"	6"	6"
12"	T	6"	6"	6"	6"	6"	6"	6"	6"	6"	6"	6"
	T-E	5-27/32"	5-13/16"	5-25/32"	5-23/32"	5-11/16"	5-19/32"	5-1/2"	5-13/32"	5-5/16"	5-7/32"	5-1/8"

1. Tolerances specified in ANSI B18.2.1. Full body diameter lag screw is shown. For reduced body diameter lag screws, the unthreaded shank diameter may be reduced to approximately the root diameter, D_r.
2. Minimum thread length (T) for lag screw lengths (L) is 6" or 1/2 the lag screw length plus 0.5", whichever is less. Thread lengths may exceed these minimums up to the full lag screw length (L).

Source: Reprinted with permission from *National Design Specification® for Wood Construction.* Copyright © 2001. American Forest & Paper Association, Inc.

A.3.3 Wood Screws[1]

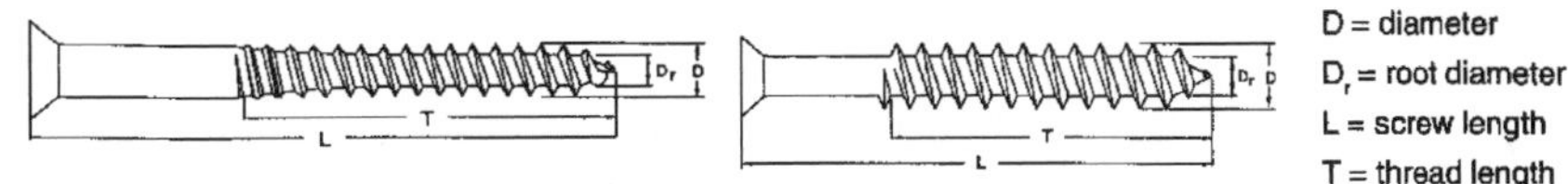

Cut Thread[2] Rolled Thread[3]

	Wood Screw Number										
	6	7	8	9	10	12	14	16	18	20	24
D	0.138"	0.151"	0.164"	0.177"	0.19"	0.216"	0.242"	0.268"	0.294"	0.32"	0.372"
D_r[4]	0.113"	0.122"	0.131"	0.142"	0.152"	0.171"	0.196"	0.209"	0.232"	0.255"	0.298"

1. Tolerances specified in ANSI B18.6.1
2. Thread length on cut thread wood screws is approximately two-thirds of the screw length.
3. Single lead thread shown. Thread length is at least four times the screw diameter or two-thirds of the screw length, whichever is greater. Screws which are too short to accommodate the minimum thread length, have threads extending as close to the underside of the head as practicable.
4. Taken as the average of the specified maximum and minimum limits for body diameter of rolled thread wood screws.

Source: Reprinted with permission from *National Design Specification® for Wood Construction.* Copyright © 2001. American Forest & Paper Association, Inc.

A.3.4 Common, Box, and Sinker Nails[2]

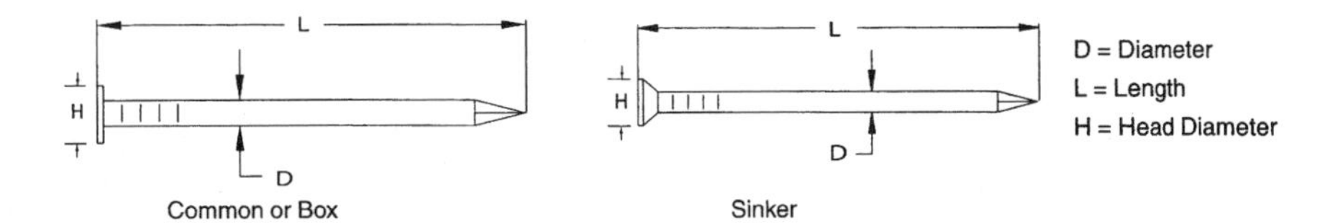

Common or Box Sinker

Type		Pennyweight										
		6d	7d	8d	10d	12d	16d	20d	30d	40d	50d	60d
Common	L	2"	2-1/4"	2-1/2"	3"	3-1/4"	3-1/2"	4"	4-1/2"	5"	5-1/2"	6"
	D	0.113"	0.113"	0.131"	0.148"	0.148"	0.162"	0.192"	0.207"	0.225"	0.244"	0.263"
	H	0.266"	0.266"	0.281"	0.312"	0.312"	0.344"	0.406"	0.438"	0.469"	0.5"	0.531"
Box	L	2"	2-1/4"	2-1/2"	3"	3-1/4"	3-1/2"	4"	4-1/2"	5"		
	D	0.099"	0.099"	0.113"	0.128"	0.128"	0.135"	0.148"	0.148"	0.162"		
	H	0.266"	0.266"	0.297"	0.312"	0.312"	0.344"	0.375"	0.375"	0.406"		
Sinker	L	1-7/8"	2-1/8"	2-3/8"	2-7/8"	3-1/8"	3-1/4"	3-3/4"	4-1/4"	4-3/4"		5-3/4"
	D	0.092"	0.099"	0.113"	0.12"	0.135"	0.148"	0.177"	0.192"	0.207"		0.244"
	H	0.234"	0.250"	0.266"	0.281"	0.312"	0.344"	0.375"	0.406"	0.438"		0.5"

1. Tolerances specified in ASTM F1667. Typical shape of common, box, and sinker nails shown. See ASTM F1667 for other nail types.

Source: Reprinted with permission from *National Design Specification® for Wood Construction.* Copyright © 2001. American Forest & Paper Association, Inc.

A.3.5 Split-Ring and Shear Plate Connectors

	Split Rings[a]	
	$2\frac{1}{2}$ in.	4 in.
Split ring		
Inside diameter at center when closed (in.)	2.5	4
Thickness of ring at center (in.)	0.163	0.193
Thickness of ring at edge (in.)	0.123	0.133
Depth (in.)	0.75	1
Groove[b]		
Inside diameter (in.)	2.56	4.08
Outside diameter (in.)	2.92	4.5
Width (in.)	0.18	0.21
Depth (in.)	0.375	0.5
Bolt		
Diameter of bolt (in.)	0.5	0.75
Diameter of bolt hole (in.)	0.56	0.81
Washers		
Round, cast or malleable iron		
Diameter (in.)	2.125	3
Square, plate		
Length of side (in.)	2	3
Thickness (in.)	0.125	0.188

	Shear Plates[c]			
	$2\frac{5}{8}$ in.	$2\frac{5}{8}$ in.	4 in.	4 in.
Shear plate				
Material	Pressed steel	Malleable iron	Malleable iron	Malleable iron
Diameter of plate (in.)	2.62	2.62	4.02	4.02
Diameter of bolt hole in plate (in.)	0.81	0.81	0.81	0.93
Depth (in.)	0.42	0.42	0.62	0.62
Groove[b]				
Inside diameter (in.)	2.29	2.31	3.54	3.54
Outside diameter (in.)	2.63	2.63	4.02	4.02
Width (in.)	0.17	0.16	0.24	0.24
Depth (in.)	0.42	0.42	0.62	0.62
Bolt				
Diameter of bolt (in.)	0.75	0.75	0.75	0.88
Diameter of bolt hole in timber or metal member (in.)	0.81	0.81	0.81	0.94
Washers				
Round, cast or malleable iron				
Diameter (in.)	3	3	3	3.5
Square, plate				
Length of side (in.)	3	3	3	3
Thickness (in.)	0.25	0.25	0.25	0.25

[a] Split-ring dimensions are from *Design Manual for TECO Timber Connectors*. Cleveland Steel Specialty Company, Bedford Heights, OH, 1997.
[b] Groove dimensions have been estimated and are provided to facilitate calculation of connection row tear-out and group tear-out failure modes.
[c] Shear plate dimensions are from ASTM D5933-96, *Standard Specification for 2-5/8 in. and 4 in. Diameter Metal Shear Plates for Use in Wood Constructions*, American Society for Testing and Materials, West Conshohocken, PA, 2001.

A.3.6 Timber Rivets

Rivet dimensions are taken from ASTM F1667.

Rivet Dimensions

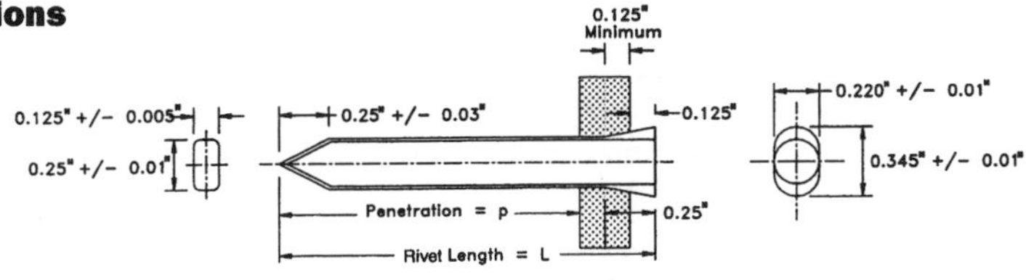

Steel Side Plate Dimensions

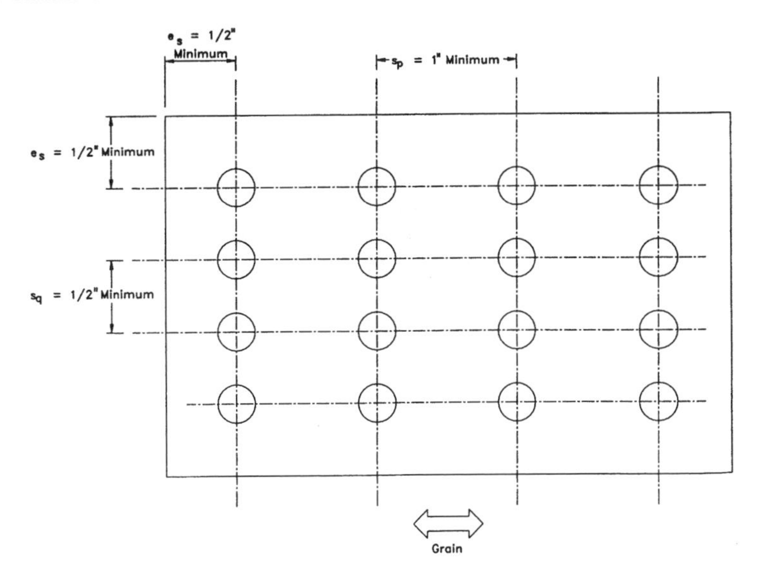

Notes:
1. Hole diameter: 17/64" minimum to 18/64" maximum.
2. Tolerences in location of holes: 1/8" maximum in any direction.
3. All dimensions are prior to galvanizing in inches.
4. s_p, and s_q, are defined in 13.3.
5. e_s is the end and edge distance as defined by the steel.
6. Orient wide face of rivets parallel to grain, regardless of plate orientation.

Source: Reprinted with permission from *National Design Specification® for Wood Construction.* Copyright © 2001, American Forest & Paper Assocation, Inc.

A.3.7 Fastener Bending Yield Strengths

Fastener Type	F_{yb} (psi)
Bolt, lag screw (with D ≥ 3/8"), drift pin (SAE J429 Grade 1- F_y = 36,000 psi and F_u = 60,000 psi)	45,000
Common, box, or sinker nail, spike, lag screw, wood screw (low to medium carbon steel)	
0.099" ≤ D ≤ 0.142"	100,000
0.142" < D ≤ 0.177"	90,000
0.177" < D ≤ 0.236"	80,000
0.236" < D ≤ 0.273"	70,000
0.273" < D ≤ 0.344"	60,000
0.344" < D ≤ 0.375"	45,000
Hardened steel nail (medium carbon steel)	
0.120" ≤ D ≤ 0.142"	130,000
0.142" < D ≤ 0.192"	115,000
0.192" < D ≤ 0.207"	100,000

Source: Reprinted with permission from *National Design Specification® for Wood Construction.* Copyright © 2001, American Forest & Paper Assocation, Inc.

REFERENCES

Chapter 1

1. American Institute of Timber Construction, *Typical Construction Details,* AITC 104, Englewood, CO, 2003.
2. American Institute of Timber Construction, *Standard Appearance Grades for Structural Glued Laminated Timber,* AITC 110, Englewood, CO, 2001.
3. American Institute of Timber Construction, *Standard for Preservative Treatment of Structural Glued Laminated Timber,* AITC 109, Englewood, CO, 1998.
4. American Wood-Preservers' Association, *Book of Standards,* Woodstock, MD, 2000.
5. U.S. Department of Agriculture, Forest Service, Forest Products Laboratory, *Wood Handbook: Wood as an Engineering Material,* Madison, WI, 1999.
6. American Institute of Timber Construction, *Designing Structural Glued Laminated Timber for Permanence,* Technical Note 12, Englewood, CO, 2001.
7. American Society of Heating, Refrigerating and Air-Conditioning Engineers, *ASHRAE Fundamentals Handbook,* Atlanta, GA, 2001.
8. American Institute of Timber Construction, *Checking in Glued Laminated Timbers,* Technical Note 11, Englewood, CO, 1987.
9. American Institute of Timber Construction, *Evaluation of Checking in Glued Laminated Timbers,* Technical Note 18, Englewood, CO, 2001.
10. American Society for Testing and Materials, *Standard Practice for Establishing Structural Grades and Related Allowable Properties for Visually Graded Lumber,* ASTM D245, West Conshohocken, PA, 2000.
11. American Institute of Steel Construction, *Manual of Steel Construction,* 9th ed., Chicago, 1989.
12. American Welding Society, *Structural Welding Code,* Dl.l, Miami, FL, 1983.

13. American Wood-Preservers' Association, *Standard for the Care of Preservative-Treated Wood Products,* AWPA M4, Woodstock, MD, 1995.
14. American Institute of Timber Construction, *Recommended Practice for Protection of Structural Glued Laminated Timber during Transit, Storage and Erection,* AITC 111, Englewood, CO, 1979.
15. American Institute of Timber Construction, *Calculation of Fire Resistance of Glued Laminated Timbers,* Technical Note 7, Englewood, CO, 1996.
16. American Forest & Paper Association/American Wood Council, *National Design Specification® for Wood Construction,* Washington, DC, 2001.
17. American Institute of Timber Construction, *Standard for Heavy Timber Construction,* AITC 108, Englewood, CO, 1993.

Chapter 2

1. U.S. Department of Agriculture, Forest Service, Forest Products Laboratory, *Wood Handbook: Wood as an Engineering Material,* Madison, WI, 1999.
2. American Forest & Paper Association/American Wood Council, *National Design Specification® for Wood Construction,* Washington, DC, 2001.
3. American Wood-Preservers' Association, *Standard for the Care of Preservative-Treated Wood Products,* AWPA M4, Woodstock, MD, 1995.
4. American Institute of Timber Construction, *Typical Construction Details,* AITC 104, Englewood, CO, 2003.
5. American Society of Heating, Refrigerating, and Air-Conditioning Engineers, *ASHRAE Fundamentals Handbook,* Atlanta, GA, 2001.

Chapter 3

1. Building Officials and Code Administrators International, now International Code Council, *National Building Code,* Falls Church, VA, 1999.
2. International Code Council, *International Building Code,* Falls Church, VA, 2003.
3. Southern Building Code Congress International, now International Code Council, *Standard Building Code,* Falls Church, VA, 1999.
4. International Conference of Building Officials, now International Code Council, *Uniform Building Code,* Falls Church, VA, 1997.
5. American Society of Civil Engineers, *Minimum Design Loads for Buildings and Other Structures,* SEI/ASCE 7-02, Reston, VA, 2003.
6. American Association of State Highway and Transportation Officials, *Standard Specifications for Highway Bridges,* Washington, DC, 2002.
7. American Institute of Timber Construction, *Glued Laminated Timber Bridge Systems,* Englewood, CO, 1999.
8. American Railway Engineering and Maintenance-of-Way Association, *Manual for Railway Engineering,* Landover, MD, 2003.
9. American Society of Civil Engineers, *Design Considerations for Fatigue in Timber Structures,* Paper 2470, Reston, VA, 1960.

10. U.S. Department of Agriculture, Forest Service, Forest Products Laboratory, *Fatigue Resistance of Quarter-Scale Bridge Stringers in Flexure and Shear,* Report 2236, Madison, WI, 1962.
11. American Forest & Paper Association/American Wood Council, *National Design Specification® for Wood Construction,* Washington, DC, 2001.
12. American Society of Civil Engineers, *Design of Structures to Resist Nuclear Weapons Effects,* Manual 42, Reston, VA, 1985.
13. American Lumber Standards Committee, Germantown, MD.
14. Southern Pine Inspection Bureau, *Standard Grading Rules for Southern Pine Lumber,* Pensacola, FL, 2002.
15. West Coast Lumber Inspection Bureau, *Grading Rules for West Coast Lumber,* Standard 17, Portland, OR, 2000.
16. Western Wood Products Association, *Western Lumber Grading Rules,* Portland, OR, 1998.
17. National Lumber Grades Authority, *Standard Grading Rules for Canadian Lumber,* New Westminster, BC, Canada, 2003.
18. American Society for Testing and Materials, *Standard Test Methods for Establishing Clear Wood Strength Values,* ASTM D2555, West Conshohocken, PA, 1998.
19. American Society for Testing and Materials, *Standard Practice for Establishing Structural Grades and Related Allowable Properties for Visually Graded Lumber,* ASTM D245, West Conshohocken, PA, 2000.
20. American Society for Testing and Materials, *Standard Practice for Establishing Allowable Properties for Visually-Graded Dimension Lumber from In-Grade Tests of Full-Size Specimens,* ASTM D1990, West Conshohocken, PA, 2000.
21. American Society for Testing and Materials, *Standard Practice for Assigning Allowable Properties for Mechanically-Graded Lumber,* ASTM D6570, West Conshohocken, PA, 2000.
22. American Institute of Timber Construction, *Standard Specifications for Structural Glued Laminated Timber of Softwood Species,* AITC 117, Centennial, CO, 2004.
23. American National Standards Institute/American Institute of Timber Construction, *American National Standard for Wood Products: Structural Glued Laminated Timber,* ANSI/AITC A190.1, Englewood, CO, 2002.
24. American Society for Testing and Materials, *Standard Practice for Establishing Allowable Properties for Structural Glued Laminated Timber (Glulam),* ASTM D3737, West Conshohocken, PA, 2001.
25. American Wood-Preservers' Association, *Book of Standards,* Woodstock, MD, 2000.
26. American Plywood Association, *Plywood Design Specification,* Tacoma, WA, 1998.
27. American Forest & Paper Association/American Society of Civil Engineers, *Standard for Load and Resistance Factor Design for Engineered Wood Construction,* AF&PA/ASCE 16-95, New York, 1996.
28. E. W. Kuenzi and B. Bohannon, Increases in Deflection and Stresses Caused by Ponding of Water on Roofs, *Forest Products Journal,* Vol. 14, No. 9, 1964, pp. 421–424.

Chapter 4

1. American Institute of Timber Construction, *Standard Specification for Structural Glued Laminated Timber of Softwood Species,* AITC 117, Englewood, CO, 2004.
2. American Forest & Paper Association/American Wood Council, *National Design Specification® for Wood Construction,* Washington, DC, 2001.
3. International Code Council, *International Building Code,* Falls Church, VA, 2003.
4. American Institute of Timber Construction, *Guidelines for the Analysis of Drilled Holes or Notches in Structural Glued Laminated Timber Beams,* Technical Note 19, Englewood, CO, 2002.
5. G. W. Thayer, and H. W. March, *The Torsion of Members Having Sections Common in Aircraft Construction,* U.S. Department of Agriculture, Forest Products Laboratory, Madison, WI, 1929.
6. American Institute of Timber Construction Technical Advisory Committee recommendations based on the proportional limit for torsion from *The Wood Handbook,* U.S. Department of Agriculture, 1999, and the relationship between torsional shear strength and shear parallel to grain reported by R. Gupta et al., Experimental evaluation of the torsion test for determining shear strength of structural lumber, *Journal of Testing and Evaluation,* Vol. 30, No. 4, July 2002, pp. 283–290; and D. S. Ryanto, and R. Gupta, A comparison of test methods for evaluating shear strength of structural lumber, *Forest Products Journal,* Vol. 48, No. 2, 1998, pp. 83–90; and the provisions of *Standard for Load and Resistance Factor Design for Engineered Wood Construction,* AF&PA/ASCE 16-95, American Society of Civil Engineers, Reston, VA, 1996, Section 5.5.
7. ASTM International, *Standard Practices for Establishing Stress Grades for Structural Members Used in Log Buildings,* D3957-90, West Conshohocken, PA, 1996.
8. American National Standards Institute, *Wood Poles,* ANSI 05.1, New York, 1992.
9. U.S. Department of Agriculture, Forest Service, Forest Products Laboratory, *Deflection and Stresses of Tapered Wood Beams,* Research Paper FPL 34, Madison, WI, 1965.
10. Colorado State University, Civil Engineering Department, *Behavior and Design of Double-Tapered Pitched and Curved Glulam Beams,* Structural Research Report 16, Ft. Collins, CO, 1976.
11. American Institute of Timber Construction, *Typical Construction Details,* AITC 104, Englewood, CO, 2003.
12. American Institute of Timber Construction, *Mathematical Solution for Design of a Three-Hinged Arch,* Technical Note 23, Englewood, CO, 1998.
13. American Institute of Timber Construction, *Timber Construction Manual,* 4th ed., Wiley, New York, 1994.
14. American Institute of Timber Construction, *Deflection of Glued Laminated Timber Arches,* Technical Note 2, Englewood, CO, 1992.

Chapter 5

1. American Institute of Timber Construction, *Typical Construction Details,* AITC 104, Englewood, CO, 2003.
2. American Forest & Paper Association/American Wood Council, *National Design Specification® for Wood Construction,* Washington, DC, 2001.

3. International Conference of Building Officials, now International Code Council, *Uniform Building Code,* Falls Church, VA, 1997.
4. American Institute of Timber Construction, *Standard Specification for Structural Glued Laminated Timber of Softwood Species,* AITC 117, Englewood, CO, 2004.
5. American Institute of Steel Construction, *Manual of Steel Construction,* 9th ed., Chicago, 1989.
6. American Society of Mechanical Engineers, *Square and Hex Bolts and Screws,* ANSI/ASME Standard B18.2.1, New York, 1997.
7. American Concrete Institute, *Building Code Requirements for Structural Concrete,* ACI 318, Chicago, 1999.
8. American Society of Mechanical Engineers, *Wood Screws,* ANSI/ASME Standard B18.6.1, New York, 1981.
9. D. E. Breyer, *Design of Wood Structures,* McGraw-Hill, New York, 1993, p. 386.
10. U.S. Department of Agriculture, Forest Service, Forest Products Laboratory, *Wood Handbook: Wood as an Engineering Material,* Madison, WI, 1999.

Chapter 6

1. International Conference of Building Officials, now International Code Council, *Uniform Building Code,* Falls Church, VA, 1997.
2. American Society for Testing and Materials, *Standard Specification and Test Method for Establishing Recommended Design Stresses for Round Timber Construction,* ASTM D3200, West Conshohocken, PA, 2000.
3. American Society for Testing and Materials, *Standard Specification for Round Timber Piles,* ASTM D25, West Conshohocken, PA, 1999.
4. American National Standards Institute, *Wood Poles,* ANSI 05.1, New York, 1992.
5. American Wood-Preservers' Association, Woodstock, MD.
6. American Wood-Preservers' Association, *Standard for the Care of Preservative-Treated Wood Products,* AWPA M4, Woodstock, MD, 1995.
7. American Forest & Paper Association/American Wood Council, *National Design Specification® for Wood Construction,* Washington, DC, 2001.
8. National Frame Builders Association, *Post-Frame Building Design Manual,* Lawrence, KS, 2000.
9. R. J. Hoyle and F. E. Woeste, *Wood Technology in the Design of Structures,* Iowa State University Press, Ames, IA, 1989.
10. T. D. Skaggs, F. E. Woeste, and D. A. Bender, A simple analysis for calculating post forces, *Applied Engineering in Agriculture,* Vol. 9, No. 2, 1993, pp. 253–259.
11. American Wood Preservers Institute, *Pile Foundations Know-How,* Vienna, VA, 1969.
12. U.S. Government, *Wood Preservation, Treating Practice,* Federal Specification TT-WW-571i (2), Washington, DC, 1972.
13. G. F. Sowers, *Introductory Soil Mechanics and Foundations: Geotechnical Engineering,* Macmillan, New York, 1979.
14. American Plywood Association, *Plywood Design Spcification,* Tacoma, WA, 1998.

15. U.S. Department of Commerce, *Construction and Industrial Plywood,* PS 1, Washington, DC, 1995.
16. U.S. Department of Commerce, *Performance Standard for Wood-Based Structural-Use Panels,* PS 2, Washington, DC, 1992.
17. American Institute of Timber Construction, *Standard for Tongue-and-Groove Heavy Timber Roof Decking,* AITC 112, Englewood, CO, 1993.
18. American Institute of Timber Construction, *Standard Specification for Structural Glued Laminated Timber of Softwood Species,* AITC 117, Centennial, CO, 2004.
19. American Institute of Timber Construction, *Standard Specifications for Structural Glued Laminated Timber of Hardwood Species,* AITC 119, Englewood, CO, 1996.
20. American Institute of Timber Construction, *Typical Construction Details,* AITC 104, Englewood, CO, 2003.
21. Western Wood Products Association, *Western Woods Use Book,* Portland, OR, 1996.
22. American Association of State Highway and Transportation Officials, *Standard Specifications for Highway Bridges,* Washington, DC, 2002.
23. American Railway Engineering and Maintenance-of-Way Association, *Manual for Railway Engineering,* Landover, MD, 2003.
24. American Institute of Timber Construction, *Glued Laminated Timber Bridge Systems,* Englewood, CO, 1999.
25. U.S. Department of Agriculture, Forest Service, Forest Products Laboratory, *Plans for Crash-Tested Bridge Railings for Longitudinal Wood Decks,* General Technical Report FPL-GTR-87, Madison, WI, 1995.
26. American Institute of Steel Construction, *Manual of Steel Construction,* 9th ed., Chicago, 1989.

Chapter 7

1. American Forest & Paper Association, *Load and Resistance Factor Design Manual for Engineered Wood Construction,* Washington, DC, 1996.
2. American Society for Testing and Materials, *Standard Specification for Evaluation of Structural Composite Lumber Products,* ASTM D5456, West Conshohocken, PA, 2003.
3. American Forest & Paper Association/American Wood Council, *National Design Specification® for Wood Construction,* Washington, DC, 2001.
4. American Institute of Timber Construction, *Standard Specification for Structural Glued Laminated Timber of Softwood Species,* AITC 117, Centennal, CO, 2004.
5. American Society for Testing and Materials, *Standard Specification for Computing the Reference Resistance of Wood-Based Materials and Structural Connections for Load and Resistance Factor Design,* ASTM D5457, West Conshohocken, PA, 1998.
6. American Society of Civil Engineers, *Minimum Design Loads for Buildings and Other Structures,* ASCE 7-02, Reston, VA, 2002.
7. American Forest & Paper Association/American Society of Civil Engineers, *Standard for Load and Resistance Factor Design for Engineered Wood Construction,* AF&PA/ASCE 16-95, New York, 1996.
8. American Forest & Paper Association, *Load and Resistance Factor Design: Example Problems for Wood Structures,* Washington, DC, 2000.

Appendix

1. American Forest & Paper Association/American Wood Council, *National Design Specification® for Wood Construction,* Washington, DC, 2001.
2. American Institute of Steel Construction, *Manual of Steel Construction Load and Resistance Factor Design,* 2nd ed., 4-190 through 4-203, Chicago, IL, 2003.

INDEX